Lecture Notes in Computer Science 7928

Commenced Publication in 1973
Founding and Former Series Editors:
Gerhard Goos, Juris Hartmanis, and Jan van Leeuwen

Ying Tan   Yuhui Shi   Hongwei Mo (Eds.)

Advances in Swarm Intelligence

4th International Conference, ICSI 2013
Harbin, China, June 12-15, 2013
Proceedings, Part I

 Springer

Volume Editors

Ying Tan
Peking University, Key Laboratory of Machine Perception (MOE)
School of Electronics Engineering and Computer Science
Department of Machine Intelligence
Beijing 100871, China
E-mail: ytan@pku.edu.cn

Yuhui Shi
Xi'an Jiaotong-Liverpool University
Department of Electrical and Electronic Engineering
Suzhou 215123, China
E-mail: yuhui.shi@xjtlu.edu.cn

Hongwei Mo
Harbin Engineering University, Automation College
Harbin 150001, China
E-mail: mhonwei@163.com

ISSN 0302-9743   e-ISSN 1611-3349
ISBN 978-3-642-38702-9   e-ISBN 978-3-642-38703-6
DOI 10.1007/978-3-642-38703-6
Springer Heidelberg Dordrecht London New York

Library of Congress Control Number: 2013939062

CR Subject Classification (1998): F.2, G.1, G.2, H.2.8, I.2.6, F.1, I.2, H.4, I.4, I.5

LNCS Sublibrary: SL 1 – Theoretical Computer Science and General Issues

Typesetting: Camera-ready by author, data conversion by Scientific Publishing Services, Chennai, India

Printed on acid-free paper

Springer is part of Springer Science+Business Media (www.springer.com)

Preface

This book and its companion volume, LNCS vols. 7928 and 7929, constitute the proceedings of the 4th International Conference on Swarm Intelligence (ICSI 2013) held during June 12–15, 2013, in Harbin, China. ICSI 2013 was the fourth international gathering in the world for researchers working on all aspects of swarm intelligence, following the successful and fruitful Shenzhen (ICSI 2012), Chongqing (ICSI 2011), and Beijing events (ICSI 2010), which provided a high-level academic forum for the participants to disseminate their new research findings and discuss emerging areas of research. It also created a stimulating environment for the participants to interact and exchange information on future challenges and opportunities in the field of swarm intelligence research.

ICSI 2013 received 268 submissions from about 613 authors in 35 countries and regions (Algeria, Australia, Austria, Bangladesh, Bonaire Saint Eustatius and Saba, Brazil, Canada, Chile, China, Czech Republic, France, Germany, Hong Kong, India, Islamic Republic of Iran, Italy, Japan, Republic of Korea, Malaysia, Mexico, Pakistan, Palestine, Romania, Russian Federation, Saudi Arabia, Singapore, South Africa, Spain, Sweden, Switzerland, Chinese Taiwan, Thailand, Tunisia, Turkey, UK, USA) across six continents (Asia, Europe, North America, South America, Africa, and Oceania). Each submission was reviewed by at least two reviewers, and on average 2.5 reviewers. Based on rigorous reviews by the Program Committee members and reviewers, 129 high-quality papers were selected for publication in this proceedings volume with an acceptance rate of 48.13%. The papers are organized in 22 cohesive sections covering all major topics of swarm intelligence research and development.

As organizers of ICSI 2013, we would like to express sincere thanks to Harbin Engineering University, Peking University, and Xian Jiaotong-Liverpool University for their sponsorship, as well as to the IEEE Computational Intelligence Society, World Federation on Soft Computing, and International Neural Network Society for their technical co-sponsorship. We appreciate the Natural Science Foundation of China for its financial and logistic support. We would also like to thank the members of the Advisory Committee for their guidance, the members of the International Program Committee and additional reviewers for reviewing the papers, and the members of the Publications Committee for checking the accepted papers in a short period of time. Particularly, we are grateful to the Springer for publishing the proceedings in the prestigious series of *Lecture Notes in Computer Science*. Moreover, we wish to express our heartfelt appreciation to the plenary speakers, session chairs, and student helpers. In addition, there are still many more colleagues, associates, friends, and supporters who helped us in immeasurable ways; we express our sincere gratitude to them all. Last but not the least, we would like to thank all the speakers, authors, and participants for their great contributions that made ICSI 2013 successful and all the hard work worthwhile.

April 2013   Ying Tan
                                                               Yuhui Shi
                                                               Hongwei Mo

Organization

General Chairs

Russell C. Eberhart   Indiana University-Purdue University, USA
Guihua Xia   Harbin Engineering University, China
Ying Tan   Peking University, China

Program Committee Chair

Yuhui Shi   Xi'an Jiaotong-Liverpool University, China

Advisory Committee Chairs

Gary G. Yen   Oklahoma State University, USA
Xingui He   Peking University, China

Organizing Committee Chair

Hongwei Mo   Harbin Engineering University, China

Technical Committee Chairs

Carlos A. Coello Coello   CINVESTAV-IPN, Mexico
Xiaodong Li   RMIT University, Australia
Andries Engelbrecht   University of Pretoria, South Africa
Ram Akella   University of California, USA
M. Middendorf   University of Leipzig, Germany
Lin Zhao   Harbin Engineering University, China

Special Sessions Chairs

Fernando Buarque   University of Pernambuco, Brazil
Benlian Xu   Changsu Institute of Technology, China

Publications Chair

Radu-Emil Precup   Politehnica University of Timisoara, Romania

Publicity Chairs

Hideyuki Takagi   Kyushu University, Japan
Shan He   University of Birmingham, UK
Yew-Soon Ong   Nanyang Technological University, Singapore
Juan Luis Fernandez
  Martinez   University of Oviedo, Spain
Jose Alfredo F. Costa   Federal University, Brazil
Kejun Wang   Harbin Engineering University, China

Finance and Registration Chairs

Chao Deng   Peking University, China
Andreas Janecek   University of Vienna, Austria

Local Arrangements Chairs

Lifang Xu   Harbin Engineering University, China
Mo Tang   Harbin Engineering University, China

Program Committee

Payman Arabshahi   University of Washington, USA
Sabri Arik   Istanbul University, Turkey
Carmelo J. A. Bastos Filho University of Pernambuco, Brazil
Walter Chen   National Taipei University of Technology,
                              Chinese Taipei
Manuel Chica   European Centre for Soft Computing, Spain
Jose Alfredo Ferreira Costa UFRN Universidade Federal do Rio Grande do
                              Norte, Brazil
Arindam K. Das   University of Washington, USA
Prithviraj Dasgupta   University of Nebraska, USA
Mingcong Deng   Tokyo University of Agriculture and
                              Technology, Japan
Yongsheng Ding   Donghua University, China
Haibin Duan   Beijing University of Aeronautics and
                              Astronautics, China
Mark Embrechts   Rensselaer Institute, USA
Juan Luis Fernández
  Martínez   University of Oviedo, Spain
Wai-Keung Fung   University of Manitoba, Canada
Luca Gambardella   Istituto Dalle Molle di Studi sull'Intelligenza
                              Artificiale, Switzerland

Dunwei Gong	China University of Mining and Technology, China
Maoguo Gong	Xidian University, China
Ping Guo	Beijing Normal University, China
Haibo He	University of Rhode Island, USA
Ran He	National Laboratory of Pattern Recognition, China
Shan He	University of Birmingham, UK
Lu Hongtao	Shanghai Jiao Tong University, China
Mo Hongwei	Harbin Engineering University, China
Jun Hu	Chinese Academy of Sciences, China
Guangbin Huang	Nanyang Technological University, Singapore
Yuancheng Huang	Wuhan University, China
Andreas Janecek	University of Vienna, Austria
Alan Jennings	University of Dayton, USA
Zhen Ji	Shenzhen University, China
Changan Jiang	RIKEN-TRI Collaboration Center for Human-Interactive Robot Research, Japan
Licheng Jiao	Xidian University, China
Colin Johnson	University of Kent, USA
Farrukh Khan	FAST-NUCES Islamabad, Pakistan
Thanatchai Kulworawanichpong	Suranaree University of Technology, Thailand
Germano Lambert-Torres	Itajuba Federal University, Brazil
Xia Li	Shenzhen University, China
Xuelong Li	University of London, UK
Andrei Lihu	Politehnica University of Timisoara, Romania
Fernando B. De Lima Neto	University of Pernambuco, Brazil
Guoping Liu	University of Glamorgan, UK
Jianhua Liu	Fujian University of Technology, China
Ju Liu	Shandong University, China
Wenlian Lu	Fudan University, China
Bernd Meyer	Monash University, Australia
Martin Middendorf	University of Leipzig, Germany
Bijaya Ketan Panigrahi	Indian Institute of Technology, Delhi, India
Thomas Potok	ORNL, USA
Radu-Emil Precup	Politehnica University of Timisoara, Romania
Yuhui Shi	Xi'an Jiaotong-Liverpool University, China
Zhongzhi Shi	Institute of Computing Technology, Chinese Academy of Sciences, China
Mohammad Taherdangkoo	Shiraz University, Iran
Hideyuki Takagi	Kyushu University, Japan
Ying Tan	Peking University, China
Ke Tang	University of Science and Technology of China, China

Ba-Ngu Vo	Curtin University, Australia
Bing Wang	University of Hull, UK
Jiahai Wang	Sun Yat-sen University, China
Lei Wang	Tongji University, China
Ling Wang	Tsinghua University, China
Lipo Wang	Nanyang Technological University, Singapore
Qi Wang	Xi'an Institute of Optics and Precision Mechanics Of CAS, China
Shunren Xia	Zhejiang University, China
Benlian Xu	Changshu Institute of Technology, China
Yingjie Yang	De Montfort University, UK
Peng-Yeng Yin	National Chi Nan University, China
Zhuhong You	Shenzhen University, China
Jie Zhang	Newcastle University, UK
Jun Zhang	Waseda University, Japan
Junqi Zhang	Tongji University, China
Lifeng Zhang	Renmin University of China, China
Qieshi Zhang	Waseda University, Japan
Qingfu Zhang	University of Essex, UK
Dongbin Zhao	Institute of Automation, Chinese Academy of Science, China
Zhi-Hua Zhou	Nanjing University, China
Zexuan Zhu	Shenzhen University, China
Xingquan Zuo	Beijing University of Posts and Telecommunications, China

Additional Reviewers

Ali, Aftab
Bo, Xing
Bova, Nicola
Dai, Wang-Zhou
Ding, Ke
Ding, Ming
Fang, Jianwu
Gambardella, Luca Maria
Hao, Pengyi
Ho, Tze-Yee
Jiesheng, Wang
Mi, Guyue
Pei, Yan
Pérez Pancho, David
Qian, Chao

Sun, Minghui
Tanoto, Andry
Wan, Wenbo
Wang, Jaidong
Wang, Li
Xing, Bo
Yeh, Ming-Feng
Yu, Chao
Yu, James
Yu, Jian
Zhang, Pengtao
Zheng, Shaoqiu
Zheng, Zhongyang
Zhou, Wei
Zhu, Guokang

Table of Contents – Part I

Applications of PSO Algorithms

Ant Colony Optimization Algorithms

Biogeography-Based Optimization Algorithms

Novel Swarm-Based Search Methods

Bee Colony Algorithms

Differential Evolution

Parameter Optimization

Neural Networks

Fuzzy Methods

Evolutionary Programming and Evolutionary Games

Table of Contents – Part II

Data Mining Methods

System and Information Security

Intelligent Control

Wireless Sensor Network

Scheduling and Path Planning

Image and Video Processing

Other Applications

Interactive Robotic Fish
for the Analysis of Swarm Behavior

Tim Landgraf[1], Hai Nguyen[1], Stefan Forgo[1], Jan Schneider[1], Joseph Schröer[1],
Christoph Krüger[1], Henrik Matzke[1],
Romain O. Clément[2], Jens Krause[2], and Raúl Rojas[1]

[1] Freie Universität Berlin, FB Mathematik u. Informatik
Arnimallee 7, 14195 Berlin, Germany
[2] Leibniz-Institute of Freshwater Ecology & Inland Fisheries
Müggelseedamm 310, 12587 Berlin, Germany
tim.landgraf@fu-berlin.de
http://biorobotics.mi.fu-berlin.de

Abstract. Biomimetic robots can be used to analyze social behavior through active interference with live animals. We have developed a swarm of robotic fish that enables us to examine collective behaviors in fish shoals. The system uses small wheeled robots, moving under a water tank. The robots are coupled to a fish replica inside the tank using neodymium magnets. The position of the robots and each fish in the swarm is tracked by two cameras. The robots can execute certain behaviors integrating feedback from the swarm's position, orientation and velocity. Here, we describe implementation details of our hardware and software and show first results of the analysis of behavioral experiments.

Keywords: biomimetic robots, biomimetics, swarm intelligence, social behavior, social networks, swarm tracking.

1 Introduction

The use of biomimetic robots that help understanding complex biological systems has several advantages over conventional methods in behavioral biology. Foremost the study of animal behavior in groups can benefit from biomimetic robots. Once the robot is accepted as a conspecific, the experimenter is in full control over the interaction with the animals, which drastically augments conventional, static setups and gives access to a whole new set of manipulations. Intriguing examples of the recent past include robotic cockroaches to explore group shelter seeking [2], robotic bees for investigating the honeybee dance communication system [3], robotic bowerbirds for the analysis of courtship behavior [4] and a robotic fish to study group decision making [1]. Similar to [6], the proposed system builds up on the results of the latter, a plotter-like positioning system under a water tank that can move a single fish replica in the tank via strong neodymium magnets. We have built a new prototype that utilizes wheeled robots advancing the system to a multi-agent platform: with our system a swarm

Y. Tan, Y. Shi, and H. Mo (Eds.): ICSI 2013, Part I, LNCS 7928, pp. 1–10, 2013.

of robotic fish can interact with a swarm of real fish. A real-time shoal tracking system is used to close the control loop and enables the robots to display specific behaviors. With more than one fish robot we can, for instance, study which morphologies or behaviors make up a better leader or a more frightening predator by comparing the effect on the shoal in a single experiment. In the following, each aspect of the system is described: the general setup, the hardware and software of the wheeled robot, the procedure to building fish replicas and the computer vision system to track the robot and the individuals of the shoal. The remainder of this contribution will describe how we analyze the swarm behavior to find patterns of group formation in time.

2   General Setup

Our focal animals, three-spined *Sticklebacks* and *Guppies*, are small in body size (1 *cm* − 5 *cm* length). Thus, we use life sized fish replicas that are moved by small wheeled robots below the water tank. The tank is positioned at about 1.40 *m* above the ground. The replicas are moulded using dead sample animals, painted and finished to obtain a fish-like appearance. They stand on a small base that integrates a neodymium magnet (see Figure 1). Below the water tank, the wheeled robots move on a second level - a transparent polycarbonate plate. Each of them holds up a magnet to the bottom side of the tank. Two infrared LEDs are attached to each robots bottom side and aligned with the robot's forward direction. An IR-sensitive camera on the ground is used to localize each robot's pair of IR-LEDs. A second camera above the tank is used to track the individuals of the shoal. The robots are controlled via a Wifi connection by a

Fig. 1. The replica is attached to the magnetic base with a thin transparent plastic stick. The picture shows a replica that was scaled up in size to investigate the acceptance in the group varying this parameter.

central computer that integrates the computer vision results to enable each robot to display various interactive behaviors in a closed control loop.

3 Fish Replica

The body of the replica is constructed from resin plaster. A mold is made using a dead template fish using Gedeo molding alginate. The fin, made from an acetate sheet, is set into the mold that is then filled using liquid resin plaster. The cast is painted with acrylic to approximately match the shading of the template fish. Afterwards, the model is coated with waterproof varnish. A thin transparent plastic rod (0.8 mm in diameter) connects the body to its plastic base, which is glued to two small block magnets (see Figure 1).

4 Robot Platform and Control

We have built a custom two-wheeled robot with a base area of 7 cm × 7 cm as depicted in Figure 2. Each robot consists of a light-weight polystyrol frame. All relevant parts are stacked in a minimal volume. Two Faulhaber gear motors are affixed to the bottom of the base plate. The battery pack (7.2 V $LiPo$) is clamped under the motors and a stack of Arduino boards (Arduino Uno, Cupperhead WifiShield and DFRobot Motor Shield) is affixed to the top plate of the plastic frame. A voltage supply board is distributing battery voltage to the Arduino, the motor board and a voltage divider which scales down the battery voltage to be read by one of the Arduino's analog pins. The topmost plate a rare-earth magnet.

The main control of the robots is executed on a central personal computer. At a frequency of 20 Hz, a command packet is sent to every robot via a UDP Wifi connection. Each robot has a unique identifier and only parses their respective packets. We use a fixed length protocol with a two bytes header, 12 bytes data and two bytes checksum. The main program runs the computer vision and controls each robot's orientation and forward speed using a PID controller with respect to a current target point in the water tank. The experimenter can either define static paths by clicking line segments in a virtual arena or choose from a set of interactive behaviors. The former will be executed without feedback from the shoal. Once the respective robot has reached the next target position the next point in the sequence is selected as the new target. The interactive behaviors use the shoal's centroid, its boundaries or the position of a certain fish. Depending on the starting conditions and the behavior of the shoal, this might result in very different trajectories. In section 6 we describe the interactive modes in more detail. Irrespective of the control mode, for each new time step a new motion is calculated, according to the PID controllers for forward and rotary motion. Each motion command is translated to motor velocities and then sent to each robot to be executed.

Fig. 2. The wheeled robots are moving on a transparent plastic plate in a space below the water tank. A rare-earth magnet is attached to the robot's top. We are using a stack of Arduino compatible extension boards ("shields") for wifi communication and motor control. The battery is positioned right under two Faulhaber DC gear motors that are directly connected to the wheels. The battery connects to the Arduino and provides voltage to the motor board. A voltage divider is used to scale down the battery voltage to a maximum of 2.56 V such that the battery level can be read out on an analog pin on the Arduino. A pair of infrared-LEDs is set in to the base plate for visual tracking of the robots from below.

5   Computer Vision

There are two computer vision systems: one for tracking the robots and one for the shoal. Without marking the replicas, the tracking from above might confuse robots and real fish. This would result in a catastrophic loss of motion control over those agents. In a previous work, we described fish replicas with in-built IR-LEDs. This was shown to work with *Sticklebacks* but failed with the smaller *Guppies* due to the resolution of our camera.

The proposed solution is to attach two IR-LEDs to the bottom sides of the robots. Running on a transparent plate, each robot is localized in the images of a camera on the ground. We use a standard webcam with the IR-block removed and an IR-pass filter added to the lens. Thus, only IR-light can reach the sensor and the IR-LEDs produce very bright spots in the camera image. The computer vision pipeline therefor is simply applying a global threshold to binarize the image. The resulting binary image is denoised by an erosion operation and gaps in the remaining blobs are filled with a dilation. Consequently, we seek connected components that likely represent IR-LEDs. Since the distance of the LEDs on each robot is known and constant, the distance of the blobs in the image is known and constant as well. With this constraint we search for pairs of blobs that, if found, define the position of the robot and two possible orientations, i.e. the real angle or the real angle plus 180 degrees. For each robot this ambiguity has to be removed manually in the graphical user interface in a setup phase prior to the experiment. Once a robot is localized, an ID will be assigned: either new IDs will be added (upon initialization) or the IDs of previously found instances will be assigned based on the distance of their projected position (using a motion model with forward and angular velocities) and the new measurement. The vision system assigns each robot a duration called time to live (TTL). If a robot is lost this property is decremented until found again in later frames or until it reaches zero. When found again the TTL is reset to the default (15 frames), otherwise the robots ID is deleted.

The shoal tracking system uses a camera above the tank which is connected to a separate PC for performance reasons. The tank is built from opaque white plastic plates. The fish appear as clear dark grey objects in the camera image. In an initialization period a background hypothesis of the scene is accumulated. In normal operation, each new camera frame is then subtracted from the

Fig. 3. A (cropped) screenshot of the shoal tracking system. Each individual of the shoal is identified by a number. An ellipse and a triangular tail is depicted to mark each fish's position. The orange tail marks the only robot in this recording. The green dots signify that those animals have been found to be in one subgroup, whose center is denoted by the white dot. The circle around the dot denotes one standard deviation.

background and the result is binarized using a global threshold. Similarly to the robot tracking algorithm, morphological operations (erosion and dilation) are used to remove noise and merge fragmented blobs. We then search for connected components, i.e. patches of foreground pixels that form a binary large object (i.e. blob) and that exceed a certain minimal size. Their orientation is determined by using the the second moments (see e.g. [5]). This again yields an ambiguous result. Since fishes usually swim forwards, the direction is extracted by integrating the motion vector over a fixed time window. The system assigns an ID to every fish object and tracks them using a motion model similar to the one described above: In the new frame, each new object is labeled with the ID of the closest projection. When fishes swim together closely they might merge into a bigger single blob. The system recognizes this and is able to assign the right IDs as follows: First, two merging blobs are recognized and the respective blob is assigned a group object label. It is continuously tracked over following frames. Once the individuals within the group object part, the new positions of the single blobs are compared to the linearly projected locations of the individuals from their last known position. Using this simple rule, most of the individuals keep the right ID after such close body contacts. Furthermore, the shoal's center is computed, as well as a number of centers of sub-groups of the shoal. For example, a shoal might divide into two sub-groups having an equal number of individuals. The global shoal center would be in the middle between the two groups. We use k-Means to find those smaller clusters of individuals. This information is sent to the main computer via an IP network connection. This information is used when an agent operates in an interactive mode that relies on the shoal center. If the shoal is divided the system chooses the bigger shoal for the consecutive path planning.

6   Interactive Behaviors

Apart from using static trajectories, our system allows to assign a number of interactive behaviors to the agents. Using the center of gravity of the shoal, a robot or a group of robots can be set to follow a preset path maintaining (or not exceeding) a certain distance to the shoal ("follower" mode). The "predator" behavior drives the robot into the shoal's centroid with maximum velocity and the "leader" behavior makes the robot approach to the shoal, swim past them, make a turn and wait for them to come closer. A direction within a $-90°$ and $90°$ range is chosen according to a normal distribution (of mean $0°$) and a straight run is executed once the shoal is within a certain proximity. This is repeatedly performed until the swarm ceases to follow. Each behavior uses a different set of parameters, like the forward velocity in "predator" mode and the frequency of the change of direction in "leader" mode. First tests confirmed that the shoal seems to have sustained interest in the robots and follows the fish model.

7   Validation and Experiments with Real Fish

We have validated the motion of the robotic fish by creating reference trajectories of various shapes. The position of the robot is measured using the robot tracker and then compared to the target trajectory. The mean positional error is approx. 1 cm (standard deviation of 0.4 cm) and the average angular error vanishes with a feasibly low dispersion (std: 2.5°). The shoal tracking system was validated in two regards: first, the positional error of static objects were calculated by placing fish-sized metal blocks on known positions and comparing the system's output to the reference position. The average error is below 1 mm but might be larger when objects are moving. Secondly, the tracking error was determined by counting the number of individuals that were either lost by the system or assigned the wrong ID (e.g. after swimming close to another individual). If an individual gets lost and found again, we would only count the loss. Five different video sequences of the same duration (1000 frames, 40 seconds) and varying number of individuals were subject to the tracker. In average 1.2 errors per minute occur - most of them produced in one sequence with many fishes overlapping. The acceptance of the replicas was shown previously in [1]. The proposed robotic platform was tested under experimental conditions with a swarm of 10 real fish. The motion of the replica is smooth and the fish did not react differently to previous observations as described in [1]. The interactive behaviors are still in improvement but first tests look promising.

8   Analysis of the Formation of Sub-groups

In experiments with only one robot we traced all individuals with the shoal tracker, as described above. The results were manually reviewed and corrected if necessary (e.g. when IDs were switched erroneously). In order to recognize patterns in their behavior, expressed in relative distances, we calculate a distance matrix for every point in time, i.e. frame in the video feed. Each row and column corresponds to an individual. The mutual distances of individual 1 to all other individuals are tabulated in row 1, and so on. Hence, the matrix is symmetric and shows low values for pairs that swim close to each other. Computing the average distance matrix over all available points in time yields the average mutual proximity. Figure 4 shows the result of such a computation. In the given example, individuals $1 - 4$ and $5 - 8$ form sub-groups that persist over time. It is also possible to recognize temporal patterns: we have binarized the distance matrix such that only close individuals are marked with a 1. The dynamics of this binary function over time can be visualized for all individuals in a three dimensional plot as depicted in Figure 5. Animals that with a mutual preference exhibit a continuous line (or recurring lines) over time. We are currently investigating how the shoal is reforming after disturbed by a robot predator.

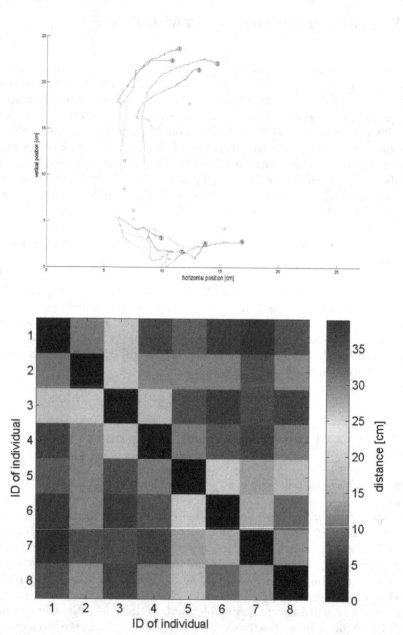

Fig. 4. Left: The figure depicts the trajectories of all individuals of the shoal in a sample sequence. Having started from roughly the same location (denoted by small blue squares), they end up as two separated groups. The end positions are marked with blue circles showing the respective animal ID. Right: The matrix shows the average mutual distance of each individual to all other members of the shoal. Individuals $1-4$ form a sub-group as well as individuals $5-8$.

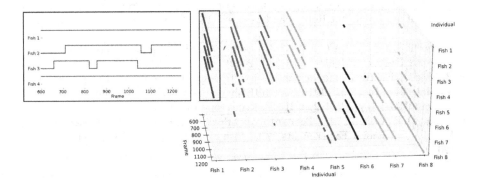

Fig. 5. By binarizing each time point's proximity matrix one can visualize each individual's social behavior over time. The 3-D plot to the right depicts a line over time for a particular pair of individuals if the two fishes were swimming closely together. In the shown sample sequence two groups of four fishes are keeping their mutual distances over a time of 600 frames, i.e. 30 seconds. The left figure shows the course of the binary proximity functions for individual 1 over time.

9 Conclusion

We have built a multi-agent platform for moving robotic fish in a large water tank. The design of the system allows biologists to investigate group decision making in fish shoals with more than one robotic agent. The system is scalable to a large number of agents. The replicas are accepted by the shoal and shown to excite following behavior [1]. The tracking of the individuals of the shoal is sufficiently robust and might only be improved by using markers on the animals. This might require a higher resolution of the camera and reduce the frame rate and therewith the smoothness of the motion control. However, the first data analysis yields promising results and is already used in conventional (non-robotic) research on collective behavior. Currently, we are preparing experiments with live *Sticklebacks* and two robotic fish to investigate the swarm behavior with robotic fish exhibiting a different morphology.

References

1. Faria, J.J., Dyer, J.R.G., Clément, R.O., Couzin, I.D., Holt, N., Ward, A.J.W., Waters, D., Krause, J.: A novel method for investigating the collective behaviour of fish. Behav. Ecol. Sociobiol. 64(8), 1211–1218 (2010)
2. Halloy, J., Sempo, G., Caprari, G., Rivault, C., Asadpour, M., Tâche, F., Saïd, I., Durier, V., Canonge, S., Amé, J.M., Detrain, C., Correll, N., Martinoli, A., Mondada, F., Siegwart, R., Deneubourg, J.L.: Social integration of robots into groups of cockroaches to control self-organised choices. Science 318(5853), 1155–1158 (2007)

3. Landgraf, T., Oertel, M., Rhiel, D., Rojas, R.: A biomimetic honeybee robot for the analysis of the honeybee dance communication system. In: Proc. of the IEEE/RSJ Int. Conf. on Intelligent Robots and Systems (IROS), pp. 3097–3102. IEEE, Taipei (2010)
4. Patricelli, G.L., Uy, J.A.C., Walsh, G., Borgia, G.: Male displays adjusted to female's response. Nature 415(6869), 279–280 (2002)
5. Bradski, G.R.: Real time face and object tracking as a component of a perceptual user interface. In: Proc. of the Fourth IEEE Workshop on Applications of Computer Vision (WACV), pp. 214–219. IEEE, Princeton (1998)
6. Swain, D.T., Couzin, I.D., Leonard, N.E.: Real-Time Feedback-Controlled Robotic Fish for Behavioral Experiments With Fish Schools. Proc. IEEE 100(1), 150–163 (2012)

The Improvement on Controlling Exploration and Exploitation of Firework Algorithm

Jianhua Liu[1,2], Shaoqiu Zheng[2], and Ying Tan[2,*]

[1] School of Information Science and Engineering, Fujian University of Technology,
Fuzhou, 350108, P.R. China
[2] Department of Machine Intelligence, School of EECS, Peking University
Key Laboratory of Machine Perception (MOE), Peking University
jhliu@fjnu.edu.cn, {zhengshaoqiu,ytan}@pku.edu.cn

Abstract. Firework algorithm (FWA) is a new Swarm Intelligence (SI) based optimization technique, which presents a different search manner and simulates the explosion of fireworks to search the optimal solution of problem. Since it was proposed, fireworks algorithm has shown its significance and superiority in dealing with the optimization problems. However, the calculation of number of explosion spark and amplitude of firework explosion of FWA should dynamically control the exploration and exploitation of searching space with iteration. The mutation operator of FWA needs to generate the search diversity. This paper provides a kind of new method to calculate the number of explosion spark and amplitude of firework explosion. By designing a transfer function, the rank number of firework is mapped to scale of the calculation of scope and spark number of firework explosion. A parameter is used to dynamically control the exploration and exploitation of FWA with iteration going on. In addition, this paper uses a new random mutation operator to control the diversity of FWA search. The modified FWA have improved the performance of original FWA. By experiment conducted by the standard benchmark functions, the performance of improved FWA can match with that of particle swarm optimization (PSO).

Keywords: Firework Algorithm, Swarm Intelligence Algorithm, Exploration and Exploitation, PSO.

1 Introduction

Firework algorithm (FWA) is a new intelligence optimization algorithm based on Swarm Intelligence (SI) developed by Y. Tan and Y. Zhu [14]. Like the other SI algorithms, such as Particle Swarm Optimization (PSO) [8], Ant System [3], Colonel Selection Algorithm [2, 13], and Swarm Robots [1], Different Evolution (DE) [11], Artificial Bee Colony (ABC) [7] etc., firework algorithm also is a population based optimization technique. Firework algorithm simulates the explosion of fireworks to search the optimal solution of problem. Compared to the other SI algorithms, firework algorithm has distinctive advantages in solving many optimization problems and presents a different search manner.

---

* Senior Member, IEEE

Y. Tan, Y. Shi, and H. Mo (Eds.): ICSI 2013, Part I, LNCS 7928, pp. 11–23, 2013.

Algorithm 1. Conventional Fireworks Algorithm.

1: Select n position for initial fireworks;
2: Calculate the number of sparks for each firework;
3: Calculate the amplitude of explosion for each firework;
4: Generate the sparks of explosion for each firework;
5: Generate the sparks of Gaussian mutation for each firework;
6: Select n location for next generation fireworks
7: If condition does not meet, algorithm turns to 2
8: Output results

Since it was proposed, fireworks algorithm has shown its significance and superiority in dealing with the optimization problems and has been seen many improvement and application with practical optimization problems. Andreas and Tan [5, 6] used FWA to compute the non-negative matrix factorization and gains a little advantages compared to SPSO, FSS, GA. Pie et al. [9] investigated the influences of approximation approach on accelerating the fireworks algorithm search by elite strategy. In [9], they compared the approximation models, sampling methods, and sampling number on the FWA acceleration performance, and the random sampling method with two-degree polynomial model gains better performance on the benchmark functions. Zheng et al. [16] proposed a hybrid algorithm between FWA and differential evolution (DE), which shows superiority to the previous FWA and DE. H. Gao and M. Diao [4] have designed a cultural firework algorithm which was used to search optimal value of filter design parameters with parallel search. Computer simulations have showed that FIR and IIR digital filters based on the cultural firework algorithm are superior to previous filters based on the other SI algorithm in the convergence speed and optimization results.

However, firework algorithm needs a policy that dynamically controls the exploration and exploitation of searching solution of problem with iteration. In the FWA, the calculation of number of explosion spark and amplitude of firework explosion have too more parameter to set. Its mutation operator cannot availably control the diversity of FWA. This paper will design a transfer function to map the rank number of firework to percentage value in order to calculate the number of explosion spark and firework explosion, and then a rand mutation is presented to generate the diversity of FWA. By the experiment, the modified techniques on FWA can improve the perfromance of original FWA. The paper is organized as follows: section 2 introduces the FWA and analyzes the drawback of FWA; the improved FWA was provided in section 3. Section 4 conducted the experiment to test the effect of improved FWA. The section 5 is the conclusion of this paper.

2 Fireworks Algorithm

Firework algorithm searches the optimal solution of problem using several fireworks explosion to generate the sparks in the space of problem, where the spark and firework are the potential solutions of problem. The procedure of original firework algorithm can be seen in Algorithm 1.

According to the idea of FWA, a good firework denotes the better fitness,which means that the firework may be close to the optimal location. Therefore, the good firework should generate more sparks in the smaller amplitude of explosion. In the contrast, for a bad firework, the search radius should be larger and it should generate less sparks in the larger amplitude of explosion. For every firework, the number of explosion sparks and amplitude of explosion must be calculated before it explodes, which formulas are as follows:

- Calculating the Number of Sparks:

$$S_i = m \frac{y_{\max} - f(x_i) + \xi}{\sum\limits_{i=1}^{n} (y_{\max} - f(x_i)) + \xi} \tag{1}$$

$$\widehat{s_i} = \begin{cases} round(a \cdot m) \ if \ s_i < am \\ round(b \cdot m) \ if \ s_i > bm \\ round(s_i) \ otherwise \end{cases} \tag{2}$$

Where S_i is the number of the spark of the ith firework explosion, m is the total number of sparks generated by the n fireworks. y_{max} is the maximum value of the objective function among the n fireworks, and is a small constant which is utilized to avoid zero-division-error. The constant a and b are the const parameters.
- Calculating the Amplitude of Explosion:

$$A_i = \tilde{A} \frac{f(x_i) - y_{\min} + \xi}{\sum\limits_{i=1}^{n} (f(x_i) - y_{\min}) + \xi} \tag{3}$$

Where $\tilde{A}$ denotes the maximum explosion amplitude and y_{min} is the minimum value of the objective function among n fireworks.

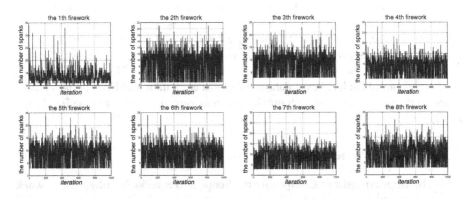

Fig. 1. The number of sparks for every firework in original FWA

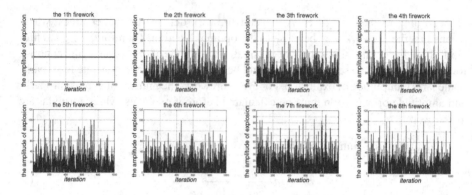

Fig. 2. The amplitude of explosion for every firework in the proposed algorithm

In order to investigate the effect which Eq.(1) and Eq.(3) impact on firework algorithm, we can observe the Fig.1 and Fig.2. Fig.1 is the number of sparks generated by eight fireworks with the FWA iteration, while Fig.2 is the amplitude of explosion generated by eight fireworks with the FWA iteration. The data in the two figures is gained with Eq.(1) and Eq.(3) provided that every iteration of FWA has only eight fireworks.

From Fig.1, different fireworks in the different iteration have no regularity to gain the number of sparks. The number of sparks generated by different firework all are vibrated almost between 5 and 15.The best firework (the first firework) has not always generated the most spark number among eight fireworks. From Fig.2, the first firework (the best firework) explodes in the smaller amplitude than the amplitude of other firework. The explosion amplitude of the first fireworks is always about 10^{-4}, while the amplitude of the other firework is about between 0 and 50. It can be found that the amplitude of first firework is too small in the early time and constant while the other firework' amplitude is variable and have not regularity.

In term of Fig.1 and Fig.2, the Eq.(1) and Eq.(3) can embody random of FWA, because the explosion number and explosion amplitude of fireworks is variable with the fitness of firework. However, spark number and amplitude of firework explosion have not dynamically changed as the algorithm iterates. Especially, the best firework (the 1th firework) has constant explosion amplitude and its sparks number of explosion does not increase with iteration. Therefore,it is difficult for the formula of Eq.(1) and Eq.(3) to effectively control the local exploration and global exploitation of FWA in the solution space. In the next section, we will provide two new equations to modify the Eq.(1) and Eq.(3).

3   The Improvement of FWA

3.1   The Improvement of Computing the Scope and Sparks Number of Firework Explosion

The above drawback of FWA is account of using the fitness of firework to compute the scope and sparks number of firework explosion. In order to improve the equations on calculating the number of firework sparks and the amplitude of firework explosion, we use the sequence number of fireworks to compute the two values. Therefore, a transfer function must be designed to map the sequence number of fireworks to function value which is used to better calculate the amplitude and spark number of firework explosion. The transfer function is cited from the sigmoid function as the Eq.(4):

$$f(x) = \frac{1}{1 + e^x} \tag{4}$$

The function is further improved and added a parameter a as the following Eq.(5):

$$f(x) = \frac{1}{1 + e^{(x-1)/a}} \tag{5}$$

Where a is a control parameter to change the shape of the above function. Eq.(5) can transfer sequence number of firework rank of fitness to different value of function, which is used to calculate the spark number and amplitude of firework explosion. The function of Eq.(5) is named as transfer function. When the parameter $a = 1, 5, 9, 13$ and 21,the figures of the transfer function with different parameter value are plotted as Fig.3 which x axis denotes sequence number. From the Fig.3, it can be found that the function fitness of different sequence number is more and more mean as the parameter a is increasing. So, the calculating number of explosion sparks is designed as the following equation:

$$S_n = m \frac{f(n)}{\sum\limits_{n=1}^{N} f(n)} \tag{6}$$

Where m is the total of number of spark, n is the sequence number of a firework. S_n denotes the spark number of the nth firework explosion. The calculating the amplitude of firework is designed as following equation:

$$A_n = A \frac{f(N - n + 1)}{\sum\limits_{n=1}^{N} f(N - n + 1)} \tag{7}$$

Where A is the maximum amplitude of firework explosion, n is the sequence number of a firework. A_n denotes the amplitude of the nth firework explosion. In the Eq.(6) and Eq.(7), N is the total number of firework in FWA. The function $f(x)$ is the Eq.(5), which parameter a is varied with the iteration from 20 to 1. With variable parameter a, the explosion number of spark and explosion amplitude of firework is dynamically changed as the iteration goes on.

In order to compare Eq.(1) and Eq.(3) to Eq.(6) and Eq.(7), Fig.4 and Fig.5 plot the number of sparks and amplitude of explosion of eight fireworks with iteration which are calculated using Eq.(6) and Eq.(7).

Compared to the Fig.1 and Fig.2, Fig.4 shows that the number of spark generated by the firework of FWA with the modified equation is very regular. For the sorting front fireworks, the number of sparks is more and more with the iteration, while for the last fireworks, the number of sparks is less and less. Fig.5 shows that the amplitude

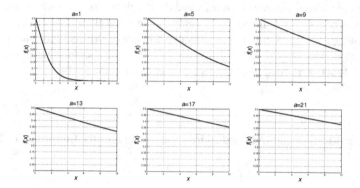

Fig. 3. The plotting figure of transfer function with different parameter

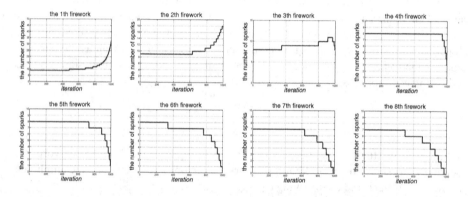

Fig. 4. The number of sparks for every firework in modifying FWA

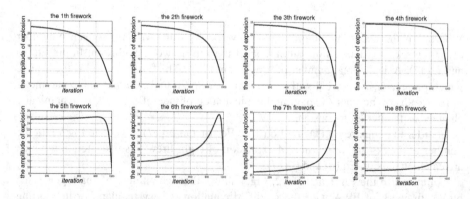

Fig. 5. The amplitude of explosion for every firework in modifying FWA

of the better firework explosion is smaller and smaller as the iteration goes on, while the amplitude of bad firework explosion is bigger and bigger with iteration. The policy can embody the idea of algorithm that a good firework has more number of generating sparks and less amplitude of explosion while a bad firework generate less number of sparks during the larger amplitude of explosion. There may be global optimal solution near the good firework, so the explosion of a good firework undertakes the local searching, while a bad firework exploding undertakes the global exploitation of solution space. In the new method, the dynamic change of number of sparks and amplitude of explosion with iteration can embody that the global exploitation is done in the early time of algorithm running, while the local exploration is enhanced during the later time of algorithm's iteration. So, the new calculating equations can better control the exploitation and exploration of firework algorithm with iteration.

3.2   The Mutation Improvement

To keep the diversity, original firework algorithm employed Gaussian mutation to generate sparks. The jth dimensions of the ith firework, x_{ij}, mutates as x_{ij} by the following equation:

$$x_{ij} = x_{ij}Gaussian(1,1) \qquad (8)$$

However, the above mutation makes original FWA easily converged to zero point of the search space, and it is difficult for FWA to generate the diversity. In order to add the diversity of FWA, the random mutation is employed to make the firework mutated. The mutation formula is as follows:

$$x_{ij} = x_{ij}^{min} + rand()(x_{ij}^{max} - x_{ij}^{min}) \qquad (9)$$

Where x_{ij} denotes the position of the jth dimensions of ith firework; x_{ij}^{min} denotes the minimal bound of the jth dimensions of the ith firework; x_{ij}^{max} denotes maximal bound of the jth dimensions of the ith firework. The function $rand()$ gains the sampling value in the interval [0, 1] with the uniform distribution.

3.3   The Selection of the Next Generation Fireworks

Original FWA selects n location for next generation fireworks by the Eq.(10) and Eq.(11):

$$R(x_i) = \sum_{j \in K} d(x_i, x_j) = \sum_{j \in K} ||x_i - x_j|| \qquad (10)$$

$$p(x_i) = \frac{R(x_i)}{\sum\limits_{j \in K} R(x_i)} \qquad (11)$$

Where the x_i is the location of ith sparks or firework, $d(x_i, x_j)$ is the distance between two sparks or fireworks. K is the set of sparks and firework generated in current generation. The $p(x_i)$ is the probability which the ith firework or spark is selected as the firework of next generation. Eq.(10) and Eq.(11) do not consider the fitness of sparks or fireworks for the selection of next generation fireworks' location. This is not consist with the idea of the equations of Eq.(6) and Eq.(7). Because Eq.(6) and Eq.(7) use the sequence number of sorting fireworks' fitness to calculate the sparks number and amplitude of fireworks explosion, but the Eq.(10) and Eq.(11) of original FWA don't consider the fitness to select the firework location. Therefore, there are two methods to be provided to modify the selection operator.

1. Fitness Selection using the roulette
   Like the original FWA, the best of the set will be selected first. Then the others are selected base on fitness proportion using the roulette. So, the selection probability of every spark or firework must be calculated with the following formula:

$$p(x_i) = \frac{y_{\max} - f(x_i)}{\sum_{i \in K} (y_{\max} - f(x_i))} \tag{12}$$

   Where y_{max} is the maximum value of the objective function among the set K which consist of the fireworks and sparks in the current generation. The other fireworks will be selected using the roulette according to the probability gained by Eq.(12).

2. Best Fitness Selection
   In [15], Zheng et al used a random selection operator to replace the previous time consuming one. It is as the following, when the algorithm has decided the number of firework of every generation, all the sparks and fireworks of the current generation are sorted according to their fitness and then select the best n sparks or fireworks with the best fitness as the location of next generation. The method is very simple and is consistent with the new calculation of explosion number and explosion amplitude of fireworks in the Eq.(6) and Eq.(7).

4   Experiment and Analysis

4.1   Experimental Design

In order to evaluate the performance of the improved FWA, fourteen benchmark functions provided by CEC 2005 are employed [12]. These benchmark functions include five unimodal functions, nine multimodal functions. The optimal fitness of these functions is not zero and is added bias. These functions are shifted and the optimal locations are shifted to different location from zero point in solution space. More details on the benchmark functions can be seen in [12].

In order to test the performance of improved FWA in this paper, the improved FWA with best fitness selection and random mutation (IFWABS) , the improved FWA with the fitness selection using the roulette and random mutation (IFWAFS), original FWA and global PSO are compared with each other. The global PSO is employed the decreasing weight w from 0.9 to 0.4 proposed in [10], and the neighbor particles of each particle is all particles. The particle population size is 100. The factor c_1 and c_2 of PSO are set as 2. The FWA and improved FWA are set the number of firework as 10, the total number of explosion sparks S as 80 and the amplitude of explosion A as the range length of problem space. The experiment is conducted in Matlab 2012b and executed in windows 7.

4.2   Experimental Results and Analysis

The experiment is conducted to compute the mean error fitness ($f(x) - f(x^*)$, $f(x^*)$ is real optimal fitness of the benchmark functions), standard square error and the best error fitness in the 25 run on the 14 benchmark functions. Each run of all algorithm is evaluated 1000, 10000, 10000 and $D * 10000$(D is the dimension of benchmark function), respectively. Each algorithm will be conducted in the 10 dimensions and 30 dimensions, respectively. Table1 is the results of mean error fitness, standard square error and the best fitness in the 25 run on 14 functions in 10 dimensions, which the Fitness Evaluated number(FEs) is 1000, 10000,100000. Table2 is the results of that in 30 dimensions, which the FEs is 10000, 100000 and 300000.

In term of Table 1, compared to FWA, the performance of IFWABS and IFWAFS is better than that of FWA. In the 10 dimensions, the mean error finesses of FWA on 14 functions all are worse than that of IFWABS or IFWAFS whether FEs is 1000, 10000 or 100000. In term of Table2, the performance of IFWABS and IFWAFS is better than that of FWA on all functions except for the 8th function in 30 dimensions. Therefore, the improved FWA has improved the performance of firework algorithm. Compared the performance of two improved FWA (IFWABS and IFWAFS), it can be found from Table1 and Table2 that IFWABS is advantage to IFWAFS. In more part of cases, the mean error fitness of function of IFWABS is outstanding to the IFWAFS. Compared to PSO, improved FWA is more optimal performance on most of functions, especially in 100000 FEs and 300000 FEs. As the FEs is more and more, the performance of improved FWA is better and better than PSO, so the improved FWA can match with PSO. Fig.6 plots the convergence process for four algorithms to optimize the 14 functions with 300000 FEs in 30 dimensions These figures are visual to illustrate the effect of four algorithms that improved FWA is excel to original FWA and can match with PSO.

5   Conclusion

Firework algorithm is a novel swarm intelligence based algorithm that can availably search the optimal solution of parameter space. FWA imitates the firework explosion to generate sparks and provide the idea that better firework can take up the local search and the bad firework do the global exploitation. This paper modifies the calculation of scope and amplitude of firework explosion, and designs a transfer function to map rank number of firework fitness to allocate the total sparks number and explosion aptitude. A parameter of transfer function was used to control the dynamical calculation of two values with iteration. This way is more effective to control the local and global search of FWA in solution space. In addition, a random mutation was presented to enhance the diversity of FWA. At last, in order to accord with the new idea of calculation of scope and amplitude of firework explosion, the best sparks selection and fitness selection was employed to improve the selection operator of FWA. By experiment conducted on 14 standard benchmark functions in CEC 2005, the improved FWA can be superior to the original FWA and can match with PSO.

Table 1. Statistical Results of Mean, Std and Best of Benchmark Functions in 10 Dimension

FES	No.		PSO	FWA	IFWAFS	IFWABS	FES	No.		PSO	FWA	IFWAFS	IFWABS
	f_1	Mean	3.32E+03	1.86E+04	4.33E+03	**3.19E+03**		f_8	Mean	2.07E+01	2.06E+01	2.06E+01	**2.06E+01**
		Std	1.45E+03	3.88E+03	1.67E+03	1.43E+03			Std	1.55E-01	1.49E-01	1.18E-01	1.16E-01
		Best	5.17E+02	1.32E+04	1.12E+03	6.31E+02			Best	2.03E+01	2.02E+01	2.04E+01	2.04E+01
	f_2	Mean	1.21E+04	3.60E+04	1.23E+04	**1.20E+04**		f_9	Mean	5.17E+01	1.21E+02	6.24E+01	**5.01E+01**
		Std	4.25E+03	1.08E+04	6.24E+03	4.51E+03			Std	1.27E+01	2.63E+01	1.88E+01	1.37E+01
		Best	4.90E+03	1.38E+04	1.32E+03	3.83E+03			Best	2.89E+01	5.54E+01	2.0E+01	2.05E+01
	f_3	Mean	**4.38E+07**	9.56E+08	7.67E+07	5.07E+07		f_{10}	Mean	8.0E+01	1.96E+02	9.81E+01	**8.66E+01**
		Std	3.11E+07	6.74E+08	4.73E+07	4.52E+07			Std	1.68E+01	3.46E+01	2.0E+01	2.11E+01
		Best	6.85E+06	2.28E+08	1.24E+07	1.42E+06			Best	4.41E+01	1.23E+02	5.82E+01	4.32E+01
1.0E+03	f_4	Mean	**1.39E+04**	3.19E+04	1.49E+04	1.51E+04	1.0E+03	f_{11}	Mean	**9.0E+0**	1.25E+01	1.06E+01	9.58E+0
		Std	6.63E+03	1.24E+04	5.41E+03	5.99E+03			Std	1.29E+0	1.03E+0	1.25E+0	1.09E+0
		Best	3.40E+03	5.74E+03	5.63E+03	3.65E+03			Best	5.55E+0	9.54E+0	8.52E+0	6.90E+0
	f_5	Mean	**1.09E+04**	2.33E+04	1.26E+04	1.12E+04		f_{12}	Mean	3.67E+04	2.84E+05	4.58709e+04	**3.50379e+04**
		Std	2.62E+03	3.48E+03	3.06E+03	2.66E+03			Std	1.79E+04	8.58E+04	2.63E+04	2.36E+04
		Best	5.90E+03	1.64E+04	5.43E+03	5.94E+03			Best	1.18E+04	1.04E+05	6.23E+03	7.83E+03
	f_6	Mean	**3.10E+08**	1.96E+10	1.17E+09	6.70E+08		f_{13}	Mean	7.81E+0	2.73E+02	1.06E+01	**7.76E+0**
		Std	3.09E+08	1.08E+10	1.23E+09	6.44E+08			Std	3.06E+0	2.95E+02	4.95E+0	2.90E+0
		Best	2.11E+06	4.18E+09	3.40E+07	3.84E+07			Best	3.76E+0	5.52E+01	4.10E+0	3.27E+0
	f_7	Mean	1.40E+03	3.56E+03	1.65E+03	**1.39E+03**		f_{14}	Mean	**4.04E+0**	4.37E+0	4.20E+0	4.13E+0
		Std	1.18E+02	4.84E+02	2.11E+02	1.11E+02			Std	2.79E-01	2.62E-01	2.64E-01	2.33E-01
		Best	1.27E+03	2.53E+03	1.31E+03	1.27E+03			Best	3.28E+0	3.63E+0	3.53E+0	3.71E+0 [b]
	f_1	Mean	2.09E+02	3.76E+03	5.97E-01	**1.08E-01**		f_8	Mean	2.05E+01	2.03E+01	**2.04E+01**	2.04E+01
		Std	3.75E+02	2.88E+03	4.58E-01	**7.24E-02**			Std	1.05E-01	1.17E-01	**1.38E-01**	1.05E-01
		Best	5.65075e-05	2.59E+02	3.99E-02	**2.0E-02**			Best	2.02552e+01	2.01E+01	**2.01E+01**	2.03E+01
	f_2	Mean	**1.30E+02**	8.07E+03	4.09E+02	9.89E+02		f_9	Mean	8.36E+0	4.61E+01	1.98E+01	**2.37E+0**
		Std	**1.31E+02**	2.62E+03	2.67E+02	5.91E+02			Std	3.27E+0	1.41E+01	6.69E+0	**1.23E+0**
		Best	**1.93164e-01**	1.76E+03	2.80E+01	1.99E+02			Best	2.75172e+00	1.73E+01	9.20E+0	**2.01E-01**
	f_3	Mean	3.0E+06	2.73E+07	1.37E+06	**7.91E+05**		f_{10}	Mean	2.05E+01	8.50E+01	4.59E+01	**5.78E+01**
		Std	6.57E+06	2.73E+07	1.25E+06	**7.35E+05**			Std	7.11E+0	1.69E+01	1.60E+01	2.42E+01
		Best	1.04099e+05	2.90E+06	3.47E+04	**3.65E+04**			Best	8.40258e+00	5.67E+01	1.45E+01	1.41E+01
1.0E+04	f_4	Mean	**1.42E+02**	1.11E+04	1.99E+03	3.13E+03	1.0E+04	f_{11}	Mean	**3.96E+0**	8.50E+0	6.45E+0	6.55E+0
		Std	**1.42E+02**	3.91E+03	1.72E+03	1.97E+03			Std	**1.38E+0**	1.36E+0	1.15E+0	1.39E+0
		Best	**1.68170e+00**	3.38E+03	3.02E+02	6.60E+02			Best	1.21034e+00	6.26E+0	3.41E+0	4.71E+0
	f_5	Mean	**7.76E-05**	9.30E+03	5.88E+02	2.04E+02		f_{12}	Mean	3.66E+03	3.12E+04	4.32E+03	**2.20E+03**
		Std	**8.06E-05**	3.73E+03	6.09E+02	4.31E+02			Std	5.64E+03	1.07E+04	5.21E+03	**2.37E+03**
		Best	**8.55801e-06**	8.95E+02	1.62E+01	1.12E+01			Best	1.08496e+04	1.11E+04	1.85E+02	**3.77E+01**
	f_6	Mean	7.75E+06	2.26E+08	2.49E+03	**2.11E+03**		f_{13}	Mean	1.71E+0	5.62E+0	1.71E+0	**8.31E-01**
		Std	1.80E+07	4.24E+08	3.56E+03	3.63E+03			Std	8.59E-01	3.51E+0	8.05E-01	**1.88E-01**
		Best	4.39794e+00	2.45E+06	1.93E+01	**1.07E+01**			Best	6.32417e-01	2.41E+0	8.59E-01	**5.84E-01**
	f_7	Mean	1.27E+03	1.27E+03	1.27E+03	**1.27E+03**		f_{14}	Mean	3.50E+0	3.86E+0	3.71E+0	3.76E+0
		Std	2.54E-01	3.23E-02	7.10E-02	**7.63E-02**			Std	**3.20E-01**	2.74E-01	2.90E-01	3.05E-01
		Best	1.26723e+03	1.27E+03	1.27E+03	1.27E+03			Best	2.8502e+00	3.34E+0	2.98E+0	3.02E+0
	f_1	Mean	4.13E+01	1.28E+02	1.44E-04	**1.01E-04**		f_8	Mean	2.03E+01	2.01E+01	2.02E+01	**2.02E+01**
		Std	5.66E+01	2.75E+02	6.73E-05	**6.56E-05**			Std	6.94E-02	7.53E-02	1.06E-01	**8.69E-02**
		Best	0.0E+0	1.72E+0	3.06E-05	**2.17E-05**			Best	2.01566e+01	2.0E+01	2.01E+01	2.0E+01
	f_2	Mean	9.11E+01	2.61E+02	**1.30E+0**	1.26E+01		f_9	Mean	5.27E+0	1.64E+01	5.86E+0	**4.14E-05**
		Std	7.93E+01	1.38E+02	**8.45E-01**	8.39E+0			Std	5.93E+0	1.04E+01	2.99E+0	**2.21E-05**
		Best	5.68E-14	1.26E+02	**1.04E-01**	9.45E-01			Best	0.000e+00	5.03E+0	2.98E+0	**5.08E-06**
	f_3	Mean	1.84E+06	5.86E+06	**2.77E+05**	3.13E+05		f_{10}	Mean	**2.03E+01**	6.36E+01	3.30E+01	5.10E+01
		Std	3.52E+06	3.52E+06	**1.53E+05**	1.60E+05			Std	**8.80E+0**	1.38E+01	1.19E+01	2.24E+01
		Best	9.96846e+04	2.30E+06	**3.24E+04**	4.15E+04			Best	8.95463e+00	4.12E+01	1.79E+01	1.29E+01
1.0E+05	f_4	Mean	2.91E+02	2.65E+03	**5.61E+0**	2.64E+02	1.0E+05	f_{11}	Mean	**3.50E+0**	6.07E+0	5.72E+0	6.73E+0
		Std	1.47E+02	1.81E+03	**3.66E+0**	2.56E+02			Std	**1.30E+0**	1.92E+0	1.69E+0	1.27E+0
		Best	5.63891e+01	1.20E+03	**8.97E-01**	1.19E+01			Best	1.22304e+00	3.91E+0	1.83E+0	4.47E+0
	f_5	Mean	**0.0E+0**	6.63E+02	6.17E+0	1.80E+02		f_{12}	Mean	1.66E+03	3.19E+03	1.33E+03	**1.20E+03**
		Std	**0.0E+0**	8.58E+02	4.82E+0	6.25E+02			Std	3.31E+03	1.38E+03	2.04E+03	**1.79E+03**
		Best	**0.000e+00**	5.69E+01	1.84E-01	1.08E-01			Best	5.41956e-02	1.82E+03	2.15E+0	**7.09E+0**
	f_6	Mean	2.35E+07	4.95E+03	1.81E+03	**3.27E+02**		f_{13}	Mean	6.90E-01	1.17E+0	7.85E-01	**4.50E-01**
		Std	6.47E+07	4.15E+03	3.46E+03	1.49E+03			Std	2.59E-01	3.87E-01	2.79E-01	**1.70E-01**
		Best	8.90593e+00	1.57E+03	1.77E-02	**1.01E+0**			Best	3.66839e-01	7.80E-01	3.79E-01	**1.88E-01**
	f_7	Mean	1.32E+03	1.27E+03	1.27E+03	**1.27E+03**		f_{14}	Mean	**2.84E+0**	3.73E+0	3.28E+0	3.46E+0
		Std	1.84E+02	6.70E-05	1.09E-04	**5.77E-05**			Std	**3.88E-01**	2.42E-01	3.11E-01	3.99E-01
		Best	1.26723e+03	1.27E+03	1.27E+03	1.27E+03			Best	1.74585e+00	3.39E+0	2.56E+0	2.47E+0

Table 2. Statistical Results of Mean, Std and Best of Benchmark Functions in 30 Dimension

FES	No.		PSO	FWA	IFWAFS	IFWABS	FES	No.	PSO	FWA	IFWAFS	IFWABS
	f_1	Mean	4.98E+03	4.63E+04	6.64E+02	**9.67E+01**		f_8	2.11E+01	**2.07E+01**	2.10E+01	2.10E+01 [t]
		Std	3.94E+03	9.20E+03	3.07E+02	**3.18E+01**			6.20E-02	**9.39E-02**	7.81E-02	8.34E-02
		Best	6.39E+02	3.47E+04	1.86E+02	**4.31E+01**			2.10E+01	**2.05E+01**	2.08E+01	2.07E+01
	f_2	Mean	**1.20E+04**	5.17E+04	4.14E+04	4.78E+04		f_9	1.52E+02	3.01E+01	1.88E+02	**8.61E+01**
		Std	**4.77E+03**	1.00E+04	1.16E+04	8.42E+03			2.52E+01	2.61E+01	2.77E+01	**1.65E+01**
		Best	**6.41E+03**	3.05E+04	2.40E+04	3.42E+04			1.01E+02	2.60E+02	1.37E+02	**5.68E+01**
	f_3	Mean	5.98E+07	4.92E+08	6.98E+07	**3.75E+07**		f_{10}	2.51E+02	6.21E+02	3.71E+02	3.58E+02
		Std	2.74E+07	2.22E+08	2.64E+07	**2.02E+07**			**2.10E+01**	6.22E+01	7.13E+01	9.29E+01
		Best	1.97E+07	1.16E+08	1.95E+07	**8.57E+06**			**2.15E+02**	5.20E+02	2.16E+02	2.07E+02
1.00E+04	f_4	Mean	**2.08E+04**	6.17E+04	6.44E+04	7.11E+04	1.00E+04	f_{11}	3.13E+01	3.93E+01	3.32E+01	**3.04E+01**
		Std	**4.25E+03**	1.14E+04	1.42E+04	2.06E+04			3.56E+01	2.48E+00	4.03E+00	**3.50E+00**
		Best	**1.29E+04**	4.28E+04	4.03E+04	3.42E+04			2.48E+01	3.29E+01	2.40E+01	**2.36E+01**
	f_5	Mean	**8.66E+03**	2.98E+04	1.61E+04	1.24E+04		f_{12}	1.15E+05	7.87E+05	1.94E+05	**1.25E+05**
		Std	**3.07E+03**	3.94E+03	2.87E+03	2.89E+03			5.83E+04	2.05E+05	9.81E+04	**6.26E+04**
		Best	**1.40E+03**	2.35E+04	1.07E+04	7.92E+03			1.66E+04	4.22E+05	4.06E+04	**3.83E+04**
	f_6	Mean	6.84E+08	1.54E+10	2.78E+07	**3.88E+05**		f_{13}	2.01E+01	5.45E+01	1.89E+01	**1.14E+01**
		Std	9.82E+08	4.39E+09	3.11E+07	**2.54E+05**			2.30E+00	3.44E+01	3.96E+00	**2.21E+00**
		Best	6.34E+06	5.25E+09	2.68E+06	**7.02E+04**			1.65E+01	2.05E+01	1.25E+01	**7.60E+00**
	f_7	Mean	4.80E+03	4.77E+03	4.76E+03	**4.72E+03**		f_{14}	1.35E+01	1.37E+01	**1.33E+01**	1.34E+01
		Std	1.31E+02	4.81E+01	4.53E+01	**2.20E+01**			2.48E-01	3.19E-01	**3.58E-01**	3.90E-01
		Best	4.70E+03	4.73E+03	4.71E+03	**4.70E+03**			1.30E+01	1.29E+01	**1.25E+01**	1.27E+01 [b]
	f_1	Mean	5.03E+03	8.81E+03	4.21E-02	**1.51E-02**		f_8	2.09E+01	2.03E+01	2.07E+01	2.07E+01 [t]
		Std	3.03E+03	5.27E+03	1.42E-02	**4.64E-03**			6.64E-02	9.01E-02	1.12E-01	8.73E-02
		Best	6.07E+02	1.32E+03	1.71E-02	**6.80E-03**			2.08E+01	2.01E+01	2.04E+01	2.05E+01
	f_2	Mean	**1.96E+03**	2.68E+04	4.00E+03	1.34E+04		f_9	6.16E+01	1.73E+02	7.16E+01	**2.88E+00**
		Std	**2.01E+03**	4.56E+03	1.22E+03	3.98E+03			1.72E+01	3.80E+01	1.16E+01	**1.54E+00**
		Best	**2.44E+02**	1.63E+04	2.52E+03	5.93E+03			2.76E+01	1.01E+02	5.39E+01	**8.99E-03**
	f_3	Mean	1.55E+07	8.68E+07	1.15E+07	**7.60E+06**		f_{10}	1.34E+02	4.66E+02	2.78E+02	3.29E+02
		Std	9.18E+06	3.47E+07	4.24E+06	**3.79E+06**			**4.97E+01**	7.26E+01	7.11E+01	8.77E+01
		Best	4.68E+06	3.57E+07	6.47E+06	**3.88E+06**			**7.14E+01**	3.14E+02	1.41E+02	2.04E+02
1.00E+05	f_4	Mean	**4.00E+03**	3.79E+04	2.85E+04	4.26E+04	1.00E+05	f_{11}	**2.23E+01**	3.52E+01	3.04E+01	2.89E+01
		Std	**3.99E+03**	5.14E+03	9.81E+03	1.34E+04			**2.92E+00**	3.52E+00	3.29E+00	4.31E+00
		Best	**1.06E+03**	2.73E+04	1.24E+04	2.35E+04			**1.77E+01**	2.51E+01	2.27E+01	2.26E+01
	f_5	Mean	8.27E+03	2.15E+04	9.46E+03	**8.08E+03**		f_{12}	7.12E+04	2.09E+05	5.29E+04	**3.55E+04**
		Std	2.17E+03	4.22E+03	2.85E+03	**2.10E+03**			6.44E+04	7.24E+04	2.80E+04	**2.75E+04**
		Best	5.25E+03	1.37E+04	4.09E+03	**4.78E+03**			1.15E+04	1.03E+05	1.82E+04	**6.33E+03**
	f_6	Mean	5.85E+08	3.30E+08	**1.50E+03**	3.36E+03		f_{13}	4.19E+00	1.17E+01	5.65E+00	**3.05E+00**
		Std	5.47E+07	4.12E+08	**3.07E+03**	4.61E+03			1.76E+00	3.95E+00	1.41E+00	**6.91E-01**
		Best	3.17E+07	6.65E+06	**3.36E+01**	2.71E+01			1.89E+00	6.50E+00	3.24E+00	**1.78E+00**
	f_7	Mean	4.82E+03	4.70E+03	4.70E+03	**4.70E+03**		f_{14}	**1.27E+01**	1.33E+01	1.31E+01	1.30E+01
		Std	1.35E+02	7.22E-03	3.02E-01	**2.80E-03**			**3.60E-01**	3.12E-01	3.78E-01	3.71E-01
		Best	4.70E+03	4.70E+03	4.70E+03	**4.70E+03**			**1.16E+01**	1.22E+01	1.21E+01	1.19E+01
	f_1	Mean	3.26E+03	3.70E+03	1.45E-03	**7.91E-04**		f_8	2.09E+01	**2.01E+01**	2.06E+01	2.05E+01 [t]
		Std	1.55E+03	3.50E+03	4.07E-04	**2.88E-04**			5.87E-02	**9.29E-02**	1.22E-01	1.05E-01
		Best	6.79E+02	3.29E+02	7.70E-04	**3.53E-04**			2.10E+01	**2.00E+01**	2.03E+01	2.03E+01
	f_2	Mean	2.50E+03	1.48E+04	**7.40E+02**	4.05E+03		f_9	6.04E+01	1.33E+02	3.99E+01	**8.01E-02**
		Std	2.38E+03	3.93E+03	**2.86E+02**	1.30E+03			1.98E+01	2.87E+01	9.30E+00	**2.76E-01**
		Best	2.74E+02	8.92E+03	**3.27E+02**	2.02E+03			3.52E+01	8.21E+01	2.59E+01	**2.51E-04**
	f_3	Mean	2.26E+07	4.52E+07	5.90E+06	**3.33E+06**		f_{10}	**1.28E+02**	4.10E+02	3.00E+02	3.34E+02
		Std	2.50E+07	1.83E+07	2.11E+06	**1.39E+06**			**2.85E+01**	7.45E+01	8.15E+01	8.13E+01
		Best	5.58E+06	2.29E+07	2.20E+06	**9.61E+05**			**7.28E+01**	2.78E+02	1.47E+02	1.82E+02
3.00E+05	f_4	Mean	**2.93E+03**	3.20E+04	1.09E+04	2.77E+04	3.00E+05	f_{11}	**2.07E+01**	3.38E+01	3.02E+01	2.94E+01
		Std	**2.79E+03**	5.03E+03	4.48E+03	1.14E+04			**3.59E+00**	3.36E+00	3.16E+00	3.12E+00
		Best	**5.02E+02**	2.39E+04	4.04E+03	9.32E+03			**1.45E+01**	2.41E+01	2.51E+01	2.43E+01
	f_5	Mean	7.61E+03	1.62E+04	8.15E+03	**7.36E+03**		f_{12}	7.35E+04	9.31E+04	2.92E+04	**1.57E+04**
		Std	1.99E+03	3.89E+03	2.18E+03	**1.66E+03**			3.75E+04	4.03E+04	1.87E+04	**1.69E+04**
		Best	3.30E+03	7.42E+03	5.24E+03	**4.29E+03**			1.48E+04	3.56E+04	1.21E+04	**6.53E+02**
	f_6	Mean	6.98E+08	2.02E+07	2.76E+03	**1.50E+03**		f_{13}	3.24E+00	6.61E+00	3.45E+00	**1.98E+00**
		Std	1.05E+09	3.72E+07	3.95E+03	**3.43E+03**			2.14E+00	4.54E+00	6.81E-01	**4.19E-01**
		Best	4.11E+07	1.02E+05	2.90E+01	**2.75E+01**			1.85E+00	2.20E+00	2.40E+00	**1.28E+00**
	f_7	Mean	4.86E+03	4.70E+03	4.70E+03	**4.70E+03**		f_{14}	**1.23E+01**	1.31E+01	1.30E+01	1.31E+01
		Std	1.62E+02	2.00E-04	1.25E-04	**1.55E-05**			**4.30E-01**	3.06E-01	3.37E-01	2.87E-01
		Best	4.70E+03	4.70E+03	4.70E+03	**4.70E+03**			**1.13E+01**	1.22E+01	1.22E+01	1.25E+01

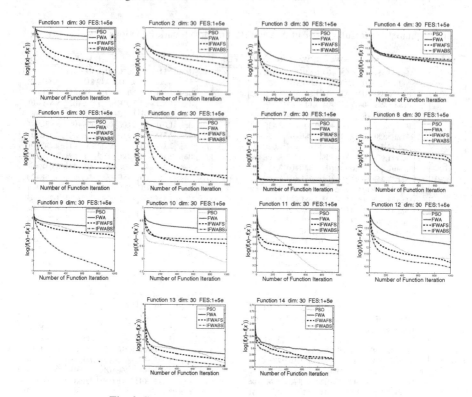

Fig. 6. Convergence cures on the benchmark functions

Acknowledgements. This work was supported by the Provincial Natural Science Foundation of Fujian under grants number 2012J01246, Foundation of supporting Province's College and University of Fujian under grants number JK2011035 and Foundation for Starting Research of Fujian university of technology under grants number E0600100.

References

1. Beni, G., Wang, J.: Swarm intelligence in cellular robotic systems. Robots and Biological Systems: Towards a New Bionics? 703–712 (1993)
2. De Castro, L.N., Von Zuben, F.J.: Learning and optimization using the clonal selection principle. IEEE Transactions on Evolutionary Computation 6(3), 239–251 (2002)
3. Dorigo, M., Maniezzo, V., Colorni, A.: Ant system: optimization by a colony of cooperating agents. IEEE Transactions on Systems, Man, and Cybernetics, Part B: Cybernetics 26(1), 29–41 (1996)
4. Gao, H., Diao, M.: Cultural firework algorithm and its application for digital filters design. International Journal of Modelling, Identification and Control 14(4), 324–331 (2011)
5. Janecek, A., Tan, Y.: Iterative improvement of the multiplicative update nmf algorithm using nature-inspired optimization. In: 2011 Seventh International Conference on Natural Computation (ICNC), vol. 3, pp. 1668–1672. IEEE (2011)

6. Janecek, A., Tan, Y.: Swarm intelligence for non-negative matrix factorization. International Journal of Swarm Intelligence Research (IJSIR) 2(4), 12–34 (2011)
7. Karaboga, D., Basturk, B.: A powerful and efficient algorithm for numerical function optimization: artificial bee colony (abc) algorithm. Journal of Global Optimization 39(3), 459–471 (2007)
8. Kennedy, J., Eberhart, R.: Particle swarm optimization. In: Proceedings of the IEEE International Conference on Neural Networks, vol. 4, pp. 1942–1948. IEEE (1995)
9. Pei, Y., Zheng, S., Tan, Y., Takagi, H.: An empirical study on influence of approximation approaches on enhancing fireworks algorithm. In: IEEE International Conference on System, Man and Cybernetics (SMC 2012), pp. 14–17. IEEE, Seoul (2012)
10. Shi, Y., Eberhart, R.: A modified particle swarm optimizer. In: The 1998 IEEE International Conference on Evolutionary Computation Proceedings: IEEE World Congress on Computational Intelligence, pp. 69–73. IEEE (1998)
11. Storn, R., Price, K.: Differential evolution–a simple and efficient heuristic for global optimization over continuous spaces. Journal of Global Optimization 11(4), 341–359 (1997)
12. Suganthan, P.N., Hansen, N., Liang, J.J., Deb, K., et al.: Problem definitions and evaluation criteria for the cec 2005 special session on real-parameter optimization. In: 2005 IEEE Congress on Evolution Computation (CEC), pp. 1–15. IEEE (2005)
13. Tan, Y., Xiao, Z.: Clonal particle swarm optimization and its applications. In: IEEE Congress on Evolutionary Computation (CEC 2007), pp. 2303–2309. IEEE (2007)
14. Tan, Y., Zhu, Y.: Fireworks algorithm for optimization. Advances in Swarm Intelligence pp. 355–364 (2010)
15. Zheng, S., Janecek, A., Tan, Y.: Enhanced fireworks algirithm. In: IEEE International Conference on Evolutionary Computation. IEEE (submitted, 2013)
16. Zheng, X.X., Y.J., H.F., L.: Differential evolution–a simple and efficient heuristic for global optimization over continuous spaces. Accepted by Neurocomputing (2013)

Diversity Analysis of Population in Shuffled Frog Leaping Algorithm

Lianguo Wang[1,2] and Yaxing Gong[2]

[1] College of Information Science and Technology, Gansu Agricultural University,
Lanzhou, China
[2] College of Engineering, Gansu Agricultural University, Lanzhou, China
wanglg@gsau.edu.cn, yaxing918@126.com

Abstract. The diversity of population is an important indicator for measuring optimal performance of swarm intelligence algorithms. The effect of three operators of Shuffled Frog Leaping Algorithm (SFLA) on the diversity of population and the average optimization results were analyzed in this paper by means of the simulation experiments. The results show that removing the global extreme learning operator will not only maintain the higher diversity of population, but also improve the operating speed and the optimization precision of the algorithm.

Keywords: swarm intelligence, shuffled frog leaping algorithm, diversity of population, function optimization.

1 Introduction

The diversity of population is an important indicator for measuring optimal performance of swarm intelligence algorithms. Lower population diversity would help local deep exploration, while higher population diversity would help global optimization to avoid the premature convergence. Nowadays, many scholars have already researched the diversity of population from different points of view and proposed many improved algorithms. Reference [1] testified the relationship between swarm diversity and global optimum capability through mathematical illation, and also presented a modified differential evolution (DE) algorithm based on a random mutation strategy for keeping the diversity. According to swarm diversity, reference [2] adopted a global distance disturbance strategy towards the worst particle which improves these particles' global searching ability. Besides, to increase the diversity of population in this paper, a probability disturbance was introduced for the best particle in the whole swarm. Reference [3] proposed a new method which integrates a diversity control strategy into quantum-behaved particle swarm optimization (PSO) to guide the particle's search and improve the capabilities of exploration. To overcome the premature convergence, exploitation ability of PSO was regulated through introducing fitness variance and position variance of the whole population in the evolutionary process to preserve the diversity of population in [4]. In reference [5], the searching behavior of

Y. Tan, Y. Shi, and H. Mo (Eds.): ICSI 2013, Part I, LNCS 7928, pp. 24–31, 2013.

PSO was effectively controlled via individual random walking and high diversity preserving to improve its optimal performance. Besides, reference [6] proposed an adaptive adjusted PSO algorithm, which could maintain the swarm diversity by adaptively adjusting uniform distribution of fitness value. The results showed that it could effectively improve the global searching ability of the algorithm. A different method to increase the particle swarm diversity was presented in [7]. As particle swarm lost its diversity, a particle outside the ultra-ball containing the swarm was selected randomly to use for disturbing the global best particle, and a disturbance was also introduced to the basic PSO formula to update each particle. Finally, reference [8] added the fuzzy controller and location hopping strategy to the algorithm for controlling the swarm diversity of PSO and improving its ability to jump out of local optimal solution.

Shuffled Frog Leaping Algorithm (SFLA) is one of biological evolutionary algorithms based on swarm intelligence presented by Eusuff and Lansey in 2003 [9]. It has the characteristics such as simple concept, fewer parameters, fast calculation speed, powerful optimal performance and easy to realize. SFLA has been successfully applied to many fields such as water distribution network, function optimization, network optimization of product oil pipeline, combination optimization, image processing, multi-user detection and power system optimization. This paper deeply studies the diversity of SFLA by using the concept of swarm diversity presented in [10], and analyzes the impact of three operators of SFLA on the swarm diversity by simulation experiments and propose some guidance advices for improvement of the algorithm.

2　Swarm Diversity

Swarm diversity refers to difference among individuals in population. It represents clustering degree of individuals in the whole searching space, and reflects the distribution of individuals to certain extent. Swarm diversity is also an important indicator to measure individual's coverage areas in searching space. Common sense suggests that evolution inevitably leads to diversity change. As diversity is higher, individual distribution is more dispersed and suitable for global detection, in addition, diversity is smaller, individual distribution is more concentrated and suitable for local exploration. It means the higher diversity is helpful for global exploration while the lower diversity is helpful to local search. However, as the swarm diversity reduces too low, there would be many similar individuals in population. Thus, the population cannot produce new individuals, searching would trap into local optimum and result in premature convergence. In this sense, maintaining higher diversity would enable algorithm to search more unknown areas and to strengthen the evolutionary ability of population. And the global searching ability and final optimal result would also be improved.

In this paper, the average distance from individual to the center of population is used to measure swarm diversity. The calculated equation was showed in eq. (1):

$$Diversity(P) = \frac{1}{N \times L} \sum_{i=1}^{N} \sqrt{\sum_{j}^{N} (x_{ij} - \overline{x}_j)^2} ; \qquad (1)$$

where P is the swarm; N is the swarm size; L is the length of the longest diagonal in the search space; n is the dimensionality of the problem; $\overline{x}_j$ is the j'th value of the average point. The quantified diversity is the real number between 0 and 1.

3   The Effects of Evolution Operator on Diversity

3.1   Evolution Operator of SFLA

SFLA treats every individual as a frog in the swarm. These frogs are divided into m subgroups and each group contains s frogs. The whole evolutionary process of the algorithm is realized by the following three operators:

1. The operator individuals learn from subgroup extremum (short for SO)

For each subgroup, new individual is generated by the best individual X_b and the worst individual X_w in current subgroup. If the fitness of this new individual is better than the previous X_w, the X_w will be updated by the new individual. The updating formula is as follows:

$$D(t+1) = rand() \times (X_b - X_w); \tag{2}$$

$$X_w(t+1) = X_w(t) + D(t+1); \tag{3}$$

where, $-D_{max} \le D \le D_{max}$; D is each moving step length; D_{max} is the maximum allowable step; t is current iteration; $rand()$ refers to a random number between 0 and 1. If the fitness of updated $X_w(t+1)$ is better than $X_w(t)$, $X_w(t)$ will be replaced by $X_w(t+1)$.

2. The operator individuals learn from global extremum (short for GO)

A new individual is generated by the global best individual X_g and the X_w of current subgroup when the solution of this group is not updated after learning from the subgroup extremum. X_w will be replaced by the new individual if the new fitness is better. The updated strategy eq.(2) in above step is changed to eq.(4) in this stage:

$$D(t+1) = rand() \times (X_g - X_w); \tag{4}$$

The update strategy (4) and (3) is executed in order. Similarly, $X_w(t)$ is updated by $X_w(t+1)$ when the fitness of $X_w(t+1)$ is better than $X_w(t)$.

3. The random operator (short for RO)

If the solution of subgroup is not improved after learning from subgroup extremum and global extremum, X_w will be replaced by a randomly generated solution which is controlled in the definition domain.

3.2   Experimental Design

In this paper, the authors took seven standard test functions of optimization problems to discuss the influence of the above three operators on swarm diversity and evolutionary process.

$$f_1(x) = \sum_{i=1}^{n} x_i^2$$

$$f_2(x) = \sum_{i=1}^{n-1} (100(x_{i+1}^2 - x_i)^2 + (1-x_i)^2)$$

$$f_3(x) = \sum_{i=1}^{n} (x_i^2 - 10\cos(2\pi x_i) + 10)$$

$$f_4(x) = \frac{1}{4000} \sum_{i=1}^{n} x_i^2 - \prod_{i=1}^{n} \cos(\frac{x_i}{\sqrt{i}}) + 1$$

$$f_5(x) = -20\exp\left(-0.2\sqrt{\frac{1}{30}\sum_{i=1}^{n} x_i^2}\right) - \exp\left(\frac{1}{30}\sum_{i=1}^{n}\cos 2\pi x_i\right) + 20 + e$$

$$f_6(x) = \sum_{i=1}^{n-1} (x_i^2 + x_{i+1}^2)^{0.25} [\sin^2(50(x_i^2 + x_{i+1}^2)^{0.1}) + 1]$$

$$f_7(x) = \frac{\pi}{n}\left\{10\sin^2(\pi y_1) + \sum_{i=1}^{n-1}(y_i - 1)^2\left[1 + 10\sin^2(\pi y_{i+1})\right] + (y_n - 1)^2\right\} + \sum_{i=1}^{n} u(x_i, 10, 100, 4)$$

$$y_i = 1 + 0.25(x_i + 1), n = 30$$

$$u(x_i, a, k, m) = \begin{cases} k(x_i - a)^m, & x_i > a \\ 0, & -a \le x_i \le a \\ k(-x_i - a)^m, & x_i < -a \end{cases}$$

In these experiments, the following 4 control strategies (S1~S4) were applied for explicitly indicating the effect of each operator in evolutionary process.

S1: SO, GO, RO exist simultaneously in the algorithm;
S2: Only GO and RO exist in the algorithm;
S3: Only SO and RO exist in the algorithm;
S4: Only SO and GO exist in the algorithm.

In these experiments, swarm size was set to 200 frogs. The swarm was divided into 20 subgroups and each group contained 10 frogs. The maximum number of iterations was 500 and the number of internal iterations in the subgroup was 10. The maximum allowable step $D_{max} = X_{max}/5$ (X_{max} refers to the maximum search range). The parameters settings of test functions were shown in table 1. Thirty trial runs were performed for each problem and the average of results was regarded as the final simulation result to reduce the random error in our simulation work.

Table 1. The parameters of benchmark test functions

Function	Dimension	Search range	Theoretical optimum
f_1	30	[-5.12,5.12]	0
f_2	30	[-30,30]	0
f_3	30	[-5.12,5.12]	0
f_4	30	[-600,600]	0
f_5	30	[-32,32]	0
f_6	30	[-100,100]	0
f_7	30	[-50,50]	0

3.3   Experiments Results

Each experiment for every test function had been repeated 30 times. Diversity in each iteration and the average of optimization results were recorded and drew as evolutionary curves to compare.

All the experiments demonstrate similar results. Due to the space limit of this paper, only the diversity curves of function f_1~f_4 varied with iterations were shown in Fig.1~Fig.4. The result shows that the highest diversity is presented when the algorithm is without GO (S3) of every function while the worst is gained when the algorithm is without RO (S4). The order of the diversity of all control strategies from high to low is: S3, S2, S1, S4.

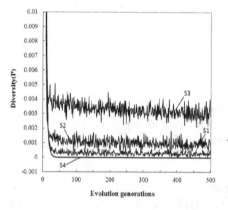

Fig. 1. The swarm diversity of f_1

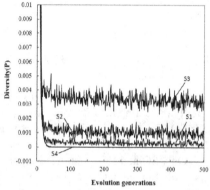

Fig. 2. The swarm diversity of f_2

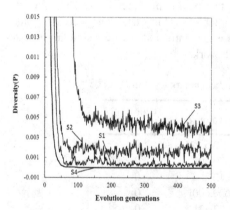

Fig. 3. The swarm diversity of f_3

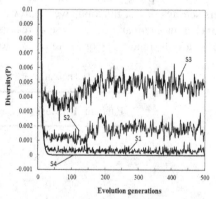

Fig. 4. The swarm diversity of f_4

Similarly, only the average evolutionary curves of function f_1~f_4 were shown in Fig.5~Fig.8. The vertical axis is the common log of the average extremum and the horizontal axis is corresponding iterations. To avoid function value being zero, 10^{-11} was added to each value as the final values. It can be seen from the illustrations that the best average optimization result of all functions is gained when the algorithm is without GO (S3), while the worst is gained when the algorithm is without RO (S4). In general, the order of the average optimization result of all control strategies from high to low is: S3, S2, S1, S4.

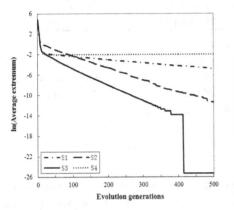

Fig. 5. Average evolutionary curve of f_1

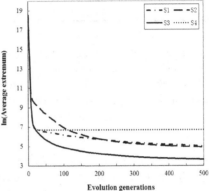

Fig. 6. Average evolutionary curve of f_2

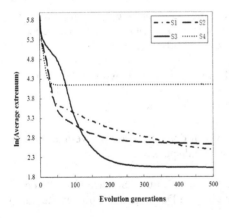

Fig. 7. Average evolutionary curve of f_3

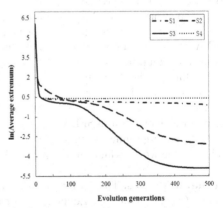

Fig. 8. Average evolutionary curve of f_4

To further verify the importance of three operators in the evolutionary process of SFLA, the authors counted the average effective improved times of each operator among SO, GO and RO which simultaneously existed in the algorithm, the result was listed in Table 2, in which the values in brackets are their success ratio (the proportion of average effective numbers in the total numbers).

Table 2. The average effective numbers of three operators

Function	Effective numbers of SO(success ratio)	Effective numbers of GO(success ratio)	Effective numbers of RO(success ratio)
f_1	479.23(95.85%)	20.24(4.05%)	0.53(0.11%)
f_2	479.43(95.89%)	20.03(4.01%)	0.53(0.11%)
f_3	471.75(94.35%)	24.52(4.90%)	3.73(0.75%)
f_4	479.25(95.85%)	20.19(4.04%)	0.56(0.11%)
f_5	475.71(95.14%)	21.69(4.34%)	2.59(0.52%)
f_6	452.21(90.44%)	24.48(4.90%)	23.30(4.66%)
f_7	478.42(95.68%)	20.85(4.17%)	0.73(0.15%)

Table 2 suggests that the success ratios of solution updated by SO were over 90%. For f_6, besides, the effective numbers of GO and RO were almost equal; for the rest functions, the success ratio of GO had remained above 4% while RO were under 1%. This demonstrates that SO plays a major role in SFLA.

4   Conclusions

1. The swarm diversity is lower as the algorithm simultaneously possesses SO, GO and RO (basic SFLA model). At the moment, its average optimal value is worse than the value when the algorithm is without GO. For individual functions, the optimum is even worse than the result when the algorithm is without SO. Therefore, the basic SFLA needs to be improved.

2. The swarm diversity and the average optimal value are the best when the algorithm only possesses SO and RO, which means that if GO could be cancelled in SFLA, it will not only enhance the optimization ability of the algorithm, but also shorten the run time.

3. Although the diversity is secondary when the algorithm is without SO, the optimization result is bad at this time. Besides, for the higher effective improved numbers of SO, it plays a main role in the evolutionary process, which shows that SO is an indispensable operator in SFLA.

4. As the algorithm is without RO, the swarm diversity and the average optimal value are the worst, which shows that RO is also indispensable in SFLA.

Acknowledgments. Thanks for the support from the Natural Science Foundation of China (61063028) and the Education Informatization Development Strategy Research Project in Gansu (2011-2).

References

1. Chen, A.H., Dong, X.M., Dong, Z., et al.: Differential evolution algorithms based on improved population diversity. Electronics Optics & Control 19(7), 80–84 (2012) (in Chinese)

2. Luo, D.S., Liu, Y.M.: Adaptive PSO based on swarm diversity for VRPSPD. Computer Engineering & Science 34(7), 160–165 (2012) (in Chinese)

3. Sun, J., Fang, W., Xu, W.B.: A Quantum-Behaved Particle Swarm Optimization With Diversity-Guided Mutation for the Design of Two-Dimensional IIR Digital Filters. IEEE Transactions on Circuits and Systems. Part II: Express Briefs 57(2), 141–145 (2010)

4. Yuan, L., Yuan, W.W.: A kind of algorithm for the improved particle swarm optimization. Journal of Shenyang Ligong University 31(3), 15–18 (2012) (in Chinese)

5. Chen, C.Y., Chang, K.C., Ho, S.H.: Improved framework for particle swarm optimization: Swarm intelligence with diversity-guided random walking. Expert Systems with Applications 38, 12214–12220 (2011)

6. Yang, Y.S.: A particle swarm optimization algorithm with adaptive adjusting. Journal of Xi'an University of Science and Technology 31(3), 356–362 (2011) (in Chinese)

7. Min, L., Liu, Q., Zhu, J.S.: An improved hybrid particle swarm optimization algorithm based on disturbance. Wireless Communication Technology 2, 43–47 (2012) (in Chinese)

8. Peng, L., Zhang, L.M., Deng, X.Y.: Particle swarm optimization based on fuzzy control of population diversity. Computer Simulation 29(4), 255–258 (2012) (in Chinese)

9. Eusuff, M., Lansey, K.E.: Optimization of water distribution network design using the shuffled frog leaping algorithm. Water Resources Planning and Management 129(3), 210–225 (2003)

10. Riget, J., Vesterstrφm, J.S.: A Diversity-Guided Particle Swarm Optimizer - The ARPSO. Technical report, Department of Computer Science, University of Aarhus (2002)

An Artificial Chemistry System
for Simulating Cell Chemistry: The First Step

Chien-Le Goh[1], Hong Tat Ewe[2], and Yong Kheng Goh[2]

[1] Faculty of Computing and Informatics,
Multimedia University, Cyberjaya, Malaysia
clgoh@mmu.edu.my
[2] Faculty of Engineering and Science, Universiti Tunku Abdul Rahman, Malaysia
{eweht,gohyk}@utar.edu.my

Abstract. Artificial chemistry is a man-made system that is similar to a real chemical system. It represents a good starting point to simulate cell processes from the bio-chemistry level. In this article, an artificial chemistry system which strikes a balance among closeness to reality, fast simulation speed and high flexibility is proposed. Preliminary results have shown that the model can simulate a general reversible reaction well.

Keywords: artificial chemistry, cell simulation.

1 Introduction

Among the techniques used in artificial life [1], there is the technique of simulating the bio-chemistry of life using artificial chemistry. Artificial chemistry is a man-made system that is similar to a real chemical system [2]. Typically, artificial chemistry simulates, in a virtual environment, molecules in a reaction space and the reactions which occur within the reaction space. The philosophy of this approach is to start from the bio-chemistry level, gradually discover the techniques to simulate cell processes and then move on to simulate higher level life behaviours within the constraints of our current knowledge of molecular cell biology and available computing power for simulation.

Various artificial chemistry models and simulators have been proposed by researchers in this field. Each is customized to fit its goals. The aim of this article is to present a new artificial chemistry model and its simulator as the first step to simulate a cell and a group of cells in the future. This work is based on existing models, integrating and extending suitable features to fit the needs to simulate cells.

The next section describes some representative artificial chemistry models. After that, the proposed model is described in section 3, experimental results in section 4 followed by the conclusion in Section 5.

Y. Tan, Y. Shi, and H. Mo (Eds.): ICSI 2013, Part I, LNCS 7928, pp. 32–39, 2013.

2   Related Work

In this section, some representative artificial chemistry models are described. Illustrations of the models can be found in Fig. 1.

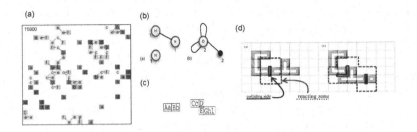

Fig. 1. (a) Squirm3, (b) synthon, (c) a string based model and (d) a tile based model

The artificial chemistry system, Squirm3, proposed by Hutton [3] to explore self-replication is a 2-D system simulating atoms and their reactions. The system has fixed six atom types (a, b, c, d, e and f) and each atom type has a variable state number (0, 1, 2, 3, ...). When two atoms collide, reactions can occur. A reaction can change the states of the atoms involved and may break or make bonds among atoms. The atoms move around in random as long as they stay within the Moore neighbourhood (eight surrounding squares) of other atoms they are bonded with. The possible reactions are predefined prior to the beginning of the simulation process. Hutton proved that his system is able to simulate self-reproduction of cells [4].

The synthon model is a model to describe molecules in detail from atoms, virtual atoms and the vertices among the atoms. Lenaerts and Bersini extended the synthon model of molecules by adding electron vertices and a method to model ionic bonds to the model [5]. Their extended model can very closely model molecules realistically. With the input of only basic reaction classes and the extended synthons, their simulator is able to generate realistic chemical reaction networks. The reactions networks can be used to predict possible reaction paths from the basic reaction classes and the possible resulting synthons from the reactions.

BioDrive [6] is an artificial chemistry model which models the reactions of molecules using differential equations. Concentration of each type of molecules in the system is used to determine the reaction rate. Reactions are expressed using differential equations and the effects of a reaction on other reactions are modelled mathematically. The precise locations of molecules are not taken into consideration in this model. Changes of concentrations of molecules against time can be calculated by solving the differential equations.

The artificial chemistry system proposed by Watanabe [7] uses inter-connected strings to represent molecules. Each string corresponds to an atom. A set of recombination rules act as reactions to transform the connected strings from one configuration to another. The strings do not only move in a compartment, which is used to represent a membrane structure, but are able to move from one compartment to another.

In [8], squares are used to represent the basic elements and to configure tiles of different shapes and sizes to represent molecules. Tiles are placed in a well mixed 'soup' from which they are randomly chosen to collide. The result of a collision are changes to the sizes and shapes of tiles. These change represent chemical reactions.

Our proposed model aims to strike a balance among closeness to reality, fast simulation speed and high flexibility to model the nature of a cell. Spheres, as shown in Fig. 2, which are less abstract than square tiles and text strings are used to represent molecules and atoms. The simulation space is in 3-D instead of 2-D. The movement and reaction of each molecule or atom are explicitly simulated instead of collectively simulated using differential equations to allow the highest level of flexibility. To counter the slowness in explicit simulation, highly realistic representation used in the synthon model is not used. Each sphere only has a name and a state number, which is used to represent the electronic state and the physical configuration of a molecule. As energy plays a major role in chemical reactions, each reaction in the model is specified together with the amount of energy consumed or released.

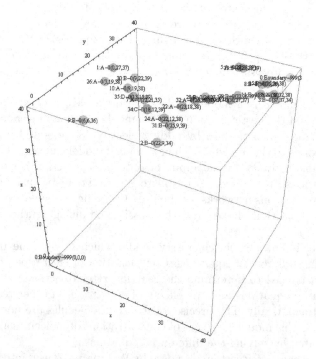

Fig. 2. Molecules in the proposed model

3   The Proposed Artificial Chemistry Model

In this section, the definition of molecules, movement of molecules and the simulation process is described.

3.1   Definition of Molecules

Spheres each with a radius of one unit length are used to represent molecules or atoms. The radius is a constant regardless of the actual size of the molecule represented. A molecule in this model can have one to n atoms. When it has only one atom, it is treated as an atom.

Each molecule has an id, a name and a state number. The id is a unique identifier in the system and the name can be a real chemical name (H_2O), a common name (water) or a generic name ($A, B, C...$). The state number accounts for the shape of the molecule and the electrical charges held by the molecule. The range of the state number is from 0 to the maximum positive integer value supported by the programming language used to implement the model. For example, in Fig. 2, there is a molecule at the coordinate (22, 9, 34) which has an id number 2, a generic name B and a state number 0.

3.2   Movement of Molecules

Molecules in this model move randomly to their neighbouring coordinates. A molecule in a coordinate (x, y, z) can move up to the coordinates $(x_1, y_1, z - 1)$, sideways to the coordinates (x_1, y_1, z) and downwards to the coordinates $(x_1, y_1, z - 1)$ where $x_1 \in \{x - 1, x, x + 1\}$ and $y_1 \in \{y - 1, y, y + 1\}$. To determine the target coordinate to move to, a random integer n between 1 to 36 is generated. If the $1 \leq n \leq 9$, the molecule will move up to one of the nine possible target coordinates, mapped to 1 to 9. Similarly, if $10 \leq n \leq 18$, the molecule moves sideways except when $n = 14$ because 14 is mapped to the current coordinate of the molecule. If $19 \leq n \leq 36$, the molecule will move downwards to the nine possible target coordinates where each coordinate is mapped to two numbers. The reason for this movement scheme is to create an environment where there is a molecule flow from the top down.

The environment where molecules exist and move is open, similar to an open environment where cells exist. At the beginning of a simulation, the environment is empty without any molecules. New molecules flow into the system from the top, gradually filling up the environment. Any molecules moving out of the simulation space are discarded from the system.

3.3   Reactions of Molecules

Reactions are specified with statements using the following syntax.

```
ID = 1
Energy = e
x1 + x2 + ... + xm -> y1 + y2 + ... + yn
```

Table 1. Examples of reaction syntax

Reaction	$H_2O \rightleftharpoons H^+ + OH^-$	$C_6H_{12}O_6 + 6O_2 \rightarrow 6CO_2 + 6H_2O$
Syntax	ID = 1	ID = 1
	Energy = 1	Energy = 5
	H2O-0 ->	C6H12O6-0 + O2-0 + O2-0 + O2-0 + O2-0 + O2-0 +
	H-1 + OH-2 .	O2-0 ->
		CO2-0 + CO2-0 + CO2-0 + CO2-0 + CO2-0 + CO2-0
	ID = 2	+ H2O-0 + H2O-0 + H2O-0 + H2O-0 + H2O-0
	Energy = -1	+ H2O-0.
	H-1 + OH-2 ->	
	H2O-0.	

ID is simply the unique identifier of reaction. xm and yn each represents a molecule or an atom with a label which includes a molecule name and a state number. e is the amount of energy needed for the reaction to occur or the amount of energy released by the reaction. A positive energy value indicates the amount of energy added to the virtual space (exothermic reaction) while a negative value indicates the amount of energy subtracted from the virtual space (endothermic reaction).

We use two examples in Table 1 to illustrate how chemical reactions are specified. The first example is the dissociation of water and the second example is the oxidation of glucose. The label of a molecule is an arbitrary text string. The state number of a molecule is normally zero. The use of positive integer values for state number is arbitrary as long as the differences in electronic and physical configurations can be expressed.

Although the amount of energy consumed or released by a reaction should be mapped appropriately to the reaction in reality, the correct mapping scheme is considered as future research work to be done. Similarly the correct mapping scheme of the state number is also considered as future work.

For a reaction to occur, sufficient local energy and the presence of all the molecules on the left hand side in close proximity is needed. Based on the total amount of global energy E, specified manually, the amount of local energy at each coordinate is calculated as E/l^3 where l is the length of cubic simulation space of volume l^3. The meaning of close proximity will be explained in the next sub-section.

3.4   The Simulation Process

Before a simulation based on the proposed model is run, several initial parameters have to be set. First is l which defines the length of a cubic simulation space. Second is the amount of global energy E. Third is the number of molecules to generate at each injection of molecules into the simulation space. Fourth is the specification of the probability of each molecule type to be generated and injected. Fifth is the uniform interval, specified as the number of simulation time step, between injections of molecules and sixth is the close proximity distance d.

At each time step of a simulation, molecules are selected one by one at random for processing until all molecules which exist in the simulation space are processed. For each selected molecule, the conditions for reactions are checked. If there is sufficient amount of local energy and all the molecules on the left hand side of a reaction, excluding the selected molecule, exist within the close proximity distance d of the selected molecule, a reaction occurs. Molecules inside the cube shown in Fig. 3 are considered as within the close proximity distance of d from a selected molecule at (x, y, z). If more than one reactions are possible, one of them is selected randomly to occur. Exothermic reactions are not checked for sufficiency of local energy.

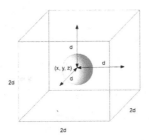

Fig. 3. The close proximity distance d from a selected molecule

If a reaction occurs, all reacting molecules including the selected molecules on the left hand side of the reaction statement will be discarded and the molecules on the right hand side of the reaction will be produced and placed randomly within the close proximity distance of the selected molecule. Any amount of energy released or consumed by the reaction will be added to or subtracted from the global energy E. This implies instant transfer or energy and energy diffusion is not considered in this model. If no reactions are possible, the selected molecule will just move randomly to its neighbouring coordinate.

4   Experimental Results

A simulator of the model has been implemented in C++. The simulator takes the necessary input and generates a series of Mathematica visualization commands as the output. Visualization of the output is done by executing the commands in Mathematica.

Using the simulator, a general reversible reaction, $A + B \rightleftharpoons C + D$, using abstract molecules with all state numbers set to zero was simulated. The parameter l was set to 40, E to 1,000,000 and d to 3. The number of molecules to generate at each injection was set to 5, the time interval between injections was set to 1 and the generation probabilities for A and B are both 0.5. Figure 4 shows the simulation space from time step 0 to time step 500.

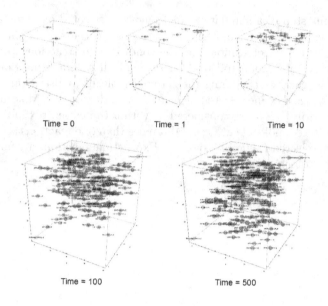

Time = 0   Time = 1   Time = 10

Time = 100   Time = 500

Fig. 4. A typical reversible reaction

The graph in Fig. 5 shows the number of molecules of A, B, C and D. It shows that the number of molecules of each molecule type eventually stabilized to an equilibrium similar to real-life chemistry.

The model and its implementation can support complex bio-chemical reactions because it does not impose a limit to the number of reactions it can simulate. However, as meaningful bio-chemical processes for cell simulation are still being studied, their simulation results are not described in this paper.

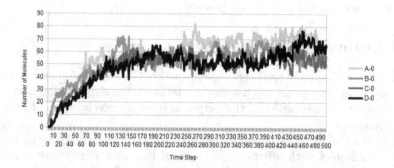

Fig. 5. The number of molecules of each molecule type

5   Conclusion

In this article, an artificial chemistry system which strikes a balance among closeness to reality, fast simulation speed and high flexibility has been proposed. It represents the first step towards a system to simulate a biological cell and a group of cells using artificial chemistry and programming code in the future. The system simulates molecules and reactions in 3-D and takes into consideration the energy aspect of chemical reactions.

For future work, complex bio-chemical pathways will be identified and simulated in the system. There are still a number of aspects of the model which need to be improved. They are the mapping scheme for energy needed for reactions, the mapping scheme for state numbers, a more realistic energy diffusion model, a model to simulate entangled molecules which move together, a model to support molecules moving at different speed and a more flexible way to inject molecules into the system.

Acknowledgement. This work is supported by the Malaysian Ministry of Higher Education under the Fundamental Research Grant Scheme.

References

1. Castro, L.: Fundamentals of natural computing: an overview. Physics of Life Reviews 4, 1–36 (2007)
2. Dittrich, P., Ziegler, J., Banzhaf, W.: Artificial Chemistries - A Review. Artificial Life 7(3), 225–275 (2001)
3. Hutton, T.J.: Evolvable Self-replicating Molecules in an Artificial Chemistry. Artificial Life 8(4), 341–356 (2002)
4. Hutton, T.J.: Evolvable Self-reproducing Cells in a Two-dimensional Artificial Chemistry. Artificial Life 13(1), 11–30 (2007)
5. Lenaerts, T., Bersini, H.: A Synthon Approach to Artificial Chemistry. Artificial Life 15(1), 90–103 (2009)
6. Kyoda, K., Muraki, M., Kitano, H.: Construction of a Generalized Simulator for Multe-cellular Organisms and its Application to SMAD Signal Transduction. In: Pacific Symposium on Biocomputing, pp. 314–325 (2000)
7. Watanabe, T., et al.: A Uniform Framework of Molecular Interaction for an Artificial Chemistry with Compartments. In: IEEE Symposium Artificial Life, pp. 54–60 (2007)
8. Yamamoto, T., Kaneko, K.: Tile Automaton in the Well-mixed Medium. Physica D (181), 252–273 (2003)

Maturity of the Particle Swarm as a Metric for Measuring the Collective Intelligence of the Swarm

Zdenka Winklerová

Brno University of Technology, Dept. of Intelligent Systems, Brno, Czech Republic
iwin@fit.vutbr.cz

Abstract. The particle swarm collective intelligence has been recognized as a tool for dealing with the optimization of multimodal functions with many local optima. In this article, a research work is introduced in which the cooperative *Particle Swarm Optimization* strategies are analysed and the collective intelligence of the particle swarm is assessed according to the proposed *Maturity Model*. The model is derived from the *Maturity Model of C2 (Command and Control)* operational space and the model of *Collaborating Software*. The aim was to gain a more thorough explanation of how the intelligent behaviour of the particle swarm emerges. It has been concluded that the swarm system is not mature enough because of the lack of the system's awareness, and that a solution would be some adaptation of particle's behavioural rules so that the particle could adjust its velocity using control parameters whose value would be derived from inside of the swarm system, without tuning.

1 Introduction

The particle swarm optimization is a population based optimization method first formulated in 1995 [1]. The *Particle Swarm Optimizer (PSO)* [2] is *a stochastic algorithm* applicable to any problem which can be characterized by an *objective function* so that the global extreme of the function has to be found. The algorithm does not require any additional information such as the derived gradient information, since the search towards the global extreme is driven by stochastic components of the particle velocity vector during the particle's flight through n-dimensional search space (hyperspace), where each particle represents a possible solution of the optimization problem. A brief description of how the algorithm works is as follows:

At the beginning of the search, some particle is identified as the best particle based on its objective function value. The swarm particles are then accelerated in the direction of this particle's position but also in the direction of their own best solutions. Each particle also has an opportunity to discover better solution during the flight, and the other particles then change direction and head for the new best particle position. The position x and velocity v of each particle in the original algorithm is updated according the following formulas:

$$v_n^i(t + \Delta t) = \omega v_n^i(t) + a_l r_1 [l_n^i(t) - x_n^i(t)]\Delta t + a_g r_2 [g_n(t) - x_n^i(t)]\Delta t \quad (1)$$

Y. Tan, Y. Shi, and H. Mo (Eds.): ICSI 2013, Part I, LNCS 7928, pp. 40–54, 2013.

$$x^i(t + \Delta t) = x^i(t) + v^i(t + \Delta t)\Delta t \tag{2}$$

with

$$n \in \mathbf{N}, r_1, r_2 \in \mathbf{U}(0, 1), \omega, a_l, a_g \in \mathbf{R}$$

where n is the number of degrees of freedom of the optimization problem, l^i is the ith particle's best position, g is the global best position of the whole swarm, ω is a real constant called inertia weight, r_1, r_2 are random numbers uniformly distributed in (0,1) to weight the local and global acceleration coefficients a_l, a_g, and Δt is a time step where the time step is assumed to be equal to one.

The equation (1) is interpreted so that the particle velocity vector consists of three components called inertia, cognitive, and social component. Both cognitive and social component contains acceleration coefficients as learning factors which influence the velocity and position of the particle. The cognitive component reflects the particle's own experience, whereas the social component expresses the interaction among particles. The inertia component keeps the particle's tendency to maintain its previous velocity.

It is evident that the *PSO* algorithm shows signs of intelligent behaviour because (i) each particle can remember its own best solution, (ii) there is an information flow among particles, by which they can communicate with the best particles depending on the topological neighbourhood structure, (iii) the swarm particles show cooperative behaviour without competition. In spite of this, the particle swarm shows unreliability in the optimization of multimodal functions with many local optima.

The *PSO* algorithm is known primarily as *a stochastic* algorithm because the particles during their flight through the hyperspace make use of a random chance. However, a swarm particle primarily orients itself using its own information about its current position and its best position it has gone through, and uses the information from the other particles about the global best position which had been previously recognized by the whole swarm. This underlying raw information is subsequently modified by the random numbers, and also by the acceleration coefficients, whose value is obtained empirically. The new direction and modulus of the particle's velocity then result from the processed information. *The behaviour of the particle is thus stochastic*, since the particle uses random elements in its decision, *but also deterministic*, since the particle learns from its own experience and/or the experience of the other particles.

Experiments have shown that depending on the specific objective function (specific problem), the predominant stochastic character of the particles' behaviour sometimes could prove advantageous for better leading to the optimum solution [3]. Thus, *could be the stochastic behaviour also taken for intelligent?* Since the particle does not have almost any information about the search space at the beginning of the search, its only opportunity is to start walking in a random direction and to gain the information. At the beginning of the search, the particle cannot rely on the experience of the other particles anyway, since their information is as poor as the particle's own information.

Is it therefore *desirable to enhance particle's cognitive abilities?* Experiments have shown that a relatively high value of the cognitive component at an early stage of the search results in extensive and, therefore, desirable particles' wanderings through the search space [3]. However, if the particle uses only its cognitive component at the later stages of the search, the performance of the algorithm is considerably worse [4]. One reason is that in the absence of the social component, no interaction among particles occurs, and thus no opportunity to use the experience of the other particles is taken. However, a relatively high value of the social component in the early stages of the search can lead to the premature convergence of the algorithm to a local optimum. The reason is that the particle should not offer the information to the others until it gains its own experience. But for some applications, on the contrary, the performance of the algorithm that uses only the social component is considerably better than the effectiveness of the original algorithm [5].

The existence of a cognitive organ is a prerequisite for individual's intelligence. Given that a cognitive ability is the ability to recognize, remember, learn, and adapt, the critical particle's ability seems (i) *the ability to adapt* its behaviour to the character of the search space and also (ii) *the ability to learn.* However, *how should we measure the collective intelligence of the swarm as a whole* if we know that the particles can act stochastically on the individual level or they can stifle their own experience completely, and in spite of it the algorithm can quickly converge to the correct solution, and thus the swarm as a whole acts intelligently?

Since the collective intelligence of the swarm emerges as a consequence of the particles' interactions, the swarm should be treated as a complex system, and the cooperative behaviour should be studied as an appropriate meta-transformation of the particle swarm system. As the first step to tackle the problem of the unreliability in terms of the meta-transformation, *it seems desirable to have a metric for measuring the collective intelligence of the swarm.* Hence, the aim of the research was (i) to introduce a *maturity model* of the particle swarm operational space as a metric for the collective intelligence of the swarm as a whole and then (ii) to assess the particle swarm intelligence according to this model. The model is proposed as a combination of the *Maturity Model of the C2 (Command and Control)* operational space and the model of *Collaborating Software.* Then, the whole particle swarm is assessed according to generic characteristics of the collaborative behaviour. While analysing the maturity of the swarm, the cooperative swarming strategies directly derived from the original version of the algorithm published in 1995, represented by equations (1) and (2), were examined.

2   Maturity Model of the C2 Operational Space

The *Maturity Model* of the *C2 (Command and Control)* operational space was suggested in 2006 [6] and further elaborated in 2010 [7]. The model is intended for distributed operational environment containing more autonomous entities constituting a system. The entities communicate through an informational exchange while solving a common task. The maturity of the operational space has to be assessed according to the following three basic indicators: (i) *the degree of allocation of decision rights* to the collective, (ii) *distribution of information,* and (iii) *patterns of interaction* among entities. The idea is illustrated in Fig. 1. Since the original diagram [7] does not evaluate the individual axes, the axes are evaluated additionally in a way the diagram is interpreted by the author of this article according to the author's own knowledge.

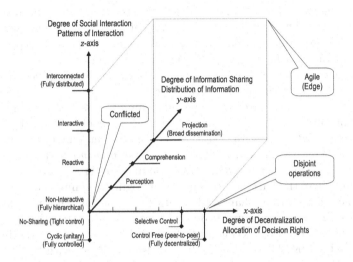

Fig. 1. *Maturity Model of the C2 (Command and Control)* operational space [6,7] as the proposed *Maturity Model of Particle Swarm Operational Space*; the axes (x, y, z) are evaluated additionally in a way the diagram is interpreted by the author.

The x-axis represents the degree of allocation of decision rights to the collective of entities, from *unitary* to *decentralized (peer-to-peer)* decision making as specified in [8] (pages 24-25). The decentralized decision making corresponds to fully autonomous entities. Initially, we could suppose that the swarm is a decentralized system, as each particle decides of its behaviour independently on the basis of the accessible information.

The y-axis represents the degree of information sharing among entities. While evaluating the y-axis, the commonly accepted *JDL (Joint Directors of Laboratories)* schema is applied [9]. The swarm particles could achieve the degree of *information sharing* from *perception* through *comprehension* to *projection* through spreading the information by means of selected topology.

The information accessible for sharing, of course, would not make any sense in case the particles would not react to this information. Hence, the z-axis indicates the degree of *social information sharing*. The evaluation of the z-axis results from the definition of the social interaction as a sequence of social action and subsequent reaction, which is accomplished by an information exchange. The purpose of the social action is to excite a reaction of one or more entities, resulting in subsequent change of their behaviour. With respect to the information exchange among entities, three levels of interactivity can be distinguished as the (i) *non-interactive* behaviour where there is no reaction on the action, (ii) *reactive* behaviour where the reaction is related only to the single action, and (iii) *interactive* behaviour where the reaction is related to more actions and to their mutual implications. The *patterns of interactions* will be thus assumed ranging from *non-interactive* through *reactive, interactive* to *integrated* (i.e. *interconnected*).

A three-dimensional *maturity vector* determines the resulting maturity of the operational space. The coordinates of the maturity vector determine the degrees of *decentralization, information sharing,* and *social interaction.* Depending on the values of the coordinates (x, y, z), the resulting *maturity of the operational space* then, according to [6], and [7] (page 80), takes one of the values (*conflicted, de-conflicted, coordinated, collaborative, agile*).

For illustration, let us give the following examples how to determine the maturity of the operational space: With *decentralized* control, shared *projection* and *integrated* (*interconnected*) entities, the operational environment is *agile.* With *unitary* control, *unshared* information, and *non-interactive* entities, the operational environment is *conflicted* (see the diagram in Fig. 1).

A swarm, where the particles use only their cognitive components, and where there is, therefore, neither shared information nor interaction among particles, seems *de-conflicted* but certainly it does not seem *collaborative.* Evidently worse behaviour of the cognitive version of the algorithm has been experimentally demonstrated in 1997 [4] and this finding is also consistent with the philosophy of the maturity model [6,7]. Hence, if a more effective swarm behaviour has to be achieved then we have to focus on the remaining indicators, namely on the *distribution of information* and the *interaction* among the particles.

If we determined the maturity values for all possible combinations of the values of (x, y, z) then we would see that not all combinations make sense. For example, with the value $(z = reactive)$, the value $(y = non\text{-}interactive)$ is not possible. The reason is that if the particle has to be reactive then the information must be delivered to it first. So we focus therefore on the question of what values should the coordinates (y, z) with the value $(x = decentralized)$ take so that the swarm could be considered as *coordinated* or *collaborative,* eventually.

3   Assessment of the Particle Swarm Given to the Generic Attributes of Collaborative Behaviour

In 2003, generic attributes of collaborative behaviour among software modules constituting a system were introduced [10]. These widely accepted attributes are: (i) an appropriate *representation* of information, (ii) the existence of *awareness,* (iii) *investigation,* (iv) *interaction,* (v) *integration,* and (vi) *coordination.*

We can note that the concepts of *interaction, integration,* and *coordination* are also discussed in the *C2 Maturity Model.* If we consider the swarm particles for the system's software modules and the *PSO* algorithm for the system then we get an opportunity to combine both models and to assess the whole particle swarm according to the characteristics of the collaborative behaviour.

3.1   Representation

Appropriate *representation* [10] of information means that the particles understand the sense of the *shared* information. The particles in their cooperative activities do not communicate directly, but they exchange information about the global best positions through shared variables within the system's environment. In case of communication among whole swarms [18], the individual swarms share the information through a context vector.

During the operation of a particle, the particle also uses other information that cannot be taken as shared information because the information does not originate from learning from the other particles. These are (i) the *random elements* r_1, r_2, (ii) the *inertia weight* ω, and (iii) the *acceleration coefficients* a_l, a_g (see equation (1)). The values of ω, a_l, a_g are supplied externally, and thus neither an observer inside of the system nor the particle itself decides on those values.

3.2   Awareness

The existence of *awareness* denotes the ability of the system to make the entities aware when something relevant to them occurs [10]. It means that the system must be able to react to any structural or functional change which can hamper the entities in its action, such as addition or loss of an entity, reducing its reliability, or its overloading, and the system still must be able to continue its operation toward the specified goal.

Within all the examined variants of the *PSO*, neither the loss nor overloading of a particle eventuates; only the particle must not leave the search space. All the examined variants of the *PSO* algorithm can detect a particle leaving the search space and return the particle back, so that the system recovers and continues in its operation. The theory of linear, discrete time dynamic systems applied for assessing the dynamic behaviour of a particle [22,23] deals with a set of tools for preventing the particle leaving the search space while ensuring its convergence and searching ability.

From the view of the system's reliability, however, the *PSO* algorithm exhibits one typical characteristic: The very nature of the particle swarm is that a particle communicates with a limited number of its neighbours, which implies *local rather than global searching ability* of the swarm. The local searching ability poses a pitfall for the multimodal object functions such that the algorithm could converge to some local optima prematurely. Since the swarm system itself is unable to decide whether its convergence is premature, its only opportunity is using the particles capabilities based on their capability to communicate.

From the communication point of view, two elemental models of the algorithm called *lbest* and *gbest* have been evolved. The difference lies in the set of the mutually influenced particles. The *gbest* model maintains only one particle as the global best solution in the whole swarm. This particle acts as an attractant causing the other particles to converge to it, which implies that the premature convergence can occur if the global best particle is not actualized properly.

The *lbest* model aims at preventing the premature convergence by maintaining multiple attractants. A subset of particles is defined for each particle as its neighbourhood, from which the local best particle is selected. The subsets are not disjoint so that a particle can be an element of more neighbourhoods, and the neighbourhoods are interconnected this way. Alternative *lbest* topologies have been designed [11,12,13], since a topology change can slow down the flow of information among particles. The slower spread of information allows particles to explore wider area of the search space, reducing the likelihood of premature convergence in the face of many local optima. The *gbest* model is a particular instance of the *lbest* model where the neighbourhood of a particle represents the whole set of the swarm particles. The *gbest* topology produces a better solution when applied to unimodal objective functions because of faster information spreading through the fully interconnected topology [13].

Another factor that affects the convergence ability of the algorithm is the *tendency of the particle to maintain its previous velocity*, from which the particle is then deflected by both cognitive and social component of particle velocity update vector (1). Therefore, the inertia weight coefficient ω has been introduced into the original *PSO* algorithm to balance the global and local searching [14]. The idea is that significantly higher inertia is necessary at the beginning of the search to explore the whole search space, whereas at later stages, when algorithm converges to the global optimum, a deceleration and thus a lower inertia is needed. The gradual deceleration is achieved so that the inertia weight coefficient ω decreases linearly for each iteration of the algorithm. The concept is known as *TVIW* (*Time Variable Inertia Weight*) [15,3]. Since it has been found that this linear concept is not effective for dynamic systems, the concept *RANDIW* (*Random Inertia Weight*) [16,3] has been introduced, where the coefficient ω was generated for several consecutive iterations using a uniform random distribution.

Although the *PSO-TVIW* algorithm localizes the optimal solution more quickly than the original version, its ability to fine-tune the optimal solution is still insufficient. The reason was understood as the *lack of particles' diversity particularly during the later stages* of the search [17,3]. Hence, another concept for enhancing the global search in early stages and for encouraging the particles to converge to the global optimum in later stages of the search has been considered, namely *TVAC* (*Time Varying Acceleration Coefficients*) [3]. It has been discovered that a relatively high value of the cognitive component at an early stage causes particles wandering through the entire search space, whereas with a relatively high value of the social component at an early stage, the particles could be trapped towards a local optimum. The conclusion is that both coefficients should pull together to make a joint effort so that both courses of searching would be balanced.

Another concept for increasing the particles' diversity is known as *CPSO* (*Cooperative Particle Swarm Optimizer*) [18] which splits the particle's position vector into components and each component is then optimized in a separate sub-swarm. To justify that the *CPSO* variant increases diversity of particles, diversity measures readily applied to *CPSO* have been proposed [24].

Even though the *PSO* modifications *TVIW*, *RANDIW*, and *TVAC* have improved the convergence ability of the original algorithm, yet they are not able to prevent *the premature convergence of the algorithm caused by stagnation of the particles*. If the current particle position is equal to the global best particle position, i.e. if $x^i = l^i = g$ in the equation (1), the particle moves only in direction of its velocity. If the particles' previous velocities are approaching zero then all the other particles cease to move in the moment they identify this particle. In doing so, it cannot be guaranteed that the algorithm converges to some local optimum – it ceases in the best position which had been previously recognized.

Specific strategies have been evolved with an intention to avoid the premature convergence caused by particles' stagnation. The *Guaranteed Convergence PSO (GCPSO)* variant of the original algorithm [18] is based on an observation whether or not the global best position has changed in comparison with the previous iteration. The *Self–organizing Hierarchical PSO (HPSO)* variant of the original algorithm [3] detects the decline of the particle's velocity and supplies the particle with a new randomly generated impulse. The *Comprehensive Learning PSO (CLPSO)* variant of the original algorithm [19] searches for particles which have been trapped in some local optima and then accelerates those weak particles towards a better particle in the swarm. This process of acceleration is associated with a comprehensive learning strategy so that the particle is allowed to learn from the best positions previously found by the other particles, thus enabling the particle to leave the local optimum.

Since the effectiveness of the *HPSO* algorithm was low with constant values of acceleration coefficients, the *HPSO* strategy was combined with the *TVAC, TVIW*, and *RANDIW* methods [3]. The best combination has proved the *HPSO-TVAC* but with one exception. To further improve the efficiency of the *CLPSO* strategy, the combination *CLPSO-TVIW* was used [19].

While assessing the *awareness ability* of the algorithm, we can learn that within all the above mentioned strategic frameworks, the global best particle tries alone to tackle its problems with the stagnation, since it does not want to become a bad learning example for the other particles. The other particles are not informed about the stagnation problem of the global best particle – the information about the stagnation is not shared.

Another inherent feature of the original *PSO* algorithm is that a particle could overshoot its target without knowing it. One of the reasons could be the fact that the particle cannot adjust the values of the inertia weight ω nor the values of the acceleration coefficients a_l, a_g using only the (shared) information accessible inside the swarm system, without the need for external control. The evidence that the control parameters could be *adapted* dynamically while the search is in progress is the *Self-adaptive Comprehensive Learning Particle Swarm Optimizer (SACLPSO)* [25].

3.3   Investigation

The *investigation* is used to quickly find information related to the activities being carried out [10]. The intelligence information is usually evaluated continuously and provided for an entity that needs it or that requests it.

All of the information shared among the particles has had a character of intelligence information virtually, since it has stemmed from the investigation of information throughout the operation of the whole swarm. The method of sharing is specified by the *gbest* or *lbest* topological structure and the intelligence information system is based on the corresponding learning strategy.

The *lbest* strategy which is based on the "intelligence" or "social intelligence" concept, has led also to the idea of making the particles "fully informed" [20]. Within the *Fully Informed Particle Swarm* (*FIPS*) variants of the *PSO*, the particle learns from the fully informed neighbourhood, where all neighbours are a source of influence. The original *PSO* can be seen as a special case of *FIPS*, which includes the selection of one particular neighbour to influence the particle, and the particle learns from two examples, i.e. from the particle's best position and from the best neighbour. Experiments with various neighbourhoods have shown better results than the original *PSO* algorithm but with increasing particle's neighbourhood, the performance of the swarm has become worse [20]. The topologies where the particle has been "truly fully informed", that is, where the particle gains the information from each particle of the swarm, the results have been very worse, as also the authors of the *FIPS* study have expected, and thus increasing the ability of information sharing up to this extreme does not seem a solution to enhance the particle in its decision making.

In the *CLPSO* algorithm [19], both cognitive and social component from (1) are replaced by a new component in which the best position is composed so that each of the particle's dimension learns from an example of the corresponding dimension of another particle called *exemplar*. A particle is allowed to learn if it has not improved its position after a certain number of iterations. A random number is generated for each dimension and if this number is greater than the particle's (empirically obtained) learning probability then the particle learns from its own best position, otherwise it learns from the personal best of another particle which is selected as the better of two randomly chosen particles. Each particle can therefore become a learning example and can determine the direction of the motion of the other particles. Moreover, each of a particle's dimension can learn from the corresponding dimension of another particle.

3.4 Interaction

The *interaction* means the ability of entities to influence each other while working on a shared task [10]. The influencing activity is manifested so that an entity changes its behaviour as a response to specific information. The effect can be only reactive, but in case of the interaction, mutual implications are essential.

The particles in the swarm interact with one another since (i) the intelligence system of the *PSO* algorithm has supplied the particles with the intelligence information, (ii) information sharing is then enabled by the topological structure (*lbest* or *gbest*) by which the particles are interconnected, and (iii) a particle updates its velocity and position following its own information and the shared information in accordance with the corresponding learning strategy.

Since the velocity vector of a particle in the original version of the *PSO* algorithm (1) directs the particle to the best position achieved in its topological neighbourhood, the particle moves to the best achieved position but the situation may change in the subsequent iterations, and the particle may itself become the best particle to which the other particles are directed. Because the velocities of the particles in the subsequent iterations have reflected both the particle's own experience and the experience of the other particles, the implications are mutual and thus the particles interact this way.

It is important to note that even if the particles interacts, the premature convergence because of stagnation of the particles in the search space may occur. The existing strategies how to lead the particle out of stagnation may use different ways from the original algorithm to update the velocity vector of the stagnating particles. In general, we can consider the behaviour of the particles as being interactive if the particles use the social components of their velocity update vectors.

3.5   Integration

The *integration* [10] lies in the ability of the entities to combine their own results with the results obtained from the other entities. It can be described as a state of the system where the entities are able to operate simultaneously and to react continuously on the intermediate results while solving a common task.

The simultaneous operation and the continuous reaction on the intermediate results is successfully simulated in a pseudo-parallel manner. The particles' activities take place sequentially in two consecutive cycles, where both cycles run over all the particles. The first cycle updates the best positions and then it is followed by the second cycle which changes the hyperspace so that it updates the particles' velocities and positions. Within the original version of the algorithm, each dimension of the n-dimensional velocity vector is updated according to the equation (1) separately, and then the position of the ith particle is updated according the equation (2). The updates are performed in the sequence that was assigned for the particles during the initialization of the swarm.

In both *HPSO* [3] and *CLPSO* [19] variants of the original *PSO* algorithm, each dimension of a particle is updated in particular so that two consecutive cycles running over n particle's dimensions are nested into the main cycle over all particles. In the first nested cycle, the best positions are updated, followed by changes of the hyperspace in the second nested cycle. The simultaneous operation and continuous reaction to the intermediate results is thus simulated on the level of the particle's particular dimensions.

The *CPSO-S* variant of the original algorithm [18] introduces n particular swarms, each of which optimizes one particular dimension of the original swarm of n-dimensional particles. The individual swarms are optimized in a sequential manner, and the information is exchanged through a context vector. The updates are performed in two consecutive cycles analogously as in the original version of *PSO* algorithm, but over the individual swarms. The simultaneous operation and continuous reaction to the intermediate results are thus simulated both on the level of the individual swarms and inside of the swarms on the level of the particles' dimensions.

When integrating the results of entities operating in mutual interactions, a novel type of control activity and organization at a higher level should emerge. A meta-system that is more complex, more intelligent and more flexible in its activity than the original system of independent entities should arise. In accordance with the *systems theory*, the integration of the parts during the meta-transformation is typically followed by their specialization, where different parts take different roles within the meta-system and undergo appropriate modifications. Each of the specialized parts remain a separate autonomous entity but with special features. The process is called differentiation of parts.

In the sense of the meta-transformation, a swarm of particles in the original *PSO* algorithm can be considered a meta-system in which, however, we cannot observe the differentiation of parts. The *CPSO-S* [18] individual swarms could be considered the specialized parts, since each swarm can be specialized to optimize one dimension. We can see a higher degree of integration in the *CPSO-H$_K$* (*Hybrid CPSO*) [18] variant of the original algorithm, where the pseudo-parallel activity and the continuous reaction to the intermediate results has been brought on the level of two algorithms.

3.6   Coordination

The *coordination* is an effort combining and guiding the entities to focus their activities on the right things at the right time [10]. The coordination of cooperating entities is carried out on the principle of *self-organization* in the absence of a central element. A relatively autonomous operation of individual entities that deal with their individual tasks and communicate in order to coordinate their activities is the essence of the synergy.

The individual task of the particle in a swarm is to move closer to the global optimum in the subsequent iterations. The common task of the particle swarm is to converge to the global optimum. The coordination of activities of the particles in a swarm is achieved (i) by regulating the speed of information flow using the *lbest* topological structure of the particles, and also (ii) by velocity control using the inertia weight and the acceleration coefficients.

As already mentioned in section 3.2, different topologies for the *lbest* model have been constructed because they allow slowing down the speed of the information flow among particles [11]. For each particle, a subset of particles as a neighbourhood is defined where each particle becomes part of the neighbourhood of one or more particles. All particles are connected this way and may exchange information indirectly. The relatively long path between distant particles from different neighbourhoods then slows down the information exchange between the distant particles, which enables the particles to explore different areas in the search hyperspace at the same time while maintaining the ability to share information. The particle is chosen in the neighbourhood according to its initially assigned index i, regardless of the particle's position in the search space. The topology is defined during the initialization of the swarm and becomes a characteristic of the environment so that the particles adapt to it naturally. The topology thus does not limit the autonomous activity of the entities.

In respect to the coordination of activities of the particles in a swarm with the velocity control based on the modification of the inertia weight and the acceleration coefficients as mentioned in section 3.2, it can be concluded that the coordination is addressed to balance the lack of awareness. While doing this, a particle does not adjust the inertia nor acceleration based on learning from the swarm search space, since the particle has no information on which it could decide what inertia and acceleration should update its velocity in its current position. The particle's velocity during its flight through the hyperspace is instead regulated artificially from outside of the system.

4   Discussion

The subject of the discussion was to determine what values should the coordinates (y, z) in the maturity model shown in Fig. 1 with the value $(x = decentralized)$ take so that the swarm could be considered as *coordinated* or *collaborative*.

The highest achieved level of *information sharing* (y-axis) can be evaluated as *shared projection*. While conducting its activity, each particle has an opportunity to learn the accessible information about the global best position (*shared perception*), to estimate the situation and to update the shared information if it becomes the global best particle (*shared comprehension*), and to move in the estimated direction. The swarm particles control their activities according to the best position yet discovered, i.e. they *project* this global information into their activities (*shared projection*).

In terms of the particle's ability to combine its own results with the results of the other swarm particles (z-axis), the swarm can be considered as an *integrated system*, since the swarm particles in all variants of the *PSO* algorithm are able to operate simultaneously and react continuously on the intermediate results.

In terms of *allocation of decision rights* (x-axis), the particles are not completely autonomous, since – except for isolated cases – the particle's velocity during its flight through the hyperspace is artificially regulated from outside the system. Thus, compared with the original assumption ($x = decentralized$), the degree of decentralization should be labelled rather as ($x = selective control$).

Hence, the answer to the question under what conditions the swarm could be considered as *coordinated* or *collaborative* is that rather than information sharing or social integration of particles, the *decision making autonomy of individual particles should be enhanced.*

5   Conclusion

It follows from the discussion that the method of the particles' coordination based on the application of tuning parameters *established externally*, namely by *external inertia weight* and *external acceleration coefficients*, is a way of control which is inconsistent with the principle of self-organization as an *autonomous activity* of the individual entities.

The conclusion from the maturity assessment of the particle swarm is that the swarm system as a whole is not mature enough, as (i) a swarm particle does not have the ability of learning *to get all the information* needed to update its velocity, and (ii) if a particle has a problem, the other particles do not know about it. Lack of the system's awareness seems to be the main cause of the persistent unreliability of the original *PSO* algorithm and its variants in optimization of multimodal functions with many local optima.

A solution would be a continual adjustment of the particle's velocity as a function of its level of confidence in its actual direction, as outlined in [21]. The confidence should be based on a particular *qualitative assessment* of the values of the objective function obtained in a given iteration. Such approach, however, would involve an adaptation of the particles' behavioural rules so that the particle could use only the coefficients (ω, a_l, a_g) whose value would be provided from inside the system, without external control.

The author's future research plan is to enhance the role of the recognized hypothetical observer within the original particle swarm – its responsibilities yet are (i) the continuous monitoring of the particles' activities and (ii) providing the particles with the information about the personal and global best positions. Enhanced observer's responsibility would be (iii) providing the particles with the set of control parameters (ω, a_l, a_g) to balance the exploration-exploitation performance, and (iv) maintaining the dynamic stability of the swarm with respect to the limitations resulting from the analyses of particle's dynamic behaviour as outlined in [22,23]. The swarming strategy would be based on time-varying adaptive control coefficients. At the beginning of the search, only local searching involving (ω, a_l) would be permitted until the hyperspace is explored as uniformly as possible. To assess the degree of the uniformity, the observer would divide the hyperspace into hyper-cubes (i.e. intervals, squares, cubes, ...) according to the particle's dimension, and would maintain a histogram that would increment the number of positions being explored by the whole swarm and falling into the disjoint hyper-cubes. The chi-squared test for assessing to what extent the observed distribution of the explored positions fits the uniform distribution could be used in each iteration. When the degree of uniformity is sufficient then the global searching (a_g) can be involved. Now, the observer could take into account that the global best (and/or worst) position could become an outlier. In such case, the observer has an opportunity to speed up the global search. To identify an outlier, a statistical method of analysis of extreme values could be used as a qualitative assessment in which the set of particles' personal best *values of objective function* would be treated as a random sample.

The author's established hypothesis is that the dependencies between the chi-squared test value at the beginning of the search, the emergence of outliers after involving the global search, and the particle's velocity control parameters exist. If the dependencies were discovered then a deeper understanding of how the intelligent behaviour of the particle swarm emerges, could be brought about.

Acknowledgements. The author thanks the anonymous reviewers for helpful comments that improved the manuscript. This work was partially supported by the *BUT FIT* grant *FIT-S-11-1* and the research plan *MSM0021630528*. Extended abstract of this article has been published by Springer–Verlag Berlin Heidelberg 2012 (`www.springer.com`). The original source [26] is available at `http://link.springer.com/chapter/10.1007/978-3-642-32650-9_40`. The development and validation of the *NATO Network Enabled Capability Command and Control Maturity Model (N2C2M2)* was a major undertaking of the *NATO* research task group *SAS-065*.

References

1. Kennedy, J., Eberhart, R.C.: Particle swarm optimization. In: Proceedings of the IEEE International Conference on Neural Networks, Perth, Australia, vol. IV, pp. 1942–1948 (1995)
2. Eberhart, R.C., Kennedy, J.: A new optimizer using particle swarm theory. In: Proceedings of the Sixth International Symposium on Micro Machine and Human Science, Nagoya, Japan, pp. 39–43 (1995)
3. Ratnaweera, A., Halgamuge, S.K., Watson, H.C.: Self-organizing Hierarchical Particle Swarm Optimizer with Time Varying Acceleration Coefficients. IEEE Transactions on Evolutionary Computation 8(3), 240–255 (2004)
4. Kennedy, J.: The particle swarm: Social adaptation of knowledge. In: Proceedings of International Conference on Evolutionary Computation, Indianapolis, USA, pp. 303–308 (1997)
5. Clearwater, S.H., Hubermann, B.A., Hogg, T.: Cooperative Problem Solving. In: Computation: The Micro and Macro View, pp. 33–70. World Scientific, Singapore (1992)
6. Moffat, J., Alberts, D.S.: Maturity Levels for NATO NEC. TR21958 V 2.0, Defence Science & Technology Laboratory, U.K (December 2006)
7. Alberts, D.S., Huber, R.K., Moffat, J.: NATO NEC C2 maturity model. DoD Command and Control Research Program (February 2010) ISBN 978-1-893723-21-4
8. Alberts, D.S., Hayes, R.E.: Power to the Edge Command.. Control.. in the Information Age, 1st printing. CCRP Publication Series, Washington, D.C. (June 2003) (reprint June 2004) ISBN 1-893723-13-5
9. Llinas, J., Bowman, C., Rogova, G., Walz, E., White, F.: Revisiting the JDL Data Fusion Model. In: Proceedings of the 7th International Conference on Information Fusion, Stockholm, Sweden (2004)
10. Corkill Daniel, D.: Collaborating Software: Blackboard and Multi-Agent Systems & the Future. In: Proceedings of the International Lisp Conference, New York (2003)
11. Kennedy, J.: Small Worlds and Mega-Minds: Effect of Neighborhood Topology on Particle Swarm Performance. In: Proceedings of the Congress on Evolutionary Computation, Washington, DC, USA, pp. 1931–1938 (July 1999)
12. Kennedy, J., Mendes, R.: Population Structure and Particle Swarm Performance. In: Proceedings of the IEEE Congress on Evolutionary Computation, Hawaii, USA, pp. 1671–1676 (2002)
13. Peer, E.S., van den Bergh, F.: Engelbrecht, A. P.: Using Neighborhoods with the Guaranteed Convergence PSO. In: Proceedings of the IEEE Swarm Intelligence Symposium, Indianapolis, USA, pp. 235–242 (2003)

14. Shi, Y., Eberhart, R.C.: A Modified Particle Swarm Optimiser. In: Proceedings of the IEEE International Conference of Evolutionary Computation, Anchorage, Alaska, pp. 69–73 (May 1998)
15. Shi, Y., Eberhart, R.C.: Empirical study of particle swarm optimization. In: Proc. of IEEE International Congress on Evolutionary Computation, vol. 3, pp. 1945–1950 (1999)
16. Eberhart, R.C., Shi, Y.: Tracking and Optimizing Dynamic Systems with Particle Swarms. In: Proc. of IEEE Congress on Evolutionary Computation 2001, Seoul, Korea, pp. 94–100 (2001)
17. Angeline, P.J.: Evolutionary Optimization Verses Particle Swarm Optimization: Philosophy and the Performance Difference. In: Porto, V.W., Waagen, D. (eds.) EP 1998. LNCS, vol. 1447, pp. 601–610. Springer, Heidelberg (1998)
18. van den Bergh, F.: An Analysis of Particle Swarm Optimizers. PhD thesis. Department of Computer Science, University of Pretoria. South Africa (2002)
19. Liang, J.J., Qin, A.K., Suganthan, P.N., Baskar, S.: Comprehensive Learning Particle Swarm Optimizer for Global Optimization of Multimodal Functions. IEEE Transactions on Evolutionary Computation 10(3), 281–295 (2006)
20. Mendes, R., Kennedy, J., Neves, J.: The Fully Informed Particle Swarm: Simpler, Maybe Better. IEEE Transactions on Evolutionary Computation 8(3), 204–210 (2004)
21. Torney, C., Neufeld, Z., Couzin, I.D.: Context – Dependent Interaction Leads to Emergent Search Behavior in Social Aggregates. PNAS 106(52), 22055–22060 (2009)
22. Clerc, M., Kennedy, J.: The Particle Swarm – Explosion, Stability, and Convergence in a Multidimensional Complex Space. IEEE Transactions on Evolutionary Computation 6(1), 58–73 (2002)
23. Fernández Martínez, J.L., García Gonzalo, E.: The PSO Family: Detection, Stochastic Analysis and Comparison. Swarm Intell. 3(4), 245–273 (2009), doi:10.1007/s11721-009-0034-8
24. Ismail, A., Engelbrecht, A.P.: Measuring Diversity in the Cooperative Particle Swarm Optimizer. In: Dorigo, M., Birattari, M., Blum, C., Christensen, A.L., Engelbrecht, A.P., Groß, R., Stützle, T. (eds.) ANTS 2012. LNCS, vol. 7461, pp. 97–108. Springer, Heidelberg (2012)
25. Ismail, A., Engelbrecht, A.P.: The Self-adaptive Comprehensive Learning Particle Swarm Optimizer. In: Dorigo, M., Birattari, M., Blum, C., Christensen, A.L., Engelbrecht, A.P., Groß, R., Stützle, T. (eds.) ANTS 2012. LNCS, vol. 7461, pp. 156–167. Springer, Heidelberg (2012)
26. Winklerová, Z.: Maturity of the Particle Swarm as a Metric for Measuring the Particle Swarm Intelligence. In: Dorigo, M., Birattari, M., Blum, C., Christensen, A.L., Engelbrecht, A.P., Groß, R., Stützle, T. (eds.) ANTS 2012. LNCS, vol. 7461, pp. 348–349. Springer, Heidelberg (2012) ISBN 978-3-642-32649-3, ISSN 0302-9743, doi:10.1007/978-3-642-32650-9

Particle Swarm Optimization in Regression Analysis: A Case Study

Shi Cheng[1,2,*], Chun Zhao[1,2], Jingjin Wu[1,2], and Yuhui Shi[2]

[1] Department of Electrical Engineering and Electronics,
University of Liverpool, Liverpool, UK
[2] Department of Electrical & Electronic Engineering,
Xi'an Jiaotong-Liverpool University, Suzhou, China
shi.cheng@liverpool.ac.uk, yuhui.shi@xjtlu.edu.cn

Abstract. In this paper, we utilized particle swarm optimization algorithm to solve a regression analysis problem in dielectric relaxation field. The regression function is a nonlinear, constrained, and difficult problem which is solved by traditionally mathematical regression method. The regression process is formulated as a continuous, constrained, single objective problem, and each dimension is dependent in solution space. The object of optimization is to obtain the minimum sum of absolute difference values between observed data points and calculated data points by the regression function. Experimental results show that particle swarm optimization can obtain good performance on regression analysis problems.

Keywords: Particle swarm optimization, regression analysis, regression models, weighted least absolute difference value.

1 Introduction

Swarm intelligence, which is based on a population of individuals, is a collection of nature-inspired searching techniques. Particle Swarm Optimization (PSO), which is one of the swarm intelligence algorithms, was introduced by Eberhart and Kennedy in 1995 [9, 13]. It is a population-based stochastic algorithm modeled on social behaviors observed in flocking birds. Each particle, which represents a solution, flies through the search space with a velocity that is dynamically adjusted according to its own and its companion's historical behaviors. The particles tend to fly toward better search areas over the course of the search process [4, 5, 10].

Optimization, in general, concerns with finding the "best available" solution(s) for a given problem, and the problem may have several or numerous optimum solutions, of which many are local optimal solutions. The goal of global optimization is to make the fastest possible progress toward the "good enough"

---

* Currently with the Division of Computer Science, The University of Nottingham Ningbo, China (Shi.Cheng@nottingham.edu.cn).

Y. Tan, Y. Shi, and H. Mo (Eds.): ICSI 2013, Part I, LNCS 7928, pp. 55–63, 2013.

solution(s). Evolutionary optimization algorithms are generally difficult to find the global optimum solutions for multimodal problems due to the possible occurrence of premature convergence [2,3,6].

The Cole-Davidson relaxation function is important in dielectric relaxation [17]. The Cole-Davidson parameters consist of τ and β. These pairs of parameters are different at different temperatures. In this paper, the proper parameters are found based on the observed data and the regression function. The regression process is formulated as a continuous, constrained, single objective problem, and each dimension is dependent in solution space. The traditional methods are difficult to solve this kind of problems due to the problem is non-separable, multimodal, and noisy. Particle swarm optimization is utilized in the regression process. The object of optimization is to obtain the minimum sum of absolute difference values between calculated value and the observed value.

The rest of this paper is organized as follows. Section 2 reviews the basic particle swarm optimization algorithm. In Section 3, the basic concept of regression analysis is introduced. The experimental setup is described in Section 4 which includes the regression problem statement, optimization problem representation, parameter setting, and regression results. Finally, Section 5 concludes with some remarks and future research directions.

2   Particle Swarm Optimization

Each particle represents a potential solution in particle swarm optimization, and this solution is a point in the n-dimensional solution space. Each particle is associated with two vectors, i.e., the velocity vector and the position vector. The position of a particle is represented as x_{ij}, i represents the ith particle, the velocity of a particle is represented as v_{ij}, $i = 1, \cdots, m$, and j is the jth dimension, $j = 1, \cdots, n$. The m represents the number of particles, and n the number of dimensions.

The particle swarm optimization algorithm is easy to implement. The velocity and the position of each particle are updated dimension by dimension, and the fitness is evaluated based on a position of all dimensions. The update equations for the velocity v_{ij} and the position x_{ij} are as follow [11,14]

$$\mathbf{v}_i(t+1) \leftarrow w_i\mathbf{v}_i(t) + c_1\text{rand}()(\mathbf{p}_i - \mathbf{x}_i(t)) + c_2\text{Rand}()(\mathbf{p}_g - \mathbf{x}_i(t)) \quad (1)$$
$$\mathbf{x}_i(t+1) \leftarrow \mathbf{x}_i(t) + \mathbf{v}_i(t+1) \quad (2)$$

where w denotes the inertia weight and usually is less than 1 [16], c_1 and c_2 are two positive acceleration constants, rand() and Rand() are two random functions to generate uniformly distributed random numbers in the range $[0,1)$, $\mathbf{x}_i$ represents the ith particle's position, $\mathbf{v}_i$ represents the ith particle's velocity, $\mathbf{p}_i$ is termed as personal best, which refers to the best position found by the ith particle, and $\mathbf{p}_g$ is termed as local best, which refers to the position found by the members in the ith particle's neighborhood that has the best fitness value so far.

Different topology structure can be utilized in PSO, which will have different strategy to share search information for every particle. Global star and local ring are two most commonly used topology structures. A PSO with global star structure, where all particles are connected to each other, has the smallest average distance in swarm, and on the contrary, a PSO with local ring structure, where every particle is connected to two near particles, has the biggest average distance in swarm [2, 15].

3   Regression Analysis

In statistics, regression analysis is a statistical technique for estimating the relationships among variables. The curve fitting is a process that apply regression analysis to data.

In general, regression models involve three parts of variables [12]:

- The unknown parameters, denoted as β, which may represent a scalar or a vector.
- The independent variables, i.e., input vector $\mathbf{x} = (x_1, x_2, \cdots, x_p)$.
- The dependent variable, i.e., an output y.

A regression model relates y to a function of $\mathbf{x}$ and β,

$$y \approx f(\mathbf{x}, \beta) \tag{3}$$

The parameters β are the goals that to be found based on the input vector $\mathbf{x}$ and real world observed output y. In this paper, we have a regression function, and a series of data points, which contains the input vector and the observed output data points. The goal of regression is to find the proper parameters for regression function.

4   Particle Swarm Optimization in Regression Analysis

4.1   Dielectric Relaxation Problem Statement

The following Cole-Davidson relaxation function is important in dielectric relaxation [17]. The input vectors are the different frequencies and the different temperatures, while the outputs are the Cacc/Cmax values. The parameters β and τ need to be tuned by the observed data. The equation is as follows:

$$\text{Cacc/Cmax} = \varepsilon^\infty + (\varepsilon^S - \varepsilon^\infty) \times (\cos(\arctan(2 \times 3.14 \times \text{frequency} \times \tau)))^\beta$$
$$\times (\cos(\beta \times (\arctan(2 \times 3.14 \times \text{frequency} \times \tau)))) \tag{4}$$

The Cole-Davidson parameters consist of τ and β. The τ is the relaxation time (related to temperature) and β is a constant for a given material (depends on the materials' physical properties). $0 \leq \beta \leq 1$ which controls the width of the distribution and $\beta = 1$ for Debye relaxation [8]. The larger the value of β is, the worst case is the dielectric relaxation for the specified high-k thin film. In our case, τ is decreasing with increasing temperature, whilst β is increased and then decreased due to the dielectric relaxation degree.

Table 1. The data of regression model

frequency	Cacc/Cmax		
	AD 150	AD 200	AD 250
100	1.0	1.0	1.0
1000	0.869281304416653	0.844311377245509	0.570121951219512
10000	0.716334230383621	0.646706586826347	0.390243902439024
100000	0.515144319560973	0.391017964071856	0.307926829268293
1000000	0.244050445886761	0.181437125748503	0.11219512195122

frequency	Cacc/Cmax	
	AD 300	AD 350
100	1.0	1.0
1000	0.778625954198473	0.720372836218375
10000	0.552671755725191	0.567243675099867
100000	0.401526717557252	0.463382157123835
1000000	0.290076335877863	0.380825565912117

Table 1 gives the observed data at different temperatures, which is from the experimental measurement. The observed data points are measured on five frequencies: which are 100, 1000, 10000, 100000, and 1000000; and five temperatures, which are AD 150, AD 200, AD 250, AD 300, and AD 350. In this optimization problem, different β and τ at different temperatures should be obtained from the 25 data points and function (4). The constrains of this optimization problem are as follows:

$$\beta_{AD\ 150} < \beta_{AD\ 200} < \beta_{AD\ 250} \tag{5}$$

$$\beta_{AD\ 250} > \beta_{AD\ 300} > \beta_{AD\ 350} \tag{6}$$

$$\tau_{AD\ 150} > \tau_{AD\ 200} > \tau_{AD\ 250} > \tau_{AD\ 300} > \tau_{AD\ 350} \tag{7}$$

4.2   Optimization Problem Representation

In this regression problem, each pair of β and τ are searched for a particular temperature. The observed data is measured at five different temperatures. There are five pairs of β and τ need to be optimized. The regression problem is formulated as a ten dimensional optimization problem.

The regression process is formulated as a continuous, constrained, single objective problem, and each dimension is dependent in solution space. The object of optimization is to obtain the minimum sum of absolute difference value between calculated value and the observed value:

$$f(\beta_i, \tau_i) = \min(\sum |\text{calculated value} - \text{observed value}|)$$

where β_i, τ_i are parameters for the regression functions.

4.3   Parameter Setting

The parameters of particle swarm optimization are set as the standard PSO [1,7]. In all experiments, the population size is 48, and $c_1 = 1.496172$, c_2 is equal to c_1, and the inertia weight $w = 0.72984$. The PSO is configured with both star and ring structures. Both PSOs are run 50 times to ensure a reasonable statistical result necessary to compare different PSOs. There are 15000 iterations for this 10 dimensional problem in every run.

4.4   Regression Results

Table 2 is the results of particle swarm optimization with star or ring structure solving regression function, respectively. The bold numbers indicate the better solutions. Three measures of performance are utilized in this paper. The first is the best fitness value attained after a fixed number of iterations. In our case, we report the best result found after 15000 iterations. The second and the last are the median and mean value of the best fitness values for all runs. It is possible that an algorithm will rapidly reach a relatively good result while becoming trapped into a local optimum. These two values reflect the algorithm's reliability and robustness.

In general, both the particle swarm optimization with star and ring structure can obtain the best solution. The particle swarm optimization with ring structure has a good mean solution than particle swarm optimization with star structure.

Table 2. Result of particle swarm optimization with star and ring structure solving regression function. All algorithms are run for 50 times, where "best", "median", and "mean" indicate the best, median, and mean of the best fitness values for all runs, respectively.

Topology	Best	Median	Mean	Std. Dev.
Star	1.053175	1.648320	29.632730	44.56709
Ring	**1.020008**	1.088310	**5.157848**	19.51159

The Table 3 gives the best results of parameter regression. The β and τ are listed while the minimum sum of absolute difference values is found by particle swarm with star or ring structure. These two groups of values are very similar to each other.

The Table 4 gives best regression results of particle swarm optimization with star structure. The Figure 1 shows the comparison of regression results and observed data. The Table 5 gives the best regression results of particle swarm optimization with ring structure. The Figure 2 shows the comparison of regression results and observed data.

From the curves in Figure 1 and Figure 2, the particle swarm optimization algorithm with star or ring structure can obtain almost the same results of regression curves. This indicates that the good solution for this kind of problems can be found through particle swarm optimization algorithm.

Table 3. The best results of parameter regression. The minimum error is 1.02000850 for PSO with ring structure, while the minimum error is 1.05317510 for PSO with star structure.

AD	Star		Ring	
	β	τ	β	τ
AD 150	0.1002494229	0.0010511750	0.1022300070	0.0009162830
AD 200	0.1416010514	0.0010099788	0.1445693227	0.0008695953
AD 250	0.2151466434	0.0009694112	0.2228354277	0.0008272211
AD 300	0.1399315195	0.0009116935	0.1431070780	0.0007834991
AD 350	0.1192440041	0.0008699401	0.1223494783	0.0007356312

Table 4. The best regression results of particle swarm optimization with star structure

frequency	Results		
	AD 150	AD 200	AD 250
100	0.9803533087200449	0.9732203075430433	0.9599714993720966
1000	0.8183140188354814	0.7531354217322795	0.6452708516352952
10000	0.6490450233171907	0.5422181905772254	0.39022384615261413
100000	0.5151534482187442	0.3911869732998386	0.2375184001008005
1000000	0.4089571613918108	0.28233432687083115	0.14471122722135724

frequency	Results	
	AD 300	AD 350
100	0.9777641605637136	0.9827987196129729
1000	0.7668024982675506	0.8039318492517724
10000	0.5542132222985895	0.6099947380705117
100000	0.4013677612982473	0.4633722900064722
1000000	0.29079687116276054	0.35210320821093655

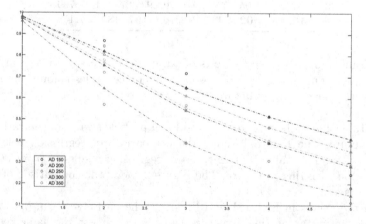

Fig. 1. The comparison of regression results and observed data. The ○ represents the observed data, and the △ represents the obtained data.

Table 5. The best regression results of particle swarm optimization with ring structure

frequency	Results		
	AD 150	AD 200	AD 250
100	0.984082908101654	0.9787492986896107	0.9681870128983179
1000	0.8264005694511332	0.7648083529054488	0.6575017199965371
10000	0.6524843293868552	0.5467651505496035	0.3901203297302975
100000	0.5155074488464255	0.3917458808543039	0.2332240862444571
1000000	0.40737442781887184	0.28080919599526033	0.1395972903770576

frequency	Results	
	AD 300	AD 350
100	0.9824993267112276	0.9868508330318926
1000	0.7785885773651047	0.8155461434535894
10000	0.5585610642007467	0.6144323526766813
100000	0.40153027594372204	0.46337978806209773
1000000	0.2887921416526203	0.34959692515051655

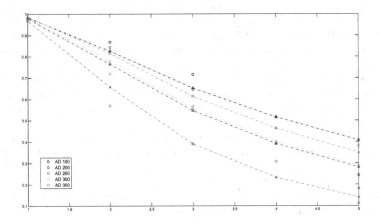

Fig. 2. The comparison of regression results and observed data. The ○ represents the observed data, and the △ represents the obtained data.

5   Conclusions

The traditionally mathematical and/or statistic methods are difficult to solve some real world problems, because these traditional methods have many requirements such as domain specific knowledge. The evolutionary computation algorithms or swarm intelligence, which does not require domain knowledge, the solution could be found by iterations of simple strategies.

In this paper, we utilized particle swarm optimization to solve a real-world regression analysis problem. Two parameters at five temperatures are optimized through the minimizing the sum of least absolute difference values. The experimental results show that particle swarm optimization can obtain good performance on solving regression analysis problems.

Acknowledgements. The authors' work is partially supported by National Natural Science Foundation of China under Grant Number 60975080, 61273367; and by Ningbo Science & Technology Bureau (Science and Technology Project Number 2012B10055).

References

1. Bratton, D., Kennedy, J.: Defining a standard for particle swarm optimization. In: Proceedings of the 2007 IEEE Swarm Intelligence Symposium (SIS 2007), pp. 120–127 (April 2007)
2. Cheng, S., Shi, Y., Qin, Q.: Experimental study on boundary constraints handling in particle swarm optimization: From population diversity perspective. International Journal of Swarm Intelligence Research (IJSIR) 2(3), 43–69 (2011)
3. Cheng, S., Shi, Y., Qin, Q.: Promoting diversity in particle swarm optimization to solve multimodal problems. In: Lu, B.-L., Zhang, L., Kwok, J. (eds.) ICONIP 2011, Part II. LNCS, vol. 7063, pp. 228–237. Springer, Heidelberg (2011)
4. Cheng, S., Shi, Y., Qin, Q.: Population diversity based study on search information propagation in particle swarm optimization. In: Proceedings of 2012 IEEE Congress on Evolutionary Computation (CEC 2012), pp. 1272–1279. IEEE, Brisbane (2012)
5. Cheng, S., Shi, Y., Qin, Q.: Population diversity of particle swarm optimizer solving single and multi-objective problems. International Journal of Swarm Intelligence Research (IJSIR) 3(4), 23–60 (2012)
6. Cheng, S., Shi, Y., Qin, Q., Ting, T.O.: Population diversity based inertia weight adaptation in particle swarm optimization. In: Proceedings of The Fifth International Conference on Advanced Computational Intelligence (ICACI 2012), pp. 395–403. IEEE, Nanjing (2012)
7. Clerc, M., Kennedy, J.: The particle swarm–explosion, stability, and convergence in a multidimensional complex space. IEEE Transactions on Evolutionary Computation 6(1), 58–73 (2002)
8. Debye, P.J.W.: Polar Molecules. Chemical Catalogue Company, New York (1929)
9. Eberhart, R., Kennedy, J.: A new optimizer using particle swarm theory. In: Proceedings of the Sixth International Symposium on Micro Machine and Human Science, pp. 39–43 (1995)
10. Eberhart, R., Shi, Y.: Particle swarm optimization: Developments, applications and resources. In: Proceedings of the 2001 Congress on Evolutionary Computation (CEC 2001), pp. 81–86 (2001)
11. Eberhart, R., Shi, Y.: Computational Intelligence: Concepts to Implementations, 1st edn. Morgan Kaufmann Publisher (2007)
12. Hastie, T., Tibshirani, R., Friedman, J.: The Elements of Statistical Learning: Data Mining, Inference, and Prediction, 2nd edn. Springer Series in Statistics. Springer (February 2009)
13. Kennedy, J., Eberhart, R.: Particle swarm optimization. In: Proceedings of IEEE International Conference on Neural Networks, pp. 1942–1948 (1995)
14. Kennedy, J., Eberhart, R., Shi, Y.: Swarm Intelligence. Morgan Kaufmann Publisher (2001)

15. Mendes, R., Kennedy, J., Neves, J.: The fully informed particle swarm: Simpler, maybe better. IEEE Transactions on Evolutionary Computation 8(3), 204–210 (2004)
16. Shi, Y., Eberhart, R.: A modified particle swarm optimizer. In: Proceedings of the 1998 Congress on Evolutionary Computation (CEC 1998), pp. 69–73 (1998)
17. Zhao, C., Zhao, C.Z., Tao, J., Werner, M., Taylor, S., Chalker, P.R.: Dielectric relaxation of lanthanide-based ternary oxides: physical and mathematical models. Journal of Nanomaterials 2012, 1–6 (2012)

Mechanical PSO Aided by Extremum Seeking for Swarm Robots Cooperative Search

Qirong Tang and Peter Eberhard

Institute of Engineering and Computational Mechanics,
University of Stuttgart, Pfaffenwaldring 9, 70569 Stuttgart, Germany
{qirong.tang,peter.eberhard}@itm.uni-stuttgart.de

Abstract. This paper addresses the issue of swarm robots cooperative search. A swarm intelligence based algorithm, mechanical Particle Swarm Optimization (PSO), is first conducted which takes into account the robot mechanical properties and guiding the robots searching for a target. In order to avoid the robot localization and to avoid noise due to feedback and measurements, a new scheme which uses Extremum Seeking (ES) to aid mechanical PSO is designed. The ES based method is capable of driving robots to the purposed states generated by mechanical PSO without the necessity of robot localization. By this way, the whole robot swarm approaches the searched target cooperatively. This pilot study is verified by numerical experiments in which different robot sensors are mimicked.

Keywords: Swarm Robotics, Mechanical Particle Swarm Optimization, Extremum Seeking, Perturbation, Cooperative Search.

1 Introduction

Swarm robotics is an area that has received a lot of attention from worldwide researchers. Using a mobile robot swarm to search targets is a typical topic in this area. Swarm robotic systems usually consist of many identical or similar simple individuals but can give super behavior in swarms. However, a swarm robotic system not only includes multiple robots but also the swarm intelligence from collaboration between the members. The methods used for controlling swarm robotics mainly boil down to two categories. The traditional ones like, e.g., artificial potential fields, or exact cell decomposition, are just inadequate when performing complex tasks. Another kind of methods is referred as non-traditional, like bacterial colony algorithms, reactive immune network, Particle Swarm Optimization (PSO), and Extremum Seeking (ES). Among them the PSO and ES are especially appealing due to their unique features, see [4], [7].

The PSO is originally only used as an optimization method, although it is extended and utilized in the robotics area, see example in [6]. The work [6] extends PSO to mechanical PSO which takes into account the mechanical properties of real robots, see also Section 2.2. Together with other strategies, encouraging results are obtained by this method. However, it requires relatively precise localization for forming the feedback loop and it is difficult to realize fast online driving due to sensor delays. Another side, Extremum Seeking is applicable as a means of navigating robots in environments where robot positions are unavailable [7]. However, the basic ES is non-cooperative, i.e., each

Y. Tan, Y. Shi, and H. Mo (Eds.): ICSI 2013, Part I, LNCS 7928, pp. 64–71, 2013.

robot is driven by ES individually. Motivated by this, this article investigates the swarm robots cooperative search by integrating mechanical PSO and ES.

2 Algorithm Design

2.1 Investigation Prerequisites

Before designing the algorithm for swarm robots, some prerequisites have to be stated. The robots used in the swarm are mobile robots. They are assumed to be relatively simple and do not have the capability of localizing their own positions. However, they are capable of sensing the relative states of their neighbors and the signal strength from the target. The simulated robots are here considered as 2D mass points without volumes.

Secondly, there are no obstacles included at the moment in the environment. The spatial distribution of the searched signal originating from the target is unknown to the robots, neither the position of the target. However, in this study the target is sending a signal which is known to decay with the distance from the source. As a usual source distribution we use the quadratic form

$$f(x, y) = f^* - q_x(x - x^*)^2 - q_y(y - y^*)^2 \tag{1}$$

to describe it. Here f is the detected signal strength, (x^*, y^*) is the maximizer while f^* represents the maximum, q_x and q_y are positive constants.

2.2 From Basic PSO to Mechanical PSO

The PSO was inspired from some biological populations, for instance, the swarm of birds. Each bird is taken as an adaptive agent and can communicate with the environment and other agents. During the process of communication, they will 'learn' or 'accumulate experience' in order to change the structure and behavior of the swarm. Such processes are the basis of PSO. In PSO, the size of the swarm is denoted by N_p and the members are called particles. The 'velocity' and position of particles are represented by $N_p \times n$ matrices $\dot{x}$ and x, respectively.

The recursion of one commonly used form of basic PSO for all N_p particles is

$$\begin{bmatrix} x^{s+1} \\ \dot{x}^{s+1} \end{bmatrix} = \begin{bmatrix} x^s \\ \omega_p \dot{x}^s \end{bmatrix} + \begin{bmatrix} \dot{x}^{s+1} \\ c_1 r_1^s \cdot (x_{self}^{best,s} - x^s) + c_2 r_2^s \cdot (\hat{x}_{swarm}^{best,s} - x^s) \end{bmatrix} \cdot \tag{2}$$

Here s denotes the iterative steps. The right-hand side of the second line of (2) contains three components, i.e., the 'inertia' which models the particles tendency of last step; the 'memory' which means moving towards the particles' self best positions, respectively; and the 'cooperation' which drives the particles to the swarm best position. In (2), c_1 and c_2 are usually non-negative constant real numbers while random effects are kept in r_1^s and r_2^s. Detailed definitions can be found in [6].

We consider one particle to represent one robot since the particles in PSO looking for the minimum (or maximum) of an objective function according to their update formulae is quite similar to the robots search scenario in which the robots are searching a target according to their cooperatively generated trajectories. Many of their correspondences are summarized in [6]. We interpret the PSO-based algorithm as providing the required forces in the view of multibody system dynamics. Namely, each robot is considered as one body in a multibody system which is influenced by forces and torques but without direct mechanical connections. In addition, the particles are replaced by mechanical robots whose motions follow physical laws. This is done in order to generate physically reasonable search trajectories. For considering the feasible dynamics, the inertia, and other physical features of the robots, the basic PSO algorithm is extended.

In a general way, if one defines k coming from Euler equations, and q contains the information of external forces and torques acting on all N_p robots, the acceleration of the entire robot swarm can be formulated by

$$\ddot{x} = \begin{bmatrix} \ddot{x}_1 \ \ddot{x}_2 \ \cdots \ \ddot{x}_i \ \cdots \ \ddot{x}_{N_p} \end{bmatrix}^T = M^{-1} \cdot (q - k) = M^{-1} \cdot F \qquad \in \mathbb{R}^{3N_p \times 1}. \qquad (3)$$

With the state vector $y_{st} = \begin{bmatrix} x \ \dot{x} \end{bmatrix}^T$, state equation

$$\dot{y}_{st} = \begin{bmatrix} \dot{x} \\ M^{-1} \cdot F \end{bmatrix}, \quad \text{and Euler forward integration} \quad y_{st}^{s+1} = y_{st}^s + \Delta t \, \dot{y}_{st}^s, \quad \text{it yields} \quad (4)$$

$$\begin{bmatrix} x^{s+1} \\ \dot{x}^{s+1} \end{bmatrix} = \begin{bmatrix} x^s \\ \dot{x}^s \end{bmatrix} + \Delta t \begin{bmatrix} \dot{x}^s \\ M^{-1} \cdot F^s \end{bmatrix}. \qquad (5)$$

We define the robot to be only influenced by forces, i.e., $l_i = 0$ at the moment. The force F^s is further determined by three parts, f_1^s, f_2^s and f_3^s, which are

$$f_1^s = -h_{f_1}^s \cdot \left(x^s - x_{self}^{best,s} \right), \quad f_2^s = -h_{f_2}^s \cdot \left(x^s - \hat{x}_{swarm}^{best,s} \right), \quad f_3^s = -h_{f_3}^s \cdot \dot{x}^s. \qquad (6)$$

Here f_1^s, f_2^s and f_3^s contain physical meanings corresponding to the 'memory', 'cooperation', and 'inertia' phenomena in basic PSO. Combining (5) and (6) yields

$$\begin{bmatrix} x^{s+1} \\ \dot{x}^{s+1} \end{bmatrix} = \begin{bmatrix} x^s \\ \left(I_{3N_p} - \Delta t \, M^{-1} \cdot h_{f_3}^s \right) \cdot \dot{x}^s \end{bmatrix}$$
$$+ \Delta t \begin{bmatrix} \dot{x}^s \\ M^{-1} \cdot h_{f_1}^s \cdot \left(x_{self}^{best,s} - x^s \right) + M^{-1} \cdot h_{f_2}^s \cdot \left(\hat{x}_{swarm}^{best,s} - x^s \right) \end{bmatrix}. \qquad (7)$$

For more detailed derivations and explanations please refer to [6]. Equation (7) is the developed mechanical PSO which is used to generate cooperatively physically reasonable trajectories for swarm mobile robots. However, it requires heavy robot localization.

2.3 Perturbation Based Extremum Seeking

Extremum Seeking (ES) had been proven to be a powerful tool in real-time non-model based control and optimization [1]. Recently, ES also has been used for swarm networked agents with each member only sensing limited local information [5]. Extremum Seeking usually is applied for questions as for seeking the maxima of objective functions. Due to its non-model based character, it is applicable to control problems which contain nonlinearity either in the plant or in its control objective. For example, in the case of cooperative search performed by swarm robots, either the robot models or the distribution of the source, or both of them can be nonlinear since ES based methods are possible to use only one external signal whose strength is detected by robots and the specific distribution form of the signal is not critical. It doesn't care about the actual positions of robots. Furthermore, the linearity or nonlinearity of the robot model is not important. Nonetheless, the signal strength f is only a single dimensional information. It is not directly sufficient for guiding robots since it lacks the 'gradient' information of the target signal. One way to solve this issue is that of equipping gradient detecting sensors on robots. However, this is a big challenge and we prefer that the robot only needs to detect one external signal. The amazing thing happens when the basic ES is varied by perturbation which is hardware free and easy for implementation. Relying on its persistence of excitation, usually a sinusoidal signal, the perturbation based ES perturbs the parameters being tuned. Through this method, the gradient information is obtained. We use the perturbation based ES scheme similarly as in [8] for guiding robots. Its control block diagram for a single robot is shown in Fig. 1.

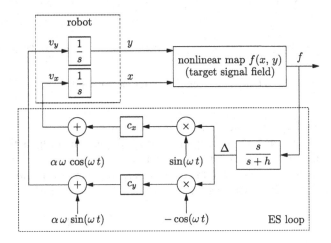

Fig. 1. Extremum Seeking scheme with x, y axes velocities as inputs for robot

In Fig. 1 the parameters α, ω, c_x, c_y and h are chosen by the designers. The washout filter $s/(s+h)$ filters out the DC component of f, then the two-channel perturbations generate gradient estimates of f in x and y directions which usually are unmeasurable by physical sensors. After the ES loop, the velocity inputs are tuned for driving the robot. The used control laws are then governed by

$$v_x = \alpha \omega \cos(\omega t) + c_x \Delta \sin(\omega t), \; v_y = \alpha \omega \sin(\omega t) - c_y \Delta \cos(\omega t), \Delta = \frac{s}{s+h}[f]. \, (8)$$

The control laws in (8) are actually optional. Dürr et al. [2] used different perturbations which also work well. Unfortunately, due to the sinusoidal perturbation, the trajectory generated from the ES based method is spiral like which artificially increases the travel distance. This is extremely serious when the distance between start position and target position is far, because the uncertainty from the AC part of f is increased. This is not acceptable for robot practical implementation. Furthermore, if this ES scheme is used for swarm robots, all robots are non-cooperative. Therefore, we try to integrate it into mechanical PSO while the localization free feature is inherited.

2.4   Mechanical Particle Swarm Optimization Aided by Extremum Seeking

Our purpose is to integrate the cooperation benefits of mechanical PSO and the localization free feature of ES. The ES is aiding the overall search algorithm of mechanical PSO. The mechanical PSO is used to generate intermediate states for guiding robots. From each state to its next adjacent state there is only a short distance. The perturbation based ES only needs to drive the robot to the next state while temporary taking the next state as its current target, see Fig. 2 for the relationship of mechanical PSO and ES.

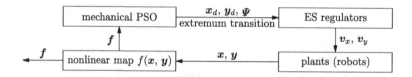

Fig. 2. Extremum Seeking aids for mechanical PSO

One should not forget that the intermediate states are not target positions. Thus, the perturbation based ES doesn't directly qualify. So, the maximum transition must be performed through which the target source (with maximal signal strength) is mathematically and temporary transited to the desired state by using information f from the actual source and Ψ from robots relative observation. This idea is also expressed by

$$f^* = f(x^*, y^*) \quad \xrightarrow[g()]{f, \Psi} \quad g^* = g(x_d, y_d, f, \Psi) \tag{9}$$

where f is a vector containing all the robots detected signal strengths. To be emphasized, f is from the actual source and it is the only signal that the robots can detect since the intermediate 'targets' (x_d, y_d) from mechanical PSO are artificial. In (9) g^* are the new maxima, g is a function which corresponds to the function f and describes the artificial targets at the intermediate states. In this study, we assume there is such a function g. By this way, each robot senses the current 'target' which locates at the corresponding state from a step of mechanical PSO. Then, our perturbation based ES drives the robot to the desired state with relatively stable trajectory due to the short distance between two adjacent states. Looking at the whole robot swarm, they are still moving in a cooperative manner since each robot traces one trajectory from mechanical PSO.

The procedures of the whole method is summarized in Algorithm 1. The method used in this study is very different to the work in [4] where PSO is still used in the view of optimization without considering the robots physical properties. The PSO generated states in [4] require a re-generation to smooth the trajectory, whereas the states from mechanical PSO in this study are ready to be traced with physically reasonable quality. Thus, the PSO in [4] has no obvious advantages compared to some other swarm intelligence based algorithms like, e.g., ant colony, bacteria foraging. In addition, [4] only handles a single robot and the particles are virtually without mapping to real robots. Some other researches, e.g., [3] and [7], provide variants of ES for swarm seeking which are basically formation control oriented. Furthermore, their control cost and energy consumption from the not well organized trajectories have restricted their applications although they are also cooperative. In contrast, the scheme in this study is more straightforward and feasible considering the implementation.

Algorithm 1. Mechanical PSO aided by ES for robots cooperative search

---
1: /* initialize: give all required control parameters, read in start positions and initial signal strength f^0 of all robots, mechanical PSO step $s = 0$, define stop criteria */
2: update mechanical PSO using (7), provides (x_d, y_d), $s = s + 1$, robot index $i = 1$
3: obtain Ψ_i^s by relative observation, perform maximum transition using (9)
4: perturbation based ES regulates robot i to $(x_{i,d}, y_{i,d})$ due to (8), $i = i + 1$
5: repeat steps 3-4 until $i > N_p$
6: measure f^s at new positions, evaluate new f^s for mechanical PSO
7: repeat steps 2-6 until a stop criterion is met

---

3 Simulation

3.1 Simulation Setup

In our simulations, the robots are assumed to run in a $3\,\text{m} \times 3\,\text{m}$ environment. The used parameters for mechanical PSO are $\omega_p = 0.6$, $c_1 = 0.1$, $c_2 = 0.8$, for ES are $\omega = 100, 125, ...$, (different for each robot) $\alpha = 0.05$, $c_x = c_y = 10$. The weights of the target field function are $q_x = q_y = 1$. We set in simulation the final maximizer at $(x^*, y^*) = (-0.5, 0.5)$. During the search the maximizers are the corresponding states generated from mechanical PSO, the final target maximum is set to $f^* = 1$.

3.2 Swarm Robots Cooperative Search by Mechanical PSO Aided by ES

We first verify a single robot to be driven by our perturbation based ES controller. The robot is actuated by x and y axes velocities and is supposed to move from $(1, 1)$ directly to $(-0.5, 0.5)$ without intermediate stops. Figure 3(a) shows the performance measured by this robot, from which one can observe the change of the detected signal. After about $10\,\text{s}$ (simulation time), it approaches the maximum which means the robot is very close to the target. Figure 3(b) demonstrates the robot trajectory which looks like a spiral curve. In Fig. 3(b) the marked R_i, R_j are the revolution radii of the robot motion from which one can see its significant change. The rotation radii of the robot motion are changing, too. In addition, both of them are unpredictable. This kind of changes will become more intense with increasing distance to the target.

After this we now set up a four robots search scenario which integrates mechanical PSO and perturbation based ES. The robot trajectories are shown in Fig. 4. In Fig. 4

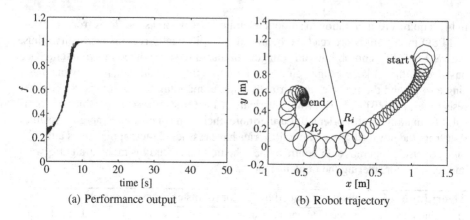

(a) Performance output   (b) Robot trajectory

Fig. 3. Extremum Seeking moves one robot from $(1, 1)$ to $(-0.5, 0.5)$

the green, red, pink and black dots (lines) are the mechanical PSO computed states (trajectories). The distances between two adjacent states are smaller. The overlapping spiral like blue curves are the ES regulated trajectories which have relatively stable spiral radii. This is very helpful for real robots implementation. From Fig. 4 one can see that the method of mechanical PSO aided by ES is feasible. Importantly, through the whole process, no robot localization is required. The blue trajectories are still longer than the ones obtained by directly connecting mechanical PSO states. This is negative but in exchange there is no localization required. From a macro point of view, the robots are still traveling cooperatively. The mechanical PSO guides the robots not to move too arbitrary as when only driven by ES while on another side the perturbation based ES frees the robots from localization.

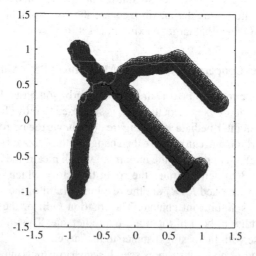

Fig. 4. Cooperative search trajectories (mechanical PSO aided by ES)

4 Open Questions and Discussion

The method gets rid of robot localization which gives the implementation a lot of freedom. However, the energy consumed from the ES trajectory is much higher than the one from mechanical PSO. How to adjust the parameters of the ES controller for energy saving is an interesting consequent investigation point.

If robot volume and obstacles are included, algorithm improvement is highly demanded. Furthermore, the maximum transition of (9) actually is a very strong assumption. Performing relative observation and building the function *g* are not easy when considering the real robots implementations.

5 Conclusion

For the swarm mobile robots cooperative search, this investigation has integrated advantages both from mechanical Particle Swarm Optimization and Extremum Seeking. The mechanical PSO provides cooperative search trajectories based on the consideration of real robots, while perturbation based ES is responsible for regulating the robots towards the purposed states from mechanical PSO. This method no longer needs the localization of the moving robots which is usually required by traditional robot navigation. This will probably open a new research window for swarm mobile robots cooperative search. The feasibility of the conducted method in this pilot study is investigated by simulation.

Acknowledgment. The authors would like to thank Dipl.-Ing. Hans-Bernd Dürr, IST, University of Stuttgart, for his valuable suggestions and helpful discussions on Extremum Seeking control.

References

1. Ariyur, K.B., Krstić, M.: Real-Time Optimization by Extremum-Seeking Control. Wiley Interscience, Hoboken (2003)
2. Dürr, H.B., Stanković, M.S., Ebenbauer, C., Johansson, K.H.: Lie bracket approximation of extremum seeking systems. Automatica (2012) (accepted for publication)
3. Ghods, N., Krstić, M.: Multiagent deployment over a source. IEEE Transactions on Control Systems Technology 20(1), 277–285 (2012)
4. Hong, C., Li, K.: Swarm intelligence-based extremum seeking control. Expert Systems with Applications 38(12), 14852–14860 (2011)
5. Stanković, M.S., Johansson, K.H., Stipanović, D.M.: Distributed seeking of Nash equilibria with applications to mobile sensor networks. IEEE Transactions on Automatic Control 57(4), 904–919 (2012)
6. Tang, Q.: Cooperative Search by Mixed Simulated and Real Robots in a Swarm Based on Mechanical Particle Swarm Optimization. Doctoral thesis, University of Stuttgart, No. 25. Aachen: Shaker Verlag (2012)
7. Zhang, C., Ordóñez, R.: Extremum-Seeking Control and Applications: A Numerical Optimization-Based Approach. Springer, London (2011)
8. Zhang, C., Siranosian, A., Krstić, M.: Extremum seeking for moderately unstable systems and for autonomous vehicle target tracking without position measurements. Automatica 43(10), 1832–1839 (2007)

Multi-swarm Particle Swarm Optimization with a Center Learning Strategy

Ben Niu[1,2,3,*], Huali Huang[1], Lijing Tan[4], and Jane Jing Liang[5]

[1] College of Management, Shenzhen University,
Shenzhen 518060, China
[2] Department of Industrial and System Engineering,
The Hong Kong Polytechnic University, Hong Kong
[3] Hefei Institute of Intelligent Machines, Chinese Academy of Sciences,
Hefei 230031, China
[4] Management School, Jinan University, Guangzhou 510632, China
[5] School of Electrical Engineering, Zhengzhou University, Zhengzhou, China
drniuben@gmail.com

Abstract. This paper proposes a new variant of particle swarm optimizers, called multi-swarm particle swarm optimization with a center learning strategy (MPSOCL). MPSOCL uses a center learning probability to select the center position or the prior best position found so far as the exemplar within each swarm. In MPSOCL, Each particle updates its velocity according to the experience of the best performing particle of its partner swarm and its own swarm or the center position of its own swarm. Experiments are conducted on five test functions to compare with some variants of the PSO. Comparative results on five benchmark functions demonstrate that MPSOCL achieves better performances in both the optimum achieved and convergence performance than other algorithms generally.

Keywords: multi-swarm particle swarm optimization, center learning strategy, particle swarm optimizer (PSO).

1 Introduction

Particle swarm optimization (PSO), as one of the most famous intelligent algorithm, was originally proposed by Kennedy and Eberhart [1][2] in 1995. It was inspired by the search behavior of particle swarm and each particle with a social behavior is regarded as a potential solution in the search space. The velocity was adjusted in the search process by a combination of its own best flying experience and the best experience of the swarm.

In the past decade, particle swarm optimization has attracted much attention and studies. However, experiments and applications for real-word problems have revealed that the PSO algorithm is not free from trapping into local optimum, especially on complex multimodal problems. Therefore a large number of relevant researches on

---

[*] Corresponding author.

Y. Tan, Y. Shi, and H. Mo (Eds.): ICSI 2013, Part I, LNCS 7928, pp. 72–78, 2013.

different variants of PSO [3-8] have been proposed to avoid the drawback of premature convergence. Based on our previous works[9][10][11], here we present a new multi-swarm scheme to balance the exploration and exploitation in PSO by using a center learning mechanism that the center position and the best position discovered so far by each particle in multiple swarms is used to update the position.

The remainder of the paper is organized as following. Section 2 gives a brief overview of original particle swarm optimization. The proposed multiple particle swarm optimization algorithm with a center learning strategy is elaborated in Section 3. Section 4 presents the comparative experimental studies and related results. Finally, some conclusions and further work are presented in Section 5.

2   An Overview of Particle Swarm Optimization

In original PSO, a particle in the swarm which represents a potential solution vector in the corresponding search boundary has a velocity to adjust its flying direction. The velocity is updated in the search process according to its own previous best experience and the best experience of the swarm. The velocity and position of each dimension of the i th particle are presented below [1][2]:

$$V_i \leftarrow V_i + c_1 * r_1 * \left(pbest - X_i \right) + c_2 * r_2 * \left(gbest - X_i \right) \tag{1}$$

$$X_i \leftarrow X_i + V_i \tag{2}$$

where $i = 1, \ldots, ps$, ps is the population size. c_1 and c_2 are the acceleration coefficients, r_1 and r_2 are two random numbers between 0 and 1, $pbest$ and $gbest$ are the best position of its own flying experience and the swarm yielding the best fitness value. V_i and X_i represent the velocity and the position of the i th particle respectively.

3   Multi-swarm Particle Swarm Optimization with a Center Learning Strategy

In this multi-swarm particle swarm optimization model, the velocity equations of the i th particle belonging to swarm k are updated as follows:

$$V_i^k = \omega * V_i^k + c_1 * r_1 * \left(P_{fi}^k - X_i^k \right) + c_2 * r_2 * \left(P_s^g - X_i^k \right) \tag{3}$$

where c_1 and c_2 are constant learning factors within one multi-swarm and between multi-swarms respectively. r_1 and r_2 are random numbers in the range [0,1]. k is the index of the multi-swarm that each particle belongs to. fi defines the exemplar which the particle i should follow. P_{fi}^k is the center position or all the particles' previous best position including its own and other particles' within swarm k. And it is decided by the center learning probability compared with a random number. If this random number is larger than the center learning probability, the corresponding particle will learn from all the individuals' best position. Otherwise it will learn from the center position found by all the particles so far. And P_s^g is the best position found so far by all the multi-swarms and g is the index of the swarm which the best position belongs to.

The center learning probability Pc_i for i th particle of multi-swarm k is calculated using[10]:

$$Pc_i = 0.05 + 0.45 * \frac{\left(\exp\left(\frac{10(i-1)}{m-1} \right) \right)}{(\exp(10) - 1)} \tag{4}$$

where m is the population size of each multi-swarm.

Table 1. Pseudo code of the MPSOCL algorithm

```
Algorithm MPSOCL
Begin
Initialize every particle and the related parameters.
While (the termination conditions are not met)
   For each multi-swarm k (do in parallel)

    For each particle i of swarm k.
      Generate a random number and compare with the center learning
              probability Pc_i in term of (4).

    Select the exemplar which ith particle should follow.
    End for
  End for (Do in parallel)
Select the fittest global individual P_s^g from all the multi-swarms.
Update the velocity and position using Eqs.(3)and(2),respectively.
Evaluate the fitness value of each particle.
End while(until a terminate-condition is met)
End
```

Through the detailed description above, the pseudo-code of MPSOCL is given as Table 1.

4   Experiments and Discussions

To compare with the performance, we choose five benchmark functions which are depicted in Table 2 and five algorithms from the literatures [3-11]. Ten-dimensional problems are tested by all optimizers.

Experiments are conducted to compare all the algorithms with the same population size of 60. Meanwhile each benchmark function is run 20 times and the maximal iteration is set at 2000 for all PSOs. The comparative algorithms LOPSO FDR-PSO, FIPS, UPSO can refer to references [4], [5], [6], [7], respectively and their parameters settings are defined as the same as used in the corresponding references.

The inertia weight is linearly decreased from 0.9 to 0.4 [3] in all PSOs except UPSO [7], LOPSO [4] which both adopt the constriction factor $\phi = 0.729$. And c_1 and c_2 are both 2.0. Meanwhile, MPSOCL has 6 multi- swarms which both include 10 particles within each multi-swarm. All the parameters used in each swarm are the same as those defined above.

Table 2. Five benchmark functions

Functions	Mathematical Representation	Search Range
Sphere(f_1)	$$f_1(\mathrm{x})=\sum_{i=1}^{n} x_i^{2}$$	[-100,100]
Ackley(f_2)	$$f_2(\mathrm{x})=20+e-20\exp(-0.2\sqrt{\frac{1}{n}\sum_{i=1}^{n}x_i^{2}})$$ $$-\exp(\frac{1}{n}\sum_{i=1}^{n}\cos 2\pi x_i)$$	[-32.768,32.768]
Rosenbrock(f_3)	$$f_3(\mathrm{x})=\sum_{i=1}^{n-1}((x_i-1)^2+100(x_{i+1}-x_i^{2})^2)$$	[-2.048,2.048]
Griewank(f_4)	$$f_4(\mathrm{x})=1+\frac{1}{4000}\sum_{i=1}^{n}x_i^{2}-\prod_{i=1}^{n}\cos(\frac{x_i}{\sqrt{i}})$$	[-600,600]
Rastrigin(f_5)	$$f_5(\mathrm{x})=\sum_{i=1}^{n}(10-10\cos(2\pi x_i)+x_i^{2})$$	[-5.12,5.12]

Table 3 and Figs 1~5 present the results on five benchmark functions in the dimension of 10. The convergence characteristics in terms of the mean values and standard deviation of the results for every benchmark function. Note that, in Table 3, optimum values obtained are in bold.

Table 3. Results on five benchmarks for 10-D

Algorithm	f_1	f_2	f_3	f_4	f_5
SPSO	6.27e-067	1.32e+000	1.73e+000	7.07e-002	2.38e+000
	±5.49e-132	±2.10e-031	±1.47e+000	±3.37e-002	±1.13e+000
LOPSO	3.26e-043	1.32e+000	2.88e+000	3.69e-002	3.03e+000
	±5.22e-085	±3.82e-031	±8.64e-001	±1.43e-002	±1.38e+000
FDR-PSO	1.35e-135	1.32e+000	**5.78e-001**	4.47e-002	2.68e+000
	±1.64e-269	±2.10e-031	**±4.80e-001**	±1.44e-002	±1.41e+000
FIPS	8.27e-021	1.32e+000	4.11e+000	1.54e-002	9.17e-001
	±3.35e-041	±2.68e-022	±1.57e-001	±1.00e-002	±6.95e-001
UPSO	7.28e-082	1.32e+000	6.23e-001	2.56e-002	5.03e+000
	±4.99e-163	±2.08e-031	±1.47e-001	±1.27e-002	±2.01e+000
MPSOCL	**6.07e-216**	**1.32e+000**	7.57e+000	**0±0**	**0±0**
	±0	**±2.07e-031**	±2.45e-001		

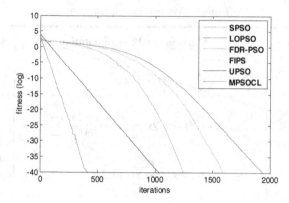

Fig. 1. Convergence characteristics on 10-dimensional Sphere function

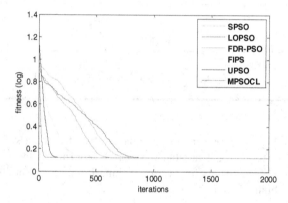

Fig. 2. Convergence characteristics on 10-dimensional Ackley function

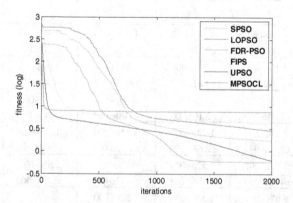

Fig. 3. Convergence characteristics on 10-dimensional Rosenbrock function

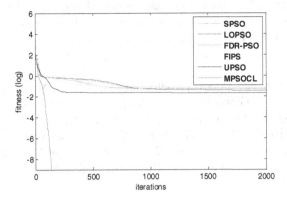

Fig. 4. Convergence characteristics on 10-dimensional Griewank function

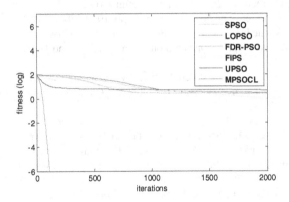

Fig. 5. Convergence characteristics on 10-dimensional Rastrigin function

As presented from Table 3 and Figs 1~5, MPSOCL is able to consistently find the better minimum within 2000 generations especially on Griewank, Rastrigin and Sphere functions for 10-D dimension. And there is no much difference of the results between MPSOCL and other optimization algorithms on the average best fitness value for Ackley and Rosenbrock functions. Overall, the proposed algorithm outperforms in the optimal values and convergence characteristics.

5   Conclusions and Further Work

In the paper, we present a multi-swarm particle swarm optimization algorithm with a center learning strategy called MPSOCL. Different from the original particle swarm optimizer and some related modified versions of this algorithm, our proposed algorithm employs new communicational scheme that each particle within one multi-swarm updates its flying direction combining historical experience from all multi-swarms with the present center position.

However, MPSOCL is not the best choice for solving any test problems. Further work may focus on effective tests of the proposed optimizer with more complicated test functions and some practical applications to evaluate the performance of MPSOCL.

Acknowledgments. This work is partially supported by The National Natural Science Foundation of China (Grants nos. 71001072, 71271140, 60905039), The Hong Kong Scholars Program 2012 (Grant no. G-YZ24), China Postdoctoral Science Foundation (Grant nos. 20100480705, 2012T50584), Science and Technology Project of Shenzhen (Grant No. JC201005280492A) and the Natural Science Foundation of Guangdong Province (Grant nos. S2012010008668, 9451806001002294).

References

1. Kennedy, J., Eberhart, R.: Particle Swarm Optimization. In: Proceedings of the 1995 IEEE International Conference on Neural Networks, pp. 1942–1948 (1995)
2. Eberhart, R., Kennedy, J.: A New Optimizer Using Particle Swarm Theory. In: Proceeding of the Sixth International Symposium on Micro Machine and Human Science, pp. 39–43 (1995)
3. Shi, Y., Eberhart, R.: A Modified Particle Swarm Optimizer. In: Proceedings of 1998 IEEE International Conference on Evolutionary Computation Proceedings, pp. 69–73 (1998)
4. Kennedy, J., Mendes, R.: Population Structure and Particle Swarm Performance. In: Proceedings of the 2002 Congress on Evolutionary Computation, pp. 1671–1676 (2002)
5. Peram, T., Veeramachaneni, K., Mohan, C.K.: Fitness-distance-ratio Based Particle Swarm Optimization. In: Proceeding of the 2003 IEEE Swarm Intelligence Symposium, pp. 174–181 (2003)
6. Mendes, R., Kennedy, J., Neves, J.: The Fully Informed Particle Swarm: Simpler, Maybe Better. IEEE Transactions on Evolutionary Computation 8(3), 204–210 (2004)
7. Parsopoulos, K.E., Vrahatis, M.N.: UPSO–A Unified Particle Swarm Optimization scheme. Lecture Series on Computational Sciences, pp. 868–873 (2004)
8. Clerc, M., Kennedy, J.: The Particle Swarm-explosion, Stability, and Convergence in a Multidimensional Complex Space. IEEE Transactions Evolutionary Computation 6(1), 58–73 (2002)
9. Niu, B., Zhu, Y., He, X.X., Wu, H.: MCPSO: A Multi-swarm Cooperative Particle Swarm Optimizer. Applied Mathematics and Computation 185(2), 1050–1062 (2007)
10. Liang, J.J., Qin, A.K., Suganthan, P.N., Baskar, S.: Comprehensive Learning Particle Swarm Optimizer for Global Optimization of Multimodal Functions. IEEE Transactions on Evolutionary Computation 10(3), 281–295 (2006)
11. Niu, B., Li, L.: An Improved MCPSO with Center Communication. In: Proceedings of 2008 International Conference on Computational Intelligence and Security, pp. 57–61 (2008)

Opposition-Based Learning Fully Informed Particle Swarm Optimizer without Velocity

Ying Gao, Lingxi Peng, Fufang Li, Miao Liu, and Waixi Liu

Department of Computer Science and Technology, Guangzhou University,
No.230 Wai Huan Xi Road, Guangzhou Higher Education Mega Center,
Guangzhou, 510006, P.R. China
falcongao@sina.com.cn

Abstract. By applying full information and employing the notion of opposition-based learning, a new opposition based learning fully information particle swarm optimiser without velocity is proposed for optimization problems. Different from the standard PSO, particles in swarm only have position without velocity and the personal best position gets updated using opposition-based learning in the algorithm. Besides, all personal best positions are considered to update particle position. The theoretical analysis for the proposed algorithm implies that the particle of the swarm tends to converge to a weighted average of all personal best position. Because of discarding the particle velocity, and using full information and opposition-based learning, the algorithm is the simpler and more effective. The proposed algorithm is applied to some well-known benchmarks. The relative experimental results show that the algorithm achieves better solutions and faster convergence.

Keywords: particle swarm optimizer, opposition-based learning, full information.

1 Introduction

Particle swarm optimization(PSO)[1] is a novel population-based evolutionary computation technique. The development of PSO was based on observations of the social behavior of animals such as bird flocking, fish schooling and swarm theory. PSO finds the global best solution by simply adjusting the trajectory of each individual toward its own best location and toward the best particle of the entire swarm at generation. Compared with other evolutionary algorithms, PSO has some attractive characteristics. It has memory, so knowledge of good solutions is retained by all particles, and there is a mechanism of constructive cooperation and information sharing between particles. Due to the simple concept, easy implementation and quick convergence, PSO has gained much attention and been successfully applied in a variety of fields mainly for continuous optimization problems.

It has been empirically investigated that standard PSO could easily being trapped in a local optimum and premature convergence in many optimization problems. Some researchers have devoted to improving its performance in various ways and developed

Y. Tan, Y. Shi, and H. Mo (Eds.): ICSI 2013, Part I, LNCS 7928, pp. 79–86, 2013.

many interesting variations. One approach is to incorporate the concept of opposition based learning in PSO to avoid such kind of situations. Opposition-based learning was first introduced by Tizhoosh[2]. It has been proved that opposition based learning process increases the convergence speed thus the evolution process accelerates. Jabeen[3] presents an algorithm called O-PSO. In O-PSO, the opposition based initialization technique is used for initialization of population in standard PSO. A method incorporating opposition based learning in PSO has been proposed by Wang[4]. The method uses opposition based learning and dynamic cauchy based mutation to avoid premature convergence in standard PSO. Shahzad[5] presents an opposition-based velocity clamping PSO algorithm(OVCPSO). OVCPSO uses opposition-based learning and velocity of particles are clamped to control the speed of particles. It avoids premature convergence and allows swarm of particles to continue search for global optima.

In this paper, a new opposition based learning fully information particle swarm optimizer without velocity is proposed for optimization problems by applying full information and employing the notion of opposition-based learning. Different from the standard PSO, O-PSO and OVCPSO, particles in swarm only have position without velocity and the personal best position gets updated using opposition-based learning in the algorithm. Besides, all personal best positions are considered to update particle position. The theoretical analysis for the proposed algorithm implies that the particle of the swarm tends to converge to a weighted average of all personal best position. Because of discarding the particle velocity and using full information, the algorithm is the simpler and more effective. The proposed algorithm is applied to some well-known benchmarks. The relative experimental results show that the algorithm achieves better solutions and faster convergence.

2   Standard Particle Swarm Optimization

In standard particle swarm optimization, the trajectory of each particle in search space is adjusted by dynamically altering the velocity of each particle, according to its own flying experience and the flying experience of the other particles in the search space. The position vector and the velocity vector of ith particle in m-dimensional search space can be represented as $\mathbf{x}_i (i = 1,2,\cdots,N)$ and $\mathbf{v}_i (i = 1,2,\cdots,N)$ respectively, N is the number of particle. In each iteration of standard PSO, the swarm is updated by the following equations:

$$\mathbf{v}_i (t+1) = w\mathbf{v}_i (t) + c_1 r_1 (\mathbf{p}_i (t) - \mathbf{x}_i (t)) + c_2 r_2 (\mathbf{p}_g (t) - \mathbf{x}_i (t)) \tag{1}$$

$$\mathbf{x}_i (t + 1) = \mathbf{x}_i (t) + \mathbf{v}_i (t + 1) \tag{2}$$

Where $\mathbf{p}_i (t)(i = 1,2,\cdots,N)$ and $\mathbf{p}_g (t)$ are given by the following equations, respectively:

$$\mathbf{p}_i (t+1) = \begin{cases} \mathbf{p}_i (t), & f(\mathbf{x}_i (t+1)) < f(\mathbf{p}_i (t)) \\ \mathbf{x}_i (t+1), & f(\mathbf{x}_i (t+1)) \geq f(\mathbf{p}_i (t)) \end{cases} \tag{3}$$

$$\mathbf{p}_g(t) \in \left\{ \mathbf{p}_1(t), \mathbf{p}_2(t), \cdots, \mathbf{p}_N(t) \middle| f(\mathbf{p}_g(t)) \right. \tag{4}$$
$$= \min\{ f(\mathbf{p}_1(t)), f(\mathbf{p}_2(t)), \cdots, f(\mathbf{p}_N(t)) \} \}$$

w is called an inertia weight. c_1 and c_2 are acceleration coefficients. r_1 and r_2 are elements from two uniform random sequences in the range $[0,1]$. $f(\mathbf{x})$ is the minimum objective function.

3   Opposition-Based Learning

The scheme of opposition-based learning was first introduced by H.R.Tizhoosh[2]. The opposition-based learning is general enough and can be utilized in a wide range of learning and optimization fields to make algorithms faster. Opposite numbers are defined as follows:

Let $\mathbf{x} = (x_1, x_2, ..., x_n)$ be an n-dimensional point, where $x_i \in [a_i, b_i], i = 1,2,..., n$. The opposite point of $\mathbf{x} = (x_1, x_2, ..., x_n)$ is defined by $\mathbf{x}' = (x_1', x_2', ..., x_n')$ where $x_i' = a_i + b_i - x_i$.

Assume $f(\mathbf{x})$ is a fitness function which is used to measure candidate's optimality. $\mathbf{x}' = (x_1', x_2', ..., x_n')$ is the opposite of $\mathbf{x} = (x_1, x_2, ..., x_n)$. Now, if $f(\mathbf{x}') \leq f(\mathbf{x})$, then point $\mathbf{x}$ can be replaced with $\mathbf{x}'$; otherwise we continue with $\mathbf{x}$. Hence, the point and its opposite point are evaluated simultaneously to continue with the fitter one.

4   Opposition-Based Learning Fully Informed Particle Swarm Optimizer without Velocity

In standard particle swarm optimization, the sharing of information among conspecifics is achieved by employing the publicly available information $\mathbf{p}_g(t)$ ('social' component). There is no information sharing among individuals except that $\mathbf{p}_g(t)$ broadcasts the information to the other particles. While the 'cognitive' component makes particles keep their own information. In fact, the position of a particle is influenced by all personal best position $\mathbf{p}_i(t)(i = 1,2,\cdots,N)$. In the paper, all personal best position is considered to update a particle position. Therefore, a particle position update is modified as:

$$\mathbf{x}_i(t+1) = \sum_{k=1}^{N} c_k r_k (\mathbf{p}_k(t) - \mathbf{x}_i(t)) \tag{5}$$

$c_k (k = 1,2,..., N)$ are non-negative constant real parameters, called acceleration coefficients which control how far a particle will move in a single iteration. $r_k (k = 1,2,..., N)$ are three independent uniform random sequences distributed in the range $[0,1]$.

The pseudocode of the opposition based learning fully informed particle swarm optimizer without velocity is as follows:

Begin
   Initial population with size N;
   Initial acceleration coefficients $c_i (i = 1,2,..., N)$;
   Initial particles $\mathbf{x}_i (0) = (x_{i,1}, x_{i,2}, \cdots, x_{i,m})$ in m-dimension space
   with $x_{i,j} \in [a_j, b_j]$, $(i = 1, \cdots, N)$, $(j = 1, \cdots, m)$;
   For i=1 to N
      $\mathbf{p}_i (0) \leftarrow \mathbf{x}_i (0)$;
      Compute opposite point of $\mathbf{p}_i (0)$: $\mathbf{op}_i (0)$;
      If $f(\mathbf{op}_i(0)) < f(\mathbf{p}_i(0))$ **Then** $\mathbf{p}_i (0) \leftarrow \mathbf{op}_i (0)$
   End
   For $t = 1$ to max iteration
      For $i = 1 : N$
         Update $\mathbf{x}_i (t)$ according to formula (5);
         Update $\mathbf{p}_i (t)$ according to formula (3);
         Compute opposite point of $\mathbf{p}_i (t)$: $\mathbf{op}_i (t)$;
         If $f(\mathbf{op}_i(t)) < f(\mathbf{p}_i(t))$ **Then** $\mathbf{p}_i (t) \leftarrow \mathbf{op}_i (t)$
      End
   End
   Output optimal solution
End

The opposition based learning fully informed particle swarm optimizer without velocity is referred to as **OFPSOV**.

It is clear that the algorithm is the simpler than the standard PSO. And the experimental results in next section show that the proposed algorithm has better convergence performance than the standard PSO.

The standard PSO algorithm has been theoretically analyzed by van den Bergh [6-7], Clerc and Kennedy[8], and Trelea[9]. Here, a theoretical analysis for the proposed algorithm is presented as follow:

When the particle swarm operates on an optimization problem, the values of $\mathbf{p}_i$ are constantly updated, as the system evolves toward an optimum. For analysis purpose, consider the situation that $\mathbf{p}_i$ keep constant during a period of time, then all particles evolve independently. Thus, only ith particle needs to be studied. For i is chosen arbitrarily, the result can be applied to all other particles.

Now, by changing (5), the following non-homogeneous recurrence relation is obtained:

$$\mathbf{x}_i (t+1) = \left(-\sum_{k=1}^{N} \alpha_k\right) \mathbf{x}_i (t) + \sum_{k=1}^{N} \alpha_k \mathbf{p}_k (t) \tag{6}$$

Where $\alpha_k = c_k r_k$ are assumed to be constant. The values α_k are thus specific instances of $c_k r_k$.

When the initial conditions $\mathbf{x}_i(0)$ has been specified, the closed form equation of (6) is given by

$$\mathbf{x}_i(t) = \left(-\sum_{k=1}^{N}\alpha_k\right)^t \mathbf{x}_i(0) + \frac{\left(1-(-\sum_{k=1}^{N}\alpha_k)^t\right)\sum_{k=1}^{N}\alpha_k\mathbf{P}_k(t)}{1+\sum_{k=1}^{N}\alpha_k} \tag{7}$$

Note that the above equations assume that $\mathbf{p}_i$ keep constant for all t. The closed form representation in Eq.(7) therefore remains valid until a better position $\mathbf{x}_i$ (and thus $\mathbf{p}_i$) is discovered. When a better position is discovered, the above equations can be used again after recalculating.

From Eq. (7), the sequence $\{\mathbf{x}_i(t)\}$ converges to the stable point when $\left|-\sum_{k=1}^{N}\alpha_k\right| < 1$. In this case,

$$\lim_{t\to\infty}\mathbf{x}_i(t) = \sum_{k=1}^{N}\alpha_k\mathbf{P}_k(t)\Big/\left(1+\sum_{k=1}^{N}\alpha_k\right) \tag{8}$$

This means that, under the condition that $\left|-\sum_{k=1}^{N}\alpha_k\right| < 1$, a particle converges to a weighted average of its all personal best position.

In the case that α_k are stochastic, c_k can be considered a supper bounds for α_k. The average behavior of the system can then be observed by considering the expected values of α_k (assuming uniform distributions):

$$E[\alpha_k] = c_k\int_0^1\frac{x}{1-0}dx = c_k\left.\frac{x}{2}\right|_0^1 = \frac{c_k}{2} \tag{9}$$

Using the expected values, the limit (15) becomes

$$\lim_{t\to\infty}\mathbf{x}_i(t) = \frac{\sum_{k=1}^{N}\frac{c_k}{2}\mathbf{p}_k(t)}{1+\sum_{k=1}^{N}\frac{c_k}{2}} = \frac{\sum_{k=1}^{N}c_k\mathbf{p}_k(t)}{2+\sum_{k=1}^{N}c_k} \tag{10}$$

From condition $\left|-\sum_{k=1}^{N}\alpha_k\right| < 1$, $\alpha_k \geq 0$, it gets $0 \leq \sum_{k=1}^{N}\alpha_k < 1$. Thus $0 \leq \sum_{k=1}^{N}E[\alpha_k] < 1$ can be obtained. Substitute (9) into $0 \leq \sum_{k=1}^{N}E[\alpha_k] < 1$, the convergence condition can be obtained.

$$0 \leq \sum_{k=1}^{N}c_k < 2 \tag{11}$$

From above analysis, we get the following conclusion:

Given $c_k \geq 0$, If $0 \leq \sum_{k=1}^{N} c_k < 2$ is satisfied, then (6) will converge to

$\sum_{k=1}^{N} c_k \mathbf{p}_k(t) / \left(2 + \sum_{k=1}^{N} c_k\right)$, that is that the particle of the swarm tends to

converge to a weighted average of its all personal best position.

5   Experimental Results from Simulations

In order test the efficiency of the opposition based learning fully informed particle swarm optimizer without velocity(**OFPSOV**). This section compares the performance of the proposed algorithm with that of the standard particle swarm optimization with weight factor. The following benchmark optimization functions have been used. For each of these functions, the goal is to find the global minimizer, formally defined as

Given $f : R^M \to R$,find $\mathbf{x}^* \in R^M$ such that $f(\mathbf{x}^*) \leq f(\mathbf{x}), \forall \mathbf{x} \in R^M$

The following functions were used:

A. Sphere function, defined as

$$f_1(\mathbf{x}) = \sum_{i=1}^{M} x_i^2 \text{ , where } \mathbf{x}^* = \mathbf{0} \text{ and } f(\mathbf{x}^*) = 0 \text{ for } -100 \leq x_i \leq 100$$

B. Rastrigin function, defined as

$$f_2(\mathbf{x}) = \sum_{i=1}^{M} (x_i^2 - 10 \cos(2\pi x_i) + 10) \text{ , where } \mathbf{x}^* = \mathbf{0} \text{ and } f(\mathbf{x}^*) = 0 \text{ for}$$

$$-5.12 \leq x_i \leq 5.12$$

C. Griewank function, defined as

$$f_3(\mathbf{x}) = \frac{1}{4000} \sum_{i=1}^{M} x_i^2 - \prod_{i=1}^{M} \cos\left(\frac{x_i}{\sqrt{i}}\right) + 1 \text{ , where } \mathbf{x}^* = \mathbf{0} \text{ and } f(\mathbf{x}^*) = 0 \text{ for}$$

$$-600 \leq x_i \leq 600$$

D. Rotated hyper-ellipsoid function, defined as

$$f_4(\mathbf{x}) = \sum_{i=1}^{M} \left(\sum_{j=1}^{i} x_j\right)^2 \text{ , where } \mathbf{x}^* = \mathbf{0} \text{ and } f(\mathbf{x}^*) = 0 \text{ for } -100 \leq x_i \leq 100$$

E.   Schwefel's Problem, defined as

$$f_5(\mathbf{x}) = \sum_{i=1}^{M} |x_i| + \prod_{i=1}^{M} |x_i| \text{ , where } \mathbf{x}^* = \mathbf{0} \text{ and } f(\mathbf{x}^*) = 0 \text{ for } -10 \leq x_i \leq 10$$

F. Schwefel's function 6, defined as

$$f_6(\mathbf{x}) = \frac{\sin^2 \sqrt{x_1^2 + x_2^2} - 0.5}{[1 + 0.001(x_1^2 + x_2^2)]^2} + 0.5, \text{ where } \mathbf{x}^* = \mathbf{0} \text{ and } f(\mathbf{x}^*) = 0 \text{ for } -100 \leq x_i \leq 100$$

G. Schwefel's function 2.2.1, defined as

$$f_7(\mathbf{x}) = \max_{i=1}^{M} \{|x_i|\}, \text{ where } \mathbf{x}^* = \mathbf{0} \text{ and } f(\mathbf{x}^*) = 0 \text{ for } -100 \leq x_i \leq 100$$

H. Ackley's function, defined as

$$f_8(\mathbf{x}) = -20 \exp\left(-0.2\sqrt{\frac{1}{M}\sum_{i=1}^{M} x_i^2}\right) - \exp\left(\frac{1}{M}\sum_{i=1}^{M} \cos(2\pi x_i)\right) + 20 + e$$

where $\mathbf{x}^* = \mathbf{0}$ and $f(\mathbf{x}^*) = 0$ for $-32 \le x_i \le 32$

I. $f_9(\mathbf{x}) = \sum_{i=1}^{M}[100(x_{i+1} - x_i^2)^2 + (1-x_i)^2]$, $|x_i| \le 300$ where $\mathbf{x}^* = (1,1,\cdots,1)$ and $f(\mathbf{x}^*) = 0$

J. $f_{10}(\mathbf{x}) = \sum_{i=1}^{M}([x_i + 0.5])^2$, $|x_i| \le 100$ where $\mathbf{x}^* = \mathbf{0}$ and $f(\mathbf{x}^*) = 0$

For the standard PSO algorithms, $w = 0.72$, and $c_1 = 1.49$ and $c_2 = 1.49$. These values have been shown to provide very good results (Clerc and Kennedy, 2002; van den Bergh, 2002; Van den Bergh and Engelbrecht, 2006). The acceleration coefficients $c_i(i = 1,2,...,N) = 0.1$ for the **OFPSOV**. For all the algorithms used in this section, N= 20. All functions were implemented in 50 dimensions except for the two-dimensional f_6 function. All experiments were repeated for 50 runs. A fixed number of maximum generations 500 was applied to all algorithm. The initial population was generated from a uniform distribution in the ranges specified below. The experimental results for each algorithm on each test function are listed in Table1. According to Table1, it can been known that **OFPSOV** can converge to global optimum with a higher precision quickly and robustly. Clearly, **OFPSOV** outperforms the standard PSO greatly for the benchmark functions.

Table 1. Comparison of **OFPSOV** and PSO for $f_1\sim f_{10}$

	OFPSOV (mean ± variances)	PSO (mean ± variances)
f_1	5.4319e-196± 0	1.8047e+3±8.5201e+4
f_2	68.1402±683.4781	1.0413e+4±1.0826e+5
f_3	0.0211± 0.3473	3.4581±46.7514
f_4	5.4973e-187± 0	2.1302e+4±3.2016e+5
f_5	8.0195e-99±1.4307e-101	2.1219e+7± 4.0146e+8
f_6	0.0012± 0.1053	0.0081± 0.1076
f_7	3.8027e-101±7.9023e-100	29.6256± 351.6197
f_8	2.9404e-16± 5.7347e-15	17.1308±270.2864
f_9	47.6095± 631.2315	4.1954e+7±7.1209e+8
f_{10}	0±0	1.0367e+4±1.3852e+5

6   Conclusions

In this paper, we proposed a new opposition based learning fully information particle swarm optimiser without velocity by using full information and opposition-based learning. In the algorithm, particles only have position without velocity, and the personal best position gets updated using opposition-based learning. Particle position is updated using all personal best positions. The theoretical analysis for the proposed algorithm implies that the particle of the swarm tends to converge to a weighted average of all personal best position. The algorithm is the simpler and effective as a result of discarding the particle velocity and using full information and opposition-based learning. The experimental results show that the algorithm has better convergence performance. In future work, we need to clarity the relationship between the algorithm performance and parameter selection to guarantee convergence.

Acknowledgments. This work is supported by the Scientific and Technological Innovation Projects of Department of Education of Guangdong Province, P.R.C. Grant No.2012KJCX0082 and Science and Technology Projects of Guangdong Province, P.R.C. under Grant No.2011B090400623, 2012B091100337, Guangzhou Science and Technology Projects under Grant No.12C42011563, 11A11020499. Authors of the paper express great acknowledgment for these supports.

References

1. Kennedy, J., Eberhart, R.C.: Particle swarm optimization. In: Proceedings of the IEEE Conference on Neural Networks, pp. 1942–1948. IEEE Service Center, Perth (1995)
2. Tizhoosh, H.R.: Opposition-Based Learning: A New Scheme for Machine Intelligence. In: Int. Conf. on Computational Intelligence for Modelling Control and Automation (CIMCA 2005), Vienna, Austria, vol. I, pp. 695–701 (2005)
3. Jabeen, H., Jalil, Z., Baig, A.R.: Opposition based initialization in particle swarm optimization O-PSO. In: GECCO Companion, pp. 2047–2052. ACM (2009)
4. Wang, H., Liu, Y., Li, C., Zeng, S.: Opposition-based Particle Swarm Algorithm with Cauchy Mutation. In: IEEE CEC, pp. 4750–4756 (2007)
5. Shahzad, F., Baig, A.R., Masood, S., Kamran, M., Naveed, N.: Opposition-Based Particle Swarm Optimization with Velocity Clamping (OVCPSO). Advances in Intelligent and Soft Computing 116, 339–348 (2009)
6. van den Bergh, F.: An analysis of particle swarm optimizers. PhD thesis, Department of Computer Science, University of Pretoria, South Africa (2002)
7. van den Bergh, F., Engelbrecht, A.: A study of particle swarm optimization particle trajectories. Information Sciences 176(8), 937–971 (2006)
8. Clerc, M., Kennedy, J.: The particle swarm-explosion, stability, and convergence in a multidimensional complex space. IEEE Transactions on Evolutionary Computation 6(1), 58–73 (2002)
9. Trelea, I.C.: The particle swarm optimization algorithm: convergence analysis and parameter selection. Information Processing Letters 85, 317–325 (2003)

GSO: An Improved PSO Based on Geese Flight Theory

Shengkui Dai, Peixian Zhuang, and Wenjie Xiang

College of Information Science and Engineering of National HuaQiao University
D.S.K.@hqu.edu.cn

Abstract. Formation flight of swan geese is one type of swarm intelligence developed through evolution by natural selection. The research on its intrinsic mechanism has great impact on the bionics field. Based on previous research achievements, extensive observation and analysis on such phenomenon, five geese-flight rules and hypotheses are proposed in order to form a concise and simple geese-flight theory framework in this paper. Goose Swarm Optimization algorithm is derived based on the Standard Particle Swarm Optimization algorithm. Experimental results show that GSO algorithm is superior in several aspects, such as convergence speed, convergence precision, robustness and etc. The theory offers the in-depth explanations for the performance superiority. Moreover, the rules and hypotheses for formation flight adhere to all five basic principles of swarm intelligence. Therefore, the proposed geese-flight theory is highly rational and has important theoretical innovations, and GSO algorithm can be utilized in a wide range of applications.

Keywords: Geese Theory, Goose Swarm Optimization, Particle Swarm Optimization, Swarm Intelligence.

1 Introduction

Swan geese often tend to line up in a J-Shape or V-Shape during their migration to the South. This extraordinary natural phenomenon was found since ancient times, but nobody thoroughly understood it. The hypothesis that the V-Shape formation flight of geese had the energy-saving advantage was firstly proposed by the German aerodynamics Carl Wieselsberger. It states that goose's winging causes wake vortices, and the air flows up outside of the vortex. If one neighbour goose is at the position with upwards gas swirling, it can save a significant amount of energy during the flight. Ever since then, studies on the mechanism of geese flight stepped into a new period.

Particle swarm optimization algorithm is a swarm intelligence optimization algorithm invented by Kennedy and Eberhart [1,2]. Compared to other algorithms, PSO requires less parameters and is easier to implement. However, it easily falls into a relative extremism of test function, and has lower convergence accuracy probably due to the easy homogenization and diversity scarcity of particles in iterative process. In order to resolve these problems, linear-decreased inertia weight is presented to perfectly balance between the local and global search abilities of PSO algorithm [3,4,5]. Then SPSO is formulated from studies on acceleration coefficients, parameter settings and boundary conditions of PSO algorithm[6,7,8]. In order to further improve

Y. Tan, Y. Shi, and H. Mo (Eds.): ICSI 2013, Part I, LNCS 7928, pp. 87–95, 2013.

the precision of SPSO algorithm, some pioneers utilized the studies on geese-flight to discover that formation flight of geese increases flight distances by 12% than one solitary goose's flight. An improved particle swarm optimization algorithm[9] based on the advantage of formation flight is proposed, and improves the optimizing precision markedly, compared with the SPSO.

To further balance between homogenization and diversity of particles in iterative process, as well as to improve the convergence precision and speed of the SPSO algorithm, a concise and simple theory of geese flight is proposed based on five rules and hypotheses which are summarized through detailed analyses on formation flight in this paper. According to these hypotheses, Geese Swarm Optimization algorithm is proposed to improve the SPSO in test indicators, such as convergence precision, convergence rate and robustness, and so on. In the rest of sections, we first provide the detailed theoretical analysis followed by the test results of the proposed algorithm.

2   GPSO Algorithm

GSO algorithm randomly initializes a particle population M, and each particle is treated as a point in the N-dimensional space. The position of the i-th particle is X_i, and velocity as V_i. Particles are sorted according to its fitness, and the best particle is chosen as the leader. After then, the personal extrema $p_{(i-1)}$ of the former particle is selected as the global extremum of the i-th particle. On the other hand, personal extrema $pbest_i$ of the i-th particle are updated by the weighted average of all personal extrema according to its fitness. Therefore, personal extrema, the position and the velocity of i-th particle are updated according to following equations:

$$V_i(k+1) = w(k) \times V_i(k) + c_1 \times r_1 \times (p_a(k) - X_i(k)) + c_2 \times r_2 \times (pbest_{(i-1)}(k) - X_i(k)) \quad (1)$$

$$X_i(k+1) = X_i(k) + V_i(k+1) \quad (2)$$

$$p_a = \sum_{i=1}^{M} pbest_i \times f(X_i) \bigg/ \sum_{i=1}^{M} f(X_i) \quad , \quad w(k) = w_{max} - \frac{(w_{max} - w_{min}) \times k}{k_{max}} \quad (3)$$

Where k for current iterative number, r_1 and r_2 are random numbers in [0,1], c_1 and c_2 are positive acceleration factors. w is linear-decreased inertia weight, w_{max} for the maximum value of w, and w_{min} for the minimum.

Experimental results show that GSPO algorithm can improve the convergence accuracy, robustness and convergence rate to a large extent [9,10]. However, GPSO also has some issues. First, in the sense of bionics for geese flight, the formula (3) means that every goose knows the status of all geese, which is incompatible with the hypotheses presented later in this paper, and the effect of formula (3) is tested to be counterproductive. Second, formula (3) is only suitable for seeking the maxima of functions, not suitable for seeking the minimum. In addition, it is inaccurate to immediately use fitness as the weight, especially for such functions with dramatically changing extremal values, which are the limitations of the GPSO algorithm.

3   Geese Theory and GSO Algorithm

3.1   Geese Flight Theory-Rules and Hypotheses

Based on other researchers' results and the in-depth analysis of geese flight, we believe that the energy-saving hypothesis is more reasonable. According to aerodynamics principles, the latter goose must be in a rear sloping position of the former, in order to make use of the vortex. With this basic idea and its extension, the paper summarizes and presents five rules and hypotheses of geese flight to attempt to form a more reasonable theory. The rules and hypotheses and their evidence are described as following:

(1) **Anosia Hypothesis: the toughness degree gradually decreases from the first goose to the last.** The strongest goose always acts as the leader with maximum labor intensity because none of front air vortex can be utilized. In case of fatigue, the leader will move to the end of the formation, and the second goose in the row replaces the position of the original leader to guide the flock flight consequentially.

(2) **View-field Hypothesis: the flying geese with limited visual field can only see the front part of the whole flock.** In order to take advantage of the air vortex produced by the front goose to save energy, a goose must follow the former at the inclined rear position. This means that the 'J-Shape' team is an oblique array. Therefore, the current goose only needs to see the front part in its visual field. Although studies show that sight range of geese may be as much as 128 degrees, which allows goose to see the entire team when flying.

(3) **Global Hypothesis: each goose adjusts its own position according to the status of all geese in its front visual field.** These adjustments may maintain the integrity of flight formation to achieve the global optimization, or one single goose utilizing the comprehensive effects produced by all geese in the front view can save more energy. Global Hypothesis is a natural extension of View Hypothesis.

(4) **Local Hypothesis: each goose quickly adjusts its positions according to the closest former.** The current goose quickly adjusts its position according to the status of the front geese in order to quickly and effectively utilize the vortex right ahead of it. This vortex is most direct, most effective and most useful in the sense of energy-saving.

(5) **Simpleness Hypothesis: Geese adopt a simple and direct method to adjust their status.** According to the above statements, all flying geese except the leader need to adjust their positions with a quick, dynamic and real-time manner. Without considering how goose intelligence is, this paper assumes that geese use a simple and effective method to quickly and timely adjust itself to a local optimal or suboptimal position.

3.2   GSO Algorithm Based on Geese Theory

As aforementioned, every goose can perceive two statuses, i.e., part of group's status global extremum, and individual status personal extrema including itself and the former's. One GSO algorithm is implemented consequentially on the basis of SPSO algorithm in this paper, which can be treated as one of improved versions of SPSO or GPSO algorithm. Of course, the proposed geese theory could have more expression forms, which means that our GSO algorithm based on SPSO is just the first example not the last. This paper realizes the GSO algorithm based on SPSO in several aspects: **Firstly**, sort all particles based on hypothesis 1 in each iteration, and then get the geese queue in accordance to each particle's fitness. **Secondly**, calculate its optimal values that each goose can perceive according to hypotheses 2 and 3, so there is more than one global extreme value. According to hypotheses 5, a simple average method is adopted to calculate its global extremum $gbest_i$ of each goose. **Finally**, adjust its individual optimal value based on hypotheses 2 and 4. A simple-substitute method is adopted to update the personal extrema, namely, the new personal extreme values of i-th goose is now $pbest_{(i-1)}$ of the former $(i-1)$ goose, which is based on hypotheses 5. As a result, in our GSO algorithm, the personal extrema, global extremum, and velocity of the i particle are updated as the following equations:

$$pbest_i = pbest_{(i-1)}, \quad gbest_i = \frac{1}{i}\sum_{m=1}^{i} pbest_m \tag{4}$$

$$V_i(k+1) = w(k)\times V_i(k) + c_1 \times r_1 \times (gbest_i\text{-}X_i(k)) + c_2 \times r_2 \times (pbest_{(i-1)}(k)\text{-}X_i(k)) \tag{5}$$

GSO algorithm can be well balanced between diversity and homogenization of particles. The global extremum and personal extrema used to update the status of each goose are different so that the particle diversity is preserved. The simple average method can prevent particles from gathering together too densely around the minority particles with higher weight, which can reduce the homogenization trend for particles. It means that every goose can learn from the flock and local goose in bionics sense. On the other hand, use better goose than the current to compute group extremum and individual extrema means that goose can maintain the homogenization trend for best result.

4   Experimental Results and Discussion

In order to verify and compare the performance of GSO algorithm, two typical benchmark functions are selected for experiments, where optimal values are both 0. Between these functions, the Sphere function is with a minimum value in the flat area. The Rastrigin function forms another category with multiple relative maxima and minima. The outer iteration number is 50 for each function, and the inner iteration

Table 1. Benchmark functions and initial parameters

Test Function	Analytic Expression	Dimension	Initial Range
Sphere	$f_1(X) = \sum_{i=1}^{N} x_i^2$	30	[-10,10]
Rastrigin	$f_3(X) = \sum_{i=1}^{N}[x_i^2 - 10\cos(2\pi x_i) + 10]$	30	[-5.2, 5.2]

number is 1000. Population size is 20, and all particles are uniformly distributed in the initial range. Linear-declined range of w is [0.9, 0.4]. c_1 and c_2 are set to 2. V_{max} is different according to the function characteristics. Benchmark test functions and initial parameters are shown in table 1.

4.1   Algorithm Validations for Fixed Dimensions and Fixed Iteration Number

According to the above parameters, four evaluation indicators for optimal fitness, including mean best fitness, optimal solution, standard deviation and excellent ratio, which are respectively short for MBF, OS, SD, ER. These indicators are adopted to evaluate the performances of the three algorithms. MBF and OS are used to measure convergence accuracy, which is on behalf of global optimization ability, SD is used to measure the robustness of all algorithms, ER means the percentage of GSO and GPSO's results superior to SPSO. The experimental results are shown in Table 2.

Table 2. Results under fixed dimensions and iterations

Test Function	Algorithm	MBF	OS	SD	ER
Sphere	SPSO	0.0072	0.0025	0.0032	
	GPSO	0.0023	0.0008	0.0009	98%
	GSO	0.0015	0.0007	0.0005	98%
Rastrigrin	SPSO	36.6122	15.2921	10.8477	
	GPSO	11.5285	5.4284	3.7324	100%
	GSO	12.1994	4.4802	3.5772	100%

The data shows that the performances of GSO are significantly improved in comparison to the SPSO, and slightly better than GPSO. Thus we can draw the conclusion that convergence precision, optimal capacity and robustness for GSO algorithm are statistically better than the SPSO and GPSO under fixed dimensions and iterations.

4.2   Experimental Comparisons on Convergence Speed and Trend

In order to show the convergence speed and the trend for these algorithms, the global optimal value is shown in a curve way during the iterative process. IGPSO is GSO. The number of iterations is 500 and the dimension is 30. Other parameters are fixed. The results are shown in figure 1.

All three algorithms drop fast in the initial iterative phase, which means that these algorithms can probably fall into the relative domain extrema. With the increase of the iteration times, SPSO and GPSO may be quickly trapped in a relative extreme value, but cannot jump out of the relative extreme area. This implies that the convergence precision of SPSO and GPSO cannot be further improved. However, GSO can approach the better extreme values, which are close to the optimal value.

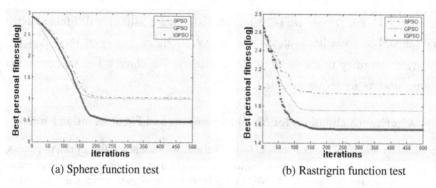

(a) Sphere function test   (b) Rastrigrin function test

Fig. 1. Fitness chart for test algorithms

4.3   Performance Verification for Respective Rule 3 and 4

Among those principles proposed above, global extremal part in SPSO algorithm is modified according to the Rule 3, and individual extremal part is updated according to the Rule 4. Several experiments are conducted to analyze the algorithm performance respectively only with rule 3 or rule 4. Rastrigrin function is used with default parameters, and the results are shown in Table 3.

Table 3. Performance Verification Results for rule 3 and/or rule 4

Rule	Algorithm	MBF	OS	SD	ER
Only Rule 3	SPSO	37.3096	18.6962	10.2504	
	GPSO	48.0027	15.2145	22.8932	34%
	GSO	14.3516	2.9932	5.5277	98%
Only Rule 4	SPSO	37.2380	21.9243	8.9809	
	GPSO	22.8587	11.5966	7.8924	92%
	GSO	22.0462	9.8606	8.9582	92%

The data shows that GSO with only rule 3 is better than SPSO, while the results of GPSO are worse than SPSO. Both GSO and GPSO with only rule 4 are similar to each other in principle, and better than SPSO.

Principle analysis: in order to solve the fault of SPSO, which easily falls into extrema, GSO algorithm with only rule 3 only with more than one global optimal extremum can avoid the risk, i.e. a single global extremum to weaken the trend that other particles approximate to single extremum, and at the same time these extrema not too far from the global extremum. GSO algorithm with only rule 4 has the ability and trend of the transition from the relative extremum to another better one. Accumulations formed from the transition effect of much iteration are finally in approximation to the theoretical extreme value for GSO algorithm. In fact, rule 4 is almost the same as the principle of the LBEST mode mentioned in [2]. However, the performance of GPSO with only rule 3 is worse than SPSO, the reason may be due to the GPSO defects pointed out in the first part of this paper. GPSO can avoid the risk single global extremum for SPSO, but the way of its weight calculation is unreasonable. Therefore, its weighted optimal value is too far from the global extreme points, which leads to worse performance.

4.4   Comparison Performances on GSO and Genetic Algorithm

In order to compare GSO with Genetic Algorithm, we take Rastrigrin function as the test model. The number of iterations is 100 and the dimension is 2, and other parameters of these algorithms are unchanged. The crossover and mutation probabilities in GA are randomly selected for each iteration. The results are shown in figure 2.

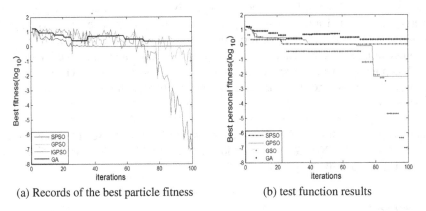

(a) Records of the best particle fitness   (b) test function results

Fig. 2. Comparison performances on GSO and GA

Figure 2 (a) shows record process of the best particle fitness for each iteration. At the iteration number from 0 to 70, all algorithms have a relative slow convergence rate. SPSO with the reason of only one global extremum easily falls into the relative domain extremum at about 40, and GA is at about 70. After the 70th iteration, GPSO and GSO can reach the better convergence rate because of multiple global extrema to increase the probability of jumping out of those extremal domain, and convergence accuracy of GSO is much better than GPSO. Figure 2 (b) shows convergence trends and results of best personal fitness, and convergence trend and accuracy of GSO are much better than GA, SPSO and GPSO. Therefore, from above data we can draw the conclusion that test performances of GA are worst, and GSO is much better than GA, SPSO and GPSO.

5   Conclusions Remarks

PSO algorithm is one type of swarm intelligence optimization algorithms invented based on bird flock foraging, which has been widely used in function global optimization, combinatorial optimization, dynamic system optimization, and other aspects. But its defect is to easily fall into a relative extremum. The geese formation flight is a natural swarm optimization phenomenon, and some geese principles have been widely used in economics and management field. However, how to combine with swarm intelligence optimization algorithm is still a new research direction. Based on in-depth observation and analysis on natural geese formation flight, five flight rules are proposed to form a concise and reasonable geese-flying theory.

This paper tends to agree with the "vortex energy-saving" hypothesis. Through extensive observations about geese formation flight, this paper presents five geese flight rules and hypotheses for the first time, in order to construct a simple and reasonable theory of geese. Similar to the SPSO algorithm, our geese theory is in line with the five basic principles of swarm intelligence[1] with better characteristics of artificial life.

Two improvements for SPSO based on geese theory are proposed to achieve a better balance between the diversity and identity of particles. The experiments results show that rule 4 has more significant influence than the other results. Moreover, it can be confirmed that GSO algorithm has been remarkably improved in convergence speed, convergence precision and robustness, etc. As a result, we draw the conclusion that these rules and hypotheses of geese formation flight are reasonable and effective.

In order to make these rules and hypotheses more reasonable and effective to have general application value, more in-depth theoretical research and verifications on geese formation flight are necessary in the future at biology, physics, mathematics and other aspects.

Acknowledgment. This research is funded by the Fundamental Research Funds for the Central Universities (No:JB-ZR1145) and High-level Talents Research Project of National HuaQiao University (No: 09BS102).

References

1. Kennedy, J., Eberhart, R.C.: Particle Swarm Optimization. In: Proceedings of the 1995 IEEE International Conference on Neural Networks, pp. 1942–1948. IEEE Press, Perth (1995)
2. Eberhart, R.C., Kennedy, J.: A New Optimizer Using Particle Swarm Theory. In: Proceedings of the Sixth International Symposium on Micro Machine and Human Science, pp. 39–43. IEEE Press, Nagoya (1995)
3. Shi, Y.H., Eberhart, R.C.: Empirical Study of Particle Swarm Optimization. In: Proceeding of Congress on Evolutionary Computation, pp. 1945–1949. IEEE Press, Piscataway (1999)
4. Shi, Y.H., Eberhart, R.C.: A Modified Particle Swarm Optimizer. In: Proceedings of the 1998 IEEE International Conference on Evolutionary Computation, pp. 69–73. IEEE Press, Anchorage (1998)
5. Fukuyama, Y.: Fundamentals of Particle Swarm Techniques. IEEE Power Engineering Society 45–51 (2002)
6. Ratnaweera, A., Halgamuge, S.: Self-organizing hierarchical particle swarm optimizer with time 2 varying acceleration coefficients. IEEE Transactions on Evolutionary Computation 8, 240–255 (2004)

7. Robinson, A., Rahamat-Samii, Y.: Particle Swarm Optimization in Electromagnetics. IEEE Transations on Antennas and Propagation 52, 397–407 (2004)
8. Shi, Y.H., Eberhart, R.C.: Parameter Selection in Particle Swarm Optimization. In: Porto, V.W., Waagen, D. (eds.) EP 1998. LNCS, vol. 1447, pp. 591–600. Springer, Heidelberg (1998)
9. Liu, J.Y., Guo, M.Z., Deng, C.: GeesePSO: An Efficient Improvement to Particle Swarm Optimization. Computer Science 33, 166–168 (2006)
10. Xiao, Z., Yuan, Y., Li, P.Y.: Learning Algorithm for Multimodal Optimization. Computers and Mathematics with Applications 57, 2016–2021 (2009)

Improved Algorithms Based
on the Simple Particle Swarm Optimization

Lei Liu, Xiaomeng Zhang, Zhiguo Shi, and Tianyu Zhang

School of Computer and Communication Engineering,
University of Science and Technology Beijing,
100083 Beijing, P.R. China
zhangxm0703@163.com

Abstract. As one of the representative algorithms in swarm intelligence, particle swarm optimization has been applied to many fields because of its several merits, such as simple concept, easy realizing and fast convergence rate in the early evolutionary. However, it still has some disadvantages such as easy falling into the local extremum, slow convergence velocity and low convergence precision in the late evolutionary. Two new algorithms based on the simple particle swarm optimization are proposed to try to improve the precision of the algorithm in a certain error range of the length of time. The algorithms have been simulated and compared with the particle swarm optimization and the simple particle swarm optimization. The simulations show that the algorithms have a higher convergence precision for some functions or a particular issue.

Keywords: Swarm Intelligence, Particle Swarm Optimization, Swarm Robots.

1   Introduction

Swarm intelligence is a method to achieve artificial intelligence by imitating biological group behavior in the natural world [1], which offers a new thought to the solutions of complex issues by using group advantage without centralized control and global model [2]. It is also a kind of soft simulation for biological group, which is different from the traditional simulation for the structure of organisms. The individual can be regarded as very simple and single, and it is also allowed to have the ability to learn to solve specific problems.

As a representative of the swarm intelligence algorithms, particle swarm optimization (PSO) algorithm is used to solve continuous optimization problems originally. Similar to the genetic algorithm, it is a kind of optimization tool based on the idea of group. But variation and cross operation which exist in genetic algorithm as we know do not exist in PSO. Particles in the solution space search the goal through following the best particle. Therefore, it achieves simply and has fewer parameters to be adjusted. But it is also easy to fall into the local extremum and has low convergence precision in the late evolutionary.

In this paper, two new algorithms are proposed based on the idea of the simple particle swarm optimization (sPSO) [3] put forward by Hu Wang et al to try to

Y. Tan, Y. Shi, and H. Mo (Eds.): ICSI 2013, Part I, LNCS 7928, pp. 96–103, 2013.

improve the precision of the algorithm. One is simulated in the Matlab simulation platform and the other is in Player/Stage. In the end, they are both compared with PSO and sPSO.

2   Research Background and Related Works

2.1   Particle Swarm Optimization

PSO is first put forward by Kennedy and Eberhart in 1995 [4-5], which is a kind of heuristic search algorithm based on population optimization. Due to the concept of it is simple, easy to be realized, and has good optimal characteristics, particle swarm optimization is rapidly developing in the short term, and has been used in many fields, such as electric power system optimization, TSP problem solving, etc.

The principle of the particle swarm algorithm is: there is only a piece of food in the region, and all the birds don't know where the food is, but they know the distance between the current location and the food. What is the optimal strategy to find the food? The most simple and effective method is to search the area around the bird whose location is nearest from the food at present.

In the PSO, each solution of the optimization problem corresponding to the location of a bird in the search space, calling the bird "Particle". D-dimensional position vector of the i-th particle is $x_{i=}$ $(x_{i1}, x_{i2}, ..., x_{iD})$. According to pre-set fitness functions (which is relevant to the problem), the current value of x_i can be calculated. Particles fly at a certain speed in the search space. $v_i=(v_{i1}, v_{i2}, ..., v_{id}, ..., v_{iD})$ is the speed of particle i, which dynamically adjust based on its own flying experience and that of companion. All the particles have a Fitness Value decided by target function, and know the best position so far (particle best, notes for pbest) $P_{i=}$ $(P_{i1}, P_{i2}, ..., P_{id}, ..., P_{iD})$ and the current position. These can be seen as their own flying experience. In addition, each particle also know the best position(global best, notes for gbest) so far in the group, which can be regarded as the companion's experience of a particle. And then the particles follow the current optimal particle to search in the solution space.

Shi[5] et al add the momentum inertia coefficient ω to improve the ability of PSO to jump out of local extrema. The velocity and position of the particle in each iteration are updated based on the following formula:

$$v_{id}^{t+1} = \omega * v_{id}^{t} + c_1 r_1 (p_{id} - x_{id}^{t}) + c_2 r_2 (p_{gd} - x_{id}^{t}) \cdot \tag{1}$$

$$x_{id}^{t+1} = x_{id}^{t} + v_{id}^{t+1} \tag{2}$$

(1) and (2) are regarded as the basic particle swarm optimization(bPSO) by many researchers..t is the number of iterations, ω is the inertia weight, r_1 and r_2 are the random numbers between [0, 1], which are used to keep the diversity of the population. c_1 and c_2 are learning factors, which make particles have the ability to summarize themselves and learn from excellent individual in the group, thus close to pbest and gbest.

2.2   The Simple Particle Swarm Optimization

PSO and most of the improved algorithms all based on the two key factors of particles, "position" and "speed". Therefore, the equations of the improved algorithms all contain a position variable and the speed variable. For most of the improved algorithms of PSO, some operators like crossover, mutation, etc. are added, which makes the description of PSO more and more complicated. This also makes the quantitative analysis of the convergence of PSO very complicated.

We can find something from the analysis of the biological model of PSO and its evolutionary iteration equation. In PSO, the velocity variable of the particle is not indispensable. Viewed from the point of the model of the basic particle swarm optimization (bPSO), the position x_i represents the solution of this problem. Because the final result of the optimization is x_i infinitesimal approaching to the optimal solution, we need to consider the direct change of x_i only. The Velocity vector v_i just represents the rapidity of the particles' movement. The speed of the movement is not able to show that the particle can approaches to the location of the optimal solution effectively. Instead, that may cause the particles to deviate from the correct direction of evolution, which is the "divergence" phenomenon, thus resulting in the phenomenon of slow convergence speed and low convergence precision in the later stage. In addition, the position and velocity directly are computed without the concept of particles' movement time in formula (2). It is not in line with the law of motion $x = vt$ in real life [3].

The theorem: bPSO evolutionary process has nothing to do with the particle velocity.

Proof: [6] has demonstrated that the speed range $v_{id} \in [-v\max, v\max]$ of the trapped particle equals the restraint factor α. Therefore, only the joint evolution equation formed by the formula (1) and formula (2) shall be considered. The update of each dimension is completely independent of one another except for the relation between p_{id}, p_{gd} and search space of every dimension. So it is without loss of generality. Since the certification process can be simplified to one-dimensional case, the subscript d can be omitted. Further, it is assumed that the particles in the population remain intact except particle i, and the subscript i can be omitted [3]. The variables are defined as follows: $\varphi_1 = r_1 c_1$, $\varphi_2 = r_2 c_2$, $\varphi = \varphi_1 + \varphi_2$, $\rho = \dfrac{\varphi_1 p_0 + \varphi_2 p_g}{\varphi_1 + \varphi_2}$ .To facilitate understanding, formula (1) and formula (2) are moved from the superscripts of variable descriptors to the parentheses behind them. Formula (1) and formula (2) can be changed to:

$$v(t+1) = \omega v(t) + \varphi(\rho - x(t)) \tag{3}$$

$$x(t+1) = x(t) + v(t+1) \tag{4}$$

Formula (5) can be obtained after the iteration of the formula (3) and the formula (4):

$$x(t+2) + (\varphi - \omega - 1)x(t+1) + \omega x(t) = \varphi \rho \tag{5}$$

Equation (5) is a classical second order differential equation without speed. The importance of Theorem 1 is to explain that bPSO can do without the concept of particle velocity and avoid the artificial determined parameter [-$v$max,$v$max], which affect the convergence speed and the convergence precision of particles.

Based on the above analysis, Hu W et al [3] proposed an optimization equation of the particle swarm without speed items, which is as follows.

$$x_{id}^{t+1} = \omega x_{id}^{t} + c_1 r_1 (p_{id} - x_{id}^{t}) + c_2 r_2 (p_{gd} - x_{id}^{t}) \ . \tag{6}$$

ωx_{id}^{k} in the formula represents the position of the particles in the previous stage, called the "historical" part. It carries the impact of the past. Through changing the value of ω, we control the degree of the impact of the next position. $c_1 r_1 (p_{id} - x_{id}^{k})$ is the difference value between the historical best position of the particle and the current position of the particle. It makes the particle have the trend to the historical best position, called the "cognition" part. It expresses the thoughts of the particle on itself. $c_2 r_2 (p_{gd} - x_{id}^{k})$ is the difference value between the historical best position of the group and the current position of the group. It makes the group have the trend to the historical best position, called the "Social Experience" part. It expresses the comparison and the imitation between the particles of the group, and achieves the information sharing and collaboration between particles.

This is the simple particle swarm optimization, abbreviated as sPSO.

3   Improved Algorithms Based on sPSO

3.1   Adding a Random Number for the "Historical" Part

In order to improve particle swarm optimization algorithm convergence precision, two algorithms are proposed based on sPSO. Firstly this paper tries to add a random number for the historical part of the formula of sPSO to reduce the degree of inheritance from the history and improve the particle's ability of exploration. The formula after adding a random number is as follows:

$$x_{id}^{k+1} = r_0 \omega x_{id}^{k} + c_1 r_1 (p_{id} - x_{id}^{k}) + c_2 r_2 (p_{gd} - x_{id}^{k}) \ . \tag{7}$$

In order to test the significance of r_0's existence for formula, bPSO, sPSO and the improved PSO with the rand (rPSO) are used to optimize the given function. It is commonly used in the comparison of the optimization algorithms. The number of particles is 16, the number of iterations is 150, $c_1 = c_2 = 2$, $\omega = 0.4 \sim 0.9$. Function form, dimension, search range and the extremum in theory are shown in Table 1.

Table 1. The function used to test the improved method

Name and code	Formula	Dim n	Range [x_{min}, x_{max}]	Optimal f
Sphere $f1$	$f_1(x) = \sum_{i=1}^{n} x_i^2$	30	$[-100,100]^{30}$	0

Experiment shows the improvement result. The length of time it takes and the final convergence value of each run of the three algorithms are shown in Table 2, Table 3 and Table 4.

Table 2. The simulation of bPSO

	1	2	3	4	5	6	7	8	9	10
Time(s)	0.05	0.02	0.06	0.02	0.02	0.02	0.04	0.02	0.02	0.02
Value	2.01	2.51	2.41	1.44	1.93	2.65	1.86	3.14	2.96	2.07

As can be seen from the table, the convergence values of the ten separate runs are all between 1 and 4. The convergence precision is not high. The length of time which it takes is between 0.02 and 0.06.

Table 3. The simulation of sPSO

	1	2	3	4	5	6	7	8	9	10
Time(s)	0.03	0.09	0.03	0.03	0.03	0.03	0.03	0.03	0.03	0.03
Value	6e-82	1e-92	5e-91	9e-101	2e-85	4e-84	5e-97	6e-90	9e-97	4e-89

The main difference between bPSO and sPSO is that latter removes the influence of the speed, and use the distance between the particles to calculate, which simplifies the computational complexity. The searching time is similar to bPSO.

Table 4. The simulation of rPSO

	1	2	3	4	5	6	7	8	9	10
Time(s)	0.09	0.08	0.03	0.03	0.03	0.03	0.03	0.03	0.03	0.03
Value	e-194	e-178	e-191	e-209	e-179	e-187	e-186	e-191	e-186	e-175

From the time point of view, they are in the same order of magnitude and not much different from each other. The average length of time of bPSO is 0.0322s, sPSO is 0.0362s, and rPSO's is 0.0405s. From the convergence value point of view, rPSO and sPSO are significantly better than bPSO, and the result of rPSO is most accurate.

In summary, adding random number has an effect on convergence precision.

3.2   The Algorithm without the Random Number for the "Cognition" and "Social Experience" Part

The researches of swarm robot systems draw lessons from optimization techniques and principles such as swarm intelligence. It is an application of swarm intelligent in the multi-robot system [7]. As a representative of the swarm intelligent algorithms, particle swarm algorithm and swarm robots searching are the instances of the smart agent searching. And there is a certain mapping relation between them. The contact between swarm robot and particle swarm is the contact of concrete and abstract, reality and model. Therefore, swarm robots in the real world can be modeled and simulated by the use of particle swarm optimization [8].

Since swarm robots are widely used in many areas of our lives, it is important to improve PSO for the robotics to make the search capacity of robots have better stability and higher optimizing precision. This paper conceives an improved PSO which deletes the random number of the "cognition" and the "social experience" part based on sPSO. The formula of the algorithm is:

$$x_{id}^{k+1} = \omega x_{id}^k + c_1(p_{id} - x_{id}^k) + c_2(p_{gd} - x_{id}^k) \ . \tag{8}$$

The fast convergence rate and the higher convergence precision are expected in the simulation. This is the algorithm called the rand number deleted PSO (drPSO).

4   Simulation

4.1   Simulation in the Obstacle-Free Environment

There are no obstacles in the simulation environment, but searching target (the larger black dots in the Fig. 1) and robots (the other small dots). The simulation environment is shown in Fig. 1. The size of the environment is set in the file called "world" in Player/Stage. The name, color, initial position and posture of the 20 pioneer robots with laser sensors are set, too.

PSO, sPSO, drPSO are simulated in this simulation environment separately. The simulation diagrams are shown below.

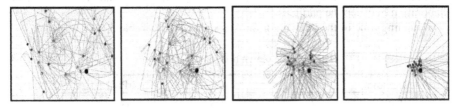

Fig. 1. The simulation of bPSO

In the beginning, the convergence speed is relatively fast and the particles can run towards the general direction of the target. Some particles can find the target in mid-run of the algorithm; the searching starts to slow down. During the running of the algorithm, the speed and position of every particle are updated according to the bPSO formula. In the end, most of the particles have found the target, and the others quickly converge to the position of the target. It takes 27seconds to finish the simulation

The simulation of sPSO is as follows.

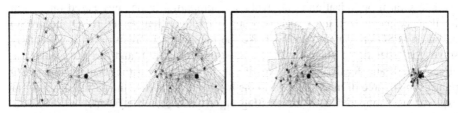

Fig. 2. The simulation of sPSO

sPSO has slow convergence, particularly in the middle of the run. Some robots run circles around the target, and then gradually close to it. The completion time of sPSO is 55 seconds. It can be learned that sPSO cannot work well   in every case.

The simulation of drPSO is showm below.

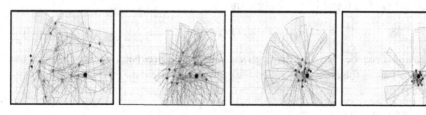

Fig. 3. The simulation of drPSO

The c_1 and c_2 in the formula (8) are the learning factors, and also called accelerating factors. They make the particles have the ability to summarize and to learn from the outstanding individuals in the groups, and close to the optimal position of the history of their own and the group. The value of the c_1 affects the movement trend of the particles to the historical best position of their own. This kind of trend is called "cognition", indicating the thinking of the particles. The value of the c_2 affects the movement trend of the particles to the historical best position of the group. This kind of trend is called the "social experience", indicating the comparison and simulating with the neighbors, which achieving the information sharing and collaboration between the particles.

The running time of the three algorithms is shown in the Table 5.

Table 5. The search time of the algorithms

Algorithm	bPSO	sPSO	drPSO
Time(s)	27s	55s	14s

From the Table 5, the shortest time of searching is drPSO, 14 seconds.

The values of the c_1 and c_2 play an important role in the multi-robot collaboration in search for the target. srPSO removes the random numbers that may limit the ability of the robots to search, which accelerate convergence speed to a certain extent.

5   Conclusion

Particle swarm optimization is an evolution computing technology based on swarm intelligence method. In this paper, two improved algorithms called rPSO and drPSO based on sPSO are put forward to make the convergence precision high when they solve the specific problems. The simulations of rPSO and drPSO show the effectiveness and feasibility of them. It can be found from the analysis that rPSO reduces the degree of inheritance from the history and improve the particle's ability of exploration. Compared with the existing algorithms, search efficiency of drPSO is increased greatly.

Acknowledgements. This work is jointly supported by NSFC under Grant No. 60903067. Beijing Natural Science Foundation under Grant No. 4122049, Funding Project for Beijing Excellent Talents Training under Grant No. 2011D009006000004, and the Fundamental Research Funds for the Central Universities(FRF-JX-12-002, FRF-TP-12-083A).

References

1. Kennedy, J., Eberhart, R.C.: Swarm Intelligence. Academic Press, USA (2001)
2. Wang, M., Zhu, Y.L., He, X.X.: Research Summarize of Swarm Intelligence. Computer Project 31, 194–196 (2005)
3. Hu, W., Li, Z.S.: A simpler and more effective particle swarm optimization algorithm. Journal of Software 18, 861–868 (2007)
4. Kennedy, J., Eberhart, R.C.: Particle swarm optimization. In: Proceedings of IEEE International Conference on Neural Networks, Washington, pp. 1942–1948 (1995)
5. Kennedy, J., Eberhart, R.C.: A new optimizer using particle swarm theory. In: Proceedings of the 6th International Symposium on Micro Machine and Human Science, Nagoya, pp. 39–43 (1995)
6. Clerc, M.: The swarm and the queen: Towards a deterministic and adaptive particle swarm optimization. In: Proceedings of the 1999 Congress on Evolutionary Computation, pp. 1951–1957 (1999)
7. Xue, S.D., Zeng, J.C.: Swarm Robotics: A Survey. Pattern Recognition and Artificial Intelligence 21, 177–185 (2008)
8. Chen, B.D.: Improved Particle Swarm algorithms based on the characteristics of swarm robots. Master thesis, Taiyuan University of Science and Technology (2009)
9. Gerkey, B., Vaughan, R., Howard, A.: The Player/Stage Project: Tools for Multi-robot and Distributed Sensor Systems. In: Proceedings of the International Conference on Advanced Robotics, pp. 317–323 (2003)

A Test of Position Determination with PSO

Walter W. Chen, Jian-An Wang, and Zhe-Ping Shen

Dept. of Civil Engineering, National Taipei University of Technology, Taipei, Taiwan
waltchen@ntut.edu.tw, petera23p@yahoo.com.tw,
fishfishfishgoo@gmail.com

Abstract. PSO has been used in combination with ultra-high resolution 360-degree panoramic images in positioning field objects at a landslide site. Although the computational efficiency was exceptional, the sum of errors was high. In order to demonstrate that the errors came from GPS readings instead of photography mistakes or erroneous computer codes, the authors designed and implemented an experiment and used it to verify the applicability of the PSO method. A conceptual layout was first sketched on paper and then tested on a rooftop of a campus building. Two sets of input data were constructed using the panoramic photos and the CAD drawing of the conceptual layout, respectively. Both data sets were computed using the brute force program and the PSO program developed in previous studies. The results showed that cm-level and sub-mm level accuracy was achieved in the experiment. Consequently, it was concluded that the PSO program was correct and the PSO method was applicable to the positioning problem. The accuracy of positioning in the field can be improved with the aid of better GPS devices.

Keywords: Surveying poles, PSO, brute force method.

1 Introduction

Particle Swarm Optimization (PSO) is an artificial intelligence technique that has been applied to many research problems including tree trunks fitting [1], landslide analysis [2, 3], slope parameters determination [4], and field positioning [5]. Among these interesting applications, one that particularly intriguing is the use of PSO in combination with ultra-high resolution 360-degree panoramic images. Chen et al. demonstrated this technique at a landslide site using bamboo sticks to denote the positions of desired objects [5]. Three sets of panoramic images at three distinct locations were taken using the GigaPan. Since the images provided the full 360-degree view of the surrounding environment, relative angles between the bamboo sticks in reference to the individual camera could be calculated by counting pixels. Using imaginary rays of lines emitting from the three camera locations, researchers successfully triangulated the locations of every bamboo stick using both the brute force method and the PSO technique. The speed-up factor was at least 546 times (7100 sec divided by 13 sec). Although PSO was proved to be extremely useful and efficient in this problem, the triangulation error, defined as the total lengths of the sides of intersection triangles, could not be reduced to less than 91.77 m. Therefore,

Y. Tan, Y. Shi, and H. Mo (Eds.): ICSI 2013, Part I, LNCS 7928, pp. 104–111, 2013.

questions remain over the use of the PSO method in position determination to achieve the desired accuracy and precision. In principle, errors could come from GPS readings, photography mistakes, erroneous computer codes, or the incorrect use of PSO. Rather than speculating on the possible/main source of errors, the authors designed a test case (for which exact solutions exist) and used it to verify the correctness of the PSO codes and therefore the applicability of the PSO technique. The results are described in the following sections.

2   Design of Test Case

With the aim of validating the suitability of the PSO method in position determination, a test layout on paper was designed in a grid manner as shown in Figure 1. Each grid was 2.1 m in both horizontal and vertical directions. Three camera locations were picked and denoted A, B and C in Figure 1. Eight target locations (numbered 1-8) were also selected in the test, and they were spaced out as evenly as possible. Note that for the test layout to be valid, no lines formed by connecting any two of the eight targets should pass through any one of the three cameras. In other words, no target was allowed to be "blocked" by other targets when it is "viewed" from any one of the three camera locations.

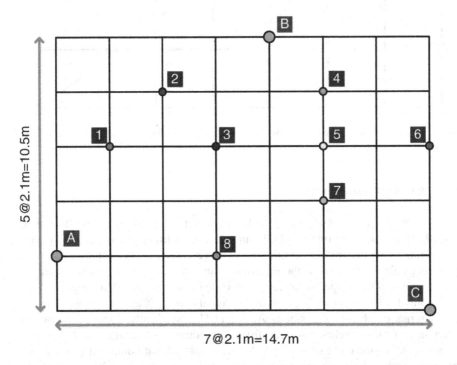

Fig. 1. Test layout of three camera locations (A, B and C) and eight target locations (1-8)

When the paper design was finished, the layout was transferred to the CAD software to create a working drawing. Lines emanating from the sources (cameras) to the targets were drawn as shown in Figure 2. The purposes are twofold: (1) to verify that no target was blocked in view by other targets and (2) to precisely measure the final rotation angles needed to triangulate the targets. As mentioned before, targets were spaced out as evenly as possible. Although the positions looked random, they were selected carefully to realize the design goal of testing the PSO method.

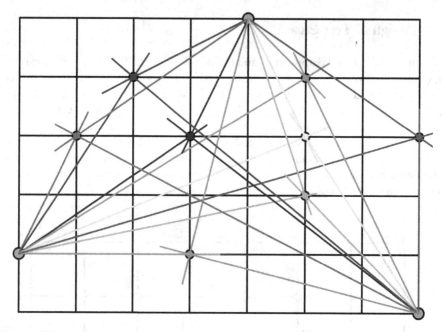

Fig. 2. CAD drawing created to triangulate eight targets using lines emanating from three sources (cameras).

3   Experimental Setup

The conceptual layout of Figures 1 and 2 were implemented on the rooftop of the Third Teaching Building on the NTUT campus. The rooftop was covered by 30 cm by 30 cm tiles, thus making the positioning of targets easy. Eight custom-made metal surveying poles were used in the experiment and they were held in place and in a vertical position using special devices with tripod bases or just dead weights (2-liter bottles filled with water). A GigaPan and a digital camera (Canon G11) were used to take a series of photos at positions A, B and C, respectively. Figure 3 shows the panoramic photos stitched together from individual photos taken at those three locations. All of the eight surveying poles were colored and numbered in the photos for subsequent analysis. The photos were imported into Photoshop where the numbers of pixels between poles were counted and divided by the total number of pixels to obtain the angles between adjacent targets.

Fig. 3. Panoramic photos taken at the rooftop of the Third Teaching Building showing the eight metal surveying poles (colored and labeled) positioned according to the design layout in Figures 1 and 2

4   Computational Results

After the angles between adjacent targets had been calculated, they were adjusted and recorded as the counter-clockwise angles from target #1. The results are shown in Table 1. The allowable values of the angles are from 0 to 360 degrees. The same procedure was repeated for each of the three photos. There were two input files in the numerical experiment. The values in Table 1 served as the first input file to the PSO and the brute force programs developed by Chen et al. [5]. For the purpose of comparison, theoretical angles between different targets as viewed from different camera locations were also computed using the CAD software. The results are shown in Table 2. The values in Table 2 formed the second input file to the two programs mentioned above. Both programs were executed on a computer with an Intel Celeron G530 2.40 GHz processor running the Ubuntu 12.04 operating system. The computational results are summarized in Table 3. The outcomes show that the brute force method took two hours to finish the computation in the experiment, which is in agreement with the results reported in [5]. Also, the PSO method took merely a fraction of the time that had been required by the brute force method. This is also in consistent with [5]. For the two input files used, speedup ratios of 906 and 1,212 were achieved, respectively. Moreover, the brute force method was only accurate to ones digits, whereas the PSO method was able to provide many significant figures. Finally, the error from the brute force method was 5.3 to 223.7 times higher than the PSO method.

Table 1. Counter-clockwise angles of eight surveying poles from target #1 computed based on the numbers of pixels in the three panoramic photos taken at positions A, B and C

No.	Angles @ point A	Angles @ point B	Angles @ point C
1	0.00	0.00	0.00
2	352.60	353.34	348.00
3	330.37	29.65	349.71
4	327.29	100.63	323.38
5	318.08	82.63	330.46
6	312.20	112.12	297.14
7	307.62	74.54	341.60
8	296.56	42.12	12.56

Table 2. Counter-clockwise angles of eight surveying poles from target #1 computed based on the precise drawing created in the CAD software

No.	Angles @ point A	Angles @ point B	Angles @ point C
1	0.00	0.00	0.00
2	352.87	352.87	347.91
3	330.26	29.74	349.70
4	327.53	101.31	323.13
5	318.37	82.87	330.26
6	312.51	112.62	296.57
7	307.88	74.74	341.57
8	296.57	42.27	12.53

Table 3. Comparison of computational results using two sets of input data derived from panoramic photos and CAD drawings, respectively

	Method	Min. sum of lengths of △ sides	# of computations	Best solution	Time spent
Photos	Brute force	309.18 cm	46,656,000	64 214 153	7249 sec = 2.01 hr
	PSO $w = 0.7$	58.23 cm	23,162	63.54976 213.58606 153.38948	8 sec
Theoretical/CAD	Brute force	346.75 cm	46,656,000	63 214 153	7273 sec = 2.02 hr
	PSO $w = 0.7$	1.55 cm	23,307	63.43220 213.68979 153.43309	6 sec

5   Discussion and Conclusions

This is an experiment designed to test the PSO method in triangulating field targets. The results were quite satisfactory. The same experimental layout was implemented in the CAD software and on a rooftop of a campus building with surveying poles. For angles calculated from panoramic photos, the sum of errors (measured as the total lengths of the sides of the eight intersection triangles) was as low as 58.23 cm (as shown in Table 3). This means that the average error of each side of the triangles was only 2.43 cm. Furthermore, if the angles from the CAD measurements (theoretically precise values) were used in the computation, the sum of errors could approach zero (1.55 cm). This is equivalent to 0.06 cm per side, a solution of extremely high accuracy. For this reason, the PSO program was proved to be correct and the PSO method was ascertained to be applicable to this kind of problems. Recall that the sum of errors reported in [5] was much higher (91.77 m) than the current study. By ruling out the possibility of photography mistakes and erroneous computer codes, it is reasonable to conclude that the errors in [5] mostly came from GPS readings, in which case the hand-held GPS and the valley terrain with poor signal cover were to blame.

Finally, Figure 4 shows the convergence curves for both sets of input data. The blue line and the red line represent the cases where the angles were calculated from panoramic photos and the CAD drawing, respectively. It has already been pointed out that the blue line converged to 58.23 cm and the red line converged to 1.55 cm. With the same inertia weight of 0.7, the red line also converged much faster than the blue line did, and both lines converged faster than the curve in [5], where w was set to 0.8. Three things are worth nothing. First, it is evident that the quality of input data affects not only the accuracy of results but also the speed of convergence. High quality data will accelerate the rate of convergence. Second, there will always be errors in field measurements even in a controlled environment of a rooftop. In this study, one likely error might come from the wind blowing over the rooftop. It might have caused the poles to move slightly. That is probably why the blue line only converged to 58.23 cm. Third and finally, the PSO method is proved to be capable of achieving cm-level (or even sub-mm level) accuracy. It is an extremely powerful and efficient algorithm for positioning field objects.

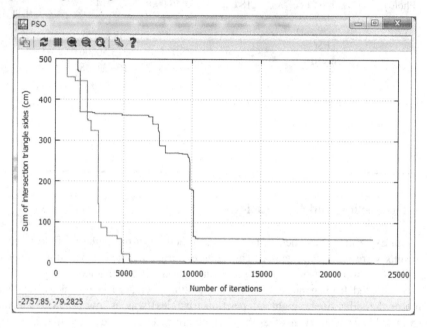

Fig. 4. Convergence curves of the triangulation process using the PSO method for angles calculated from panoramic photos (blue) and precise CAD drawings (red).

Acknowledgments. This study was partially supported by grant numbers NSC 101-2218-E-027-001, NSC 100-2218-E-027-002, and NSC 99-2218-E-027-008 from the National Science Council of Taiwan (ROC). The financial support is highly appreciated. Thanks are also due to Miss Ting-Yun Huang for proof-reading the manuscript.

References

1. Chen, W.W., Chen, P.: Reconstructing Tree Trunks from Point Clouds using PSO. In: Proc. 2010 International Conference on Intelligent Computation Technology and Automation, pp. 996–999. IEEE Computer Society (2010)
2. Chen, W.W., Chen, P.: PSOslope: a Stand-alone Windows Application for Graphical Analysis of Slope Stability. In: Tan, Y., Shi, Y., Chai, Y., Wang, G. (eds.) ICSI 2011, Part I. LNCS, vol. 6728, pp. 56–63. Springer, Heidelberg (2011)

3. Wen, J.-C., Chen, W.W.: Finding Most Likely Sliding Surfaces using PSO. In: Proc. 2011 Conference on Technologies and Applications of Artificial Intelligence, pp. 309–312. Conference Publishing Services (2011)
4. Chen, W.W., Shen, Z.-P., Wang, J.-A., Tsai, F.: Scripting STABL with PSO for Analysis of Slope Stability. Accepted for Publication in Neurocomputing
5. Chen, W.W., Shen, Z.-P., Chen, H.-C., Wang, J.-A.: Position Measurement Assisted by PSO and 360-degree Images. In: Proc. the 33rd Asian Conference on Remote Sensing, PS1 2012. Asian Association on Remote Sensing (2012)

Initial Particles Position for PSO, in Bound Constrained Optimization

Emilio Fortunato Campana[1], Matteo Diez[1], Giovanni Fasano[2], and Daniele Peri[1]

[1] National Research Council-Maritime Research Centre (CNR-INSEAN), Via di Vallerano, 139, 00128 Rome, Italy
{emiliofortunato.campana,matteo.diez,daniele.peri}@cnr.it
[2] University Ca'Foscari of Venice, Department of Management
fasano@unive.it

Abstract. We consider the solution of bound constrained optimization problems, where we assume that the evaluation of the objective function is costly, its derivatives are unavailable and the use of exact derivative-free algorithms may imply a too large computational burden. There is plenty of real applications, e.g. several design optimization problems [1,2], belonging to the latter class, where the objective function must be treated as a 'black-box' and automatic differentiation turns to be unsuitable. Since the objective function is often obtained as the result of a simulation, it might be affected also by noise, so that the use of finite differences may be definitely harmful.

In this paper we consider the use of the evolutionary Particle Swarm Optimization (PSO) algorithm, where the choice of the parameters is inspired by [4], in order to avoid diverging trajectories of the particles, and help the exploration of the feasible set. Moreover, we extend the ideas in [4] and propose a specific set of initial particles position for the bound constrained problem.

Keywords: Bound Constrained Optimization, Discrete Dynamic Linear Systems, Free and Forced Responses, Particles Initial Position.

1   Introduction

Applied sciences offer several challenging applications of bound constrained optimization, where the computational cost of the objective function is remarkably large. In this regard, optimization tools combining the theoretical properties of exact methods and the fast progress of heuristics represent an active research area. Furthermore, on large scale real problems, which are typically more difficult and require correspondingly larger computational resources, both practitioners and theoreticians claim for robust methods, often endowed also with theoretical properties. Moreover, in many cases the derivatives are unavailable. In the latter case, the use of 'black-box' simulations for computing the objective function makes the adoption of automatic differentiation impossible, due to the unavailability of the source code. In addition, simulations represent an essential tool,

Y. Tan, Y. Shi, and H. Mo (Eds.): ICSI 2013, Part I, LNCS 7928, pp. 112–119, 2013.

but often introduce an unexpected artificial noise, which unavoidably imposes strong care when adopting finite differences.

This paper considers PSO [6], with a specific choice of the parameters, for the solution of the bound unconstrained *global optimization problem*

$$\min_{x \in \mathcal{F}} f(x), \qquad f : \mathbb{R}^n \to \mathbb{R}, \tag{1}$$

where $\mathcal{F} = \{x \in \mathbb{R}^n : l \leq x \leq u\}$, $l, u \in \mathbb{R}^n$ and without loss of generality $l < u$. Obviously, in case $l_i = -\infty$, $i \in \{1, \ldots, n\}$ and $u_i = +\infty$, $i \in \{1, \ldots, n\}$ problem (1) reduces to an unconstrained optimization problem. At present $f(x)$ is assumed to be a nonlinear and non-convex *continuous function*.

This paper has a twofold purpose. First we propose some novel rules for the selection of parameters in PSO, using the reformulation of PSO iteration described in [4]. Then, we suitably adapt the choice of particles position/velocity studied in [4] for the unconstrained case, to the feasible set $\mathcal{F}$ of (1). The latter adaptation requires some geometric insight and involves a negligibly small algebra, even when the scale n is large.

As regards the symbols used in this paper, the subscripts identify the particles in a PSO scheme, while we use the superscript to indicate the iteration. I_k is the identity matrix of order k. If σ is a real random unknown and $u \in \mathbb{R}^n$, the symbol $\sigma \otimes u$ indicates an n-real vector, whose j-th and i-th entries are respectively given by $\sigma_j u_j$ and $\sigma_i u_i$, where σ_j and σ_i are different occurrences of σ. Finally, $\|A\|_F$ indicates the Frobenius norm of matrix A, i.e. $\|A\|_F = tr(A^T A)^{1/2}$, where $tr(\cdot)$ indicates the *trace* of a matrix.

In Section 2 we propose a reformulation of PSO iteration, which is essential for our proposal, then Section 3 suggests some basics on the choice of parameters in PSO, and Section 4 proposes some indications to properly choose the initial position/velocity of particles for problem (1).

2   A Reformulation of PSO

Consider the trajectory of the j-th PSO iteration ($k \geq 0$)

$$v_j^{k+1} = \chi_j^k \left[w_j^k v_j^k + c_j r_j \otimes (p_j^k - x_j^k) + c_g r_g \otimes (p_g^k - x_j^k) \right], \tag{2}$$

$$x_j^{k+1} = x_j^k + v_j^{k+1},$$

where $j = 1, \ldots, P$ indicates the j-th *particle* and P is a positive integer. The vectors v_j^k and x_j^k are n-real vectors representing respectively the *velocity* (i.e. the search direction) and the *position* of the j-th particle at step k. Moreover, the n-real vectors p_j^k and p_g^k satisfy

$$f(p_j^k) \leq f(x_j^\ell), \qquad \text{for any } \ell \leq k, \ \ p_j^k \in \{x_j^\ell\},$$

$$f(p_g^k) \leq f(x_j^\ell), \qquad \text{for any } \ell \leq k \text{ and } j = 1, \ldots, P, \ \ p_g^k \in \{x_j^\ell\}, \tag{3}$$

while $\chi_j^k, w_j^k, c_j, r_j, c_g, r_g$ are positive bounded coefficients. As well known, p_j^k represents the 'best position' in the trajectory of the j-th particle up to step k, while p_g^k is the 'best position' among all the particles up to step k. The choice of the coefficients as well as the number of particles P is often problem dependent (see also [7]), and here we consider the choice [4], which is very general. The latter choice also includes the case where both the *inertia* coefficient w_j^k and the *constriction* coefficient χ_j^k are used. Finally, as usually, we can assume without loss of generality that r_j and r_g are uniformly distributed random parameters, with $r_j \in [0,1]$ and $r_g \in [0,1]$.

After some simplifications, for each particle j, assuming for brevity that $w_j^k = w_j$ and $\chi_j^k = \chi_j$, for any $k \geq 0$ the iteration (2) is equivalent to the *discrete stationary (time-invariant) system* (see also [4])

$$X_j(k+1) = \begin{bmatrix} \chi_j w_j I_n & -\chi_j(c_j r_j + c_g r_g)I_n \\ \chi_j w_j I_n & [1 - \chi_j(c_j r_j + c_g r_g)]I_n \end{bmatrix} X_j(k) + \begin{bmatrix} \chi_j\left(c_j r_j p_j^k + c_g r_g p_g^k\right) \\ \chi_j\left(c_j r_j p_j^k + c_g r_g p_g^k\right) \end{bmatrix}$$

$$(4)$$

where

$$X_j(k) = \begin{pmatrix} v_j^k \\ x_j^k \end{pmatrix} \in \mathbb{R}^{2n}, \qquad k \geq 0. \tag{5}$$

From a geometric perspective the sequence $\{X_j(k)\}$ represents the trajectory of the j-th particle in the state space $\mathbb{R}^{2n}$. Moreover, using a standard notation for linear systems, we can split $X_j(k)$ into the *free response* $X_{jL}(k)$ and the *forced response* $X_{jF}(k)$ (see also [8]). Thus, on summary for any $k \geq 0$ the $2n$-real vector $X_j(k)$ may be rewritten as

$$X_j(k) = X_{jL}(k) + X_{jF}(k), \tag{6}$$

where

$$X_{jL}(k) = \Phi_j(k)X_j(0), \qquad X_{jL}(k) = \sum_{\tau=0}^{k-1} H_j(k-\tau)U_j(\tau), \tag{7}$$

and after some computation we obtain (see also [4])

$$\Phi_j(k) = \begin{pmatrix} \chi_j w_j I_n & -\chi_j(c_j r_j + c_g r_g)I_n \\ \chi_j w_j I_n & [1 - \chi_j(c_j r_j + c_g r_g)]I_n \end{pmatrix}^k \in \mathbb{R}^{2n \times 2n}. \tag{8}$$

$$(9)$$

We urge to recall that from the expressions (6)-(7), unlike the vector $X_{jF}(k)$, the free response $X_{jL}(k)$ only depends on the initial point $X_j(0)$, and not on the vectors p_j^τ, p_g^τ, with $\tau \geq 0$. As described in the next section, the latter observation plays a keynote role, in order to design efficient PSO schemes for solving (1).

3   Issues on Parameters Assessment in PSO

Observe from (8) that $\Phi_j(k) = \Phi_j(1)^k$, for any $k \geq 0$, and the $2n$ eigenvalues of the unsymmetric matrix $\Phi_j(1)$ are real (see also [4]). Setting for simplicity in (8)

$$a_j = \chi_j w_j, \qquad w_j = \chi_j(c_j r_j + c_g r_g), \qquad j = 1, \ldots, P, \qquad (10)$$

after some computation we see that the matrix $\Phi_j(1)$ has only the two distinct eigenvalues λ_{j1} and λ_{j2} given by

$$\lambda_{j1} = \frac{1 - w_j + a_j - \left[(1 - w_j + a_j)^2 - 4a_j\right]^{1/2}}{2}$$

$$\lambda_{j2} = \frac{1 - w_j + a_j + \left[(1 - w_j + a_j)^2 - 4a_j\right]^{1/2}}{2}, \qquad (11)$$

each of them having algebraic multiplicity n. A necessary (but possibly not sufficient) condition for $\{X_j(k)\}$ to be non-diverging (which implies that also $\{x_j^k\}$ and $\{v_j^k\}$ in (2) are non-diverging), is

$$|\lambda_{j1}| < 1, \qquad |\lambda_{j2}| < 1, \qquad (12)$$

which affect the choice of PSO parameters as described in the next proposition (the next conditions are simplified with respect to [4]).

Proposition 1. *Consider the position (10) in (2), with $\chi_j^k = \chi_j$ and $w_j^k = w_j$, $j = 1, \ldots, P$. Suppose for $k \geq 0$*

$$0 < a_j < 1, \qquad 0 < w_j < 2(a_j + 1), \qquad j = 1, \ldots, P, \qquad (13)$$

with $w_j^k \neq (1 \pm a_j^{1/2})^2$. Then, for any $k \geq 0$ and $j = 1, \ldots, P$, conditions (12) are fulfilled. $\diamond$

Observe that conditions (12) imply $\lim_{k \to \infty} X_{jL}(k) = 0$, $j = 1, \ldots, P$, and most of the typical settings for PSO parameters proposed in the literature (see e.g. [7,9]) satisfy (13). Moreover, from relations (7), (8), (10) and considering that $\Phi_j(1)$ is unsymmetric, we have also that for any j

$$\|\Phi_j(k)\|_F \leq \|\Phi_j(1)\|_F^k = tr\left[\Phi_j(1)^T \Phi_j(1)\right]^{\frac{k}{2}}, \qquad (14)$$

and

$$tr\left[\Phi_j(1)^T \Phi_j(1)\right]^{\frac{1}{2}} = tr \begin{bmatrix} 2a_j^2 I_n & a_j(1 - 2w_j)I_n \\ a_j(1 - 2w_j)I_n & [w_j^2 + (1 - w_j)^2]I_n \end{bmatrix}^{\frac{1}{2}}$$

$$= \left[2a_j^2 + w_j^2 + (1 - w_j)^2\right]^{\frac{1}{2}}.$$

Using Fact 9.12.1 in [10] (where $B = I_n$ and $\|B\|_F = \sqrt{n}$) we have that

$$\frac{1}{\sqrt{n}} |tr\, [\Phi_j(1)]| \leq \|\Phi_j(1)\|_F \qquad (15)$$

where

$$tr\, [\Phi_j(1)] = a_j + (1 - \omega_j).$$

Now, from (7) and (14)

$$\|X_{jL}(k)\|_F \leq \|\Phi_j(1)\|_F^k \cdot \|X_j(0)\|_F,$$

and though $\lim_{k \to \infty} X_{jL}(k) = 0$, $j = 1, \ldots, P$, we would like $\|X_{jL}(k)\|_F$ not to be attenuated when the index k is still relatively small. On this purpose, given the coefficients c_j, $j = 1, \ldots, P$ and c_g, we propose to set χ_j and w_j by solving for each $j = 1, \ldots, P$ one of the following two programs, inspired by Proposition 1 and, respectively, relation (14) and relation (15):

$$\max_{\chi_j, w_j} 2a_j^2 + \omega_j^2 + (1 - \omega_j)^2$$
$$0 < a_j < 1, \qquad\qquad (16)$$
$$0 < \omega_j < 2(a_j + 1),$$

$$\max_{\chi_j, w_j} |a_j + (1 - \omega_j)|$$
$$0 < a_j < 1, \qquad\qquad (17)$$
$$0 < \omega_j < 2(a_j + 1).$$

The programs (16)-(17) attempt to possibly force larger values of $\|X_{jL}(k)\|_F$ for k small. In Section 4 we give more motivations about the latter issue.

Now, in the light of (7), (12) and the results in Proposition 1, we think that the following question still deserves special consideration: *can we properly choose the initial points $X_j(0)$, $j = 1, \ldots, P$, for problem (1), so that the trajectories $\{x_j^k\}$ span as much as possible the feasible set $\mathcal{F}$?* Section 4 addresses the latter issue, in order to give indications on the choice of the initial point and velocity of particles.

4   Initial Particles Position and Velocity in PSO, for Bound Constrained Optimization

In this section we study some proposals of initial particles position and velocity, for the bound constrained optimization problem (1). To this aim let us consider the feasible set $\mathcal{F}$ in (1); we remind that possibly we allow $l_i = -\infty$ and/or $u_i = +\infty$ for some indices $i \in \{1, \ldots, n\}$. In the previous section we studied settings for PSO parameters, such that the free response $X_{jL}(k)$ associated to particle j is possibly not attenuated too early, i.e. when k is still relatively small. In this section we show a method to exploit the latter property, in order to possibly improve the overall performance of PSO on bound constrained optimization. In particular, we want to give indications for the choice of the vectors $X_j(0)$, so that possibly the orthogonality conditions (or similar properties)

$$X_{j_i L}(k)^T X_{j_h L}(k) = 0, \qquad 1 \le i \ne h \le m, \tag{18}$$

among the free responses of the first m particles (with $m \le n$), are satisfied. Observe that conditions (18) do not impose the trajectories of PSO particles to be orthogonal; however, they guarantee that *part of particles trajectories* (i.e. the free responses in $\mathbb{R}^{2n}$) are orthogonal, as long as they do not fade. This explains why in Section 3 we studied conditions on PSO parameters, in order to prevent a premature extinction of $X_{jL}(k)$ when k increases.

In particular, our first proposal for the choice of $X_j(0)$, $j = 1, \dots, P$, is the following:

1. If $l < 0 < u$ then set $X_j(0)$ such that $x_j^0 \in \mathcal{F}$, randomly for $j = n+1, \dots, P$, and $v_j^0 \in \mathbb{R}^n$ for $j = n+1, \dots, P$. On the other hand, for $j = 1, \dots, n$ set

$$t_j = \left[\frac{\sqrt{n}}{n} \sum_{i=1}^{n} -\frac{\sqrt{n}}{2} e_j \right] \in \mathbb{R}^n, \qquad X_j(0) = \begin{pmatrix} \alpha_j t_j \\ \beta_j t_j \end{pmatrix} \in \mathbb{R}^{2n}, \tag{19}$$

where α_j is any real value such that $\alpha_j t_j \in \mathcal{F}$, $j = 1, \dots, n$.

2. Otherwise, set $X_j(0)$ such that $x_j^0 \in \mathcal{F}$, randomly for $j = n+1, \dots, P$, and $v_j^0 \in \mathbb{R}^n$ for $j = n+1, \dots, P$. Then, for $j = 1, \dots, n$ consider the vertex $\hat{u} \in \mathcal{F}$ which is the closest to the origin; take

$$X_j(0) = \begin{pmatrix} \hat{u}_j \\ z_j \end{pmatrix}, \qquad j = 1, \dots, n, \tag{20}$$

$\hat{u}_j$ being the j-th vertex of $\mathcal{F}$ adjacent to $\hat{u}$ (i.e. such that an edge of $\mathcal{F}$ connects $\hat{u}$ and $\hat{u}_j$), and $z_j \in \mathbb{R}^n$ is randomly chosen.

Observe that while (19) satisfies (18) and α_j is very easy to compute, the choice (20) simply ensures that the vectors $X_j(0)$, $j = 1, \dots, n$, are at least linearly independent (though in general not orthogonal). Now, in order to force condition (18) (or similar conditions) in a more general framework, let us consider the geometry of the feasible set $\mathcal{F}$ (shaded area) in Fig.1. Suppose the point c is the intersection of the diagonals of $\mathcal{F}$, i.e. $c = (u + l)/2$, and the segment a_i is given by $a_i = (u_i - l_i)/2$, $i = 1, \dots, n$. We want to compute the equations of the dashed hyperellipsoids E_0, E_1 and E_2 in Fig.1, E_0 being a sphere. It is not difficult to realize that

$$\begin{aligned} E_0 &: (x - c)^T A_0 (x - c) = 1, & A_0 &= diag_{1 \le i \le n} \left\{ \left(\textstyle\sum_{i=1}^{n} a_i^2 \right)^{-1} \right\}, \\ E_1 &: (x - c)^T A_1 (x - c) = n, & A_1 &= diag_{1 \le i \le n} \left\{ a_i^{-2} \right\}, \\ E_2 &: (x - c)^T A_1 (x - c) = 1; \end{aligned} \tag{21}$$

indeed, it suffices to consider that E_0 is a sphere, the extreme points v_ℓ, $\ell = 1, \dots, 2^n$, in the corners of $\mathcal{F}$ have coordinates $c_i \pm a_i$, $i = 1, \dots, n$ (which satisfy

Table 1. We list the results for 6 test functions from the literature (n is the number of unknowns and f^* is the value of f at a global minimum). The results are over 25 PSO runs, $f\_bst/f\_wst/f\_av$ is the *best/worst/average* value of f over the 25 runs, while *st. dev.* indicates the *standard deviation*. x_{rand} indicates random initial choice for particles position, while x_{orth} indicates initial choice for particles position as in (19).

Function		x_{rand}	x_{orth}	Function		x_{rand}	x_{orth}
Griewank (n=10)	f^*	0.0000		Levy 10^n loc.min. (n=30)	f^*	0.0000	
	f_bst	0.5562	**0.0057**		f_bst	53.8192	**1.1428**
	f_av	0.8485	**0.0332**		f_av	107.3033	**3.4678**
	f_wst	1.1650	**0.0731**		f_wst	299.5744	**3.9709**
	st. dev.	0.0067	0.0004		st. dev.	0.0001	0.0000
Griewank (n=20)	f^*	0.0000		Levy 15^n loc.min. (n=30)	f^*	0.0000	
	f_bst	1.2360	**0.0016**		f_bst	14.4646	**3.1471**
	f_av	1.3872	**0.0022**		f_av	31.7934	**3.3890**
	f_wst	1.7438	**0.0653**		f_wst	60.5632	**3.5046**
	st. dev.	0.0001	0.0000		st. dev.	0.0002	0.0000
Levy 5^n loc.min. (n=30)	f^*	0.0000		Griewank (n=30)	f^*	0.0000	
	f_bst	3.0273	**0.0268**		f_bst	1.5631	**0.0007**
	f_av	8.9546	**0.0483**		f_av	2.1459	**0.0389**
	f_wst	13.6678	**0.0942**		f_wst	2.8092	**0.0710**
	st. dev.	0.0000	0.0000		st. dev.	0.0000	0.0000

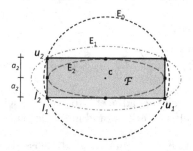

Fig. 1. The feasible set $\mathcal{F} \subset V$ of (1), V is the region inside E_0 or E_1, and $V \supset E_2$

the first two equations (21)), and the centers of the facets of $\mathcal{F}$ have entries in the sets $\{c_i, l_i, u_i\}$, $i = 1, \ldots, n$. We would like to show that for problem (1) it is possible to set $X_j(0)$, $j = 1, \ldots, n$ (other than (19)-(20)), so that conditions (18) at least in some cases are satisfied, with $X_j(0)$ such that $x_j^0 \in V$, where V is the region inside either of the hyperellipsoids E_0, E_1 or E_2. The importance of the latter property relies on the fact that it *tries to force orthogonality* among particles trajectories, while particles move within $\mathcal{F}$. Thus, we expect that PSO will be able to explore the feasible region of interest $\mathcal{F}$, as accurately as possible, while possibly ignoring the exploration in the set $\mathbb{R}^n \setminus V$.

In a more general scheme where $\mathcal{F}$ is treated in a penalty framework (i.e. PSO is used for the unconstrained minimization of a penalty function, which is the sum of $f(x)$ and a term penalizing the constraints violation), then we can set

$X_j(0)$, $j = 1, \ldots, P$, in a different fashion with respect to **1.** and **2.** Indeed, we can consider the choice:

$\hat{\mathbf{1}}$. If $l < 0 < u$ then set $X_j(0)$ such that $x_j^0 \in \mathcal{F}$, randomly for $j = n+1, \ldots, P$, and $v_j^0 \in \mathbb{R}^n$ for $j = n+1, \ldots, P$. On the other hand, from (19) set

$$X_j(0) = \begin{pmatrix} \alpha_j t_j \\ \beta_j t_j \end{pmatrix} \in \mathbb{R}^{2n}, \qquad j = 1, \ldots, n,$$

where now α_j *is any real value such that* $\alpha_j t_j \in V$, $j = 1, \ldots, n$, and V is the region inside either of the hyperellipsoids E_0, E_1 or E_2 in Fig.1.

$\hat{\mathbf{2}}$. Otherwise, take the choice **2.**

We still have to complete in a separate paper a numerical experience, giving full evidence of the effectiveness of the proposals above, in a framework where exact penalty methods are adopted. However, Table 1 summarizes a few preliminary results on six test problems from the literature (the caption describes the setting adopted), indicating that our proposal might be effective and efficient.

References

1. Mohammadi, B., Pironneau, O.: Applied Shape Optimization for Fluids. Clarendon Press, Oxford (2001)
2. Haslinger, J., Mäkinen, R.A.E.: Introduction to Shape Optimization. In: Advances in Design and Control. SIAM, Philadelphia (2003)
3. Pinter, J.D.: Global Optimization in Action. In: Continuous and Lipschitz Optimization: Algorithms, Implementations and Applications. Kluwer Academic Publishers, The Netherlands (1996)
4. Campana, E.F., Fasano, G., Pinto, A.: Dynamic analysis for the selection of parameters and initial population, in particle swarm optimization. Journal of Global Optimization 48, 347–397 (2010)
5. Campana, E.F., Fasano, G., Peri, D.: Globally Convergent Modifications of Particle Swarm Optimization for Unconstrained Optimization. In: Olsson, A.E. (ed.) Particle Swarm Optimization: Theory, Techniques and Applications. Advances in Engineering Mechanics, pp. 97–118. Nova Publishers Inc., South Africa (2011)
6. Kennedy, J., Eberhart, R.C.: Particle swarm optimization. In: Proceedings of the 1995 IEEE International Conference on Neural Networks IV, pp. 1942–1948. IEEE Service Center, Piscataway (1995)
7. Clerc, M., Kennedy, J.: The Particle Swarm - Explosion, Stability, and Convergence in a Multidimensional Complex Space. IEEE Transactions on Evolutionary Computation 6, 58–73 (2002)
8. Sarachik, P.E.: Principles of linear systems, Cambridge University Press, Cambridge (1997)
9. Zheng, Y.L., Ma, L.H., Zhang, L.Y., Qian, J.X.: On the convergence analysis and parameter selection in particle swarm optimization. In: Proceedings of the Second International Conference on Machine Learning and Cybernetics, Xi'an, November 2-5 (2003)
10. Bernstein, D.S.: Matrix Mathematics: Theory, Facts, and Formulas, 2nd edn. Princeton University Press, NJ (2009)

Visualizing the Impact of Probability Distributions on Particle Swarm Optimization

Tjorben Bogon, Fabian Lorig, and Ingo J. Timm

Business Information Systems 1
University of Trier
{bogon,s4falori,ingo.timm}@uni-trier.de

Abstract. In this paper we present a simulation tool for the visualization of the impact of different probability distributions on Particle Swarm Optimization (PSO). PSO is influenced by a high number of random values in order to simulate a more nature like behaviour. Based on these random numbers the optimization process may vary. Usually the uniform distribution is chosen but regarding certain underlying fitness functions this may not the best choice. To test the influence of different probability distributions on PSO and to compare the different approaches, the presented simulation system consist of a simple user interface and allows the integration of own distribution formulas in order to test their impact on PSO.

Keywords: Particle Swarm Optimization, Probability Distributions, Random Numbers, Simulation System, Visualization.

1 Motivation

Particle Swarm Optimization (PSO) is a nature inspired metaheuristic based on the movement behaviour of bird flocks and fish shoals and was introduced by Kennedy and Eberhardt in 1995 [1]. To integrate the exploration factor of the flocks and to simulate the "natural" behaviour within this optimization technique, random factors are integrated into the PSO-formula. The PSO technique is based on a swarm of particles which influence themselves and try to find the global optimum of an underlying fitness function. Every particle receives a fitness value out of the actual position in the search space and calculates the next point among others regarding the best found place of the other particles within the swarm. Hereby random factors are being used to vary the influence of these independent variables on the calculation of each particle's next position and the scattering of the swarm.

According to the standard definition for computing and comparing PSO introduced by Bratton and Kennedy in 2007 [2], a total amount of 600 000[1] random

---

[1] As defined by Bratton and Kennedy in 2007, the standard PSO algorithm includes 50 particles and 6 000 epochs. The calculation of each particle's next position contains two random values (ϵ_1, ϵ_2) which are integrated into the inertia-weight-update-formula [3].

Y. Tan, Y. Shi, and H. Mo (Eds.): ICSI 2013, Part I, LNCS 7928, pp. 120–128, 2013.
© Springer-Verlag Berlin Heidelberg 2013

numbers is required for the calculation during one optimization process. This demonstrates that the probability distribution, which the random numbers are based on, influences the movement of the particles. In Bratton and Kennedy's standard for PSO, a uniform distribution is being used to generate random numbers. This works fine on the most of the underlying fitness functions. But regarding certain fitness landscapes, different PSO-configurations will lead to a better optimization results [4] just as other probability distributions do. In this work we are going to develop a simulation tool for the examination and testing of other probability distribution's impact on the movement of a particle.

Therefore we introduce a simulator for random distribution on PSO. The simulation system can be used to artificially generate possible movements of a particle. The resulting positions of the particle will be visualized in order to illustrate the way the search space is being swept using a certain probability distribution. This enables the user to compare how different distributions make a swarm scatter on a simulated fitness landscape.

This work is structured as follows: First we give a brief introduction into PSO and the state of the art in random distribution on PSO. Afterwards we describe the developed simulation system. In section 4 the simulation system is being evaluated and finally we discuss our results and give a short look on the future aspects.

2   State of the Art

Random distributions are being used in most of the PSO variations. In this section we give a brief introduction into the PSO-technique and describe some different types of random distributions used in PSO variations.

2.1   Particle Swarm Optimization

Particle Swarm Optimization is inspired by the social behaviour of flocks of birds and shoals of fish. A number of simple entities, the particles, are placed in the domain of definition of some function or problem. The fitness (the value of the objective function) of each particle is evaluated at its current location. The movement of each particle is determined by its own fitness and the fitness of particles in its neighbourhood [1]. The results of one decade of research and improvements in the field of PSO have recently been summarized by Bratton and Kennedy [2], also recommending standards for comparing different PSO methods. Therefore the definition used in this paper is based on Bratton and Kennedy as well. We aim at continuous optimization problems in a search space S defined over the finite set of continuous decision variables $X_1, X_2, \ldots, X_n$. Given the set Ω of conditions to the decision variables and the objective function $f : S \to \mathbb{R}$ (also called fitness function) the goal is to determine an element $s^* \in S$ that satisfies Ω and for which $f(s^*) \leq f(s)$, $\forall s \in S$ takes effect. $f(s^*)$ is called a global optimum.

Given a fitness function f and a search space S the standard PSO initializes a set of particles, the swarm. In a D-dimensional search space S each particle

P_i consists of three D-dimensional vectors: its position $\vec{x}_i = (x_{i1}, x_{i2}, \ldots, x_{iD})$, the best position the particle visited in the past $\vec{p}_i = (p_{i1}, p_{i2}, \ldots, p_{iD})$ (particle best) and a velocity $\vec{v}_i = (v_{i1}, v_{i2}, \ldots, v_{iD})$. Usually the position is being initialized uniformly distributed over S and the velocity is also uniformly distributed depending on the size of S. The movement of each particle takes place in discrete steps using an update function. In order to calculate the update of a particle we need a supplementary vector $\vec{g} = (g_1, g_2, \ldots, g_D)$ (global best), the best position of a particle in its neighbourhood. The update function, called inertia weight, consists of two parts. The new velocity of a particle P_i is calculated for each dimension $d = 1, 2, \ldots, D$:

$$v_{id}^{new} = w \cdot v_{id} + c_1 \epsilon_{1d} (p_{id} - x_{id}) + c_2 \epsilon_{2d} (g_d - x_{id}). \tag{1}$$

As a next step, the position is being updated: $x_{id}^{new} = x_{id} + v_{id}^{new}$. The new velocity depends on the global best (g_d), particle's best (p_{id}) and the old velocity (v_{id}) which is weighted by the inertia weight w. The parameters c_1 and c_2 provide the possibility to determine how strong a particle is attracted by the global and the particle best. The random vectors $\vec{\epsilon}_1$ and $\vec{\epsilon}_2$ are uniformly distributed over $[0, 1)^D$ and produce the random movements of the swarm. According to Pan et al. this extension is required to be able to use the PSO for stochastic optimization purposes. [5]

2.2   Probability Distribution in PSO

Only in infinite time the specific scattering of numbers using the uniform distribution will be notable. It is possible, that the first numbers generated using a uniform distribution are equal to those generated by an extreme value distribution. In conjunction with the PSO update formula the computation of the next position of a particle seems to be set randomly. In order to avoid this, Kennedy developed the *Bare Bone PSO*[6], where the next position of the particle is being computed using a Gaussian distribution and not with the introduced update-formula. An analysis of this type of PSO shows that the Bare Bone PSO returns adequate results but not as good as those of the original PSO [7]. Exchanging the probability distribution leads to improved results [8]. One problem of PSO, the oscillation around a certain point, emanates from the chosen distribution. In [9] the exchange of the normal distribution for an dynamic probabilistic distribution leads to a better result and shows that the chosen distribution is a valid factor for the optimization process.

PSO using different random distributions is widely spread. For example the Cauchy distribution is used in [10] and the exponential distribution has been PSO is introduced by Krohling and dos Santos Coelho in [11]. But not only during the optimization process different distribution types are being used. The initialization of the starting position depends on the underlying probability distribution as well, as the random distribution is being used to specify the position. To gain a better optimization process this initialization effect has been probed in [12].

These different approaches show that random distribution is an important factor for the optimization process and that it is not trivial to choose the right distribution for a given fitness function. In order to try out different distributions and to analyse the chosen distribution's impact on the swarm a tool is needed to illustrate the scattering of the particles. With a simulator it is possible to see how the swarm behaves and to examine whether this behaviour could lead to a better optimization process. In the next section we will describe our approach by simulating a swarm choosing different random distributions and show how we developed an adequate user interface for the comparison of different distributions.

3 Concept

In this approach we aim to visualize the occurring differences by varying the probability distribution used for the generation of the random numbers. Therefore we intended the development of a simulation system visualizing the effect of different probability distributions on the scattering of the particles and the way the search space is being swept. Hereby we do not only want to simulate the first possible steps of a particle. As well each particle's next few steps shall be examined to receive an impression of how the particle moves though the search space.

The developed simulator is based on a given particle, where a certain amount (set by the user) of possible next positions will be calculated. In order to keep the results comparable, every repetition of the simulation-process will be executed outgoing from the same initially given particle. By generating new random factors (ϵ_1, ϵ_2) for each repetition the calculated next position of the initial particle will vary. This leads to possible positions inside the search space the particle can reach applying a given probability distribution. By increasing the number of repetitions, a first impression of the search space's coverage will be given.

Using the simulation system the user can arbitrary set the initial particle's position as the origin of the calculated particles. The p_{best} and g_{best} position and as it may be required the particle's old speed v_i can be set as well. This settings are used to scatter the points. The user can choose between the two main update-formulas (inertia weight and constriction update formula) and configure these formulas as desired. To compare different distribution functions a *SEED*-value can be set as well.

Each of the calculated particle's movements can be traced over a number of defined epochs. For this purpose one of the calculated possible positions is randomly picked and will be used as an origin for the next repetition of the simulation. The resulting sweep of the search space can be viewed in a 2d-model illustrating the possible scatter of the particles based on the chosen probability distribution.

In order to simulate a random fitness function as well two random points out of the amount of possible new positions can be set as new p_{best} and g_{best}. This is not a good fitness function but it demonstrates some variance of a very cliffy function. If the user decides not to reset the p_{best} and g_{best} values the underlying fitness is a very specific function. Hereby the points p_{best} and g_{best} represent very small local minimas surrounded by a large flat area.

In addition to our predefined common probability distributions integrated into this simulator the user can easily add own distribution algorithms to analyse their impact as well using a Java-interface which is being provided by the simulation system. This enables the user to use all possible number generators for examining their impact on PSO.

3.1   The Implementation of the Simulator

Due to its simpleness, the high spreading and its platform independence, the programming language Java has been chosen for the implementation of the tool.

Simulating PSO Behaviour with one Particle. As the tool is meant to be easy to use, the PSO's preconditions have been simplified for the user by using a point and click behaviour. The simulation is based on a single particle whose starting position can be set by the user. Furthermore the global best position (g_{best}), the personal best position (p_{best}) and the old speed of the particle (v_i) have to be chosen as well. For this purpose a panel is given on which the user can freely set these positions. If no positions are set by the user a set of standard positions[2] will be used. In case specific positions shall be chosen, a coordinate system can be placed over the panel to simplify the accurate choice. The standard PSO-configuration for the parameter set (v_i, c_1 and c_2) is automatically being set in the tool, too.

Based on the given parameters a possible next position of the initial particle will be simulated a given number of times. The calculated positions will be displayed as coloured dots within the panel the user has set the initial state in. As a result, a scatter plot of all calculated possible points enfolded by a rectangle will be shown as seen in fig. 1.

All choosen settings within the visualization can be replaced with simple point and click. The configuration of the particle (update-formula and parameter) can be set by the user as well. With this possibilities of setting up a particle, all PSO-behaviours can be simulated.

Simulating an Artificial Fitness Function. In order to simulate more than one epoch of the particle's movement, a fitness function is required for the re-calculation of the p_{best} and g_{best} value. As the handling of the tool is meant to be simple, we try to simulate an artificial fitness function being used for the determination of p_{best} and g_{best}. As a first possible fitness function, the swarm does not find a better place for g_{best} and p_{best}, the initially given p_{best} and g_{best} will be the same the for the further calculation. An alternative fitness function

---

[2] The standard position settings are (x/y): p = (150/20), g_{best} = (80/165), p_{best} = (400/15), v_i = (180/85).

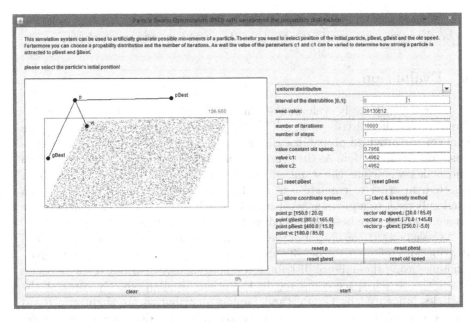

Fig. 1. Screenshot of the simulation system

is that p_{best} and/or g_{best} can be reset to a new position after each the termination of each epoch. By doing so, one of the new positions calculated within the last epoch will be randomly chosen as a new g_{best} for the following epoch. Analogously the particle's p_{best} will be set to the latest position calculated for each repetition.

Implementation of the Probability Distributions. The random numbers needed within the calculation are provided by five pre-implemented common probability distributions. These are uniform, discrete uniform, Cauchy, Laplace and Gumbel distribution, which represent different types of density functions. The user can choose between these functions and examine their impact on PSO.

In case the user intends to use a distribution which has not yet been implemented or an own distribution, the tool as well provides the possibility to integrate self-written Java-classes containing the distribution-algorithm. Therefore the so-called factory method pattern [13] has been used in order to simplify the process of creating an object of the user-included distribution-class. For that reason the implementation of the tool's *generator*-interface is required for these classes, which enforces the existence of a *getRandom()*-method. In case the class satisfies these preconditions and is added to the *generatorFactory*-class, the new distribution-algorithm can be selected by the user within the tool.

In addition the distribution-classes need to provide a method for setting a *SEED*-value, which is required in order to ensure the identical processing of two or more simulation-runs for the purpose of testing and reproduction.

Furthermore the boundaries out of which the generated numbers emerge need
to be changeable to specify the range of the resulting random numbers.

4   Evaluation

In order to conclusively evaluate the simulation system, the output based on different probability distributions will be analysed and compared. For this purpose
some of our provided probability distributions have been exemplarily chosen as
basis for the simulation. As the density of these distributions depends on location
and scale parameters, the standard parameters[3] for each distribution have been
selected. In order to emulate a fitness function, the position of g_{best} has been
randomly reset to a previously generated particle after each epoch. Furthermore
all simulations have been initialized with the same particles and pre-settings[4].

As seen in fig. 2, the output of the different probability distribution varies. The
distributions influence the density and the scattering of the particles. In the given
example, using the uniform distribution homogenously distributed the particles
inside the search space compared to the other distributions. By contrast the
particles influenced by Gumble and Laplace distribution are slightly focussed
on the side of the search space next to the initial particle. Applying Cauchy
distribution leads to a linear expansion of the search space, which results in an
elongated but narrow scattering of the particles.

Table 1. Surface area of the search space

distribution	1st epoch	2nd epoch	3rd epoch
uniform	106 650	386 835	661 182
Gumbel	106 594	366 300	743 562
Cauchy	105 925	399 620	1 052 580
Laplace	105 702	452 412	834 960

In order to estimate the way the different probability distributions cover the
search space, the surface area can be considered as a measurement parameter.
The spread of the space being covered during the 1st epoch of the simulation
using different distributions does not differ significantly (see table 1). However,
considering the subsequent epochs, variations regarding the surface of the covered
space can be observed. Using a uniform distribution covers results in a small, but
evenly distributed search space, whereas the other distribution's search space is
up to 60% bigger. For a better understanding the calculation of the convex-hull
is a better measurement but not implemented yet.

---

[3] Standard parameters: Cauchy $[\mu = 0, \beta = 0.5]$, Laplace $[\mu = 0, \beta = 1]$, Gumbel $[\mu = 0.5, \beta = 2]$

[4] Initial Particles: p $= (150/20)$, $g_{best} = (80/165)$, $p_{best} = (400/15)$, $v_i = 180/85$.
Presettings: interval of the distribution $=]0,1[$, seed-value $= 20\,130\,612$, number of
iterations $= 20\,000$, number of epochs $= 3$, $w_i = 0.7968$, $c_1 = 1.4962$, $c_2 = 1.4962$.

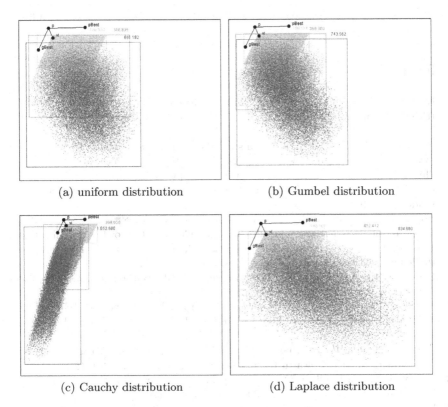

(a) uniform distribution   (b) Gumbel distribution

(c) Cauchy distribution   (d) Laplace distribution

Fig. 2. Scattering of the results using different probability distributions

5   Discussion

In this paper we presented a simulator for random distributions in PSO. The simulation system allows the user to define a specific configuration of one particle and to examine the scattering of this particle based on the chosen probability distribution. In order to use any possible distribution, a Java-interface is defined to implement own distributions. The user-interface allows the user to easily configure the epochs, PSO-settings and the amount of possible positions being simulated. As a result we compare different random distributions and show some specific scatter plots of the possible PSO behaviour. Using the Cauchy-distribution we got more slowing down particles compared to the uniform distribution, but on the other hand the uniform distribution explores a greater part of the search space.

Finally this approach gives the chance to analyze the PSO more detailed and visualize the behaviour of different random distributions.

As next steps a real fitness function based on the provided functions from Bratton and Kennedy [2] should be implemented into the simulation system to view the results of the distributions in a real environment. Another feature is to configure the chosen distribution in order to set new parameters for the random number generator.

References

1. Kennedy, J., Eberhart, R.: Particle swarm optimization. In: Proceedings of the 1995 IEEE International Conference on Neural Network, Perth, Australia, pp. 1942–1948 (1995)
2. Bratton, D., Kennedy, J.: Defining a standard for particle swarm optimization. In: Swarm Intelligence Symposium, pp. 120–127 (2007)
3. Shi, Y., Eberhart, R.: Parameter selection in particle swarm optimization. In: Porto, V.W., Waagen, D. (eds.) EP 1998. LNCS, vol. 1447, pp. 591–600. Springer, Heidelberg (1998)
4. Bogon, T., Poursanidis, G., Lattner, A.D., Timm, I.J.: Automatic Parameter Configuration of Particle Swarm Optimization by Classification of Function Features. In: Dorigo, M., et al. (eds.) ANTS 2010. LNCS, vol. 6234, pp. 554–555. Springer, Heidelberg (2010)
5. Pan, J.S., Huang, H.C., Jain, L.C.: Intelligent watermarking techniques. World Scientific, River Edge (2004)
6. Kennedy, J.: Bare bones particle swarms. In: Proceedings of the 2003 IEEE Swarm Intelligence Symposium, SIS 2003, pp. 80–87. IEEE Servoce Center, Piscataway (2003)
7. Feng, P., Xiaohui, H., Eberhart, R.C., Yaobin, C.: An analysis of Bare Bones Particle Swarm. In: IEEE Swarm Intelligence Symposium. IEEE, Piscataway (2008)
8. Richer, T.J., Blackwell, T.M.: The Lévy Particle Swarm. In: IEEE Congress on Evolutionary Computation, CEC 2006, pp. 808–815 (2006)
9. Kennedy, J.: Dynamic-probabilistic particle swarms. In: Proceedings of the 2005 Conference on Genetic and Evolutionary Computation, GECCO 2005, pp. 201–207. ACM, New York (2005)
10. Li, C., Liu, Y., Zhou, A., Kang, L., Wang, H.: A fast particle swarm optimization algorithm with cauchy mutation and natural selection strategy. In: Kang, L., Liu, Y., Zeng, S. (eds.) ISICA 2007. LNCS, vol. 4683, pp. 334–343. Springer, Heidelberg (2007)
11. Krohling, R., dos Santos Coelho, L.: Pso-e: Particle swarm with exponential distribution. In: IEEE Congress on Evolutionary Computation, CEC 2006, pp. 1428–1433 (2006)
12. Thangaraj, R., Pant, M., Deep, K.: Initializing pso with probability distributions and low-discrepancy sequences: The comparative results. In: World Congress on Nature Biologically Inspired Computing, NaBIC 2009, pp. 1121–1126 (December 2009)
13. Gamma, E., Helm, R., Johnson, R., Vlissides, J.: Design Patterns: Elements of Reusable Object-Oriented Software. Addison-Wesley Professional (1994)

Local and Global Search Based PSO Algorithm

Yanxia Sun[1,*], Zenghui Wang[2], and Barend Jacobus van Wyk[1]

[1] Department of Electrical Engineering, Tshwane University of Technology,
Pretoria 0001, South Africa
[2] Department of Electrical and Mining Engineering, University of South Africa,
Florida 1710, South Africa
{sunyanxia,wangzengh,vanwykb}@gmail.com

Abstract. In this paper, a new algorithm for particle swarm optimisation (PSO) is proposed. In this algorithm, the particles are divided into two groups. The two groups have different focuses when all the particles are searching the problem space. The first group of particles will search the area around the best experience of their neighbours. The particles in the second group are influenced by the best experience of their neighbors and the individual best experience, which is the same as the standard PSO. Simulation results and comparisons with the standard PSO 2007 demonstrate that the proposed algorithm effectively enhances searching efficiency and improves the quality of searching.

Keywords: Local search, global search, particle swarm optimisation.

1   Introduction of PSO

PSO is an evolutionary computation technique developed by Kennedy and Eberhart [1] in 1995: it is a population-based optimisation technique, inspired by the motion of bird's flocking, or fish schooling. The particle swarms are social organizations whose overall behavior relies on some sort of communication amongst members, and cooperation. All members obey a set of simple rules that model the communication within the flock, between the flocks and the environment. Each solution is a "bird" in the flock and is referred to as a "particle". PSO is not largely affected by the size and nonlinearity of the problem, and can converge to the optimal solution in many problems [2-5] where most analytical methods fail to converge. It can, therefore, be effectively applied to different optimisation problems.

The standard particle swarm algorithm works by iteratively searching in a region and is concerned with the best previous success of each particle, the best previous success of the particle swarm as a whole, the current position and the velocity of each particle [4]. The particle searches the domain of the problem, according to

$$V_i(t+1) = \omega V_i(t) + c_1 R_1 (P_i - X_i(t)) + c_2 R_2 (P_g - X_i(t)), \tag{1}$$

$$X_i(t+1) = X_i(t) + V_i(t+1) \tag{2}$$

---

* Corresponding author.

Y. Tan, Y. Shi, and H. Mo (Eds.): ICSI 2013, Part I, LNCS 7928, pp. 129–136, 2013.
© Springer-Verlag Berlin Heidelberg 2013

where $V_i = \left[v_i^1, v_i^2, \cdots v_i^n \right]$ is the velocity of particle i ; $X_i = \left[x_i^1, x_i^2, \cdots x_i^n \right]$ represents the position of particle i ; P_i represents the best previous position of particle i (indicating the best discoveries or previous experience of particle i); P_g represents the best previous position among all particles (indicating the best discovery or previous experience of the social swarm); ω is the inertia weight that controls the impact of the previous velocity of the particle on its current velocity and is sometimes adaptive. R_1 and R_2 are two random weights whose components r_1^j and r_2^j ($j = 1, 2 \cdots, n,$) are chosen uniformly within the interval [0,1] which might not guarantee the convergence of the particle trajectory; c_1 and c_2 are the positive constant parameters. Generally the value of each component in V_i should be clamped to the range $[-v_{max}, v_{max}]$ to control excessive roaming of particles outside the search space.

Among these parameters, the inertia weight ω plays an important role and affects the global and local search ability of PSOs. If the value of ω is too big, the global search ability of PSO will be improved, but its local search ability will not be adequate. Otherwise, if the value of ω is small, the global search ability will decrease and the particles easily fall in premature. Some parameters of adaptive PSOs has been proposed but these usually change the inertia weight: ω is large at the beginning of the search procedure and ω decreases as time increased [7, 13]. However, there is a similar problem with the fixed inertia weight method: 1) at the beginning, the local search ability is not effective as ω is big; while 2) the global search ability is not satisfactory at the end of the search procedure as ω becomes small. To balance the local search and global search ability at the same time, a new particle swarm optimisation algorithm is proposed which can perform the local and global search es simultaneously.

In the proposed algorithm, the particles are divided into two groups. The velocity of the first group of particles is only influenced by the best experience of its neighbors. And the velocity of the second group is influenced by both the best experience of its neighbors and its own best experience. The rest of this paper is arranged as follows: In Section 2, the proposed algorithm is described. Section 3 describes the problems used to evaluate the new algorithm and the results are obtained. Finally, the concluding remarks appear in Section 4.

2   Local and Global Search Based PSO Algorithm

Referring to equation (1), the right side consists of three parts: the first is the previous velocity of the particle; the second and third are those parts contributing to the change in the velocity of a particle. As explained in [7], without these two parts, the particles will keep on flying at the current speed in the same direction until they hit the boundary. PSO will not find an acceptable solution unless there are acceptable solutions on their flying trajectories. But this is a rare case. On the other hand, referring to equation (1) without the first part, the flying particles' velocities are only determined by their current positions and their best positions in its history. At the same time, each other particle will be flying toward its weighted centroid of its own best position and the global best position of the population [8]. Some authors have suggested adjustments to the parameters of the PSO algorithm: adding a random component to the inertia weight [9, 10], using a secondary PSO to find the optimal parameters of a primary PSO [11], and adaptive critics [12]. From our literature study and simulation experience, the optimum is often found near the global best experience in various optimisation problems. To help the particles to enhance searching the region around the global best experience, the first group particles are separated from the whole set of particles to search the area around the global best experience. Then equations (1) and (2) will be altered to

$$V_i(t+1) = 0.5 \times \omega \times V_i(t) + c_1 R_1 (P_g - X_i(t)) + c_2 R_2 (P_g - X_i(t)), \tag{3}$$

$$X_i(t+1) = X_i(t) + V_i(t+1). \tag{4}$$

As can be seen from equation (3), the particles will focus on searching the area around the best experience among their neighbors.

The particles in the second group will continue to the search the global experience of the swarm and its own best experience according to equations (1) and (2), which are the same as the standard PSO.

The following procedure can be used for implementing the proposed particle swarm algorithm:

1) Initialize the swarm, assign a random position in the problem hyperspace to each particle and calculate the fitness function which is yielded by the optimisation problem whose variables are corresponding to the elements of particle position coordinates.

2) The particles in the first group search the area according to equations (3) and (4). Meanwhile, those in the second group search the area according to equations (1) and (2).

3) Evaluate the fitness function for each particle.

4) For each individual particle, compare the particle's fitness value with its previous best fitness value. For each individual particle, compare the particle's fitness value with its previous best fitness value. If the current value X_i is better than the previous best value P_i, then set P_i as X_i.

5) Repeat steps 2) - 4) until the criterion for stopping is met (e.g., maximum number of iterations or a sufficiently good fitness value).

3   Numerical Simulation

To demonstrate the efficiency of the proposed technique, six well-known benchmarks are used to compare the proposed method and the standard PSO 2007 (Matlab version compiled in 2011) [14]. These six optimisation problems were used as shown in Table 1. Their parameters are given in Table 2. These six are famous test functions for minimization methods; each of them has several local minima. In the numerical simulation of the proposed LGPSO method and standard PSO, the particle swarm population size is set floor($10 + 2\sqrt{D}$). Here D is the dimension of the optimisation problems and the function floor(A) rounds the elements of float number A to the nearest integers less than or equal to A. The rest of the parameters are as follows: inertia weight $\omega = \dfrac{1}{(2\log 2)} \approx 0.7213$, learning rates $c_1 = c_2 = 0.5 + \log 2$, and velocity Vmax set to the dynamic range of the particle in each dimension. The topology of LGPSO is the same as the standard PSO 2007 (SPSO 2011) [14]. It should be noted that the initial variables are set random float numbers in the range [0, 1] to check the effect of big search range. The maximum number of function evaluations is 2000 for these two methods with 100 independent runs. The optimisation statistical analysis of these two algorithms is reported in Table 3. The evolutionary curves of LGPSO and the standard PSO 2007 are depicted in Figures 1-6.

Table 1. Functions used to test the effects of the LGPSO method

Problem	Objective functions
Rosenbrock	$f(x) = \sum_{i=1}^{D}(100(x_{i+1} - x_i^2)) + (x_i - 1)^2)$
Ackley	$f(x) = 20 + e - 20e^{-\frac{1}{5}\sqrt{\frac{1}{D}\sum_{i=1}^{D}x_i^2}} - e^{-\frac{1}{D}\sum_{i=1}^{D}\cos(2\pi x_i)}$
Griewank	$f(x) = \frac{1}{4000}\sum_{i=1}^{D}(x_i - 100)^2 - \prod_{i=1}^{D}\cos(\frac{(x_i - 100)}{\sqrt{i}}) + 1$
Salomon	$f(x) = \cos(2\pi\sqrt{\sum_{i=1}^{D}x_i^2}) + 0.1\sqrt{\sum_{i=1}^{D}x_i^2} + 1$
Rotated-hyper-ellipsoid	$f(x) = \sum_{i=1}^{D}(\sum_{j=1}^{i}x_j)^2$
Quartic function	$f(x) = \sum_{i=1}^{D}ix_i^4 + rand()$
Alpine	$f(x) = \sum_{i=1}^{D}\lvert x_i\sin(x_i) + 0.1x_i\rvert$
Levy	$f(x) = \sin^2(\pi x_1)$ $+\left(\begin{array}{l}\sum_{i=1}^{D-1}(x_i - 1)^2\left(1 + 10\sin^2(\pi x_i + 1)\right)\\ +(y_{D-1} - 1)^2(1 + \sin^2(2\pi x_{D-1}))\end{array}\right)$

Table 2. Functions parameters for the test problems

Problem	Dimension	Search range	Initial range
Rosenbrock	30	±500	[0, 1]
Ackley	30	±500	[0, 1]
Griewank	30	±500	[0, 1]
Salomon	30	±500	[0, 1]
Rotated-hyper-ellipsoid	30	±500	[0, 1]
Quartic function	30	±500	[0, 1]
Alpine	30	±500	[0, 1]
Levy	30	±500	[0, 1]

Table 3. Comparison between standard PSO 2007 and LGPSO

Problem	Method	best	Mean	Std.dev	Worst
Rosenbrock	Standard PSO 2007	122.3422	222.7063	31.1602	343.6297
Rosenbrock	LGPSO	122.0898	183.5776	24.3594	261.8783
Ackley	Standard PSO 2007	1.8158	2.3669	0.2861	3.0259
Ackley	LGPSO	1.2924	2.0563	0.3337	2.9711
Griewank	Standard PSO 2007	0.0411	0.0788	0.0226	0.1827
Griewank	LGPSO	0.0247	0.0584	0.0172	0.1110
Salomon	Standard PSO 2007	0.2999	0.2999	1.0235e-004	0.3005
Salomon	LGPSO	0.2999	0.2999	6.9229e-006	0.2999
Rotated hyper-ellipsoid	Standard PSO 2007	38.1115	144.2199	67.4747	432.1835
Rotated hyper-ellipsoid	LGPSO	12.1226	48.0499	30.0913	143.8851
Quartic function	Standard PSO 2007	1.8093	5.8692	2.5883	15.6572
Quartic function	LGPSO	1.3892	3.7318	1.5546	8.2722
Alpine function	Standard PSO 2007	0.9011	1.9667	0.6130	4.0312
Alpine function	LGPSO	0.4702	1.4589	0.5381	3.5732
Levy function	Standard PSO 2007	0.4363	0.7213	0.1312	1.1353
Levy function	LGPSO	0.2605	0.5791	0.1267	0.8630

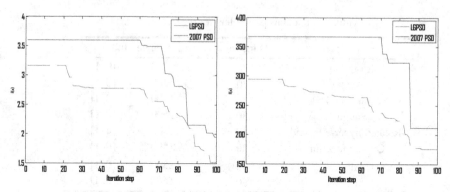

Fig. 1. Comparison for Rosenbrock function   **Fig. 2.** Comparison for Ackley function

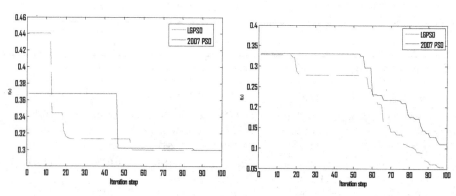

Fig. 3. Comparison for Griewank function   **Fig. 4.** Comparison for Salomon function

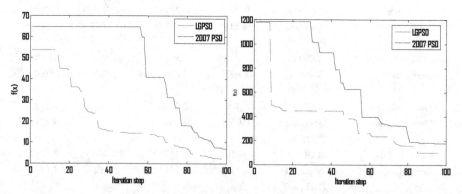

Fig. 5. Comparison for Rotated hyper-ellipsoid function   **Fig. 6.** Comparison for Quartic function

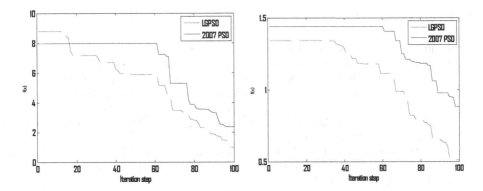

Fig. 7. Comparison for Alpine function   **Fig. 8.** Comparison for Levy function

As can be seen from Table 3, for the test functions, the best, mean, standard deviation and worst results obtained by the LGPSO are better than the results gained from the standard PSO 2007. The optimization performance of the proposed method is also more stable than the standard PSO 2007 according to the statistical analysis of mean and standard deviation. From Figures 1-8, the optimisation performance is better when the procedure begins, as the local search is added into the algorithms. The simulation results obtained by the LGPSO are better than the results from the standard PSO 2007, which means the final solutions obtained from the LGPSO are more closely focused on the best solution than those from the standard PSO 2007.

4   Conclusion

In this paper, a local and global search based particle swarm optimisation (LGPSO) method was proposed to improve the optimisation performance of the PSO. In this new model, the first group of particles focused on the search around the global best experience while the second group particles are influenced by both the best experience of their group and their own best experience. The simulations showed that the proposed method can achieve good optimisation performance no matter whether at the beginning or at the end of the search period. Moreover, the complexity of the proposed algorithm is not increased over that of the Standard PSO 2007 while the performance of the proposed FCPSO is more stable and more accurate than the Standard PSO 2007.

Acknowledgements. This work was supported by China/South Africa Research Cooperation Programme (No. 78673 & CS06-L02), South African National Research Foundation Incentive Grant (No. 81705), SDUST Research Fund (No. 2010KYTD101) and Key Scientific Support Program of Qingdao City (No. 11-2-3-51-nsh).

The authors thank the Matlab version codes of standard PSO 2007 compiled by Mahamed Omran in 2011, which are available at the Particle Swarm Central website [14].

References

1. Kennedy, J., Eberhart, R.C.: Particle Swarm Optimisation. In: Proceedings of IEEE International Conference on Neural Networks, Perth, Australia, pp. 1942–1948 (1995)
2. Hu, X., Shi, Y., Eberhart, R.: Recent Advances in Particle Swarm. In: Congress on Evolutionary Computation, pp. 90–97. IEEE Service Center, Piscataway (2004)
3. Huang, C.M., Huang, C.J., Wang, M.L.: A Particle Swarm Optimisation to Identifying the ARMAX Model for Short term Load Forecasting. IEEE Transactions on Power Systems 20, 1126–1133 (2005)
4. Clerc, M.: Particle Swarm Optimisation. ISTE Publishing Company (2006)
5. Nedjah, N., Mourelle, L.D.M.: Systems Engineering Using Particle Swarm Optimisation. Nova Science Publishers (2007)
6. Aihara, K., Takabe, T., Toyoda, M.: Chaotic Neural Networks. Physics Letter A 144(6-7), 333–340 (1990)
7. Shi, Y., Eberhart, R.: A Modified Particle Swarm Optimiser. In: IEEE International Conference on Evolutionary Computation, pp. 69–73. IEEE Press, Piscataway (1998)
8. Zhang, W.J., Xie, X.F.: DEPSO: Hybrid Particle Swarm with Differential Evolution Operator. In: Proceedings of IEEE International Conference on Systems, Man and Cybernetics, Washington DC, USA, pp. 3816–3821 (2003)
9. Mohagheghi, S., Del Valle, Y., Venayagamoorthy, G., Harley, R.: A Comparison of PSO and Back Propagation for Training RBF Neural Networks for Identification of a Power System with STATCO. In: Proceedings of IEEE Swarm Intelligence Symposium, pp. 381–384 (June 2005)
10. del Valle, Y., Venayagamoorthy, G.K., Mohagheghi, S., Hernandez, J., Harley, R.G.: Particle Swarm Optimisation: Basic Concepts, Variants and Applications in Power Systems. IEEE Transactions On Evolutionary Computation 12(2), 171–195 (2008)
11. Doctor, S., Venayagamoorthy, G., Gudise, V.: Optimal PSO for Collective Robotic sSearch Applications. In: Proceeding IEEE Congress on Evolutionary Computation, Portland, Oregon, USA, pp. 1390–1395 (2004)
12. Venayagamoorthy, G.: Adaptive Critics for Dynamic Particle Swarm Optimisation. In: Proceedings of IEEE International Symposium on Intelligence Control, Taipei, Taiwan, pp. 380–384 (September 2004)
13. Liu, B., Wang, L., Jin, Y.H., Tang, F., Huang, D.X.: Improved Particle Swarm Optimisation Combined with chaos. Chaos, Solitons and Fractals 25, 1261–1271 (2005)
14. Kennedy, J., Clerc, M., et al.: Particle Swarm Central (2012), http://www.particleswarm.info/Programs.html

Cask Theory Based Parameter Optimization for Particle Swarm Optimization

Zenghui Wang[1,*] and Yanxia Sun[2]

[1] Department of Electrical and Mining engineering, University of South Africa,
Florida 1710, South Africa
[2] Department of Electrical engineering, Tshwane University of Technology,
Pretoria 0001, South Africa
{wangzengh,sunyanxia}@gmail.com

Abstract. To avoid the bored try and error method of finding a set of parameters of Particle Swarm Optimization (PSO) and achieve good optimization performance, it is desired to get an adaptive optimization method to search a good set of parameters. A nested optimization method is proposed in this paper and it can be used to search the tuned parameters such as inertia weight ω, acceleration coefficients c_1 and c_2, and so on. This method considers the cask theory to achieve a better optimization performance. Several famous benchmarks were used to validate the proposed method and the simulation results showed the efficiency of the proposed method.

Keywords: PSO, Parameter Optimization, Try and Error method, Nested Optimization method, Cask theory.

1 Introduction

Particle Swarm Optimization (PSO) was developed by Kennedy and Eberhart [1]. This algorithm is inspired by the social behavior of a flock of migrating birds trying to reach an unknown destination. All members obey a set of simple rules that model the communication within the flock, between the flocks and the environment. Each solution is a "bird" in the flock and is referred to as a "particle". PSO has attracted a lot of attention as it makes few or no assumptions about the problem being optimized and can search very large spaces of candidate solutions [2, 4-7]. The formula of PSO is realized by two update functions:

$$V_i(t+1) = \omega V_i(t) + c_1 R_1 (P_i - X_i(t)) + c_2 R_2 (P_g - X_i(t)), \tag{1}$$

$$X_i(t+1) = X_i(t) + V_i(t+1). \tag{2}$$

Here $V_i = \left[v_i^1, v_i^2, \cdots v_i^n \right]$ is the velocity of particle i; $X_i = \left[x_i^1, x_i^2, \cdots x_i^n \right]$ represents the position of particle i; P_i represents the best previous position of particle i

---

* Corresponding author.

Y. Tan, Y. Shi, and H. Mo (Eds.): ICSI 2013, Part I, LNCS 7928, pp. 137–143, 2013.

(indicating the best discoveries or previous experience of particle i); P_g represents the best previous position among all particles (indicating the best discovery or previous experience of the social swarm); ω is the inertia weight that controls the impact of the previous velocity of the particle on its current velocity and is sometimes adaptive. R_1 and R_2 are two random weights whose components r_1^j and r_2^j ($j=1,2\cdots,n,$) are chosen uniformly within the interval $[0,1]$ which might not guarantee the convergence of the particle trajectory; c_1 and c_2 are the positive constant parameters. Generally the value of each component in V_i should be clamped to the range $[-v_{max}, v_{max}]$ to control excessive roaming of particles outside the search space.

The generalized procedure of applying standard PSO 2011 (SPSO 2011) [8] is

1) Initialize the swarm and assign a random position in the problem hyperspace to each particle and calculate the fitness function which is given by the optimization problem whose variables are corresponding to the elements of particle position coordinates; and set the topology of the whole particles.

2) The particles search the area according to equations (1) and (2); check the velocity and position of particles to find whether they violate the boundaries.

3) Evaluate the fitness function for each particle.

4) For each individual particle, compare the particle's fitness value with its previous best fitness value. For each individual particle, compare the particle's fitness value with its previous best fitness value. If the current value X_i is better than the previous best value P_i, then set P_i as X_i.

5) Change the topology if the optimization performance is not improved in a certain number of iterations.

6) Repeat steps 2)-5) until a stopping criterion is met (e.g., maximum number of iterations or a sufficiently good fitness value).

As can be seen from (1) and (2), there are several parameters which should be determined before PSO was applied. Similar as most of the evolutionary optimization algorithms, the parameters of PSO need to be chosen carefully to achieve good optimization performance. The parameters are often chosen based try and error method as different optimization problems have different characteristics and the parameters should not be same to achieve good optimization results. Hence, it is desired to find a suitable set of parameters of PSO without using the bored try and error method. For the evolutionary optimization algorithms, there are some methods optimizing the parameters of the optimization algorithms which are called meta-optimization. Meta-optimization is reported to have been used as early as in the late 1970s by Mercer and Sampson for finding optimal parameter settings of a genetic algorithm [9]. There are some meta-optimizations [10], [11], [12]. For different meta-optimizations, there are different performance indexes.

In this paper, an automatic parameters searching method is proposed based on the particle swarm optimization algorithm and the cask theory. The rest of this paper is arranged as follows: Section 2 presents the proposed algorithm with details. Simulations and comparison are given in Section 3. Finally, the concluding remarks appear in Section 4.

2   Cask Theory Based Parameter Optimization

As the optimization performance depends on the optimization problems, the parameters of optimization algorithms should also depend on the optimization problems, which means different optimization problems should have different sets of parameters of optimization algorithms. As the optimization algorithms can find optimal or sub-optimal solution for the optimization problems, the optimization algorithms can also be used to find the optimal or sub-optimal parameters for PSO. Similar as the optimization procedure, the objective function or criteria related to the parameters of PSO must be defined firstly. There is an important theory is cask theory or barrel theory in Management Science [3]. The cask theory describes that the cubage of a cask is dependent on the shortest wood plate as shown in Fig. 1. This method takes the worst case as the performance criteria and it is possible to make the optimization performance not worse than the achieved one.

Fig. 1. Cask theory (www.baike.com)

The parameter optimization procedure is same with the standard one as mentioned in Section 1. The core of the parameter optimization is defining the objective function or criteria. The followings are the factors, which should be considered, when design the objective function for PSO parameter optimization:

1)   Important parameters of PSO should be chosen and they will be the inputs of the objective function.
2)   The optimization problem should be considered as the implicit objective as the parameters are used to achieve good optimization performance for the optimization problem.
3)   The optimization performance should be stable when the obtained parameters are implemented.
4)   The output of the objective function should follow cask theory to guarantee the worst optimization performance is not too bad.

Here, without loss of generality, the algorithm of the SPSO 2011 [8] is chosen as the optimization algorithm whose parameters (inertia weight ω, and acceleration coefficients c_1 and c_2) will be optimized and the SPSO 2011 with fixed parameter is used to optimize these parameters. Hence, for 1) the inputs of the objective function are the inertia weight ω, the acceleration coefficients c_1 and c_2. For 2), the optimization problem will be the target of the SPSO 2011 with variant parameters (VSPSO 2011). For 3) and 4), the optimization problem should be optimized several runs by VSPSO to make sure the results are not stochastic; and the worst fitness value is chosen as the output of the objective function which follows the   cask theory.

After the set of parameters are obtained, the normal procedure of PSO will be used to optimize the optimization problems.

3   Numerical Simulation

To demonstrate the efficiency of the proposed technique, eight well-known benchmarks are used to compare the proposed method and the standard PSO 2011 (Matlab version) [8]. The eight optimization problems were used as shown in Table 1. The parameters of these optimization problems are given in Table 2. These eight optimization problems are famous test functions for minimization methods and each of them has high dimension and several local minima. In the numerical simulation of SPSO 2011 with fixed parameters, the particle swarm population size is set floor(10 + $2\sqrt{D}$). Here D is the dimension of   the optimization problems and   function floor(A) rounds the elements of float number A to the nearest integers less than or equal to A.

The rest of the parameters are as follows: inertia weight $\omega = \dfrac{1}{(2\log 2)} \approx 0.7213$,

learning rates $c_1 = c_2 = 0.5 + \log 2$, and velocity Vmax set to the dynamic range of the particle in each dimension. For VSPSO 2011, the inertia weight ω, the acceleration coefficients c_1 and c_2 are the parameters to be optimized and all the initial ranges of ω, c_1 and c_2 are [0.2, 3]. To reduce the run time, the maximum number of function evaluations is 500 with 10 independent runs. The maximum number of function evaluations is 500 for these VSPSO 2011 using the parameters obtained and SPSO 2011 with 100 independent runs.

The optimized parameters were given in Table 3. The optimization statistical analysis of proposed method and SPSO 2011 with fixed parameters was given in Table 4. As can be seen from Table 3, the parameters are totally different from the fixed parameters of SPSO 2011 and there are no rules to follow as the optimization problems are totally different. As can be seen from the Table 4, the optimization performance of VSPSO 2011 is more stable and it can guarantee the worst results are not worse than the worst results of SPSO 2011 as the proposed parameter optimization method is cask theory based parameter optimization method.

Table 1. Functions used to test the effects of the LGPSO method

Sphere	$f(x) = \sum_{i=1}^{D} x_i^2$
Rastrigin	$f(x) = 10D + \sum_{i=1}^{D}(x_i^2 - 10\cos(2\pi x_i))$
Step	$f(x) = \sum_{i=1}^{D}[(x_i + 0.5)^2]$, [.] is rounding function
Rosenbrock	$f(x) = \sum_{i=1}^{D}(100(x_{i+1} - x_i^2)) + (x_i - 1)^2)$
Ackley	$f(x) = 20 + e - 20e^{-\frac{1}{5}\sqrt{\frac{1}{D}\sum_{i=1}^{D}x_i^2}} - e^{-\frac{1}{D}\sum_{i=1}^{D}\cos(2\pi x_i)}$
Griewank	$f(x) = \frac{1}{4000}\sum_{i=1}^{D}(x_i - 100)^2 - \prod_{i=1}^{D}\cos(\frac{(x_i - 100)}{\sqrt{i}}) + 1$
Salomon	$f(x) = \cos(2\pi\sqrt{\sum_{i=1}^{D}x_i^2}) + 0.1\sqrt{\sum_{i=1}^{D}x_i^2} + 1$
Rotated hyper-ellipsoid	$f(x) = \sum_{i=1}^{D}(\sum_{j=1}^{i}x_j)^2$

Table 2. Functions parameters for the test problems

Functions	Dimension	Initial range
Sphere	30	±500
Rastrigin	30	±500
Step	30	±500
Rosenbrock	30	±500
Ackley	30	±500
Griewank	30	±500
Salomon	30	±500
Rotated hyper-ellipsoid	30	±500

Table 3. Optimized parameters for the test problems

Functions	Inertia weight ω, and	Acceleration coefficient c_1	Acceleration coefficient c_2
Sphere	0.5728	0.6336	0.8422
Rastrigin	0.5908	0.6726	0.9059
Step	0.6539	0.5442	0.6911
Rosenbrock	0.6392	1.2737	0.5954
Ackley	3.0000	3.0000	2.9441
Griewank	0.5901	0.9769	0.7857
Salomon	0.5424	0.3778	0.5264
Rotated hyper-ellipsoid	0.5360	0.8172	0.6147

Table 4. Comparison between standard PSO 2011 and VSPSO 2011

Problem	Method	best	Mean	Std.dev	Worst
Sphere	Standard PSO 2011	1.0482e+005	2.2438e+005	5.6333e+004	4.1521e+005
Sphere	VSPSO 2011	2.4199e+004	7.0329e+004	2.1287e+004	1.2790e+005
Rastrigin	Standard PSO 2011	1.1813e+005	2.2891e+005	4.7518e+004	3.5182e+005
Rastrigin	VSPSO 2011	3.1282e+004	7.6292e+004	2.2688e+004	1.3040e+005
Step	Standard PSO 2011	114834	2.1631e+005	4.3216e+004	345796
Step	VSPSO 2011	27158	7.2645e+004	2.2680e+004	151432
Rosenbrock	Standard PSO 2011	9.3528e+010	4.8421e+011	2.1182e+011	1.2005e+012
Rosenbrock	VSPSO 2011	4.8036e+009	3.5983e+010	2.4238e+010	1.4794e+011
Ackley	Standard PSO 2011	20	20.2424	0.1278	20.5651
Ackley	VSPSO 2011	20	20	0	20
Griewank	Standard PSO 2011	27.3732	57.4863	11.4453	90.7654
Griewank	VSPSO 2011	7.7910	16.8858	4.8584	35.0051
Salomon	Standard PSO 2011	37.5720	47.5551	4.7485	59.1266
Salomon	VSPSO 2011	17.2006	29.1404	4.0430	37.6110
Rotated hyper-ellipsoid	Standard PSO 2011	4.5205e+005	7.6140e+005	1.7838e+005	1.2665e+006
Rotated hyper-ellipsoid	VSPSO 2011	6.8559e+004	1.4094e+005	4.2328e+004	2.7582e+005

4   Conclusion

In this paper, a cask theory based parameter optimization based particle swarm optimization was proposed to find a good set of parameter of. This method can find sets of optimized parameters and using the obtained parameters can achieve better optimization performance than the standard set of parameters. No prior experience is needed for this method. The simulations showed that the proposed method can achieve good optimization performance comparing with the SPSO 2011. Moreover, the simulations show that it can make sure the worst results are not worse than the worst results of SPSO 2011 as this is cask theory based parameter optimization method. This method can also be used to find the parameters of other optimization algorithms.

Acknowledgements. This work was supported by China/South Africa Research Cooperation Programme (No. 78673 & CS06-L02), South African National Research Foundation Incentive Grant (No. 81705), SDUST Research Fund (No. 2010KYTD101) and Key scientific support program of Qingdao City (No. 11-2-3-51-nsh).

The authors thank the Matlab version codes of standard PSO 2011 compiled by Mahamed Omran and the codes are available at the particle swarm central website [8].

References

1. Kennedy, J., Eberhart, R.C.: Particle Swarm Optimization. In: Proceedings of IEEE International Conference Neural Networks, Perth, Australia, pp. 1942–1948 (1995)
2. Clerc, M.: Particle Swarm Optimization. ISTE Publishing Company (2006)
3. Ju, J., Wei, S.: Endowment versus Finance: A Wooden Barrel Theory of International Trade, CEPR Discussion Papers 5109, C.E.P.R. Discussion Papers (2005)
4. Xinchao, Z.: A perturbed particle swarm algorithm for numerical optimization. Applied Soft Computing 10, 119–124 (2010)
5. Yang, X.S.: Nature-Inspired Metaheuristic Algorithms. Luniver Press (2008)
6. Roy, R., Dehuri, S., Cho, S.B.: A Novel Particle Swarm Optimization Algorithm for Multi-Objective Combinatorial Optimization Problem. International Journal of Applied Metaheuristic Computing 2(4), 41–57 (2012)
7. Chen, W., Zhang, J.: A novel set-based particle swarm optimization method for discrete optimization problem. IEEE Transactions on Evolutionary Computation 14, 278–300 (2010)
8. Kennedy, J., Clerc, M., et al.: Particle Swarm Central (2012), http://www.particleswarm.info/Programs.html
9. Mercer, R.E., Sampson, J.R.: Adaptive search using a reproductive metaplan. Kybernetes 7(3), 215–228 (1978)
10. Keane, A.J.: Genetic algorithm optimization in multi-peak problems: studies in convergence and robustness. Artificial Intelligence in Engineering 9(2), 75–83 (1995)
11. Pedersen, M.E.H., Chipperfield, A.J.: Simplifying particle swarm optimization. Applied Soft Computing 10(2), 618–628 (2010)
12. Birattari, M., Stützle, T., Paquete, L., Varrentrapp, K.: A racing algorithm for configuring metaheuristics. In: Proceedings of the Genetic and Evolutionary Computation Conference (GECCO), pp. 11–18 (2002)

A Piecewise Linearization Method of Significant Wave Height Based on Particle Swarm Optimization

Liqiang Liu, Zhichao Fan, and Xiangguo Wang

College of Automation, Harbin Engineering University, Harbin 150001, China
xgwangcq@gmail.com

Abstract. A piecewise linearization method of significant wave height is proposed based on particle swarm optimization. Piecewise linearization model is used to approximate significant wave height inversion model, minimum radius of neighborhood is used to eliminate wild value in the sample data and sparse the data, and then the particle swarm optimization algorithm is applied for piecewise area division and parameter optimization of the model. Simulation result shows that compared with traditional inversion method, better practicability and the higher significant wave height inversion precision are obtained by the proposed method.

Keywords: particle swarm optimization, significant wave height, piecewise linearization, X-band radar.

1 Introduction

Ocean wave is the closest sea phenomenon related to human beings. Wave height, wave direction, and other factors of wave play an important role in shipping and harbor as well as the security of marine oil platform. The sea clutter image obtained by marine X-band navigation radar echo forming contains rich ocean wave information, so radar's echo intensity can be used for the inversion of ocean wave spectrum and ocean wave parameter. In 1985, an ocean wave information extraction method was first proposed based on the sea clutter radar image sequence by Young [1]. As soon as this method was discovered, it attracted people's enormous interest. In the following 10 years, Zimer, Rosenthal, Günther and colleagues [2]-[6] have also focused on ocean waves information inversion based on X-band navigation radar. The ocean wave information extraction method currently has become a hot research topic in the area of ocean dynamical environment monitoring.

The significant wave height is one kind of the ocean wave information. Because of the nonlinearity ocean waves imaging mechanism, when the ocean wave parameters are inverted with X-band radar image, only the relative value of ocean wave spectrum energy can be obtained. In 1982, method of synthetic aperture radar (SAR) estimating the significant wave height was proposed by Alpers and Hasselmann [7]. In 1994, this method was extended to X-band navigation radar image by Ziemer and Günther to calculate the significant wave height. This method assumed significant wave height had a linear relationship with the square root of the radar image signal-to-noise (SNR)

Y. Tan, Y. Shi, and H. Mo (Eds.): ICSI 2013, Part I, LNCS 7928, pp. 144–151, 2013.

ratio and the significant wave height can be calculated by the linear model. But in reality, due to different computational methods of SNR, differences between radar systems, variable marine environments and other factors, significant wave height doesn't exactly have a linear relationship with the square root of radar image SNR. Regarding this problem, in 2009, H.M. Duan and J. Wang [8] divided the significant wave height into two regions of the low wave height and the high wave height, each region uses the standard inversion method to obtain A and B respectively. In 2011, L.Q. Liu et al [9] proposed a significant wave height inversion method based on radial basis function neural network. These methods have improved inversion precision of significant wave height, but there still exist some problems. In these methods, the significant wave height is divided into two regions by manually selecting a piecewise point, which can't ensure the optimal solution. The model constructed by radial basis function neural network is relatively complicated, which takes long computing time and is infeasible for practical engineering. In view of this, a piecewise linearization method of significant wave height based on particle swarm optimization is proposed in this paper, in which the piecewise area is automatically divided and it is suitable for practical engineering.

2   Problem Description

The piecewise linearization model of significant wave height is shown in Fig. 1.

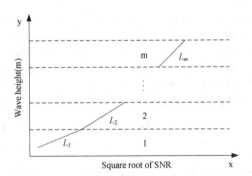

Fig. 1. Piecewise linearization model

Assume significant wave height has a piecewise linearization relationship with the square root of radar image SNR in this model. For any region L_i (i=1,2,$\cdots$, m), its model is given in equation (1).

$$H_{L_i} = A_{L_i} + B_{L_i} \cdot \sqrt{SNR} \ . \tag{1}$$

where H_{Li} is the significant wave height, A_{Li} and B_{Li} are the undetermined coefficients, SNR is the radar image signal-to-noise ratio.

The piecewise linearization model needs to meet the flowing three heuristic requirements.

1. The intersection point of any line segment L_i (i=1,2,···, m) is required to be in the region of the sample data.
2. The slope of any line segment L_i (i=1,2,···, m) is a positive value, and increases as i increases.
3. For any line segment L_i (i=1,2,···, m), its intersection point on y-axis decreases as i increases.

3   Data Preprocessing

3.1   Wild Value Elimination

The aim of the wild value elimination is to find out and abandon the abnormal point in the sample data sets. Each sample data have two values, which are the significant wave height and the square root of SNR. Firstly, definition is given as follows:

Definition 1. Minimum Neighborhood Radius

For any sample data point i (i=1,2,···, N), there exists a radius r, with the neighborhood points are distributed on the circle centered at point i and with the radius r. Radius r is represented as the minimum neighborhood radius of point i. In reality, radius r is calculated by the length between point i and its nearest neighborhood point j ($j \neq i$).

The concentration of the sample data is reflected through the minimum radius of neighborhood. The smaller the radius r is, the higher concentration the data near the point i is. If the minimum neighborhood radius of point i is significantly greater than other sample data, point i is regarded as a wild value. According to the analysis above, a method of wild value elimination based on the minimum neighborhood radius is proposed in this paper, the process of which is given as follows.

Step 1. Calculate the minimum neighborhood radius value for each sample data.

Step 2. Rearrange the sample data points from small to large according to the minimum neighborhood radius value.

Step 3. Search the jump points of the minimum neighborhood radius value from the minimum point to the maximum point.

Step 4. Set the jump point's minimum neighborhood radius value R as the threshold, and the sample data with the minimum neighborhood radius value greater than R are eliminated as wild value.

3.2   Data Sparseness

The inversion model of significant wave height is mainly affected by the dense region of sample data. However, the sample data acquired by test is not uniform distributed in reality. If the sample data is directly used in the inversion model, it will lead to inaccurate modeling. In view of this problem, the sample data is sparse before modeling the significant wave height. The process is given as follows.

Step 1. Determine the minimum neighborhood radius R_0 as the threshold of the data sparseness.

Step 2. Draw a circle which the center is the any point in the sample data and the radius is R_0. The points in the circle are eliminated, and then the center point is saved in the sparse data set.

Step 3. Judge whether there are any points in the sample data, if not, terminate the sparse process, otherwise return to step 2.

Assume the range of minimum neighborhood radius is $[R_{min}, R_{max}]$ after eliminating the wild value in the sample data, then we have $R_{min} \leq R_0 \leq R_{max}$. The range of minimum neighborhood radius is $[R_0, R_{max}]$ after data sparse. If $R_0 = R_{max}$, the intensive degree of the data set is basically identical.

4   Modeling Method Design

4.1   Particle Encoding

Piecewise linearization model is used to approximate the significant wave height inversion model, which is shown in Fig. 1. Each line is given in equation (1), and the particle in PSO is encoded as equation (2).

$$X_i^t = \{x_{i,j,A}^t, x_{i,j,B}^t\} \ . \tag{2}$$

where $i = 1, 2, \cdots, n$ represents n particles; $j = 1, 2, \cdots, m$ represents the number of piecewise in the model; A is the intercept parameter; B is the slope parameter; t is the iteration number of the PSO.

4.2   Initialize Settings

According to the sample data distribution, it can be found that slope of L_i in the piecewise linearization model increase as square root of SNR increase. In order to improve the algorithm's modeling speed, the position X_i^0 and velocity V_i^0 of particle i are initialized by equation (3) and equation (4) respectively.

$$X_i^0 = \begin{cases} x_{i,j,A}^0 = A_{max} - \dfrac{(rand + j - 1) \cdot (A_{max} - A_{min})}{m} \\ x_{i,j,B}^0 = B_{min} + \dfrac{(rand + j - 1) \cdot (B_{max} - B_{min})}{m} \end{cases} . \tag{3}$$

$$V_i^0 = \begin{cases} v_{i,j,A}^0 = rand \cdot v_{max} \\ v_{i,j,B}^0 = rand \cdot v_{max} \end{cases} . \tag{4}$$

where rand is a random number in [0, 1]; A_{max} and A_{min} represent the maximum intercept and the minimum intercept respectively, and $\min(Hs) \leq A_{max} \leq \max(Hs)$,

Hs is the real values of significant wave height; B_{max} and B_{min} represent the maximum and the minimum value of slope respectively, and $0 \leq B_{min} \leq B_{max}$; $v_{max} > 0$ is the particle's maximum velocity.

4.3   Fitness Function

The fitness of swarm particles is computed and evaluated. The particle is improved when the present fitness value is smaller than the last fitness value. The fitness function is shown in equation (5).

$$F(X_i^t) = \sum_{j=1}^{m} \sum_{k=1}^{N_j} \left(\left(x_{i,j,A}^t + x_{i,j,B}^t \cdot \sqrt{SNR_{jk}} \right) - Hs_{jk} \right)^2 + \delta \cdot P . \tag{5}$$

where N_j is the amount of data, which belong to the j-th line region in the sample data; $\sqrt{SNR_{jk}}$ and Hs_{jk} represent the square root of signal-to-noise ratio and the real values of significant wave height respectively; δ is the penalty coefficient; P is the penalty value.

The penalty value P is calculated by equation (6), which account for the heuristic requirements that are given in section 2.

$$P = p_1 + p_2 + p_3 . \tag{6}$$

where p_1, p_2, and p_3 represents the violation degree of heuristic requirements, which is evaluated through the violation number.

5   Simulation and Analysis

The proposed modeling method is performed with Matlab environment and basic PSO. Simulation is performed with the data acquired from the experiment in Fujian Pingtan on October, 2009. SNR of the radar image is obtained by calculating the radar image acquired and real values of significant wave height is set as the values obtained by WamosII in the experiment.

5.1   Data Preprocessing

There are 1386 groups of available SNRs of radar image and significant wave heights from WamosII. The relationship between real values of significant wave height and the square root of SNR is shown in Fig. 2(a). According to values of minimum neighborhood radius, the sample data points are rearranged from small to large. The relationship is shown in Fig. 2(b).

The threshold of minimum neighborhood radius is set according to section 3.1, and the wild value is eliminated while the threshold $R = 0.1225$ from Fig. 2(b). Then set $R_0 = R_{max} = 0.1225$ and spare the sample data. The wild value elimination result and the data sparseness result are shown in Fig. 3(a) and Fig. 3(b) respectively.

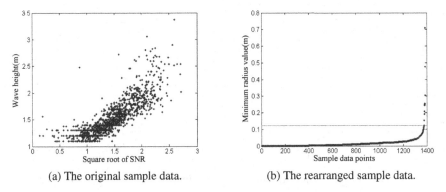

(a) The original sample data.   (b) The rearranged sample data.

Fig. 2. Relation diagram of the sample data

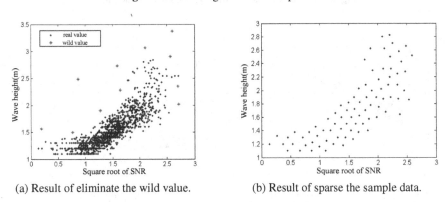

(a) Result of eliminate the wild value.   (b) Result of sparse the sample data.

Fig. 3. Result of the data preprocessing

5.2   Simulation Modeling

In order to compare the efficiency of the proposed inversion method with the traditional inversion method, simulation experiment is performed under the preprocessed data. The parameters used in the experiments are shown in Table 1.

Table 1. Parameters in experiments

Parameter	Value
N	100
c_1	1.4962
c_2	1.4962
ω_{min}	0.4
ω_{max}	0.9
v_{max}	3
A_{min}	1.5
A_{max}	-1.5
B_{min}	0
B_{max}	3
δ	100

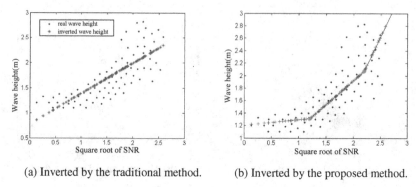

(a) Inverted by the traditional method.　　(b) Inverted by the proposed method.

Fig. 4. Relation diagram of the significant wave height

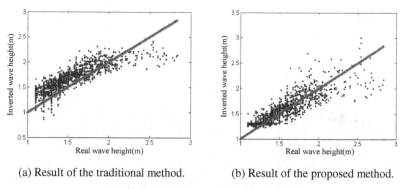

(a) Result of the traditional method.　　(b) Result of the proposed method.

Fig. 5. Regression chart of the wave height

The significant wave height inverted by the traditional method and the proposed method are shown in Fig. 4(a) and Fig. 4(b) respectively. Regression charts between the inversion wave height and the real wave height of both the traditional method and the proposed method are shown in Fig. 5(a) and Fig. 5(b).

From the simulation results, we can see that the significant wave height is represented as three piecewise linearization regions. Compared with the traditional method, stronger consistency between the inverted wave height and the real wave height can be obtained by the proposed method.

Inversion precisions of the two methods are evaluated by standard deviation and correlation coefficient. Calculation results are shown in Table 2.

Table 2. Performance comparison result

Inversion method	Standard deviation	Correlation coefficient
The tradition	0.2196	0.8490
This article	0.1777	0.8544

From the comparison of standard deviation and the correlation coefficient, we can see that smaller standard deviation and larger correlation coefficient can be obtained in this paper.

6   Conclusions

For the inversion method of significant wave height, a piecewise linearization method of significant wave height based on PSO is proposed. The concept of minimum neighborhood radius is proposed, and then the method of wild value elimination and data sparseness is given. Piecewise area division and parameter optimization of the model based on PSO are designed. Simulation performed and the results show that a better adaption and higher inversion precision are obtained in this paper.

Acknowledgment. This work has been financially supported by the National Natural Science Foundation of China under the Grant 51009036 and 51109041.

References

1. Young, R., Rosenthal, W., Ziemer, F.: A Three Dimensional Analysis of Marine Radar Images for the Determination of Ocean Waves Directionality and Surface Currents. J. Geophys. Res. 90, 1049–1059 (1985)
2. Ziemer, F., Rosenthal, W.: On the Transfer Function of a Shipborne Radar for Imaging Ocean Waves. In: Proc. IGARSS 1987 Symp., Ann Arbor, Michigan (May 1987)
3. Ziemer, F.: Directional Spectra from Shipboard Navigation Radar during LEWEX, Directional Ocean Wave Spectra. The John Hopkins University
4. Ziemer, F., Rosenthal, W.: Measurement of Two-Dimensional Wave Energy Spectra during SAXON-FPN 1990. In: Ocean 1993 Conference Proceedings, vol. 2, pp. 326–331 (1993)
5. Ziemer, F., Rosenthal, W., Richter, K., Schrader, D.: Two-Dimensional Wave Field Measurements as Sea Truth for ERS-1 SAR. In: Wooding, M. (ed.) Proceedings of Workshop ERS-1 Geophysical Validation, Penhors, Bretagne, France. ESA WPP-36, April 27-30, pp. 35–40 (August 1992)
6. Ziemer, F., Günther, H.: A System to Monitor Ocean Wave Fields. In: Proceeding of the Second International Conference on Air-Sea Interaction and Meteorology and Oceanography of the Coastal Zone, Lisbon, Portugal, pp. 22–27 (September 1994)
7. Alpers, W., Hasselmann, K.: Spectral Signal to Clutter and Thermal Noise Properties of Ocean Wave Imaging Synthetic Aperture Radars. International Journal of Remote Sensing 3, 423–446 (1982)
8. Duan, H.M., Wang, J.: An Improved Method on the Wave Height of Ocean Surface Based on X-Band Radars. Marine Science Bulletin 28(2), 103–107 (2009)
9. Liu, L., Fan, Z., Tao, C.: An Inversion Method of Significant Wave Height Based on Radial Basis Function Neural Network. In: 2011 Fourth International Joint Conference on Computational Sciences and Optimization, KunMing, China, pp. 965–968 (June 2011)

Design Fuzzy Logic Controller
by Particle Swarm Optimization for Wind Turbine

Nasseer K. Bachache and Jinyu Wen

College of Electrical and Electronic Engineering,
Huazhong University of Science and Technology (HUST), Wuhan 430074, China
tech_n2008@yahoo.com

Abstract. In this work the Particle Swarm Optimization (PSO) is utilized to framing the optimal parameters of Fuzzy Logic Controller FLC, this parameter is (centers and width) of triangle membership functions, the proposed method can design a robust controller to govern the speed of wind turbine WT, adjusting pitch angle of blade can regulate the output power of WT at a wide range of wind speed, the mean objective of this work is to make the operation of WT works as like as traditional motivator used in power system. By SIMULINK-MATLAB we implement the complete mathematical model of the system. The simulation results demonstrate that the Optimized Fuzzy Logic Control (OFLC) gets a better parameters of fuzzy sets using PSO, and realizes a good dynamic behavior compared with conventional FLC.

Keywords: Particle Swarm Optimization PSO, Fuzzy Logic Control FLC, Pitch Angle of Wind Turbine PAWT.

1   Introduction

Nowadays, there are widely adopted on bulky wind energy systems in power system, but its stochastic generation is a big problem. Pitch-controlled system is normally used in a large wind turbine to adjust blades in the rotor hub. It tends to decrease the angle attack causing a reduction in the pressure difference between (front and back) around the blade, that leads to lifting force of the blade is reduced too, vice versa, Figure(1) shows the cross section area of wind turbine blade aerodynamics and angle of attack. This system provides mechanical power of the turbine operating tightly controlled. Therefore, a controllable power generation can obtain during normal operation (rated wind speed). Adjustable-pitch drive system consisted of two parts: first (mechanism actuators) which is achieved either by an electric device (servo-motor) or hydraulic equipment; second using (regulation drive controller) to regulate the pitch angle of a rotor blade, fundamentally its control strategies have a tendency to solve the nonlinearity, time variation, randomness and uncertainties, fuzzy logic controller FLC is a powerful techniques in control applications of the complex system [1]. Recent year there are many solutions devoted to design evolutions of aerodynamic control; (variable-speed adjustable-pitch) is used in variety wind velocity to raise the output power. An adaptive sliding mode control is combined with

Y. Tan, Y. Shi, and H. Mo (Eds.): ICSI 2013, Part I, LNCS 7928, pp. 152–159, 2013.

neural network in a high wind velocity condition proposed in [2], and adaptive Fuzzy Sliding-mode control in variable speed in [3], the validation of H∞ control adjustable-pitch at a rated value was studied in [4], and a self-tuning fuzzy-PID position controller in [5], and an Elman Neural Network based control algorithm [6], certainly a few can used in practical industrial applications. In this paper the FLC is proposed to solve a highly nonlinear system. This controller suffers from the tuning of its parameters (number of membership functions and its type, rule number, and formulating rules). The tuning of scaling factors for this parameter is done either by interactively method (trial and error) or by human expert [7]. Therefore, the tunings of the FLC parameters are necessitated need to effective tuning process preferable similar to new intelligent optimization techniques, such as Genetic Algorithms (GA), Ant Colony Optimization (ACO), Simulated Annealing (SA), and Bacteria Foraging Optimization (BFO). But a well-studied and has a proven optimizer in a high potential and global optimization is Particle Swarm Optimization (PSO) algorithm [8], In this paper the generating of fuzzy controller parameters is designed by a modern intelligent algorithm PSO.

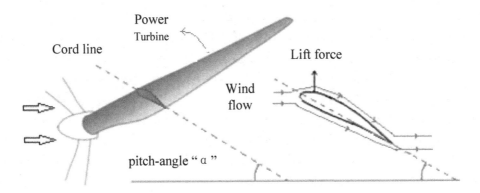

Fig. 1. Wind turbine blade aerodynamics Lift Force "FL" and the pitch-angle "α" and power turbine

2   Turbine Mechanism Model

A mechanical wind turbine aerodynamic formulated on "Betz theory", the wind power received from air (P_T) and pneumatic torque turbine (T_T) are prearranged by equations (1, 2) respectively.

$$P_T = \frac{\rho}{2} . v^3 . A_T . C_p(\lambda, \alpha) \tag{1}$$

$$T_T = \frac{\rho}{2.\lambda} . v^3 . A_T . C_p(\lambda, \alpha) \tag{2}$$

Where ρ is the air density, v is the wind velocity, A_T is the rotor disk area of blade rotation and C_p is a turbine characteristic, "α" is the pitch-angle on pitch-angle which illustrated in figure 1, ratio λ between the blade-point speed ω and the wind velocity represented in equation (3) [9], where: K is the relationship between blade tip speed and turbine rotor speed. There are many expressions for C_p equation (4) one of them [7].

$$\lambda = \frac{K}{v} \tag{3}$$

$$C_p(\lambda, \alpha) = c1 \left(\frac{c_2}{\lambda + c_3 \alpha} - \frac{c_2 c_9}{1 + \alpha^3} - c_3 - c_4 \alpha^{c_5} - c_6 \right)^{\left(\frac{c_7}{\lambda + c_8 \alpha} \frac{c_7 c_9}{1 + \alpha^3} \right)} + c_{10} \lambda \tag{4}$$

The values of C1 to C10 usually depend on turbine manufacturing. In this work we take the expression of $C_p(\lambda, \alpha)$ enthusiastically as a nonlinear function.

$$C_p(\lambda, \alpha) = 0.22 \left(\frac{116}{\lambda_i} - 0.4\alpha - 5 \right) e^{\left(\frac{-125}{\lambda_i} \right)} \tag{5}$$

$$\frac{1}{\lambda_i} = \frac{1}{\lambda + 0.08\alpha} - \frac{0.035}{1 + \alpha^3} \tag{6}$$

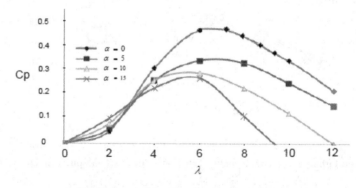

Fig. 2. Torque coefficient characteristics

The power coefficient C_p, is a highly non-linear function of the tip speed ratio λ and blade-pitch angle as shown in Figure (2) and equation (5), the large wind turbine depends on adjusting blade angle to maintain the rotational speed of the wind turbine; and output power will be around its rated value. Equation (7) is the dynamic equation of wind turbine where: (T_g) Is the electromagnetic torque of generating power and J is the moment of inertia.

$$J \frac{d}{dt} = T_T - T_g \tag{7}$$

The adjustable-pitch actuator model is achieved by electric servo motor apparatus, which can be expressed as a first order system equation (8), where α_c is the desired angle and τ is the time delay.

$$\alpha_c = \tau\alpha + \frac{d\alpha}{dt} \tag{8}$$

3   Particle Swarm Optimization

Particle swarm optimization (PSO) is a computation technique first proposed in 1995 by Kennedy and Eberhart [8]. This method has been found to be a robust method in solving non-linearity or non-differentiability problems, the PSO algorithm didn't use evolutionary operators (mutation or crossover to manipulate algorithms). However, it simulates a dynamic population behavior of (fish swarm or bird flocks), where the social sharing of information takes place and individuals can profit from the discoveries and previous experience of all the other companions during the search for food. Thus, each companion is called particle and the population is called a swarm, it is assumed to fly in many directions over the search space in order to meet the demand fitness function. For n-variables of the problem need to get its optimum; a flock of particles is put into the n-dimensional search space with randomly chosen velocities and positions knowing their best values, then (Pbest) is the best position in the n-dimensional space. The velocities of each particle are adjusted accordingly to its own flying experience and the other particles flying experience. For the *nth* particle and n-dimensional space can represented as an equation (9), the best previous position of its particle is recorded as equation (10).

$$x_i = \left(x_{i,1}, x_{i,2}, \ldots \ldots x_{i,n}\right) \tag{9}$$

$$P_{best_i} = \left(P_{best_{i,1}}, P_{best_{i,2}}, \ldots \ldots P_{best_{i,n}}\right) \tag{10}$$

The velocity is an essential part of how PSO work so as modified velocity and position of each particle can be calculated using the current velocity and distance from $(P_{best_{i,d}})$ to (g_{best_d}) as shown in equations. (11, 12).

$$V_{i,m}^{(It.+1)} = W * V_{i,m}^{(It.)} + c1 * r * \left(P_{best_{i,m}} - x_{i,m}^{(It.)}\right) + c2 * r * \left(g_{best_m} - x_{i,m}^{(It.)}\right) \tag{11}$$

$$x_{i,m}^{(It.+1)} = x_{i,m}^{(It.)} + v_{i,m}^{(It.)} \tag{12}$$

Where i:Number of particles, m: Dimension, It.: Iterations pointer, $V_{i,m}^{(It.)}$:Velocity of particle no. i at iteration It, W:Inertia weight factor, c1, c2: Acceleration constant, r: Random number between(0-1), $x_{i,m}^{(It.)}$: Current position of particle i at iteration It., P_{best_i} : Best previous position of ith particle, g_{best_m}: Global best particle among all the particles in the population.

4   PSO Implementation Adapts FLC

The selection of Membership Functions (MFs) of the input and output variables and the determination of fuzzy rules are not available; for a sensitive system it is very difficult to designing that parameter. There is no formal framework for the choice of the parameters of FLC. The conventional method (trial-and-error method) can be used in such situations. In general the tuning and learning models have become an important subject of fuzzy control. There are two input signals to the fuzzy controller: $\Delta\omega$ and ΔP_T, the function of the fuzzy controller is to observe the pattern of the power loop error signal to updating the pitch control signal, A simple fuzzy logic controller of two inputs and one output. This FLC has three triangle memberships for each input and five memberships for output and nine "if" statement rules. PSO was utilized off line to design positions of triangle shape for input/output memberships. The complete system simulation using MATLAB/SIMULINK program is presented including simulated systems with servomechanism model as shown in figure (3); But, the optimization algorithm is implemented by using MATLAB/m-file program and linked with the system simulation program MATLAB/SIMULINK the performance of the system must be examined in each iteration and particles position during the optimization algorithm, the flow chart of this work shown in figure 4. The optimization criteria (Integrated of Time Weight Square Error ITSE) equation (13) is used to evaluate accuracy performance of the fuzzy controller.

$$FF = ITSE = \int_0^t t * e^2(t)dt \qquad (13)$$

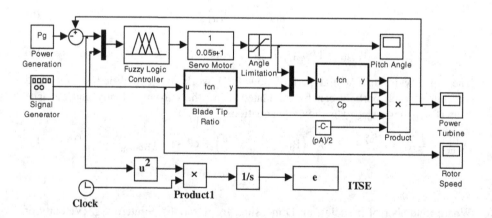

Fig. 3. Simulation of pitch angle system controlled by FLC

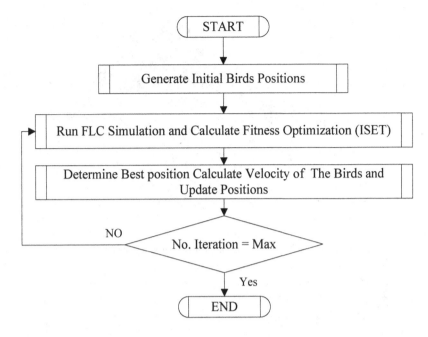

Fig. 4. Flowchart of PSO algorithm

5   Simulation and Result

In figure (5) the two inputs and one output FLC memberships designed by PSO, and its surface shown in figure (6), in figure (7) the power turbine and the pitch angle at a variable speed controlled by FLC designed by traditional and in Figure (8) the optimize FLC designed by PSO algorithm is shown.

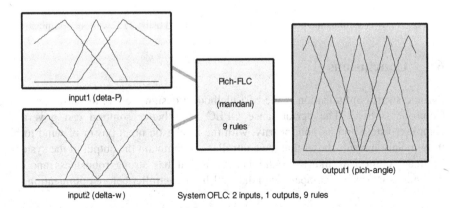

Fig. 5. FLC memberships designed by PSO

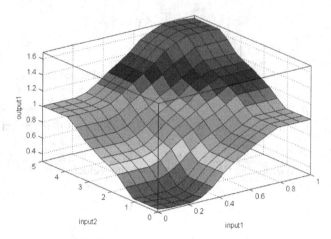

Fig. 6. FLC surface for two inputs one output

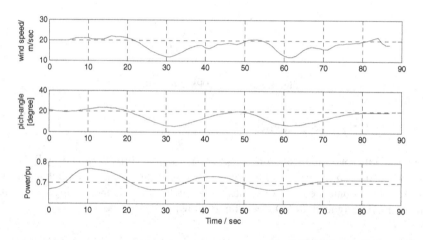

Fig. 7. Power turbine and pitch angle at a variable speed using FLC without optimization

6 Conclusions

Particle swarm optimization is the best method to design the Optimized Fuzzy Logic Controller OFLC, The performance OFLC as a better solution can govern the complex system and its nonlinearity, with the stochastic input power of wind turbine the simulation result shows that the controller can maintain the output of the system to be under control as like as traditional turbine, it has strong robustness and good dynamic performance compare with the well-built controller before optimization.

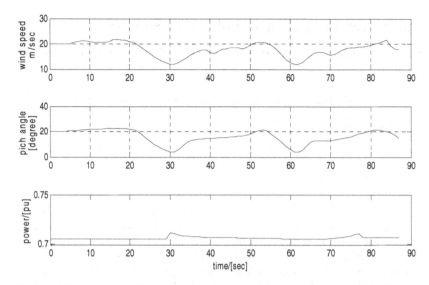

Fig. 8. Power turbine and pitch angle at a variable speed using FLC optimized by PSO

References

1. Wu, B., Lang, Y., Navid, Z., Samir, K.: Power Conversion and Control of Wind Energy System. A John Wiley & Sons, Inc., Canada (2011)
2. Jiao, B., Wang, L.: RBF Neural Network Sliding Mode Control for Variable Speed Adjustable Pitch System of Wind Turbine. In: International Conference on Electrical and Control Engineering, China (2010)
3. Yau, X., Liu, Y., Guo, C.: Adaptive Fuzzy Sliding-mode Control in Variable Speed Adjustable Pitch Wind Turbine. In: IEEE International Conference on Automation and Logistics, China (2007)
4. Guo, H., Guo, Q.: H∞ Control of Adjustable-Pitch Wind Turbine Adjustable-Pitch System. In: IEEE 5th International Conference Power Electronics and Motion Control, Slovenia (2006)
5. Dou, Z.L., Cheng, M.Z., Ling, Z.B., Cai, X.: An Adjustable Pitch Control System in a Large Wind Turbine Based on a Fuzzy-PID Controller. In: International Symposium on Power Electronics, Electrical Drives, Automation and Motion, Italy (2010)
6. Lin, W., Hong, C.: A New Elman Neural Network Based Control Algorithm for Adjustable-Pitch Variable-Speed Wind-Energy Conversion Systems. IEEE Trans. on Power Electronics 26(2) (2011)
7. Isaac, I.A., Cabrera, D., Pizarro, H., Giraldo, D., Gonzalez, J.W., Biechl, H.: Fuzzy Logic Based Parameter Estimator for Variable Speed Wind Generators PI Pitch Control. In: International Conference on Fuzzy Systems, Spain, pp. 18–23. IEEE Press (2010)
8. Lawrence, K.L., Josiah, L.M., Alex, H.: Particle Swarm Optimized T-S Fuzzy Logic Controller for Maximum Power Point Tracking in a Photovoltaic System. In: 35th Photovoltaic Specialists Conference, pp. 89–94. IEEE Press, Hawai'i (2010)
9. Olimpo, A.L., Janaka, E., Phill, C., Mike, H.: Wind Energy Generation Modelling and Control, pp. 4–6. A John Wiley and Sons, Ltd. (2009)

Parameter Identification of RVM Runoff Forecasting Model Based on Improved Particle Swarm Optimization

Yuzhi Shi, Haijiao Liu, Mingyuan Fan, and Jiwen Huang

Water Resources Research Institute of Shandong Province, Jinan 250013, China
{syz101066,liuhaijiao2005}@163.com,
{fantina715,sdskyhjw}@126.com

Abstract. Runoff forecasting which subjects to model pattern and parameter optimization, has an important significance of reservoir scheduling and water resources management decision-makings. This paper proposed a new forecasting model coupled phase space reconstruction technology with relevance vector machine, and its model parameters is optimized by an improved PSO algorithm. The monthly runoff time series from 1953 to 2003 at Manwan station is selected as an example. The results show that the improved PSO has efficient optimization performance and the proposed forecasting model could obtain higher prediction accuracy.

Keywords: Improved PSO algorithm, Relevance vector machine, Phase space reconstruction, parameter identification, Runoff forecasting.

1   Introduction

Runoff forecasting is very important for reservoir control, water resources planning and management. However, the hydrology system is a highly complex nonlinear system composed of uncertain and deterministic parts under the influence of rainfall system and underlaying surface system[1]. It is difficult to describe it in terms of rigorous physical model. So the data-driven model has become the important model in practice. Many innovated models, such as uncertain reasoning model(RM)[1], Artificial Neural Network (ANN)[2], support vector machine (SVM)[3] are gradually introduced into the hydrological forecasting, and further develop its applications. Tipping[4]puts forward sparse probability model (Relevance Vector Machine, RVM) on basis of SVM and Bayesian theory, this method has been used in the fields of image analysis [5,6], channel equalization [7],etc. and obtained effective performance. The researches show that there two aspects should be mainly involved as: (1) is runoff relevance vector machine choice; (2) is model parameters optimization identification.

Generally, relevance vector based on time series is built in sequence, which is lack of physical basis, this article applied phase space reconstruction technique[1,8] to construct the relevance vectors, and its model parameters is identified by the improved PSO algorithm, which firstly proposed by Kennedy and Eberhart [9] based on the social behavior metaphor, and has been widely applied in global optimization problems as

Y. Tan, Y. Shi, and H. Mo (Eds.): ICSI 2013, Part I, LNCS 7928, pp. 160–167, 2013.
© Springer-Verlag Berlin Heidelberg 2013

well as GA, EA, DE, ACO optimization algorithms. There are many researches[10-13] illustrated PSO algorithm has effectively optimization performance, however, the standard PSO like the others' is easy to entrap the local best fitness, so the crossover and mutation algorithms is developed to expand the search space and applied to the proposed model parameters identification.

This article mainly includes five parts: In Sect.2 Briefly introduces improved PSO algorithms. In Sect.3 Build relevance vector machine runoff forecast model. In Sect.4 Identify RVM model parameter. In Sect.5 Application. In Sect.6 Conclusions.

2   PSO Algorithm

The PSO algorithm is initialized with a population of random candidate solutions, conceptualized as particles. Each individual in PSO algorithm is assigned random velocity in search space, and is iteratively updating according to its own local best fitness and its global best fitness, which is attracted by the particle locations. Each individual in the particle swarm is composed of D-dimensional vector $x_i = [x_{i1}, x_{i2}, \cdots, x_{id}]$ and the ith particle velocity $v_i^k = [v_{i1}^k, v_{i2}^k, \cdots, v_{id}^k]$. During each iteration, the ith particle is updated by the following two best values: $p_i^{best} = [p_{i1}, p_{i2}, \cdots, p_{id}]$, which is the local best value of the ith particle has been achieved so far, and $g^{best} = [g_1, g_2, \cdots, g_d]$, which is the global best value obtained in the swarm so far. Each particle is updated iteratively by

$$v_{id}^k = w_i \cdot v_{id}^{k-1} + c_1 \cdot rand_1 \cdot (p_{id}^{k-1} - x_{id}^{k-1}) + c_2 \cdot rand_2 \cdot (g_d^{k-1} - x_{id}^{k-1}) \tag{1}$$

$$x_{id}^k = x_{id}^{k-1} + v_{id}^k \tag{2}$$

Where $c_1 = c_2$ are acceleration coefficients, w_i is the ith weight, $rand_1$ and $rand_2$ are two independent uniform random number within the range of [0,1]. To ensure convergence and be made much more stable, appropriate values is proposed by Kennedy and Eberhart with $c_1 = c_2 = 2$ and $w_i \in [0.5, 1.4]$, $v_{id} \in [-v_d^{max}, v_d^{max}]$.

Generally, PSO algorithm likes as other evolutionary optimization algorithms, which is easily fall into local optimum. In order to solve the problem and expand particles search space, the crossover and mutation algorithm are applied to improve the search space. r pair of particles are selected randomly from the k-1th iteration to crossover each other, The crossover algorithm is represented as

$$\begin{cases} x_i = rand \cdot x_i + (1 - rand)x_j \\ x_j = rand \cdot x_j + (1 - rand)x_i \end{cases} \tag{3}$$

Where $rand$ is the uniform random number within the range of [0,1].

While the mutation algorithm of the particle velocity is written as[15]:

$$v_{id}^k = \begin{cases} w_i \cdot v_{id}^{k-1} + c_1 \cdot rand_1 \cdot (p_{id}^{k-1} - x_{id}^{k-1}) + c_2 \cdot rand_2 \cdot (g_d^{k-1} - x_{id}^{k-1}) & rand_3 < c_3 \\ 0 & rand_3 \geq c_3 \end{cases} \tag{4}$$

Where $rand_3$ is the uniform random number in the range of $[0,1]$, c_3 is mutation rate. When $c_3 = 1$, the proposed method is the same as original PSO.

3   RVM Runoff Forecasting Model

Given the relevance vectors $\{X_i : i = 1, 2, \cdots, N\}$, Set training sample $\{X_i, y_i\}_{i=1}^N$, the relevance vector machine makes predictions based on regression function of the form:

$$\hat{y} = \sum_{i=1}^N w_i K(X^*, X_i) + w_0 \tag{5}$$

Where $\{w_i\}$ are model weights and $K(\cdot, \cdot)$ is kernel function.

Given runoff time series $\{x_i : i = 1, 2, \cdots, n\}$, according to phase space reconstruction method[1,8], the relevance vector $X_i = [x_i, x_{i+\tau}, x_{i+2\tau}, \cdots, x_{i+(m-1)\tau}]^T$ is made by selecting appropriate values of time lag τ and embedding dimension m, which meets the condition of $m \geq 2D + 1$, where D is saturated correlation dimension calculated by formula(6). The reconstruction system is composed of N relevance vectors $\{X_i : i = 1, 2, \cdots, N\}$, where $N = n - (m-1)\tau$. The saturated correlation dimension D is calculated by

$$D = \lim_{r \to 0} \lim_{m \to \infty} \frac{\partial \ln C(r, m)}{\partial \ln r} \tag{6}$$

Where $C(r, m) = \dfrac{2}{(N+1)N} \sum_{i=1}^N \sum_{j=i+1}^N H(r - \| X_i - X_j \|)$, $H(x) = \begin{cases} 0 & if \quad x \leq 0 \\ 1 & if \quad x > 0 \end{cases}$, r is scale, $\| \cdot \|$ is Euclidean norm.

Set the training sample $\{X_i, y_i\}_{i=1}^N$, and assume $P(y \mid X)$ is Gaussian $N(y \mid \hat{y}, \sigma^2)$. The $\hat{y}$ is defined as formula (5), the likelihood of the dataset [4] is written as

$$p(y \mid w, \sigma^2) = (2\pi\sigma^2)^{-N/2} \exp\{-\| y - \Phi w \|^2 / (2\sigma^2)\} \tag{7}$$

Where $y = (y_1, y_2, \cdots, y_N)$, $w = (w_0, w_1, \cdots, w_N)$, Φ is $N \times (N+1)$ demension kernel function matrix, $\Phi_{n,n+1} = K(X_n, X_n)$, $\Phi_{n,1} = 1$. It is indicated that maximum likelihood estimation of w, σ will lead to overfitting, so the Gaussian prior of w is shown as

$$p(w \mid \alpha) = \prod_{i=0}^N N(w_i \mid 0, \alpha_i^{-1}) \tag{8}$$

Then on the base of Bayesian rule, the posterior probability distribution of w is denoted as

$$p(w \mid y,\alpha,\sigma^2) = (2\pi)^{-(N+1)/2} \mid \Sigma \mid^{1/2} \exp\{-(w-\mu)^T \Sigma^{-1}(w-\mu)/2\} \tag{9}$$

Where $\Sigma = (\sigma^2\Phi^T\Phi + A)^{-1}, \mu = \sigma^2 \Sigma\Phi^T y$ are posterior covariance and mean respectively, $A = diag(\alpha_0,\alpha_1,\cdots,\alpha_N)$.

The EM algorithm[4]is used to maximize the marginal likelihood distribution function as follows

$$p(y \mid \alpha,\sigma^2)=N(0,\sigma^2 I + \Phi A^{-1}\Phi^T) \tag{10}$$

And then hyper-parameters $\{\alpha_n\},\sigma$ are calculated iteratively by

$$\alpha_i^{new} = \frac{1}{\Sigma_{ii} + \mu_i^2} \tag{11}$$

$$(\sigma^2)^{new} = \parallel y - \Phi\mu \parallel^2 /(N - \sum_i \gamma_i) \tag{12}$$

Where $\gamma_i = 1 - \alpha_i\Sigma_{ii}$ and Σ is covariance matrix.

It is noted that the hyper-parameter $\alpha_{MP},\sigma^2_{MP}$ is obtained when reaches the convergence and $w = \mu_{MP}$, and input vector X^* into the optimized RVM model, the mean and variance of RVM forecasting will be achieved, its formula is described as

$$\mu^* = \Phi(X^*)\mu_{MP}, \sigma^2_* = \sigma^2_{MP} + \Phi(X^*)^T \Sigma\Phi(X^*) \tag{13}$$

4   RVM Parameters Identification

In order to optimize the RVM model parameters, three important aspects should be mentioned: (1) kernel selection; (2) parameter optimization objective function; (3) parameters in the model.

(1)The kernel function of RVM model must satisfy Mercers' condition[2]. Many researches indicated that Radial basis kernel and Sigmoid Kernel are more effective in regression and classification problems. Radial basis kernel has the same function as sigmoid kernel, however, it has less parameters and so radial basis kernel is selected in the RVM model.

(2) In general, the objective function of runoff forecasting is formed on the basis of training sample fitting error evaluated the desired optimization fitness, however, the convergence process will make training fitting error close to zero and leads to serious overfitting. It is indicated that using parameters optimized to runoff forecasting during the validation period will be obtained unreasonable results, thus a new method is proposed, which comprehensively considers training and test data set errors to establish objective function, which is described by

$$minimize F = (\frac{N_2}{N_1 + N_2})\sqrt{(\sum_{i=1}^{N_1} (y_i - \hat{y}_{train})^2)} + (\frac{N_1}{N_1 + N_2})\sqrt{(\sum_{i=1}^{N_2} (y_i - \hat{y}_{test})^2)} \qquad (14)$$

Where N_1, N_2 are number of training and test sets respectively.

(3) RVM model contains phase reconstruction parameter m, τ, kernel function scale ε and model hyper-parameters $\{\alpha_n\}, \sigma$. The hyper-parameters are optimized iteratively by EM algorithm, and m, τ, ε are calculated by improved PSO. The general steps of model parameters optimization identification as follows:

Step1: Set m, τ, ε parameter ranges and initialize the particle population size N, Maximum iterations T, maximum updating velocities v_d^{max}, particle random positions x_i^0 and velocities v_i^0;

Step2: In the kth particle loop, reconstruct the relevance vector, and initialize hyper-parameters $\{\alpha_n\}, \sigma$, adopt formula (11) and (12) to evaluate iteratively the hyper-parameters α_i, σ in the EM interior loop, if the EM algorithm convergence criterion is met, end the kth loop, $w = \mu_{MP}^T$ is obtained, and then go to the $k+1$th outside loop.

Step3: For each particle, calculate the fitness functions $F(x_i)$ with formula(4),(2),(14), and compare to local best fitness $F(p_i^{best})$ achieved so far, if $F(x_i) < F(p_i^{best})$, then set the $F(p_i^{best})$ equal to the current valve, and p_i^{best} equal to the current location, that is to say $F(p_i^{best}) = F(x_i)$, $p_i^{best} = x_i$;

Step4: Compare the local best fitness $F(p_i^{best})$ to the global best fitness $F(g^{best})$ achieved so far, if $F(p_i^{best}) < F(g^{best})$, and then set the $F(g^{best})$ equal to the current value, and g^{best} equal to the current location, that is to say $F(g^{best}) = F(p_i^{best})$, $g^{best} = p_i^{best}$;

Step5: If a criterion is not met the sufficient good fitness or a maximum number of iteration, go to step2, or else end loop.

5   Application

The monthly runoff time series from January, 1953 to December, 2003 is selected to evaluate the proposed method at Manwan station. It covers the basin area of 114,500km^2, and the years average monthly runoff is 1251m^3/s, variation coefficient is 0.717, the maximum and minimum monthly runoff are 5000m^3/s, 248m^3/s respectively. Runoff time series is divided into three parts: (1) training data set, 41 years from 1953 to 1993; (2) test data set: 5 years from 1994 to 1998; (3) validation data set, 5 years from 1999 to 2003. Both of training and test data set are applied to determine model parameters, and the remaining data set (validation) is used to test model forecasting performance. The indexes of mean absolute relative error (MARE), correlation coefficient (CR), deterministic coefficient R2 are selected to evaluated forecasting accuracy.

Set ε , m , τ ranges of $[0.1,100]$, $[1,20]$ and $[1,10]$ respectively and initialize $c_1 = c_2 = 2$, $v_1^{max} = v_2^{max} = v_3^{max} = 0.1$, hyper-parameters $\alpha(0) = (0.25, 0.25, \cdots, 0.25)$, $\sigma_0^2 = \mathrm{var}(y)*0.01$, $\alpha_{max} = 1 \times 10^5$, maximum iterations $T=50$, population size $N=20$, weight $w_1^t = w_2^t = w_3^t = 0.9 - t*0.5/T$, where t is the current tth iteration. Finally parameters ε , m , τ optimized by PSO equal to $(2.1723,14,4)$. Hyper-parameter σ^2 , $\{\alpha_n\}$ are basically achieved after a few iterations. The results indicate that the EM algorithm has an effective performance and some researchers also directly make hyper-parameter as a fixed value in prediction.

The proposed model (RVM) is compared with least squares support vector machine model (LSSVM) widely applied in regression problems, relevance vector machine model (RVM*) without phase space reconstruction ($m = 12, \tau = 1$) and the automatic regression moving average model (ARMA) to analyze the runoff forecasting accuracy shown in Table 1, runoff hydrographs of forecasted and observed with RVM during training, test and validation periods are presented as Fig. 1-Fig. 2.

Table 1. Prediction accuracy of monthly flow from various methods

Model	Training period			Test period			Validation period		
	MARE /%	CR	R2	MARE /%	CR	R2	MARE /%	CR	R2
RVM	12.34	0.94	0.88	11.92	0.95	0.91	13.59	0.93	0.81
RVM*	15.62	0.92	0.83	17.58	0.89	0.79	14.93	0.92	0.75
LSSVM	13.76	0.93	0.86	13.37	0.93	0.85	14.31	0.92	0.74
ARMA(5,6)	18.27	0.91	0.72	17.75	0.89	0.72	17.07	0.93	0.70

Table 1 shows that forecasting accuracy of RVM with phase space reconstruction is higher than that of RVM* without phase space reconstruction with evaluation indexes of MARE, CR, R2 during training, test and validation periods. Moreover, compared with the other LSSVM and ARMA (5,6) models, it is also denoted that RVM has higher forecasting accuracy than the others', so it is proved actually that the proposed RVM model has a better prediction performance in runoff forecasting problem.

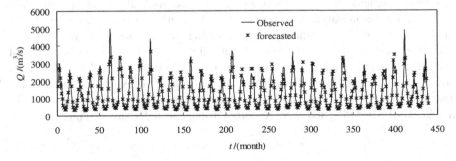

Fig. 1. Comparison between observed and RVM forecasted hydrograph during train period

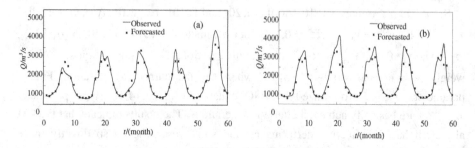

Fig. 2. Hydrographs of observed and predicted: (a) test period, (b) validation period

In order to further describe uncertainty of runoff forecasting, prediction interval is achieved under probability of 80%, as shown in Fig.3. It is indicated that the forecasting interval could be basically covers the observed, so the probability interval prediction with RVM model is reliable.

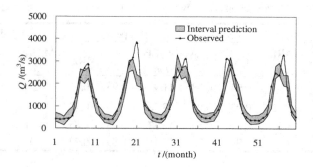

Fig. 3. Interval prediction during validation period

6   Conclusions

This article adopted an improved PSO to identify runoff forecasting model parameters, an improved PSO expand search space with crossover and mutation algorithm and facilitate the scheme to find an optimal parameter set efficiently, moreover, the RVM optimized by the improved PSO is performed efficiently to runoff prediction. A new runoff forecast model is established by coupling phase space reconstruction with relevance vector machine model, and compares it with other existing models, LSSVM and ARMA, its prediction accuracy was evaluated with indexes of MARE, CR, R2, it is proved that proposed model has higher forecasting accuracy, moreover, it could quantitatively represent forecasting uncertainty, and provide more information for flood control scheduling and water resources management decision making.

References

1. Shi, Y.Z., Zhou, H.C.: Research on monthly flow uncertain reasoning model based on cloud theory. Sci. China Tech. Sci. 53, 2408–2413 (2010)
2. Sivakumar, B., Jayawardena, A.W., Fernando, T.M.K.G.: River flow forecasting: use of phase-space reconstruction and artificial neural networks approaches. Journal of Hydrology 265, 225–245 (2002)
3. Yu, X.Y., Liong, S.Y., Babovic, V.: EC-SVM approach for real time hydrologic forecasting. Journal of Hydroinformatics 6(3), 209–223 (2004)
4. Tipping, M.E.: The relevance vector machine. Advances in Neural Information Processing System 12, 652–658 (2000)
5. Agarwal, A., Triggs, B.: 3D human pose from silhouettes by relevance vector regression. Computer Vision and Pattern Recognition 2, 882–888 (2004)
6. Bowd, C., Medeiros, F.A., et al.: Relevance Vector Machine and Support Vector Machine Classifier Analysis of Scanning Laser Polarimetry Retinal Nerve Fiber Layer Measurements. Investigative Ophthalmology & Visual Science 46, 1322–1329 (2005)
7. Chen, S., Gunn, S.R., Harris, C.J.: The relevance vector machine technique for channel equalization application. IEEE Trans on Neural Networks 12(6), 1529–1532 (2002)
8. Kantz, H., Schreiber, T.: Nonlinear time series analysis. Cambridge University Press, Cambridge (1997)
9. Kennedy, J., Eberhart, R.C.: Particle swarm optimization. In: Proc IEEE Conf. on Neural Networks, pp. 1942–1948. IEEE Press, Piscataway (1995)
10. Shi, Y., Eberhart, R.C.: A Modified Particle Swarm Optimizer. In: Proceedings of the IEEE International Conference on Evolutionary Computation, pp. 69–73. IEEE Press (1998)
11. Clerc, M., Kennedy, J.: The Particle Swarm: Explosion, Stability, and Convergence in a Multi-dimensional Complex Space. IEEE Transactions on Evolutionary Computation 6, 58–73 (2002)
12. Eberhart, R.C., Shi, Y.: Comparing Inertia Weights and Constriction Factors in Particle Swarm Optimization. In: Proceedings of the Congress on Evolutionary Computation, pp. 84–88 (2000)
13. Trelea, I.C.: The particle swarm optimization algorithm: convergence analysis and parameter selection. Information Processing Letters 85(6), 317–325 (2003)
14. Jiang, H.M., Xie, K., Ren, C., et al.: A Novel Particle Swarm Optimization with Stochastic Stagnation. Journal of Sichuan University (Engineering Science Edtion) 38(4), 118–121 (2006)

An Approach Based on Evaluation Particle Swarm Optimization Algorithm for 2D Irregular Cutting Stock Problem

Yan-xin Xu, Gen-Ke Yang, and Chang-chun Pan

Department of Automation, Shanghai JiaoTong University
and Key Laboratory of System Control and Information Processing,
Ministry of Education of China, 800 DongChuan Rd, MinHang District, Shanghai, China
{iamxuyanxin,gk_yang,pan_cc}@sjtu.edu.cn

Abstract. Cutting stock problem is an important problem that arises in a variety of industrial applications. An irregular-shaped nesting approach for two-dimensional cutting stock problem is constructed and Evolution Particle Swarm Optimization Algorithm (EPSO) is utilized to search optimal solution in this research. Furthermore, the proposed approach combines a grid approximation method with Bottom-Left-Fill heuristic to allocate irregular items. We evaluate the proposed approach using 15 revised benchmark problems available from the EURO Special Interest Group on Cutting and Packing. The performance illustrates the effectiveness and efficiency of our approach in solving irregular cutting stock problems.

Keywords: Cutting Stock Problem, EPSO, Grid Approximation.

1 Introduction

Cutting and Packing Problem (C&PP) are a large family of problems arising in a wide variety of industrial applications, including the cutting of standardized stock units in the wood, steel and glass industries, packing on shelves or truck beds in transportation and warehousing, and the paging of articles in newspapers. The objective of the packing process is to maximize the utilization of material. There are many classic cutting and packing problems, including cutting stock, trim loss, bin packing, strip packing, pallet loading, nesting, and knapsack problems etc. Figure 1 provides an example of a layout from the garment manufacturing industry.

In this paper, we focus on cutting stock problem (CSP) with convex and concave shapes. In CSP, a number of two-dimensional items must be cut from a couple of same stocks. The objective is to minimize the number of stocks. Using the typology of Wäscher, this is a Two-Dimensional Single Stock-Size Cutting Stock Problems (2DCSP) [1]. The 2DCSP has been proved to be NP-hard [2].

This paper is organized as follows. A brief review of previous work in the field is presented in Section 2. The proposed approach based on EPSO is introduced in detail in section 3. Section 4 gives experimental results on benchmark problems from the literatures that demonstrate the capabilities of the proposed approach. In Section 5, the research is concluded and possible issues for future work are suggested.

Y. Tan, Y. Shi, and H. Mo (Eds.): ICSI 2013, Part I, LNCS 7928, pp. 168–175, 2013.

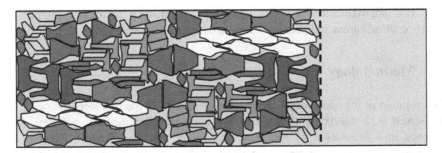

Fig. 1. An example layout from garment manufacturing

2   Literature Review

Different strategies for producing solutions to the irregular cutting stock problem have been presented according to their type and size. They include optimization approaches, (e.g. linear programming, column generation), heuristic and meta-heuristic approaches, and the emerging approaches combining these methods into a solution approach.

Due to the NP-complete nature of the problem, published solution approaches focus on heuristic and meta-heuristics methodologies. Heuristic placement strategy such as bottom-left (BL) and bottom-left-fill is proposed to supply a rule for shapes to be placed on sheet [3, 4]. Meta-heuristics are general frameworks for heuristics in solving combinatorial optimization problems. They include simulated annealing, tabu search, neural networks, genetic and particle swarm optimization algorithm [5-7]. The applications of PSO are considerably less in 2DC&PP. D. S. Liu et al 2006 presented a two-objective mathematical model with multiple constraints for two-dimensional bin packing problem and solve the problem with a hybrid multi-objective PSO algorithm [8]. Instead of using only optimization or heuristic approaches as a solution, many researchers have investigated the possibility of combining these methods into a solution approach to overcome the disadvantages of each of them. A hybrid algorithm to solve Irregular Strip Packing problems is presented in Gomes and Oliveira (2006), where the meta-heuristic simulated annealing is used to guide the search over the solution space while linear programming models are solved to generate neighborhoods during the search process [9]. A combination of these techniques that utilizes the advantages of each may produce a better solution to the problem.

In recent years, some effective approaches are proposed and used to check/generate feasible regions to pack items of irregular shape. A detailed tutorial about the geometry of nesting problems is given in Bennell and Oliveira (2008) [10]. Among these approaches, NFP (No-Fit Polygon) is definitely the most successful one, which can be gained in two ways: orbital sliding and Minkowski Sum [11, 12]. Phi Function has also get some focus and been use in some literature [13]. Furthermore, rectangle enclosure and grid approximation are also indirectly used to solve geometric problem of irregular shapes [14, 15]. In this paper, we adopt the grid approximation method to represent the shapes with two dimensional matrices for its less computer time. By using grid approximation, it's not necessary to introduce additional routines to identify enclosed areas and geometric tool to detect overlap.

3   Methodology

The operation of the approach is divided into two steps. In the first step, the items are represented as a matrix by taking the grid approximation method, and the initial sequence and orientation of the items based on their geometrical features are determined. Then the items are allocated on stock sheet one by one according to the initial sequence and orientations. Then the second step utilizes EPSO to search optimal solution by means of Bottom-Left-Fill heuristic (BFL) [16]. The flow of the approach is shown in Fig.2.

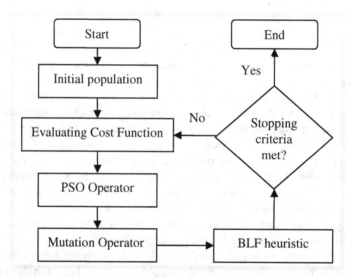

Fig. 2. Flowchart of EPSO for cutting stock problem

3.1   Initialization

In this research, each item is represented by a list of vertex coordinates $[(x_1,y_1),\cdots,(x_M, y_M)]$, where M is the number of vertices. During the allocation process, the degree of overlap among items on the stock sheet should be tested. The grid approximation, a digitized representation technique, is used to test overlaps among items. This matrix representation method was proposed by Dagli (1990) [17]. Each item is represented by a matrix, which is the smallest rectangular enclosure of the irregular-shaped item.

The detailed technique applied in the paper is referred to W.K. Wong and Z.X. Guo (2010) [18]. By using this technique, each item is enclosed by an imaginary rectangle for the sake of obtaining the reference points during the nesting process. Then, this particular rectangular area is divided into a uniform grid called pixel. In the case, the value of a pixel is '1' when the material of the sheet is occupied. Otherwise the value of the pixel is '0'. $P_L^{(k)}$ and $P_W^{(k)}$ denote the length and the width of an enclosing rectangle corresponding to the item p_k. R denotes the square side of a pixel.

The item with a two-dimensional matrix of size $A_W^{(k)} \times A_L^{(k)}$ is represented as follows:

$$A^{(k)} = \begin{pmatrix} a_{11}^{(k)} & a_{12}^{(k)} & \cdots & a_{1A_L^{(k)}}^{(k)} \\ a_{21}^{(k)} & a_{22}^{(k)} & \cdots & a_{2A_L^{(k)}}^{(k)} \\ \vdots & \vdots & \ddots & \vdots \\ a_{A_W^{(k)}1}^{(k)} & a_{A_W^{(k)}1}^{(k)} & \cdots & a_{A_W^{(k)}A_L^{(k)}}^{(k)} \end{pmatrix}$$

Where
$$A_W^{(k)} = \frac{P_W^{(k)}}{R} \quad and \quad A_L^{(i)} = \frac{P_L^{(k)}}{R} ;$$

For each entry,
$$a_{i,j}^{(k)} = \begin{cases} 1, & \text{if pixel } (i,j) \text{ is occupied} \\ 0, & \text{otherwise} \end{cases}.$$

Similar to the item representation, the stock sheet is able to be discretized as a finite number of equal-size pixels of size R^2. Hence, the stock sheet with the length L and the width W are characterized by a matrix U of size $U_W \times U_L$ as follows:

$$U = \left[u_{i,j} \right]_{U_W \times U_L}, \qquad \text{Where,} \qquad U_W = \frac{W}{R} \quad and \quad U_L = \frac{L}{R}$$

For each entry,
$$u_{i,j} = \begin{cases} 1, & \text{if pixel } (i,j) \text{ is occupied} \\ 0, & \text{otherwise} \end{cases}.$$

A value in the stock sheet matrix which is greater than one is an indication of an overlap. Although the overlap test can be performed easily and quickly, there is a need for a large memory for this representation scheme. Consequently, only required matrices are generated by the algorithm while the overlap test is performed. The initial sequence of the items is firstly determined in non-increasing order according to their area. In the case that area of items are same, the sequence of these items is decided in non-increasing order according to their length. The initial orientation for each packing item is confirmed by the MRE (Minimum Rectangular Enclosure (Jakobs 1996)) of each item. After the initial sequence and orientations of items are confirmed, the placement of the items follows a single-pass placement strategy and takes place in a sequential manner according to initial sequence and orientation by means of BFL heuristic.

3.2   EPSO

The proposed EPSO is a PSO-based algorithm which incorporates EA concepts such as the use of mutation operator as a source of diversity. It is referred to D. Liu et al. (2008) [19]. EPSO is characterized by the fact that particle movement is directed by either personal best or global best only in each instance. This is contrary to existing works where particle movement is influenced by both personal and global best at the same time. For the success of EPSO, an appropriate particle representation can be easily handled by PSO operator, mutation operator and BLF heuristic. Once initialization has finished, the particles are evaluated against the fitness function. After selecting either their personal best stock or global best stock as the velocity vector, particles are updated by inserting the stock represented by the velocity vector as the first stock and deleting duplicate items in other stocks (PSO operation). At their new positions, the particles then undergo specialized mutation operations. If any constraint is violated during mutation, the stocks violating constrains and the two least filled stocks will be selected for repacking using BLF.

3.2.1   Solution Coding

An effective encoding strategy for PSO operation and mutation operation is important for the success of EPSO. An order-based variable length particle structure is adapted as the representation of a solution for cutting stock problems in proposed EPSO. The solution including the number of stocks used and the sequence of the items packed into the stock is encoded as a particle. Each stock must be allocated at least one item which must not be found in any other stocks. And the permutations encoded are unique for every stock. This encoding structure allows the algorithm to manipulate the permutation of items in each stock without affecting other stocks.

Specifically, each particle consists of several strings of distinct integers. Each string represents a stock and each integer represents an item. The pbest and gbest are encoded in the form of extra stocks stored in the particle memory. The pbest is defined as the particles' own best stock due to their own best values in the solution space. The gbest is defined as the best stock found by the population. We use the best stock instead of best solution as pbest and gbest in proposed EPSO in order to help the algorithm avoid falling into random search by transmitting good stock among generations. Furthermore, a state variable is attached to every index in the permutation so as to encode for orientation of item.

3.2.2   PSO Operator

In PSO, each individual in the population will try to emulate the gbest and pbest solutions through updating by PSO equations [20]. How to emulate the pbest and gbest is also the key question in 2DCSP and our strategy is to set best stock as pbest and gbest, and let other stocks being repacked. In this way, best stock information is kept. The proposed PSO operator is shown as followed:

$$V_{id} = \alpha * P_{id} \oplus (1-\alpha) * P_{gd}, \quad \alpha = 0,1, \tag{1}$$

$$X_{id} = X_{id} + V_{id}. \tag{2}$$

The velocity is governed by either pbest or gbest as shown in Eq. (1). Which one is the governor is determined by the value of α. Then the movement of solution particle is confirmed by inserting the stock into the solution particle as the first stock and any duplicate items in other stocks will be deleted.

3.2.3   Mutation Operators

The purpose to adapt mutation operators in EPSO is to change the internal structure of the particle so as to further optimize the solution. The proposed mutation operators consist of two modes: partial swap and merge stocks mechanism. In the first mode, sequences of items in two stocks can be randomly cut and exchanged. The operation is used to search for more closely packed items. The items in the two least filled stocks will be merged into one stock in the second mode. The operation may help reduce the number of stocks used. Firstly, which mode is applied is decided by a random-generated number. Then the other two modes of mutation are employed. The first one randomly rotates one item in the stock, and the other one shuffle the items of the stock in order to improve the packing configuration and the stability of the stocks. Then a feasibility check is carried out to ensure the satisfaction of all the constraints. Whenever a constraint is violated, the particle eliminates two emptiest stocks, in which the items miss from the solution and be inserted back in a random order By BLF heuristic. This is helpful to confirm that all particles are valid solutions of 2D-CSP.

4   Performance Evaluation

All algorithms are implemented in Visual C++ and we also use the library CGAL to do some geometrical operations. The tests are performed on a computer with processor Intel Pentium Dual 1.8 GHz, 2 GB of RAM, and Windows XP operating system.

Few work for packing problems with pieces of irregular shape are done in the literature, and especially we can hardly find any related work for the 2DCSP and 2DBPP when pieces have irregular shape. As a result, we adapt some other known instances for packing problems with one open dimension to test our algorithms, referring to AM Del Valle .et .al 2012 [21]. The generated instances are adapted from the Two-Dimensional Irregular Strip Packing problem, and they can be found at the ES-ICUP website.

Table 1 presents solutions computed by our algorithm. We use the total area of the pieces divided by the area of one stock as a lower bound for the optimal solution. The rows in this table contain the following information: instance name, the same name as the original benchmark problem of SPP (Instance); shapes, the total number of polygons; vertices, the average number of the vertices of different shapes; solution value computed by our algorithm (Solution); the lower bound (LB); the difference (in percentage) on number of stocks computed by Solve 2DCSP and LB; the time spent in seconds (Time).

The results of the algorithm should be much closer to the optimal solutions and these differences are mainly due to the weakness of the lower bound. It is worth to mention that all the instances are executed 100 times and all the results presented on Table 1 are the average values of all executions. Such results show that the algorithm

Table 1. Results obtained for the 2CS adapted from strip packing problem

Instance	Shapes	Vertices	Solution	LB	Difference%	Time(s)
FU	12	3.58	70	56	22.079	254.60
JACKOBS1	25	5.60	46	38	25.556	7056.40
JACKOBS2	25	5.36	43	30	49.730	7035.76
SHAPES0	43	8.75	51	30	75.000	27124.81
SHAPES1	43	8.75	52	32	67.094	59703.37
SHAPES2	28	6.29	59	50	21.395	9574.17
DIGHE1	16	3.87	57	32	83.421	301.07
DIGHE2	10	4.70	42	29	50.595	87.85
ALBANO	24	7.25	80	65	25.765	6918.84
DAGLI	30	6.30	54	43	29.532	9463.45
MAO	20	9.22	45	33	41.051	7241.92
MARQUES	24	7.37	49	44	14.295	8462.74
SHIRTS	99	6.63	41	37	14.000	1235476.94
SWIM	48	21.90	56	34	69.259	467213.87
TROUSERS	64	5.06	49	42	20.361	274526.36

returns good solutions for the cutting stock problem with pieces of irregular shape. However, it requires high CPU time when solving instances with several pieces of completely irregular shape.

5   Conclusion

In this paper, a novel approach, based on EPSO, for two-dimensional irregular cutting stock problem is presented. Furthermore, the proposed approach combines a grid approximation method with BLF heuristic to allocate irregular items. The adhibition of EA concepts such as the use of mutation operator make EPSO a flexible optimization algorithm. Specifically, the advantages include the following aspects: firstly, the placement approach based on grid approximation provides the system designers with an easier way to detect whether overlap occurs. Secondly, BLF heuristic improves the quality of packing pattern for its ability to fill in the gaps in the partial layout. Thirdly, the searching method based on EPSO gets better performance compared with other meta-heuristics. Numerical investigations are performed on 15 test instances and the results demonstrate the effectiveness and efficiency of the proposed approach.

Acknowledgements. The work of this paper was supported by National Natural Science Foundation of China, No.61074150 and No. 61203178.

References

1. Wäscher, G., Haußner, H., Schumann, H.: An improved typology of cutting and packing problems. Eur. J. Oper. Res. 183(3), 1109–1130 (2007)
2. Garey, M., Johnson, D.: Computers and Intractability: A Guide to the Theory of NP-Completeness. W. H. Freeman and Company, New York (1979)

3. Oliveira, J.F., Gomes, A.M., Ferreira, J.S.: TOPOS – a new constructive algorithm for nesting problems. OR Spektrum 22(2), 263–284 (2000)
4. Burke, E.K., Hellier, R.S.R., Kendall, G., Whitwell, G.: A new bottom-left-fill heuristic algorithm for the two-dimensional irregular packing problem. Operations Research 54(3), 587–601 (2006)
5. Gonçalves, J.: A hybrid genetic algorithm-heuristic for a two-dimensional orthogonal packing problem. European Journal of Operational Research 183(3), 1212–1229 (2007)
6. Alvarez-Valdes, R., Parreño, F., Tamarit, J.M.: A tabu search algorithm for two-dimensional non-guillotine cutting problems. European Journal of Operational Research 183(3), 1167–1182 (2007)
7. Burke, E.K., Kendall, G., Whitwell, G.: A simulated annealing enhancement of the best-fit heuristic for the orthogonal stock-cutting problem. INFORMS Journal on Computing 21(3), 505–516 (2009)
8. Liu, D.S., Tan, K.C., Goh, C.K., Ho, W.K.: On solving multi-objective bin packing problems using particle swarm optimization. In: IEEE Congress on Evolutionary Computation, Vancouver, pp. 7448–7455 (2006)
9. Gomes, A.M., Oliveira, J.F.: Solving irregular strip packing problems by hybridizing simulated annealing and linear programming. European Journal of Operational Research 171(3), 811–829 (2006)
10. Bennell, J.A., Oliveira, J.F.: The geometry of nesting problems: A tutorial. Eur. J. Oper. Res. 184, 397–415 (2008)
11. Burke, E.K., Hellier, R.S.R., Kendall, G., Whitwell, G.: Complete and robust no-fit polygon generation for the irregular stock cutting problem. Eur. J. Oper. Res. 179(1), 27–49 (2007)
12. Burke, E.K., Hellier, R.S.R., Kendall, G., Whitwell, G.: Irregular Packing Using the Line and Arc No-Fit Polygon. Oper. Res. 58(4), 1–23 (2010)
13. Bennell, J., Scheithauer, G., Stoyan, Y., Romanova, T.: Tools of mathematical modelling of arbitrary object packing problems. Annals of Operations Research 179, 343–368 (2010)
14. Jakobs, S.: On genetic algorithms for the packing of polygons. European Journal of Operational Research 88(1), 165–181 (1996)
15. Poshyanonda, P., Dagli, C.H.: Genetic neuro-nester. Journal of Intelligent Manufacturing 15(2), 201–218 (2004)
16. Hopper, E., Turton, B.C.H.: An empirical investigation on meta-heuristic and heuristic algorithms for a 2d packing problem. European Journal of Operational Research 128, 34–57 (2001)
17. Dagli, C.H., Hajakbari, A.: Simulated annealing approach for solving stock cutting problem. In: Proceedings of IEEE International Conference on Systems, Man, and Cybernatics, pp. 221–223 (1990)
18. Wong, W.X., Guo, Z.X.: A hybrid approach for packing irregular patterns using evolutionary strategies and neural network. International Journal of Production Research 48(20), 6061–6084 (2010)
19. Liu, D., Tan, K., Huang, S., Goh, C., Ho, W.: On solving multi-objective bin packing problems using evolutionary particle swarm optimization. European Journal of Operational Research 190, 357–382 (2008)
20. Srinivasan, D., Seow, T.H.: Particle swarm inspired evolutionary algorithm (PS-EA) for multi-objective optimization problems. In: Proceedings of IEEE Congress on Evolutionary Computation, pp. 2292–2297 (2003)
21. Del Valle, A., De Queiroz, T., Miyazawa, F., Xavier, E.: Heuristics for twodimensional knapsack and cutting stock problems with items of irregular shape. Expert Systems with Applications 39(16), 12589–12598 (2012)

Optimization Analysis of Controlling Arrivals in the Queueing System with Single Working Vacation Using Particle Swarm Optimization

Cheng-Dar Liou

Department of Business Administration, National Formosa University,
64,Wunhua Rd., Huwei, Yunlin County, 63201, Taiwan
cdliou@nfu.edu.tw

Abstract. A cost function in the literature of queueing system with single working vacation was formulated as an optimization problem to find the minimum cost. In the approach used, a direct search method is first used to determine the optimal system capacity K and the optimal threshold F followed by the Quasi-Newton method to search for the optimal service rates at the minimum cost. However, this two stage search method restricts the search space and cannot thoroughly explore the global solution space to obtain the optimal solutions. In overcoming these limitations, this study employs a particle swarm optimization algorithm to ensure a thorough search of the solution space in the pursuit of optimal minimum solutions. Numerical results compared with those of the two stage search method and genetic algorithms support the superior search characteristics of the proposed solution.

Keywords: Direct search method, Quasi-Newton method, Particle swarm optimization, Genetic algorithms.

1 Introduction

For clarity, we briefly repeat the description of the queueing model and adopt the notations and assumptions of Yang et al. [1].

The server takes a single vacation whenever the system becomes empty. During a vacation period, the server remains working at a different service rate rather than completely terminating the service. Such a vacation is called a working vacation [2]. When the number of customers in the system reaches its capacity K (i.e. the system becomes full), no further arriving customers are allowed to enter the system until the queue length decreases to a certain threshold value F ($0 \leq F \leq K-1$). At that time, the server requires an exponential startup time to restart allowing customers to enter the system. This queueing system is referred to as F-policy M/M/1/K/WV queueing system with an exponential startup time.

Yang et al. [1] used the state-transition-rate diagram to set up the steady-state equations. However, two specific conditions appeared in the queueing model, that is $F = 1$ and $F = K-1$, seem to be neglected. This makes the steady-state equations

Y. Tan, Y. Shi, and H. Mo (Eds.): ICSI 2013, Part I, LNCS 7928, pp. 176–183, 2013.

cannot represent the real characteristics of the queueing model and incurs miscalculations as the specific conditions become visible. In addition, Yang et al. [1] proposed a direct search method (DSM) and Quasi-Newton method (QNM) for the optimization problem to determine the joint optimal values at the minimum cost. In their two-stage approach, they first employed the DSM to find the optimal capacity, K^*, and the optimal threshold value, F^*, while the other two continuous variables μ_B and μ_V were assumed to be constant and fixed. However, this assumption considerably restricts the search space, resulting in the final solution falling into a local optimal solution. In the second stage, QNM is used to search the best combinations of two continuous variables (μ_B, μ_V), while the number of capacity K and the threshold value F are assumed to be fixed according to the search result of DSM in the first stage. This method restricts the search space in a similar manner, causing the searching result to fall into local extrema. Also, in the numerical examples (see p.554 of [1]), there are 2 out of 6 instances that the service rate in a working vacation is faster than that in a normal busy period (i.e. $\mu_V > \mu_B$). This seems to be a contradiction with the definition of working vacation (see p.48 of [2]). On the basis of the above observations, in this article, we reset the state-state equations and use the particle swarm optimization (PSO) algorithm to simultaneously determine the four optimum variables, F^*, K^*, μ_B^*, and μ_V^*, to minimize the cost function. We conducted the numerical experiments using QNM, genetic algorithms (GA) and PSO, respectively and compared the search results. Note that GA is used to ensure the searching quality of the PSO algorithm.

2   Steady-State Equations

Referring to the state-transition-rate diagram for the F-policy M/M/1/WV queueing system with an exponential startup time shown in Fig.1, we can see that $F = 1$ and $F = K-1$ are two specific conditions due to the link between $P_2(1)$ and $P_3(0)$ as $F = 1$, and the different transition conditions as $F = K-1$. Therefore, the following four conditions have to be considered:

Case 1: $F = 1$ and $F = K-1$ Case 2: $F = 1$ and $F < K-1$
Case 3: $F \neq 1$ and $F = K-1$ Case 4: $F \neq 1$ and $F < K-1$.
The steady-state equations developed by Yang et al. [1] are given by:

$$(\theta + \gamma)P_0(0) = \mu_V P_0(1), \tag{1}$$

$$(\mu_V + \theta + \gamma)P_0(n) = \mu_V P_0(n+1), \quad n = 1, 2, ..., F, \tag{2}$$

$$(\mu_V + \theta)P_0(n) = \mu_V P_0(n+1), \quad n = F+1, F+2, ..., K-1 \tag{3}$$

$$(\mu_V + \theta)P_0(K) = \lambda P_3(K-1), \tag{4}$$

$$\gamma P_1(0) = \mu_B P_1(1) + \theta P_0(0), \tag{5}$$

$$(\mu_B + \gamma)P_1(n) = \mu_B P_1(n+1) + \theta P_0(n), \quad n = 1, 2, ..., F, \tag{6}$$

$$\mu_B P_1(n) = \mu_B P_1(n+1) + \theta P_0(n), \quad n = F+1, F+2, ..., K-1, \tag{7}$$

$$\mu_B P_1(K) = \lambda P_2(K-1) + \theta P_0(K), \tag{8}$$

$$\lambda P_2(0) = \gamma P_1(0) + \theta P_3(0), \tag{9}$$

$$(\lambda + \mu_B)P_2(n) = \gamma P_1(n) + \lambda P_2(n-1) + \mu_B P_2(n+1) + \theta P_3(n), \quad n = 1, 2, ..., F, \tag{10}$$

$$(\lambda + \mu_B)P_2(n) = \lambda P_2(n-1) + \mu_B P_2(n+1) + \theta P_3(n), \quad n = F+1, F+2, ... K-2, \tag{11}$$

$$(\lambda + \mu_B)P_2(K-1) = \lambda P_2(K-2) + \theta P_3(K-1), \tag{12}$$

$$(\lambda + \theta)P_3(0) = \gamma P_0(0) + \mu_B P_2(1) + \mu_V P_3(1), \tag{13}$$

$$(\lambda + \mu_V + \theta)P_3(n) = \gamma P_0(n) + \lambda P_3(n-1) + \mu_V P_3(n+1), n = 1, 2, ..., F \tag{14}$$

$$(\lambda + \mu_V + \theta)P_3(n) = \lambda P_3(n-1) + \mu_V P_3(n+1), n = F+1, F+2, ..., K-2 \tag{15}$$

$$(\lambda + \mu_V + \theta)P_3(K-1) = \lambda P_3(K-2). \tag{16}$$

Note that the above steady-state equations can only cover the situations for case 2 and case 4.

For case 1, Eq. (10) has to be modified as

$$(\lambda + \mu_B)P_2(n) = \gamma P_1(n) + \lambda P_2(n-1) + \theta P_3(n), \quad n = 1, 2, ..., F, \tag{17}$$

For case 3, Eqs. (12) and (16) have to be replaced by Eqs. (18) and (19), respectively.

$$(\lambda + \mu_B)P_2(K-1) = \lambda P_2(K-2) + \theta P_3(K-1) + \gamma P_1(K-1), \tag{18}$$

$$(\lambda + \mu_V + \theta)P_3(K-1) = \lambda P_3(K-2) + \gamma P_0(K-1) \tag{19}$$

Using the matrix-analytic method developed by Yang et al. [1], the steady-state probabilities for the F-policy M/M/1/WV queueing system with an exponential startup time can be obtained.

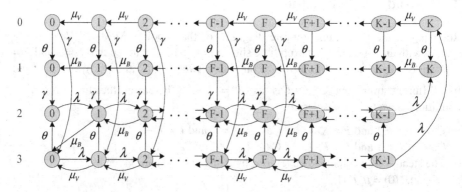

Fig. 1. The state-transition-rate diagram for an F-policy M/M/1/WV queueing system with an exponential startup time [1]

3   Basic Concept of PSO Algorithm and Encoding Scheme

The particle swarm optimization (PSO) algorithm is a stochastic search technique, motivated by the social behavior simulation of bird flocking or fish schooling, developed by Kennedy and Eberhart [3]. As an evolutionary algorithm, PSO conducts a search through updating a population (called a swarm) of individuals (called particles). The relationship between the swarm and particles in PSO algorithm is similar to the relationship between a population and its chromosomes in GA. In PSO algorithm, the problem solution space is described as a search space and each position in the search space is a possible solution for the problem. It is known that PSO combines local search (by self-experience) and global search (by neighboring experience), and has been introduced as an optimization technique in continuous and discrete spaces. More details for the applications of PSO algorithm are referred to Kennedy et al. [4], and Clerc [5].

To clearly explain the definition of particles, positions, best position, and population of the PSO, a small example using the same cost parameters with those of Yang et al. [1] is illustrated. Consider the test example with $(\lambda, \gamma, \theta) = (4, 3, 2)$, we use a population with 5 particles to explore the solution space. We set the threshold value F on the interval $[0, K-1]$, and the system capacity K on the interval $[2, 12]$. Both μ_B and μ_V are continuous on the interval of 0 to 10 as well as $\mu_V \leq \mu_B$. For an initial particle $P(R_1, R_2, R_3, R_4)$ with four random variables on the interval of 0 to 1, we can use the following encoding scheme, $F = \lfloor (11-0+1)*R_1 - \varepsilon \rfloor$, $K=2+\lfloor (12-2+1)*R_2 - \varepsilon \rfloor, \mu_B = 10*R_3$, and $\mu_V = 10*R_4$, to transform a particle to its position. The notation $\lfloor d \rfloor$ denotes the largest integer smaller than or equal to $|d|$ and $\varepsilon = 0.0001$ is used to avoid F and K falling into 12 and 13, respectively. Note that the constraint conditions $F \leq K-1$ and $\mu_V \leq \mu_B$ can be overcome by using a penalty value to delete the infeasible solutions. Each position has its own objective value shown in Table 1. The best position is the position with the best objective value among all positions. Obviously, the best position is $F = 2$, $K= 5$, $\mu_B = 4.608$, and $\mu_V = 1.803$, at this iteration in the example.

Table 1. The particles, current position, and objective values

Particle	Current position	Objective value
(0.1987, 0.3461, 0.4608, 0.1803)	(2, 5, 4.608, 1.803)	**843.2516**
(0.0153, 0.4225, 0.3663, 0.2760)	(0, 6, 3.663, 2.760)	892.2494
(0.4451, 0.4902, 0.8518, 0.0129)	(5, 7, 8.518, 0.129)	880.7553
(0.8998, 0.9048, 0.4551, 0.1573)	(10, 11, 4.551,1.573)	857.3484
(0.2897, 0.3352, 0.8678, 0.3844)	(3, 5, 8.678, 3.844)	859.5493

4   Numerical Results

4.1   Direct Search Method and Exact Results

Yang et al. [1] presented a numerical example for finding the joint optimal values (F^*, K^*) by using direct search method. However, due to their steady-state equations without the considerations of specific conditions, it seems to be possible for miscalculation. We redo the same example and obtain the numerical results shown in Table 2. We can find that the minimum expected cost $792.76 per day is obtained with $F^* = 6$ and $K^* = 8$, but not the cost $794.92 with $F^* = 5$ and $K^* = 7$ obtained by Yang et al. [1]. To show the calculation process, a small example with the specific conditions $F = 1$ and $K = 2$ shown in Fig. 2 is considered. Obviously, the steady-state equations developed by Yang et al. [1] cannot satisfy this situation. The details related to the steady-state probabilities and various system parameters are illustrated in Table 3. After the calculation, we can obtain that the cost is $1034.98 but not $983.54 shown in Table 1 of Yang et al. [1]

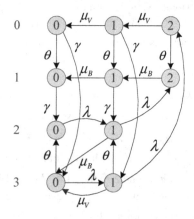

Fig. 2. The state-transition-rate diagram for an F-policy M/M/1/WV queueing system with $F = 1$ and $K = 2$

4.2   Comparison of Three Heuristic Methods

To compare the search results of QNM, GA, and PSO algorithm for the cost minimization problem, we tested 7 examples with the following cost parameters:

C_h=$15/unit, C_b=$450/day, C_f=$400/day, C_s=$500/day, C_w=$60/day, C_k=$20/unit, C_1=$30/unit, C_2=$20/unit.

The objective function can be presented mathematically as

$$\underset{F,K,\mu_B,\mu_V}{Minimize}\, TC(F,K,\mu_B,\mu_V)$$

$$s.t. \quad 0 \le F \le K-1, \quad 2 \le K \le 12,$$

$$0 \le \mu_B \le 10, \quad 0 \le \mu_V \le 10,$$

$$\mu_V \le \mu_B.$$

where

$$TC(F,K,\mu_B,\mu_V) = C_h L_S + C_b P_B + C_l \lambda P_L + C_s P_S + C_w W_S + C_k K + C_1 \mu_B + C_2 \mu_V$$

Although the GA and PSO algorithm appear to be sensitive to the tuning of various weights and parameters, we took advantage of the experiences gained through many previous experiments, adopting the following GA and PSO parameters (see Kennedy et al. [4]; Clerc [5]; Gen and Cheng [6]; Eberhart and Shi [7]; Liou [8]).

PSO Algorithm Parameters

- population size = 100;
- generations = 200 ;
- inertia weight factor is set to vary linearly from 0.9 to 0.4 ;
- the limit of change in velocity of each member in an individual is as $V_{max} = 0.3$,
- acceleration constant $c_1 = 2.0$ and $c_2 = 2.0$.

GA Parameters

- population size $n = 100$
- generation number $t_c = 200$;
- crossover rate $P_c = 0.8$;
- mutation rate $P_m = 0.05$;
- bit length of a variable $l_b = 60$.

GA and PSO were coded in MATLAB 7.0 and all results were computed using an Intel-Pentium 2.4GHz PC with 1.97GB RAM. Table 4 shows the numerical results of QNM and 100 independent experiments for each example by GA and PSO algorithm, respectively. Note that, for the convenience of comparison, the mean ratio was utilized. The ratio of the solution produced by the heuristic algorithm is calculated by V/V*, where V is the solution generated by the heuristic algorithm and V* is the minimum solution among 100 independent experiments by the heuristic algorithm.

From Table 4, we observe:

(1) QNM was unable to acquire the global minimum value in the 7 instances. This implies that the QNM is not well suited to searching for optimal values (F^*, K^*, μ_B^*, μ_V^*) to minimize the cost function.

(2) in the case $(\lambda, \gamma, \theta) = (4.0, 3.0, 1.0)$, the QNM cannot converge to any solution because the converge condition $Max(|\partial TC / \partial \mu_B|, |\partial TC / \partial \mu_V|) < \varepsilon$ (i.e. Step 5 of QNM, see Yang et al. [1], p.551) cannot be met even though the tolerance ε is set as 0.1. This implies that the QNM cannot be operated well as the extreme point of a concave/convex is not included in the feasible solutions.

(3) the mean values (V/V*) of PSO algorithm all lead to 1.0000 for all instances. This implies that the PSO algorithm is robust for all test instances.

(4) both GA and PSO can obtain the similar optimal solutions. This implies that the searching quality of PSO is conceivable.

(5) the CPU times to attain solutions per run for GA and PSO was approximately 15-18 and 2-4 seconds, respectively, depending on the selection of parameters (Table 4). This implies that the PSO is capable of solving the test instances within a reasonable time and more efficient than GA.

Table 2. The expected cost $TC(F, K)$ for $\lambda = 4.0$, $\mu_B = 5.0$, $\mu_V = 3.0$, $\gamma = 2.0$, $\theta = 3.0$

K	0	1	2	3	4	5	6	7	8	9	10	11
2	1330.28	1034.98	—	—	—	—	—	—	—	—	—	—
3	935.75	925.76	974.63	—	—	—	—	—	—	—	—	—
4	879.02	867.53	860.02	897.40	—	—	—	—	—	—	—	—
5	844.79	834.76	826.36	821.59	849.98	—	—	—	—	—	—	—
6	825.18	817.09	809.78	804.20	801.43	822.99	—	—	—	—	—	—
7	815.45	809.12	803.21	798.24	794.68	793.21	809.67	—	—	—	—	—
8	812.58	807.67	803.01	798.92	795.64	793.44	**792.76**	805.43	—	—	—	—
9	814.53	810.73	807.10	803.83	801.07	798.96	797.65	797.45	807.27	—	—	—
10	819.94	817.00	814.17	811.59	809.35	807.53	806.20	805.46	805.54	813.22	—	—
11	827.86	825.58	823.38	821.35	819.56	818.06	816.88	816.07	815.71	815.94	821.98	—
12	837.63	835.85	834.13	832.54	831.12	829.90	828.92	828.18	827.73	827.60	827.92	832.69

Table 3. The steady-state probabilities and various system parameters for the considered example

Probability	$P_0(0)$	$P_1(0)$	$P_2(0)$	$P_3(0)$	$P_0(1)$	$P_1(1)$	$P_2(1)$	$P_3(1)$	$P_0(2)$	$P_1(2)$
	0.0072	0.2404	0.2056	0.1139	0.0120	0.0918	0.1278	0.0479	0.0320	0.1214
Parameters	μ_B	μ_V	λ	θ	γ	L_S	P_B	P_L	P_S	W_S
	5.0	3.0	4.0	3.0	2.0	0.5863	0.4329	0.5048	0.3514	0.2960

Table 4. The searching results of QNM, GA, and PSO

	QNM[1]				GA					PSO				
$(\lambda, \gamma, \theta)$	(F^*, K^*)	μ_B^*	μ_V^*	TC^*	(F^*, K^*)	μ_B^*	μ_V^*	TC^*	Mean	(F^*, K^*)	μ_B^*	μ_V^*	TC^*	Mean
(3.0,3.0,2.0)	(5, 7)	6.6109	4.8106	703.3821	(3, 5)	7.0335	5.7286	692.2813	1.0001	(3, 5)	7.0338	5.7290	692.2813	1.0000
(4.0,3.0,2.0)	(6, 8)	7.7606	6.1081	816.0133	(4, 6)	8.2183	7.1967	805.3591	1.0001	(4, 6)	8.2181	7.1956	805.3591	1.0000
(5.0,3.0,2.0)	(7, 9)	8.8917	7.2314	919.3518	(5, 7)	9.8389	8.4327	908.1817	1.0000	(5, 7)	9.3579	8.4333	908.1817	1.0000
(4.0,1.0,2.0)	(8, 10)	7.7013	5.8614	850.8664	(5, 7)	8.4485	7.3459	827.6232	1.0001	(5, 7)	8.4491	7.3367	827.6232	1.0000
(4.0,6.0,2.0)	(6, 8)	7.6555	5.9112	812.2468	(4, 6)	8.0211	6.8459	794.6115	1.0000	(4, 6)	8.0211	6.8452	794.6115	1.0000
(4.0,3.0,1.0)	(7, 9)	—	—	—	(4, 6)	7.9054	7.9047	806.9839	1.0003	(4, 6)	7.9784	7.9784	806.9338	1.0000
(4.0,3.0,3.0)	(6, 8)	8.0858	3.7665	798.3834	(5, 7)	8.3238	4.4014	790.9199	1.0000	(5, 7)	8.3231	4.4026	790.9199	1.0000

5   Computational Results and Conclusions

This study investigated the optimization problem for the cost function in the F-policy M/M/1/K queueing system with working vacation and an exponential startup time. This problem was previously studied by Yang et al. [1]. Our approach involved the use of the PSO algorithm to search thoroughly for the optimal solutions within a limited period of time. The numerical results indicate that the two stage method is capable of searching only a local solution space, and thus causing the searching result to fall into local extrema. Moreover, the QNM cannot be operated well for the constrained optimization problem. The ability of the PSO to thoroughly explore the solution space supports the superior search characteristics of the proposed solution.

References

1. Yang, D.Y., Wang, K.H., Wu, C.H.: Optimization and Sensitivity Analysis of Controlling Arrivals in the Queueing System with Single Working Vacation. J. Comput. Appl. Math. 234, 545–556 (2010)
2. Servi, L.D., Finn, S.G.: M/M/1 Queues with Working Vacations (M/M/1/WV). Perform. Evaluation 50, 41–52 (2002)
3. Kennedy, J., Eberhart, R.C.: Particle Swarm Optimization. In: Proceedings of IEEE International Conference on Neural Networks, pp. 1942–1948. Piscataway, NJ (1995)
4. Kennedy, J., Eberhart, R.C., Shi, Y.: Swarm Intelligence, CA. Morgan Kaufmann, New York (2001)
5. Clerc, M.: Particle Swarm Optimization. Wiley-ISTE (International Scientific and Technical Encyclopedia) (2006)
6. Gen, M., Cheng, R.: Genetic Algorithms & Engineering Optimization. John-Wiley & Sons, Inc., New York (2000)
7. Eberhart, R.C., Shi, Y.: Computational Intelligence Concepts to Implementations. Morgan Kaufmann, New York (2007)
8. Liou, C.D.: Note on "Cost Analysis of the M/M/R Machine Repair Problem with Second Optional Repair: Newton-Quasi Method". J. Ind. Manag. Optim. 8(3), 727–732 (2012)

Anomaly Detection in Hyperspectral Imagery Based on PSO Clustering

Baozhi Cheng and Zongguang Guo

College of Physics and Electricity Information Engineering,
Daqing Normal University, Daqing 163712, China
chengbaozhigy@163.com

Abstract. In this paper, we propose a novel anomaly targets detection algorithm baesd on information processing method and KRX anomaly detector. It use fully nolinear feature and decrease bands redundancy for hyperspectral imagery. Firstly, the original hyperspectral imagery is clustered by a new clustering method, i.e. k-means clustering of particle swarm optimization. Then, we extract a largest fourth-order cumulant value in every class, and constitute a optimal band subset. Finally, the KRX detector is used on the band subset to get anomaly detection results. The simulation results demonstrate that the proposed PSOC-KRX algorithm outperforms the other algorithm, it is higher precision and lower false alarm rate.

Keywords: hyperspectral anomaly detection, particle swarm optimization, clustering.

1 Introduction

The hyperspectral imagery are composed of hundreds of contiguous and narrow spectral bands, which cover a wide spectral range with nanometer spectral resolution that large amounts of detailed characteristic features are provided for the target objects [1]. Therefore, it plays an important role in the area of geological exploration, military and agriculture [2]. In recent years, the question of anomaly targets detection becomes a hot because it isn't need prior information to the actual hyperspectral images.

The classic algorithm of hyperspectral anomaly target detection is the RX [3], which detects anomaly targets by mahalanobis distance measure for estimating the background mean vector and covariance matrix [4]. The detection performance of RX algorithm is low because it isn't use nolinear property between hyperspectral imagery bands [5], and it is invalid for the local gaussian assumption in some cases. In [4], [6], kernel method is regarded as an effective method to processing anomaly targets issues. Kwon [4] presents a nonlinear RX-algorithm, which is the kernel RX anomaly target detection algorithm, through kernel mapping the original hyperspectral image to a high dimensional feature space, and exploit the higher order relationships in bands. KRX algorithm apply nolinear property between hyperspectral bands, the performance of algorithm get enhancing, but the complexity of algorithm is higher because there is complex to computing the covariance matrix and its inverse. Thus, we propose a new

Y. Tan, Y. Shi, and H. Mo (Eds.): ICSI 2013, Part I, LNCS 7928, pp. 184–191, 2013.

anomaly targets detector which is the extended KRX algorithm. The new anomaly algorithm preprocesses hyperspectral image by using clustering method and extract targets feature by the fourth-order cumulant, the KRX detector is applying to the optimal band subsets.

This paper is organized as follows: section 2 proposes the new anomaly detection algorithm based on a new clustering method and KRX detector, which is the extended KRX algorithm, and improve the classics KRX detection performance. Section 3 provides simulation results and analysis. Conclusion is provided in Section 4.

2 Anomaly Detection Algorithm Model

In this section, we propose a new anomaly detection algorithm for hyperspectral imagery. The algorithm will detect anomaly targets more than the other algorithm, the proposed new anomaly detection algorithm preprocess the hyperspectral images by using the new information processing such as PSO, clustering, PCA, et al, we will analyze them in detail.

2.1 Particle Swarm Optimization

PSO is a stochastic optimization technique recently introduced by Kennedy and Eberhart, which is inspired by social behavior of bird flocking and fish schooling [7], [8]. Similar to other evolutionary computation algorithms such as genetic algorithms [9], [10], PSO is a population-based search method [11], on PSO, each individual called particle, which determines a fitness based on objective function. During the search process in the solution space, each particle will adjust their own position and velocity based on these good positions. The velocity adjustment is based upon the historical behaviors of the particles themselves as well as their neighbors, all particle will flying the best position, i.e., the global optimization [12].

Let us consider a swarm of size m, where is D-dimensional search spaces, $X_i = (x_{i1}, x_{i2}, ...x_{id}...x_{iD})$ ($i = 1,..., M$) denotes position vector of i th particle; $V_i = (v_{i1}, v_{i2}, ...v_{id}...v_{iD})$ defines velocity of particle flying; $Pbest_i = (pbest_{i1}, pbest_{i2}, ...pbest_{id}...pbest_{iD})$ denotes the best local position; $Gbest = (Gbest_1, Gbest_2, ...Gbest_d...Gbest_D)$ denotes the best global position. The searching procedure based on this concept can be described by (1), (2).

$$V_{id}^{(t+1)} = v_{id}^{(t)} + c_1 rand_1 (pbest_{id} - x_{id}^t)$$
$$+ c_2 rand_2 (Gbest_d - x_{id}^t) \tag{1}$$

$$x_{id}^{(t+1)} = x_{id}^{(t)} + v_{id}^{(t+1)} \tag{2}$$

where $i = 1, 2, ...m, d = 1, 2, ...D$, t denotes iterations, $rand_1$ and $rand_2$ denote random variables drawn from a uniform distribution in the range [0,1], c_1 and c_2 is two acceleration constants that regulate the relative velocities with respect to the best global and local positions, respectively. The inertia weight w is used as a tradeoff between global and local exploration capabilities of the swarm. Large values of this parameter permit better global exploration, whereas small values lead to a fine search in the solution space [11],[12].

2.2   k-means Clustering

Clustering analysis is applied in many fields such as the natural sciences, the medical sciences, economics, marketing, etc [13]. So far, there are essentially two types of clustering methods: hierarchical algorithms and partitioning algorithms [13]. Now, partitioning clustering algorithms is a hot, there are essentially three types of partitioning clustering methods: graph theoretic, k-means, and FCM. By contrast, we use k-means clustering algorithms in the paper, it is a iterative clustering algorithm, it is $X = (x_1, x_2, \ldots x_i, \ldots x_n)$ data set, the idea behind k-means that it is find k clustering center, such as $c_1, c_2, \ldots c_j, \ldots c_k$, it would the minimum distance between x_i and c_j, c_j is the nearest clustering center at the range of x_i, given by

$$J_c = \sum_{i=1}^{n} \sum_{j=1}^{k} \left\| x_i - c_j \right\| \tag{3}$$

k-means must determine the k number of clustering at first, and set k clustering center, clustering algorithm will stop until clustering centers aren't change. The clustering result has related to k. However, the k value is difficulty to setting in realistic, so, the result isn't the ideal.

2.3   KRX Algorithm

Reed and Yu [4] developed a generalized-likelihood ratio test (GLRT) for hyperspectral imagery data, the method is called RX anomaly detector, which assumed that the spectrum of received signal and the covariance of the background are unknown [4]. RX algorithm is anomaly detection operator of pure pixels based on local normal model [14], the local parameters (local mean and covariance) is obtained by sliding a double concentric window, namely, called the inner window region and the outer window region. The outer window is larger than the inner window, the mean of spectral pixels is calculated falling within the outer window; the size of the inner window is assumed to be size of the typical target of interest in image [4].

Let each input spectral signal consist of 1 spectral bands be denoted by

$$x(n) = [x_1(n), x_2(n), \ldots x_1(n)]^T \tag{4}$$

Where define X_B to be a $1 \times p$ matrix of the p reference background clutter pixels. Each observation spectral pixel is represented as a column in the sample matrix X_B [4]

$$X_B = [x_1, x_2, \dots x_P] \tag{5}$$

The two competing hypotheses that the RX need distinguish are given by [8]

$$
\begin{aligned}
& H_0 : x = n \quad \text{(Target absent)} \\
& H_1 : x = \alpha s + n \text{ (Target present)}
\end{aligned}
\tag{6}
$$

where $\alpha = 0$ under H_0 and $\alpha > 0$ under H_1. n is a vector that represents the process of background clutters noise, and $s = [s_1, s_2, \dots s_J]^T$ is the spectral signature of anomalous target [4].

The two competing hypotheses of Kernel RX algorithm is the same as RX, the original data X_B is mapped into a potentially much higher dimensional feature space by a nonlinear mapping function Φ.

$$\Phi(X_B) = [\Phi(x_1), \Phi(x_2), \dots \Phi(x_P)] \tag{7}$$

$\Phi(X_B)$ is consists of two Gaussian distributions, thus modeling the two hypotheses as

$$
\begin{aligned}
& H_{0\Phi} : x = n_\Phi \quad \text{(Target absent)} \\
& H_{1\Phi} : x = \beta \Phi(s) + n_\Phi \quad \text{(Target present)}
\end{aligned}
\tag{8}
$$

Where $\beta = 0$ under H_{0_Φ} and $\beta > 0$ under H_{1_Φ}, n_Φ represents a noise process in the feature space, $\Phi(s)$ represents target spectral signature[4]. Thus the Kernel RX-algorithm is represented as

$$KRX(\Phi(r)) = (\Phi(r) - \hat{\mu}_{B\Phi})^T \hat{C}_{B\Phi}^{-1} (\Phi(r) - \hat{\mu}_{B\Phi}) \tag{9}$$

Where $\hat{\mu}_{B\Phi}$ 和 $\hat{C}_{B\Phi}$ are the estimated covariance and mean of the background in the feature space, respectively, given by

$$\hat{\mu}_{B\Phi} = \frac{1}{M} \sum_{i=1}^{M} \Phi(x_i) \tag{10}$$

$$\hat{C}_{B\Phi} = \frac{1}{M} \sum_{i=1}^{M} (\Phi(x(i)) - \hat{\mu}_{B\Phi})(\Phi(x(i)) - \hat{\mu}_{B\Phi})^T \tag{11}$$

Using the well-known "kernel trick", given by

$$h(x, y) = <\Phi(x), \Phi(y)> = \Phi(x) \cdot \Phi(y) \tag{12}$$

Substituting (12) into (9) yields

$$KRX(\Phi(r)) = (h_r^T - k_{\hat{\mu}_B}^T)^T \hat{H}_B^{-1} (h_r^T - h_{\hat{\mu}_B}^T) \tag{13}$$

Where $\hat{H}_B = H_B - H_B I_M - I_M H_B + I_M H_B I_M$.

Equation (13) is be implemented with no knowledge of the mapping function Φ. The only requirement is a good choice for the kernel function h, we use the gaussian radial basis function as the kernel function defined as

$$h(x, y) = \exp(-\|x - y\|^2 / \hat{\sigma}) \tag{14}$$

2.4   Proposed Algorithm

As shown in Fig.1, The proposed new anomaly detection algorithm consists of four principal steps:

1) At first, the hyperspectral imagery is identified by clustering the data into different class. clustering use k-means method, but k-means hasn't enough that the shortage had discussed in preamble, so, the PSO method is used to improve the search process of the clustering center so that we achieve globally optimal solution.

2) Secondly, during clustering, PSO method determines the clustering center first, i.e., find the optimal value. Then, we determine the clustering results by k-means algorithm. Thus, the total hyperspectral imagery bands are divided for the k classes with different bands numer.

3) Thirdly, the fourth-order cumulant are used to measure the singularity of each nonlinear principal component one by one in clustering hyperspectral images, we can acquire the optimal band subsets.

4) At last, the bands subset will input to KRX algorithm, we will acquire detection results, the result will be converted for binary image.

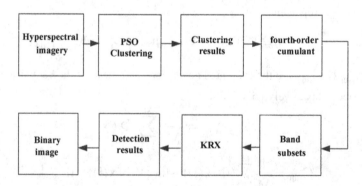

Fig. 1. Flow chart of the proposed algorithm for anomaly detection

3 Analysis and Experiments on Simulated Data

In this section, we apply a real hyperspectral images to do anomaly target detection experiments, which collected by the Airborne Visible/Infrared Imaging Spectrometer (AVIRIS). The image is a naval air station in San Diego, California, USA[15], Fig.2(a) shows the pseudo-color image of data set (composed of band 10, 55 and 100), which consist of a scene of size 100 by 100 pixels, and contain 38 panels of anomaly targets. Fig.2(b) shows truth case of anomaly target distribution, respectively. The proposed PSOC-KRX algorithm was only compared with the other detection algorithms including the RX algorithms and the kernel RX algorithm, respectively.

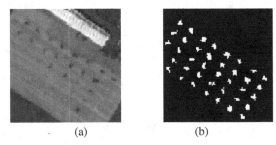

(a)　　　　　　　(b)

Fig. 2. Real hyperspectral image and corresponding distribution of anomaly targets (a) pseudo-color image. (b) truth anomaly targets distribution.

At first, we use PSO k-means clustering method to the original hyperspectral images, and there are get five different classes by setting the numbers of clustering center for five. Secondly, we extract five bands for those classes through fourth-order cumulant value, the value of five bands are largest in every class, which have the largest singularity, i.e. contain the most anomaly target feature. Then, the bands subset is composed of five bands. Finally, the KRX algorithm are used on the bands subset, and get anomaly targets detection results, kernel function is RBF, the parameter is 0.07. At the same time, we test also the other two algorithms for RX and KRX, Fig.3 show the detection performance of all algorithms, the proposed PSOC-KRX algorithm is demonstrated higher performance for detecting the possible targets.

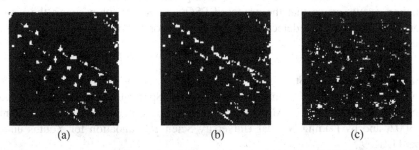

(a)　　　　　　　(b)　　　　　　　(c)

Fig. 3. Anomaly detection results of three algorithms : (a) the Proposed PSOC-KRX algorithm, (b) KRX algorithm, and (c) RX algorithm

The Receiver Operating Characteristic (ROC) curves provide a quantitative performance comparison by plotting the probability of detection. ROC represents the varying relationship of detection probability and false alarm rate [16]. The ROC curves of Fig.4 show the performance of PSOC-KRX anomaly targets detection is better than the others, the proposed algorithm significantly outperforms the kernel RX and the conventional RX at lower false alarm rate, Which improves detection perf- ormance of the conventional anomaly algorithms.

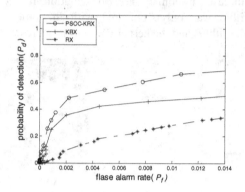

Fig. 4. ROC comparison of three anomaly detection algorithms

4   Conclusion

This paper proposes a novel anomaly detection algorithm by preprocessing hyperspectral imagery, which is extended the KRX algorithm to using fully the nolinear feature and decreasing redundancy between bands. The proposed PSOC- KRX algorithm for anomaly detection has several key differences. First, through particle swarm optimization clustering method, the original hyperspectral imagery are divided for a few different classes, which are similar in every classes, so, we use the fourth-order cumulant value to estimate the singularity of bands in every classes, and extract the band its value of cumulant is the largest. Then, the computation time is short because of bands reduction and without large kernel matrix computation. The simulation results prove that the proposed PSOC-KRX algorithm shows the well s performance for anomaly detection, which is superior to the conventional RX and kernel RX detector.

Acknowledgment. This Study was Partially Supported by Science Foundation of Heilongjiang Educational Committee (Research on Spectral Feature Based Nolinear Anomaly Target Detection Algorithms in Hyperspectral Imagery) under Grants 12533002, and by Daqing Normal University Science Foundation for Youths under Grants 11ZR09.

References

1. Du, B., Zhang, L.: Random-Selection-Based Anomaly Detector for Hyperspectral Imagery. IEEE Transactions on Geoscience and Remote Sensing 49, 1578–1589 (2011)
2. Qi, B., Zhao, C.H., Youn, E., Nansen, C.: Use of Weighting Algorithms to Improve Traditional Support Vector Machine Based Classifications of Reflectance Data. Optics Express 19, 26816–26826 (2011)
3. Reed, I.S., Yu, X.: Adaptive Multiple-Band CFAR Detection of an Optical Pattern With Unknown Spectral Distribution. IEEE Trans. Acoust., Speech Signal Process. 38, 1760–1770 (1990)
4. Kwon, H., Nasrabad, N.M.: Kernel RX-Algorithm: A Nonlinear Anomaly Detector for Hyperspectral Imagery. IEEE Transactions on Geoscience and Remote Sensing 43, 388–397 (2005)
5. Matteoli, S., Diani, M., Corsini, G.: Improved Estimation of Local Background Covariance Matrix for Anomaly Detection in Hyperspectral Images. Opt. Eng. 49, 046201-1- 046201-16 (2010)
6. Banerjee, A., Burlina, P., Dieh, C.: A Support Vector Method for Anomaly Detection In Hyperspectral Imagery. IEEE Transactions on Geoscience and Remote Sensing 44, 2282–2291 (2006)
7. Kennedy, J., Eberhart, R.C.: Swarm Intelligence. Morgan Kaufmann, San Mateo (2001)
8. Bazi, Y., Melgani, F.: Semisupervised PSO-SVM Regression for Biophysical Parameter Estimation. IEEE Trans. Geosci. Remote Sens. 45, 1887–1895 (2006)
9. Bazi, Y., Melgani, F.: Toward An Optimal SVM Classification System for Hyperspectral Remote Sensing Images. IEEE Trans. Geosci. Remote Sens. 44, 3374–3385 (2006)
10. Ghoggali, N., Melgani, F., Bazi, Y.: A multiobjective genetic SVM approach for classification problems with limited training samples. IEEE Trans. Geosci. Remote Sens. 47, 1707–1718 (2009)
11. Paoli, M., Melgani, F.: Clustering of Hyperspectral Image Based on Multiobjective Particle Swarm Optimization. IEEE Transactions on Geoscience and Remote Sensing 47, 4175–4178 (2009)
12. Niknam, T., Amiri, B.: An Efficient Hybrid Approach Based on PSO, ACO and K-Means for Cluster Analysis. Applied Soft Computing, 183-197 (2010)
13. Härdle, W., Simar, L.: Applied Multivariate Statistical Analysis, pp. 303–322. Springer, Heidelberg (2007)
14. Matteoli, S., Diani, M., Corsini, G.: A Tutorial Overview of Anomaly Detection in Hyperspectral Images. IEEE Aerosp. Electron. Syst. Mag. Tutorials 25, 5–28 (2010)
15. Hazai, S.K., Safari, A., Mojaradi, B., Homayouni, S.: A Fast-Adaptive Support Vector Method for Full-Pixel Anomaly Detection in Hyperspectral Images. In: 2011 IEEE Geoscience and Remote Sensing Society, Vancourer, Canada, pp. 1763–1766 (2011)
16. Gu, Y.F., Liu, Y., Zhang, Y.: A Selective Kpca Algorithm Based on High-Order Statistics For Anomaly Detection in Hyperspectral Imagery. IEEE Geoscience and Remote Sensing Letters 5, 43–47 (2008)

Transcribing Bach Chorales
Using Particle Swarm Optimisations

Somnuk Phon-Amnuaisuk

Music Informatics Research Group,
Faculty of Business and Computing,
Brunei Institute of Technology, Brunei Darussalam
somnuk.phonamnuaisuk@itb.edu.bn

Abstract. This paper reports a novel application of particle swarm optimisation to polyphonic transcription task. The system transforms an input audio into activation strength of pitches in the desired range. This transformation begins with audio information in time-domain to frequency-domain and finally, to activation strength of pitches (a.k.a. piano-roll representation). We can infer the likely sounding pitches by comparing the observed activation strength of input audio to reference Tone-models. Although each Tone-model is learned offline from the pitches one wish to perform transcription with, this process often only approximates the Tone-model characteristics due to the variations in volume and other effects introduced from the manner of note executions. Hence, predicting sounding notes based solely on Tone-models gives inaccurate predictions. Here, we apply PSO to search for an optimum aggregation of different predicted pitches that best represents the input activation strength. We describe our problem formulation and the design of our approach. The experimental results show our approach to be of potential in the task of polyphonic transcription.

Keywords: Particle swarm optimisation, Polyphonic transcription, Tone-models, Transcribing Bach's Chorales.

1 Introduction

Polyphonic transcription is an important research problem since it could enable applications such as score following, automatic transcription of sheet music and other related downstream applications in music education. One of the pioneer works by [1], approached this problem using a blackboard expert system approach. The system employed many knowledge sources to search for polyphonic notes based on rules. Knowledge elicitation, knowledge engineering and depth of inference chain pose challenges to the rule-based approach. In contrast to the rule-based approach, a Bayesian approach [2,3,4] summarises rules as conditional probability table (CPT). This provides the means to handle uncertainty as well as reducing the depth of the inference chain since the number of CPT in a Bayesian network tends to be a lot smaller than the number of rules in an expert system.

Y. Tan, Y. Shi, and H. Mo (Eds.): ICSI 2013, Part I, LNCS 7928, pp. 192–199, 2013.

Supervised learning techniques can lift the burden of knowledge elicitation and knowledge engineering by facilitating model construction through learning from training examples. Researchers have employed artificial neural networks [5] and graphical models [6] to build their transcription models. Although the learning can be automated, the supervised learning approach requires enough training data points to construct a good model. This implies that the model must be trained with enough polyphonic sound examples. This is ineffective for our problem since all possible combinations of only four pitches in the pitch range between C2 to B5 (four octaves) would already be $\frac{48!}{44! \times 4!} = 194580$ patterns.

Instead of learning a complete example set, if we could learn only basis vectors in which a complete example set could be constructed from the basis vectors, then this should be a much more effective method. Recently, factoring techniques [7,8] have been applied to polyphonic transcription task. The factoring technique can be interpreted as solving for aggregations of basis vectors from a given input. The learning of basis vectors, where all observations could be reconstructed based on the basis vectors, has received a lot of interest recently [9,10]. The idea of the basis vector is appealing and in this work, we explore the concept using PSO to search for a good aggregation of Tone-models (basis vectors).

This paper is organised into the following sections. Section 2 discusses our proposed concept and gives the details of the techniques behind it. Section 3 provides the output of the proposed apprach. Finally, the conclusion and further research are presented in section 4.

2   Problem Formulation

Let $\mathbf{x}$ be a vector of length N and $x_1, ..., x_n$ are elements of the vector representing time-domain samples of polyphonic audio input. Let $\mathbf{X}$ be a vector representing frequency domain contents of the corresponding time-domain $\mathbf{x}$, Fourier transformation states that the element k of $\mathbf{X}$ is:

$$X_k = \sum_{n=0}^{N-1} h_n x_n e^{-j2\pi kn/N} \tag{1}$$

where h_n is the hamming window defined as $0.54 - 0.46 cos(2\pi \frac{n}{N})$ and $k = 0, 1, ..., N/2$. In our implementation, X_k is further binned according to the musical pitches. The binning transforms X_k coefficients into a piano roll of activation strength with the center frequency f_c of pitch $i, i \in \{40, ..., 99\}$ [1] at:

$$f_c(i) = 440 \times 2^{(i-69)/12} \tag{2}$$

The activation magnitude of the pitch i is the average of the magnitude of X_k coefficients in the range of $k_l = 0.99 f_c(i)$ to $k_u = 1.01 f_c(i)$ where, elements of $\mathbf{X}$ is the average of X_k coefficients between k_l and k_u for $i \in \{40, ..., 99\}$.

The vector $\mathbf{X}$ represents resulting activation strength aggregated from unknown pitches. In order to infer a suitable combination of pitch, we need to have

---

[1] This corresponds to pitches E2, ..., D#7.

activation information of each pitch. Let $\mathbf{X}_{pn}$ denotes the piano-roll vector for pitch pn. Hence, an audio input of a sounding note C4 (MIDI note #60) will be notated as $\mathbf{X}_{60}$. Theoretically, this vector should have activation activities peak at its overtone series; the following MIDI note numbers: 60, 72, 79, 84, 88, 91, 94, 96, 98. The harmonic pattern of each pitch provides a useful *Tone-model* signature of the pitch.

2.1   Framing Polyphonic Transcription as a PSO Search

Formally, let a particle $\mathbf{p} = (x_{11}, ..., x_{ij})$ represents weighted factors of j plausible Tone-models where $x_{ij} \in \mathcal{R}$ denotes the weight of particle i for Tone-model j. The particles at time point t, $\mathbf{p}(t)$ could be transformed across time to $t + 1$ as a new vector $\mathbf{p}(t + 1)$ as follow:

$$x_{ij}(t + 1) = x_{ij}(t) + v_{ij}(t + 1) \tag{3}$$

where v_{ij} is the velocity of particle i of a Tone-model j.

$$v_{ij}(t + 1) = wv_{ij}(t) + c_1 r_{1j}(p_{ij}(t) - x_{ij}(t)) + c_2 r_{2j}(g_{ij}(t) - x_{ij}(t)) \tag{4}$$

where w is the inertia weight for PSO. The v_{ij} is the velocity of a particle i in a j dimension. The $c_1 r_{1j}$ and $c_2 r_{2j}$ are weight parameters that combine the influence of the local best position p_{ij}, global best position g_{ij} and the current velocity to determine the velocities of the particles in the next step. The inertia w, parameters c_1 and c_2 are modified in each run according to the information below:

Parameters	Value		
Swarm size	40		
Max-iteration for each run	50		
Time step t	$t = 1, ...,50$		
$w \in [0.5\ 1.2]$	$w \leftarrow w_{max} - t(w_{max} - w_{min})/maxIteration$		
$c_1 \in [0.3\ 0.9]$	$c_1 \leftarrow c_{1max} - t(c_{1max} - c_{1min})/maxIteration$		
$c_2 \in [0.7\ 1.5]$	$c_2 \leftarrow c_{2max} - t(c_{2max} - c_{2min})/maxIteration$		
r_1, r_2	random values $\in [0\ 1]$		

Let $\mathbf{P}$ denotes a vector of p_{ij} and $\mathbf{G}$ denotes a vector of g_{ij} where p_{ij} represents the activation strength of the pitch j of the particle i, and g_{ij} represents the most probable activation strength of the pitch j of the whole swarm. The objective function of each particle in the swarm is to minimise

$$f_{obj} = argmin|\mathbf{X} - \mathbf{X}_{\{pn\}}\mathbf{P}| \tag{5}$$

where $\mathbf{G}$ will take the values of the best $\mathbf{P}$ at each step. The final $\mathbf{G}$ suggests the optimum aggregation of Tone-models which is actually the transcribed notes, i.e., $\mathbf{X} \approx \mathbf{X}_{\{pn\}}\mathbf{G}$.

In this implementation, we implement two PSO variations. In the first implementation, 15 sub-swarms, each with the size of 40, searches for the optimum weighted aggregation of four pitches. In the second implementation, a swarm size of 40 searches for the optimal weighted aggregation of pitches from a given six possible choices. This is motivated by the need to reduce the computational expense of the sub-swarm approach.

3   Results and Discussion

3.1   Data Preparation

All Bach chorales MIDI files used in this experiment were downloaded from http://www.jsbchorales.net/bwv.shtml. The choice of chorales were randomly picked and it was ascertained that all the notes did not fall out of the range between E2 to G5. We recorded the wave files from these MIDI files using a standard MIDI sound card. The recording parameters were set as follows: sound patch - acoustic grand piano, sampling rate - 44100 Hz, bit-depth - 16 bits, and channel - mono. Figure 1 shows the first five bars of Bach chorale titled *Aus tiefer Not schrei ich zu dir* and its corresponding piano roll representation. The piano-roll representation provided a ground truth for us to evaluate the performance of the system.

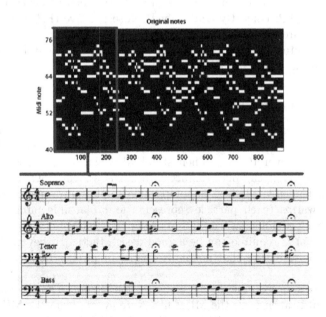

Fig. 1. A piano roll representation indicating note on (white) and note off (black). The area inside the red rectangle corresponds to the music notations below.

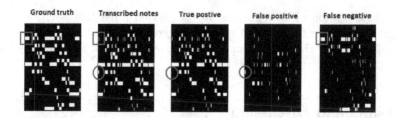

Fig. 2. This figure illustrates our pixel based evaluation scheme. This approach calculate TP, FP, and FN from the number of pixels. The three squares highlight a false negative example and the three circles highlight a false positive example.

3.2 Evaluation Scheme

From the literature, we found that different researchers may define the notion of correctly transcribed notes differently. For example, should a note with pitch value of C4 and duration of 2 seconds, be considered as correctly transcribed if the pitch is correctly identified but the duration was erroneously identified as 1.5 seconds? Due to this issue, we do not count the note, rather we count the pixel. This was also the approach taken by [10,11]. This approach represented a ground truth and the transcription output as images of note on/off (see Figure 2). Overlaying the transcribed notes on the ground truth revealed pixels corresponding to true positive (TP: correctly transcribed), false positive (FP: transcribed as notes while they are not notes) and false negative (FN: not transcribed as notes while they are notes). We measured the performance of the system using the standard precision, recall and F-score measures [11].

3.3 Experimental Results

In this experiment, a total of eight chorales were tested. The precison, recall and F score of all chorales are summarised in Table 1. The values reported here are the average values over 10 runs. Their standard deviations are less than 0.5%. The results show that the performance of a single swarm PSO (each particle represents six plausible pitches) is comparable to the performance from PSO with 15 sub-swarms (each particle represents four plausible pitches). In this problem, running a single swarm and allowing each particle to represent more pitches than what is required can be a good design tactic. However, there must be some process to ensure that not more than four pitches be included in the final output.

Figure 3 shows an example of the transcription output. The top most row is the ground truth taken from the MIDI file. Note events from the MIDI file are scaled to fit the numbers of the frames in the output transcriptions on the second row. The third, fourth and fifth rows are the TP, FP and FN respectively. The TP shows correctly identified pitches. The FP appears to be from a slight mismatch of correctly transcribed notes. The FN often appears in the low pitch region. This pattern appears in all the ten chorales and this decreases the performance

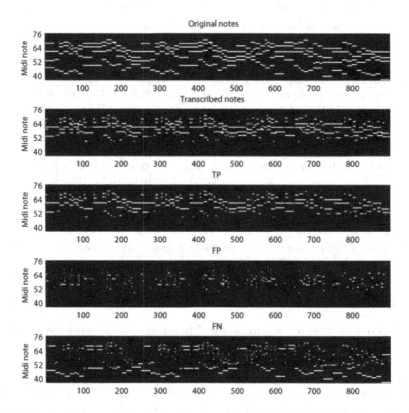

Fig. 3. Transcription output of chorale *Aus tiefer Not schrei ich zu dir*, see text for more detailed discussion

Table 1. Summary of PSO performances with 15 sub-swarms, and PSO with single swarm in transcribing Bach Chorales

Chorales	PSO 15 sub-swarms			PSO single swarm		
	Prec	Recall	F	Prec	Recall	F
Aus tiefer Not schrei ich zu dir	0.76	0.56	0.64	0.76	0.56	0.64
Nun komm, der Heiden Heiland	0.70	0.60	0.65	0.71	0.60	0.65
Ach wie nichtig, ach wie flüchtig	0.78	0.62	0.69	0.77	0.62	0.69
Wär' Gott nicht mit uns diese Zeit	0.68	0.59	0.63	0.67	0.59	0.63
Herr Jesu Christ, du höchstes Gut	0.71	0.59	0.65	0.70	0.60	0.65
Ach Gott und Herr	0.78	0.61	0.68	0.78	0.61	0.68
Es ist das Heil uns kommen her	0.76	0.65	0.69	0.74	0.65	0.69
Nun ruhen alle Wälder	0.73	0.62	0.67	0.74	0.62	0.68

Table 2. Summary of performance comparison between PSO and NMF. Numbers in bold signify the higher F-measures between the two experiments.

Chorales	NMF			PSO		
	Prec	Recall	F	Prec	Recall	F
Aus tiefer Not schrei ich zu dir	0.63	0.63	0.63	0.76	0.56	**0.64**
Nun komm, der Heiden Heiland	0.66	0.70	**0.68**	0.71	0.60	0.65
Ach wie nichtig, ach wie flüchtig	0.60	0.78	0.67	0.77	0.62	**0.69**
Wär' Gott nicht mit uns diese Zeit	0.66	0.69	**0.67**	0.67	0.59	0.63
Herr Jesu Christ, du höchstes Gut	0.65	0.69	**0.67**	0.70	0.60	0.65
Ach Gott und Herr	0.67	0.68	**0.68**	0.78	0.61	0.68
Es ist das Heil uns kommen her	0.67	0.71	**0.69**	0.74	0.65	0.69
Nun ruhen alle Wälder	0.55	0.70	0.62	0.74	0.62	**0.68**

of the system substantially. Upon inspection of the sound files, we believe this could be due to the nature of the bass parts which tends to be lower in volume as well as in their strength in FT coefficients.

3.4 Comparison with Non-negative Matrix Factorisation Technique

Table 2 provides the results from the PSO and the Non-negative Matrix Factorisation (NMF) techniques. NMF factorises an input matrix V into two non-negative components i.e., $\mathbf{V} \approx \mathbf{WH}$. The non-negative constraint gives an interpretation of part-based aggregations [7]. NMF has been successfully applied to many applications, including to polyphonic transcription problem. In [8], the authors showed that the technique was applicable to polyphonic transcription tasks. Recently, NMF has been applied to polyphonic transcription problem [12,13]. In this comparison, NMF factors an input activation strength $\mathbf{V}$ to transcription output $\mathbf{H}$ based on the Tone-model $\mathbf{W}$. We would like to point out the similarity between $\mathbf{V} \approx \mathbf{WH}$ and $\mathbf{X} \approx \mathbf{X}_{\{pn\}}\mathbf{G}$. The experiment shows comparable performance between the NMF and our proposed approach.

4   Conclusion and Future Work

In this paper, we have shown the application of PSO in polyphonic transcription task. The problem domain is formulated as the optimisation of Tone-models aggregation, where each particle searches for a good aggregation factor. The final aggregation is the global best information of the swarm. The experimental results demonstrate that our proposed technique has a good potential. The same data (i.e., chorales, Tone-models) are also employed with the competing NMF technique. The results from both techniques are comparable. In future work, we will improve on the transcription accuracy of notes in the low frequency region which we believe is not due to the limitation of the PSO, rather from the signal processing part of the system.

Acknowledgement. We wish to thank Brunei Institute of Technology for their partial financial support given to this research. We would also like to thank anonymous reviewers for their constructive comments and suggestions.

References

1. Martin, K.D.: A blackboard system for automatic transcription of simple polyphonic music. M.I.T. Media Lab, Perceptual Computing. Technical Report. 385 (1996)
2. Kashino, K., Nakadai, K., Kinoshita, T., Tanaka, H.: Application of bayesian probability network to music scence analysis. In: Proceedings of IJCAI Workshop on CASA, Montreal, pp. 52–59 (1995)
3. Walmsley, P.J., Godsill, S.J., Rayner, P.J.W.: Bayesian graphical models for polyphonic pitch tracking. In: Proceedings of Diderot Forum on Mathematics and Music, Vienna, Austria, December 2-4, pp. 1–26 (1999)
4. Davy, M., Godsill, S.J.: Bayesian Harmonic Models for Musical Signal Analysis. In: Bernardo, J.M., Bayarri, M.J., Berger, J.O., Dawid, A.P., Heckerman, D., Smith, A.F.M., West, M. (eds.) Bayesian Statistics 7, pp. 105–124. Oxford University Press (2003)
5. Marolt, M.: A connectionist approach to automatic transcription of polyphonic piano music. IEEE Transactions on Multimedia 6(3), 439–449 (2004)
6. Vincent, E., Rodet, X.: Music transcription with ISA and HMM. In: Proceedings of the Fifth International Conference on Independent Component Analysis and Blind Signal Separation (ICA 2004), Gradana, Spain, pp. 1197–1204 (2004)
7. Lee, D.D., Seung, H.S.: Learning the parts of objects by non-negative matrix factorozation. Nature 401, 788–791 (1999)
8. Smaragdis, P., Brown, J.C.: Non-negative matric factorization for polyphonic music transcription. In: Proceedings of IEEE Workshop Applications of Signal Processing to Audio and Acoustics, pp. 177–180. New Paltz, NY (2003)
9. Hoyer, P.O.: Non-negative sparse coding. In: Proceedings of IEEE Workshop on Neural Networks for Signal Provcessing XII, Martigny, Switzerland (2002)
10. Plumbley, M.D., Abdullah, S.A., Blumensath, T., Davies, M.E.: Sparse representation of polyphonic music. Signal Processing 86(3), 417–431 (2005)
11. Phon-Amnuaisuk, S.: Transcribing Bach chorales: Limitations and potentials of non-negative matrix factorisation. EURASIP Journal on Audio, Speech and Music Processing 2012, 11 (2012)
12. Phon-Amnuaisuk, S.: Polyphonic transcription: Exploring a hybrid of tone models and particle swarm optimisation. In: Machado, P., Romero, J., Carballal, A. (eds.) EvoMUSART 2012. LNCS, vol. 7247, pp. 211–222. Springer, Heidelberg (2012)
13. Bertin, N., Badeau, R., Vincent, E.: Enforcing Harmonicity and Smoothness in Bayesian Non-Negative Matrix Factorization Applied to Polyphonic Music Transcription. IEEE Transactions on Audio, Speech, and Language Processing 18(3), 538–549 (2010)

Deadline Constrained Task Scheduling
Based on Standard-PSO in a Hybrid Cloud

Guoxiang Zhang[1] and Xingquan Zuo[2]

[1] Institute of Microelectronics of Chinese Academy of Sciences, Beijing, China
[2] Beijing University of Posts and Telecommunications, Beijing, China
zhgx87@163.com, zuoxq@bupt.edu.cn

Abstract. Public cloud providers provide Infrastructure as a Service (IaaS) to remote users. For IaaS providers, how to schedule tasks to meet peak demand is a big challenge. Previous researches proposed purchasing machines in advance or building cloud federation to resolve this problem. However, the former is not economic and the latter is hard to be put into practice at present. In this paper, we propose a hybrid cloud architecture, in which an IaaS provider can outsource its tasks to External Clouds (ECs) without establishing any agreement or standard when its local resources are not sufficient. The key issue is how to allocate users' tasks to maximize its profit while guarantee QoS. The problem is formulated as a Deadline Constrained Task Scheduling (DCTS) problem which is resolved by standard particle swarm optimization (PSO), and compared with an exact approach (CPLEX). Experiment results show that Standard-PSO is very effective for this problem.

Keywords: IaaS cloud, task scheduling, hybrid cloud, Standard-PSO.

1 Introduction

Cloud computing attracts an increasing number of individual and corporations to rent cloud service due to its convenience and economy. As an important part of cloud service, Infrastructure as a Service (IaaS) becomes much more popular because it is the foundation for higher-level service such as Platform as a Service (PaaS) and Software as a Service (SaaS) [1]. Many giant IaaS providers such as Amazon EC2, IBM Smart Cloud Enterprise and Sun Grid have emerged. They can offer different VM instance types, which are characterized with different device configuration, Quality of Service (QoS) and pricing models. However, the arriving of users' requests is aperiodic and the type and amount of required VM instance are uncertain. Thus, it may occur that an IaaS provider cannot meet all the demands while guaranteeing users' QoS when peak demands happen.

To tackle this problem, in this paper an effective approach is proposed to utilize hybrid cloud model to allocate limited resources from IaaS provider's perspective. In

Y. Tan, Y. Shi, and H. Mo (Eds.): ICSI 2013, Part I, LNCS 7928, pp. 200–209, 2013.
© Springer-Verlag Berlin Heidelberg 2013

our approach, an IaaS cloud is regarded as a private cloud and is able to outsource to its tasks to other clouds when its local resources are not enough to satisfy user's requirements. The resource allocation problem in the hybrid cloud model is a kind of Deadline Constrained Task Scheduling (DCTS) problem, in which each task has a strict deadline and the objective is to maximize the profit of an IaaS cloud provider under the premise of guaranteeing each task's deadline constraint. To achieve this, a particle swarm optimization (PSO) based scheduling algorithm is proposed to get an optimal or suboptimal solution in a shorter computational time.

The rest of the paper is organized as follows: Section 2 establishes an integer programming (IP) model for the DCTS. Standard-PSO based scheduling algorithm is proposed for the DCTS in Section 3. The experimental results are given in Section 4. Section 5 reviews relative works. We conclude this paper in Section 6.

2   Problem Definition

From an IaaS provider's perspective, the private cloud refers to itself, and ECs to other public IaaS Clouds. Our model focuses on batch types of workloads that consist of a bag of independent instances. For example, an enterprise wants to understand its customers in order to develop a better marketing plan, or examine its supply chain to look for opportunities to improve efficiency, or analyze sensor data to predict machines failure and prevent revenue lost earlier [2]. In our problem, each application submitted by users consists of a number of parallel and independent tasks and has a strict deadline, before which all of its tasks must be finished. Each task requires one specific type of VM instance.

Suppose that $CP = \{CP_1, CP_2, \ldots, CP_n\}$ is a set of cloud providers. Assume CP_1 is the private cloud and $CP_2, \ldots, CP_n$ are ECs. $A = \{a_1, a_2, \ldots, a_w\}$ is a set of applications. Each application a_j ($j \in \{1, 2, \ldots, w\}$) has a strict deadline d_j and runtime r_j, and consists of a task set $Task_j = \{t_{j1}, t_{j2}, \ldots, t_{jT_j}\}$. Time is explicitly represented in the IP model by introducing time slots with a granularity of one hour. Let S be the maximum number of time slots in the schedule process, we have $S = \max_{j \in \{1,2,\ldots,w\}} (d_j)$. The objective is to allocate the w applications to $CP_k (k = 1, 2, \ldots, n)$ to maximize the profit of CP_1. Each task must be allocated to one unique $CP_k (k \in \{1, 2, \ldots, n\})$. Once a task begins to execute, its running slots are consecutive. In each slot $s (s \in \{1, 2, \ldots, S\})$, resources used by all task executed in CP_1 cannot exceed the total resources of CP_1.

Problem parameters and decision variables are defined in Tables 1 and 2, respectively.

Table 1. Problem Parameters

parameter		parameter	
n	Number of cloud providers	I	Number of VM types
w	Number of applications	p_v	Price of the vth VM type in CP_1
c_{kv}	Cost of the vth VM type in CP_k	d_j	Deadline of the jth application
r_j	Runtime of the jth application	S	Maximum deadline
T_j	Number of tasks in the jth application	b_{jv}	If $b_{jv}=1$, application a_j use VM_v; if $b_{jv}=0$, it does not use this type.
cpu_v	Number of CPUs for the vth VM type in CP_1	mem_v	Size of memory for the vth VM type in CP_1
$total\_cpu$	Total number of CPUs in CP_1	$total\_mem$	Total size of memory in CP_1

Table 2. Decision Variables

y_{jlk}	Binary decision variable, such that $y_{jlk}=1$ if task l in application j is allocated to CP_k; otherwise $y_{jlk}=0$
st_{jl}	Integer decision variable, start time slot of task t_{jl}
z_{jls}	Binary decision variable, such that $z_{jls}=1$ if task t_{jl} is allocated to time slot s of CP_1

The problem can be formulated as the following IP model.

Maximize

$$\text{Profit} = \sum_{j=1}^{w}\sum_{v=1}^{I} T_j b_{jv} p_v r_j - \sum_{j=1}^{w}\sum_{l=1}^{T_j}\sum_{v=1}^{I}\sum_{k=1}^{n} y_{jlk} b_{jv} c_{kv} r_j \quad (1)$$

Subject to:

$$\sum_{k=1}^{n} y_{jlk} = 1, \ \forall \ j \in \{1,2,\cdots,w\}, \ l \in \{1,2,\cdots,T_j\} \quad (2)$$

$$\sum_{s=1}^{d_j} z_{jls} = y_{jl1} r_j, \ \forall \ j \in \{1,2,\cdots,w\}, \ l \in \{1,2,\cdots,T_j\} \quad (3)$$

$$st_{jl} \geq 1, \ \forall \ j \in \{1,2,\cdots,w\}, \ l \in \{1,2,\cdots,T_j\} \quad (4)$$

$$st_{jl} \leq d_j - r_j + 1, \forall \ j \in \{1,2,\cdots,w\}, \ l \in \{1,2,\cdots,T_j\} \quad (5)$$

$$(s \leq st_{jl} - 1) \vee (s \geq d_j - r_j) \vee ((s \geq st_{jl}) \wedge (s \leq st_{jl} + r_j - 1) \wedge (z_{jls} = y_{jl1})) \quad (6)$$
$$\forall \ s \in \{1,2,\cdots d_j\}, \ j \in \{1,2,\cdots,w\}, \ l \in \{1,2,\cdots,T_j\}$$

$$\sum_{j=1}^{w}\sum_{l=1}^{T_j}\sum_{v=1}^{I} z_{jls}b_{jv}cpu_v \le total\_cup, \quad \forall \ s \in \{1,2,\cdots S\} \tag{7}$$

$$\sum_{j=1}^{w}\sum_{l=1}^{T_j}\sum_{v=1}^{I} z_{jls}b_{jv}mem_v \le total\_mem, \quad \forall \ s \in \{1,2,\cdots S\} \tag{8}$$

$$y_{jlk} \in \{0,1\}, \forall \ j \in \{1,2,\cdots,w\}, \ l \in \{1,2,\cdots,T_j\}, k \in \{1,2,\cdots n\} \tag{9}$$

$$z_{jls} \in \{0,1\}, \forall \ j \in \{1,2,\cdots,w\}, \ l \in \{1,2,\cdots,T_j\}, s \in \{1,2,\cdots S\} \tag{10}$$

$$st_{jl} \in \{1,2,\cdots,S\}, \forall \ j \in \{1,2,\cdots,w\}, \ l \in \{1,2,\cdots,T_j\} \tag{11}$$

The first term of the objective function (1) represents the income of CP_1 and the second one means its cost. Constraint (2) guarantees that each task is allocated to exactly one cloud. Constrain (3) ensures that each task must be finished before its deadline. Constraints (4)-(6) guarantee that each task is non-preemptable. Constraints (7) and (8) are only applied to CP_1 and restrict the number of CPUs and amount of memory used for each slot. Finally, Eq. (9)-(11) describe the decision variable definitions.

3   PSO Based Scheduling Approach

The problem is a task allocation and sequence one. Solving such a problem using exact approach will take a large amount of computational time for a large size problem. Hence, a Particle Swarm Optimization (PSO) [3] based heuristic scheduling approach is used to solve it. In this approach, each particle represents a set of priorities of all tasks. The purpose of PSO is to produce an optimal particle (a set of priorities) to allocate cloud resources in the most effective manner.

3.1   Standard-PSO

In Standard-PSO, each individual treated as a particle in D-dimensional search space, and represented by a three triple $\{X_i, V_i, P_i\}$. $X_i = (x_{i1}, x_{i2}, \ldots, x_{iD})$ and $V_i = (v_{i1}, v_{i2}, \ldots, v_{iD})$ denote the position and velocity of particle i, respectively. $P_i = (p_{i1}, p_{i2}, \ldots p_{iD})$ represents the personal best (*pbest*) of particle i. $G = (g_1, g_2, \ldots, g_D)$ denotes the global best (*gbest*). The quality of a position (particle) is evaluated by a fitness function. The value of each component in V_i can be clamped to the range of $[-v_{\max}, v_{\max}]$ to control excessive roaming of particle, and updated by

$$v_{id}(t+1) = \omega v_{id}(t) + c_1 r_1 \left[x_{id}(t) - p_{id}(t)\right] + c_2 r_2 \left[x_{id}(t) - g_d(t)\right] \tag{12}$$

where $i = 1, 2, \ldots, M$ denotes the number of particles and $d = 1, 2, \ldots, D$ is the dimension of particles. r_1 and r_2 are uniformly distributed random number whose range is [0, 1]. c_1 and c_2 are learning factors. c_1 is the individual cognition component and c_2 is the social communication component. ω is the inertia weight to avoid unlimited growth of particle's velocity. The particle flies toward a new position according to Eq. (13), and each value of the new position should not exceed the range of $[\min X, \max X]$.

$$x_{id}(t+1) = x_{id}(t) + v_{id}(t+1) \tag{13}$$

The procedure of Standard-PSO is as follows.

Step 1) Initialize position and velocity of all particles randomly in the search space.

Step 2) Evaluate fitness value of all particles; let each particle's *pbest* and its fitness value equal to the current position and fitness value, respectively; let *gbest* be the best one among all particles.

Step 3) Updated each particle's velocity and position using (12) and (13).

Step 4) Calculate the fitness value of all particles;

Step 5) Update *pbest*. For each particle, if fitness value of its new position is better than that of its *pbest*, then replace the *pbest* by the new position.

Step 6) Update *gbest*. For each particle, if fitness value of its new position is better than that of its *gbest*, then replace its *gbest* by the new position.

Step 7) If the stopping criterion is satisfied, then output *gbest* and its fitness value; otherwise go to Step 3).

3.2 Solution Representation

Each application contains many tasks, such that all applications can be considered as a set of tasks, i.e., $T = \{t_1, t_2, \ldots, t_{TN}\}(TN = \sum_{j=1}^{w} T_j)$. The purpose is to allocate these *TN* tasks to *n* clouds. A particle is expressed as a *TN* (that is, $D = TN$) dimensions vector and each dimension (position) represents a task. The ranked-order value rule [4] is used to decode a particle $X_i = (x_{i1}, x_{i2}, \ldots, x_{iD})$ into a permutation of tasks $T = \{t_1, t_2, \ldots, t_D\}$ [5] to evaluate this particle. For example, for a problem with 5 tasks (*D*=5), the *i*th particle is denoted by $X_i = (0.94, 3.46, 2.78, 4.83, 3.67)$. The position x_{i4} has the greatest value, such that the task represented by x_{i4} is assigned a rank value *one*. Similarly, the rank values of 2, 3, 4 and 5 are assigned to x_{i5}, x_{i2}, x_{i3} and x_{i1}, respectively. Thus, a priority sequence of the tasks, $Pri = \{5, 3, 4, 1, 2\}$ is obtained.

3.3 Evaluation Function

Each task in a particle is allocated to $CP_k (k = 1, 2, \ldots, n)$ according to its priority. If CP_1 has available resources to meet a task's demand during its runtime, then the task is allocated to CP_1; otherwise, an EC with minimal cost is chosen by

$$EC_l = \arg \min_{k \in \{2, \ldots, n\}} \{ \sum_{v=1}^{l} b_{app(l)v} c_{kv} \}, \, l \in \{1, 2, \ldots, D\} \tag{14}$$

where *app(l)* means the application that the *l*th task belongs to.

The pseudo-code of evaluation process for each particle is as follows.

Step 1) Initialization

$total\_cost=0$, $avail\_cpu_s = total\_cpu$, $avail\_mem_s = total\_mem$, $s \in \{1,2,...,S\}$
Calculate the total income by

$$total\_income = \sum_{l=1}^{D}\sum_{v=1}^{I} b_{app(l)v} P_v r_{app(l)} \qquad (15)$$

Step 2) Calculate the total cost:
Sort the task set T in a descending order according to the code of the particle.
Let the lth task in T be t_l, its start time be st_l and let $st_l = 1$, $l \in \{1,2,...,D\}$;

For each task t_l in T

    While $st_l \le d_{app(l)} - r_{app(l)} + 1$

        $IsPC$= true;

        For each $s \in \{st_l,...,st_l + r_{app(l)} -1\}$

            If $\sum_{v=1}^{I} cpu_v b_{app(l)v} \ge avail\_cpu_s$ or $\sum_{v=1}^{I} mem_v b_{app(l)v} \ge avail\_mem_s$

                Set $IsPC$ = false; break for;

            End If

        End For

        If $IsPC$ = true

            Calculate the cost for task t_l by

$$cost_l = \sum_{v=1}^{I} c_{1v} b_{app(l)v} r_{app(l)} \qquad (16)$$

            Update $avail\_cpu_s$ and $avail\_mem_s$ for each $s \in \{st_l,...,st_l + r_l -1\}$;

            Break while;

        End If

        $st_l = st_l +1$;

    End While

    If $IsPC$==false

        Select the EC using (14) and calculate the cost of task t_l by

$$cost_l = \sum_{v=1}^{I} c_{(EC_l)v} b_{app(l)v} r_{app(l)} \qquad (17)$$

    End If

    $total\_cost = total\_cost + cost_l$;

End for

Step 3) Output $Profit = total\_income - total\_cost$.

4   Experimental Results and Analysis

To verify the effectiveness of our approach, it is applied to several problem instances. The private cloud and EC's prices and instance types are set according to practice, and are given in Table 3-5. CPU and memory are chosen here because of their importance for a VM instance and common use in many researches [6-7].

Table 3. Instance Types

Name	CPUs	Memory
Small	1	1.7
Large	4	7.5
Xlarge	8	15

Table 4. Private Cloud's Cost and Price

	Small	Large	Xlarge
cost	0.03	0.12	0.24
price	0.08	0.32	0.64

Table 5. ECs' Price

EC	Small	Large	Xlarge
A	0.085	0.34	0.68
B	0.070	0.30	0.70
C	0.100	0.40	0.72

Three problem instances are designed. Problem instance 1 consists of 8 applications. Problem parameters are shown in Table 6. VM instance type requested by each application is randomly selected from the above VM instance types. Problem instance 1 is a small size one, which is used to compare Standard-PSO with the exact approach (CPLEX v12.0) to show accuracy of our approach. Problem instances 2 and 3 are large size ones, and are used to test the ability of our approach to handle large size problem instances. Their parameters are shown in Table 7.

Table 6. Parameters of Problem Instance 1

Applications		Cloud Resources	
Number of tasks	$\sim$unif[1,5]	CPU	20
VM instance type	$\sim$unif[1,3]	Memory	40GB
Deadline (hours)	$\sim$unif[1,5]		
Runtime (hours)	$\sim$unif[1,Deadline]		

Table 7. Parameters of Problem Instances 2 and 3

Applications		Cloud Resources	
Number of tasks	$\sim$unif[1,50]	CPU	512
VM instance type	$\sim$unif[1,3]	Memory	1024GB
Deadline (hours)	$\sim$unif[1,168]		
Runtime (hours)	$\sim$unif[1,Deadline]		

Standard-PSO based heuristic approach (SPSO-HA) is compared with CPLEX v12.0. SPSO-HA's parameters are shown in Table 8. To make the number of particles in the swarm increase with the number of tasks, the swarm size *pop_size* is set to 2*D*. c_1 and c_2 are given by 2.0 as suggested in [8]. ϖ is assigned as 0.4. For all problem instances, the termination criteria of Standard-PSO are to reach *maxGen*.

Table 8. Parameters of SPSO-HA

pop_size	maxGen	v_{max}	minX	maxX	c_1	c_2	ϖ
2D	1000	$\lfloor D/2 \rfloor$	0	D	2	2	0.4

SPSO-HA is coded in Matlab 7.0 and run on a PC with 64-bit intel core i5 CPU and 4 GB memory using Windows 7 operation system. For all problem instances, SPSO-HA carries out 10 independent runs. CPLEX is a mathematical programming solver, which is able to solve the formulation (Equations (1)-(11)) of this problem to obtain optimal solution.

For problem instance 1, CPLEX can obtain its optimal solution within reasonable time. The average profits obtained in the 10 runs of SPSO-HA and their average runtime are given in Table 9. We can see that the average profit obtained by SPSO-HA is 4.84, which is very close to the optimal profit found by CPLEX.

Table 9. Comparison of Average Profit and Runtime of Standard-PSO

Algorithms	Average profit	Average runtime(second)
SPSO-HA	4.8400	14.72
CPLEX	**4.9100**	**0.98**

Since we obtain the optimal solution, we can calculate the offline error (OE) and standard deviation (SD) of solutions got by SPSO-HA by (18) and (19), respectively.

$$OE = 1/(rn \sum_{i=1}^{rn} | prof_i - prof |) \tag{18}$$

where *rn* represents the number of runs; $prof_i$ is the profit of the solution obtained in the *i*th run and $prof$ is the optimal profit obtained by CPLEX.

$$SD = \sqrt{\sum_{i=1}^{rn} (prof_i - mean\_prof)^2 / rn} \tag{19}$$

where *mean_prof* is the average profit obtained in the 10 runs.

The OE and SD obtained by the SPSO-HA are given in Table 10. We can observe that the profit achieved by the SPSO-HA is very close to the optimal profit.

Table 10. Comparison of OE and SD for Problem Instance 1

Algorithm	offline error	standard deviation
SPSO-HA	0.0700	0.0733

For lager size problem instance 2 and 3, since CPLEX cannot obtain their optimal solutions with a reasonable computational time, the maximal runtime of CPLEX is restricted as 5 hours. Table 11 gives the average profits and runtime in the 10 runs of SPSO-HA and profits obtained by CPLEX for the two instances. It is observed that although CPLEX runs much longer time, solutions found by CPLEX are worse than those of SPSO-HA.

Table 11. Comparison of Average Profit and Runtime for Problem Instances 2 and 3

Algorithms	Problem instance 2		Problem instance3	
	Ave profit	Avetime (second)	Ave profit	Avetime (second)
SPSO-HA	**3516.5840**	**2790.60**	**2846.0600**	**4211.59**
CPLEX (5 hrs)	3027.8400	18000.00	2830.8800	18000.00

5   Related Work

At present, many researches have been studied to effectively allocate cloud resources for a single cloud [9-11]. Nathani et al. [9] proposed an algorithm to deal with deadline-sensitive task scheduling problem. This scheduling approach solves the limited resources problem by rejecting user requests, resulting in decreasing users' QoS. Zhao et al. [10] proposed an approach to schedule independent and divisible tasks to minimize the maximal complement time of all tasks, and did not consider the case of resource limitation. In [11], Li suggested that the job scheduling system should use cloud resources as few as possible to reduce the cost while meeting the requirement of QoS. All above researches focused on the local resources allocation in a single IaaS cloud, and did not consider scheduling tasks amongst different clouds. An IaaS provider will reject task requests when its resources are not sufficient; however, rejecting task requests may lose the reputation and decrease QoS [12].

In order to enhance clouds' elasticity and reliability, resource and task scheduling among multiple clouds has been studied. The concept of cloud federation was proposed in [13] and [14], which supplies local resources by integrating multiple clouds. However, cloud federation is not easy to be applied to practice before standardization and cooperation agreements are published and signed by cloud federation members.

From cloud providers' perspective, effectively allocating limited resources is important to maximize its profit and guarantee the QoS. To the best of our knowledge, our research is the first one on building a framework to outsource tasks to external clouds from an IaaS provider's perspective to maximize profit and guarantee user-level QoS.

6   Conclusion

In the paper, an integer programming model is established for the problem of resource allocation in a hybrid IaaS cloud environment. Standard-PSO based approach is proposed to solve this problem. In Standard-PSO, each dimension of a particle represents a task and a set of tasks' priorities is obtained by sorting the dimensions according to their values in a descending order. This approach can make cloud resource allocation more elastic and guarantee user-level QoS. Experimental results show that our approach is able to produce scheduling solutions that are better than those obtained by CPLEX for large size problems.

References

1. Bhardwaj, S., Jain, L., Jain, S.: Cloud computing: a study of infrastructure as a service (IaaS). International Journal of Engineering and Information Technology 2(1), 60–63 (2010)
2. Liu, H., Orban, D.: GridBatch: cloud computing for large-scale data-intensive batch applications. In: IEEE International Symposium on Cluster Computing and the Grid, Lyon, France, pp. 295–305 (2008)
3. Kennedy, J., Eberhart, R.C.: Particle swarm optimization. In: IEEE Conference on Neural Networks, Piscataway, NJ, pp. 1942–1948 (1995)
4. Liu, B., Wang, L., Jin, Y.: An effective PSO-based memetic algorithm for flow shop scheduling. IEEE Transactions on System, Man, and Cybernetics, Part B: Cybernetics 37(1), 985–997 (2007)
5. Bean, J.C.: Genetic algorithms and random keys for sequencing and optimization. ORSA Journal on Computing 6(2), 154–160 (1994)
6. Bossche, R.V., Vanmechelen, K., Broeckhove, J.: Cost-optimal scheduling in hybrid IaaS clouds for deadline constrained workload. In: IEEE International Conference on Cloud Computing, Miami, Florida, pp. 228–235 (2010)
7. He, S., Guo, L., Guo, Y.: Real time elastic cloud management for limited resources. In: IEEE International Conference on Cloud Computing, Washington D. C., USA, pp. 622–629 (2011)
8. Doctor, S., Venayagamoorthy, G.K., Gudise, V.G.: Optimal PSO for collective robotic search applications. In: IEEE Congress on Evolutionary Computation, San Diego, CA, USA, pp. 1390–1395 (2004)
9. Nathani, A., Chaudhary, S., Somani, G.: Policy based resource allocation in IaaS cloud. Future Generation Computer System 28(1), 94–103 (2012)
10. Zhao, C., Zhang, S., Liu, Q., Xie, J., Hu, J.: Independent tasks scheduling based on genetic algorithm in cloud computing. In: International Conference on Wireless Communications, Networking and Mobile Computing, Marrakech, Morocco, pp. 1–4 (2009)
11. Li, L.: An optimistic differentiated service job scheduling system for cloud computing service users and providers. In: International Conference on Multimedia and Ubiquitous Engineering, Qingdao, China, pp. 295–299 (2009)
12. Li, C., Li, L.: A distributed multiple dimensional QoS constrained resource scheduling optimization policy in computational grid. Journal of Computer and System Science 72(4), 706–726 (2006)
13. Toosi, A.N., Calheiros, R.N., Thulasiram, P.K., Buyya, R.: Resource provisioning policies to increase IaaS provider's profit in a federated cloud environment. In: IEEE International Conference on High Performance Computing and Communications, Banff, Canada, pp. 279–287 (2011)
14. Breitgand, D., Maraschini, A., Tordsson, J.: Policy-driven service placement optimization in federated cloud. IBM Research Report (2011)

An Enhanced Node Repeatable Virtual Network Embedding Algorithm Based PSO Solution

Cong Wang[1], Ying Yuan[2], Ying Yang[3], and Xi Hu[1]

[1] School of Computer and Communication Engineering,
Northeastern University at Qinhuangdao, 066004, China
[2] School of Information Science and Engineering, Northeastern University,
Shenyang, 11004, China
[3] Liren College of Yanshan University, Qinhuangdao, 066004, China
{congw1981,yuanying1121,huxi214}@gmail.com

Abstract. The major challenge in network virtualization is the efficient mapping of virtual nodes and links of virtual networks onto substrate network. In this paper we propose ENR-VNE, an algorithm which can achieves high VN request acceptance ratio in the same time. We modeled VNE problem as an optimal problem to minimize the substrate resource utilization degree. Leverage the advantage of ram data switch between virtual machines host on same physical machine instead of using physical link bandwidth, our algorithm allow repeatable node mapping for same VN. Because the initial value of PSO algorithm is crucial, we present an initial position assign method to accelerate convergence and achieve more repeatable features. Simulation results show that our algorithm achieve high acceptance ratio on same substrate network than unrepeatable approach and initial position assign method can further improve the algorithm performance.

Keywords: virtual network embedding, resource allocation, repeatable node mapping, particle swarm optimization.

1   Introduction

The major challenge in network virtualization is the efficient mapping of virtual networks with constraints on both nodes and links onto substrate network, which known as the Virtual Network Embedding (VNE) problem. However, the VN embedding problem is known to be NP-hard even in the offline case. With constraints on virtual nodes and links, the VNE problem can be reduced to the NP-hard multi-way separator problem [1]. Therefore, its solutions mainly rely on heuristic algorithms. Particularly, there are some technique about ram data switch between VMs co-resident on the same single physical machine [2], such approach can provide a fully transparent and high performance data switch through ram channel instead of the traditional link switch and the major advantage of such technique is that it can save much bandwidth between VMs co-resident on the same physical machine.

Y. Tan, Y. Shi, and H. Mo (Eds.): ICSI 2013, Part I, LNCS 7928, pp. 210–217, 2013.

2   Related Work

Many algorithms have been proposed for the VNE problem. They can be classified to one-stage VNE algorithm and two stage VNE algorithm. In one-stage VNE solution, Houidi et al. [3] propose a distributed VNE algorithm that simultaneously maps virtual nodes and virtual links without any centralized controller, which using a multi-agent system to improve robustness of system and reduce costs. They assume unlimited resources to accept all the VN requests. Lischka et al. [4] proposed a back-tracking-based VN embedding algorithm using subgraph isomorphism detection that extensively searches the solution space in a single stage.

In two-stage VNE solutions, Minlan Y et al. [5] have provided a two stage algorithm for embedding the VNs. Firstly, they embedding the virtual nodes. Secondly they proceed to map the virtual links using shortest paths and multi-commodity flow (MCF) algorithms in order to increase the acceptance ratio and the revenue. Y. Zhu and M. Ammar [6] proposed an algorithm greedily chooses the substrate nodes that are lightly loaded to map the virtual nodes and uses the shortest path between the selected nodes to map the virtual links. Xiang, C et al. [7] present a Particle Swarm Optimization based algorithms named VEN-R-PSO, as a heuristic algorithm they did not consider the repeatable node mapping and so there is make no sense to improve the efficiency by adjust the position.

The difference between our solution from previous studies are that: Firstly, our algorithm bring into play the repeatable features so as to save more physical bandwidth to accept more VN; Secondly, because of the position initial and the position update phases in PSO is crucial for convergence of the algorithm and we want a virtual network node will be mapped onto the same physical host as far as possible, so we present a position enhanced algorithm to afford more efficiency and VNR accept ratio.

3   Enhanced Node Repeatable Embedding Algorithm

In this section, we will firstly introduce the particular model for VNE problem, and then describe the discrete PSO algorithm in details to solve the problem.

3.1   VNE Problem Description

We model the substrate network as a weighted undirected graph and denote it by $G^S = (N^S, E^S, A_N^S, A_E^S)$, where N^S is the set of substrate nodes and E^S is the set of substrate link. We denote the set of loop-free substrate paths by P^S. The notations A_N^S denote the attributes of the substrate nodes, including CPU capacity, storage, and location. The notations A_E^S denote the attributes of the substrate edges, including bandwidth and delay. In this paper, each substrate node $n^s \in N^S$ is associated with the CPU capacity. Each substrate link $e^s(i, j) \in E^S$ between two substrate nodes i and j is associated with the bandwidth.

Similar to the substrate network, a virtual network can be represented by a weighted undirected graph $G^V = (N^V, E^V, C_N^V, C_E^V)$, where N^V and E^V denote the set of virtual nodes and virtual link, respectively. Virtual nodes and edges are associated with constraints on resource requests, denoted by C_N^V and C_E^V, respectively.

The goal is to minimize the usage of the substrate resources. Leveraging the advantage of ram switch between VMs host on same physical machine, we can map virtual nodes of same VN onto same machine as far as possible instead of allocation them a physical link capacity for their communication, i.e. try to embed each VNR to the least number of physical machines to save physical bandwidth. Thus the object of our optimization problem just needs to calculate link cost. It is defined as follows:

$$\text{Minimize} \sum_{(i,j) \in P^s} \varphi_{ij}^w \times bw(e^w)$$

$$\text{s.t.} \quad \forall n \in N^V, \forall j \in N^S \quad Cpu(j) - \sum_{n^v \to j} Cpu(n^v) \geq Rcpu(n) \tag{1}$$

$$\forall w \in E^v, \quad \forall (i,j) \in p^s, \ w \to p^{ij} \quad \min_{e^s \in p^{ij}} Cbw(e^s) \geq Rbw(w)$$

Where φ_{ij}^w is a binary variable:

$$\forall w \in E^v, w \to p^{ij}, \forall i, j \in N^s \quad \varphi_{ij}^w \begin{cases} 0 & i = j \\ 1 & i \neq j \end{cases} \tag{2}$$

The first qualification is node resource constraints, where $\sum Cpu(n^v)$ is the total amount of CPU capacity which has already been allocated; $Cpu(j)$ is the total amount of CPU capacity of the substrate node j.

The second qualification is link resource constraints, where $\min Cbw(e^s)$ is the minimum bandwidth of links in the path p^{ij}, that means if a virtual link w is embedded onto a substrate path p^{ij}, the capacity of each link in this path must be higher than the request bandwidth of virtual link w.

3.2   Repeatable DPSO Solution

We use Disperse Particle Swarm Optimization (DPSO) to solve this optimal problem described in previous subsection. For a VNR, the search space is N-dimensional, where N is the number of node of the VN. Then a particle swarm is used to search the optimal position $X^i = [x_1^i, x_2^i, ..., x_N^i]$ to map the virtual nodes of a VN. To VNE problem the position and velocity of particles are determined according to the following velocity and position update recurrence relations:

$$V^{k+1} = \varphi_1(X_p^k - X^k) + \varphi_2(X_g^k - X^k)$$
$$X^{k+1} = X^k \oplus V^{k+1} \tag{3}$$

Where $V^{k+1} = [v_1^i, v_2^i, ..., v_N^i]$ is velocity of a particle, where v_i^k is a binary variable. For each v_i^k, if $v_i^k = 1$, the corresponding virtual node's position in the current VNE solution should be preserved; otherwise, should be adjusted by selecting another substrate node.

Because we use DPSO to calculate the optimal position, we must give the relevant discrete quantity operation definitions:

Definition 1. Subtraction of Position $X_* - X$ If X_* and X have the same values at the same dimension, the resulted value of the corresponding dimension is 1, otherwise, the resulted value of the corresponding dimension is 0.

Definition 2. Addition of Multiple $\varphi_1 X' + \varphi_2 X''$ a new velocity that corresponds to a new virtual network embedding solution, where $\varphi_1 + \varphi_2 = 1$. If X' and X'' have the same values at the same dimension, the resulted value of the corresponding dimension will be kept; otherwise, keep X' with probability φ_1 and keep X'' with probability φ_2.

Definition 3. Addition of Position and Velocity $X^k \oplus V^k$ a new position that corresponds to a new virtual network embedding solution. If the value of v_i^k equals to 1, the value of x_i^k will be kept; otherwise, the value of x_i^k should be adjust by selecting another substrate node.

We use equation (3) to update position and velocity in our DPSO algorithm to calculate optimal position for each VN. After every update, particles calculate their fitness according to equation (1). In each round, if the position cannot match the two qualifications the fitness will be set to be $+\infty$. In our solution, link mapping is implement in fitness calculate phase simultaneously, according to the position we use FloydWarshall shortest path algorithm to calculate virtual link mapping between every virtual node pairs, if all shortest path of the node pairs are exit, the fitness is gain by the object function in (1); otherwise, the fitness is set to be $+\infty$. Then for each VNR, after several rounds implementation, a particle swarm can find an optimal mapping solution for it. The detail of our algorithm is shown as follows:

```
program DPSO_Mapping (G_v, G_s) {
   Generate particles for G_v
   remove nodes and links which capacity is less than mi-
nimal request of G_v
   for(int i=0;i<ParticleCount;i++){
      particle[i].initial_position()
   }
   for(int i=0;i<MaxItCount;i++){
      gbestpre=particles.getgBest()
      for(int j=0;j<particle numbers;j++){
         particle[j].calculateFitness()
         particle[j].updateSpeed()
         particle[j].updatePosition()}
      if(gbestpre==particles.getgBest()){
         numfoit++}else{
```

```
                  numfoit=0}
        if(numfoit==10){
             break}
  }
    if(particles.getgBest()!=+∞){
        this.solution=Particle.getGbsolution()}
    return this.Solution
  }
```

In the algorithm, remove nodes and links which capacity is less than minimal node and link capacity request of VNR is to reduce the search space for particles. The terminate condition of the algorithm is it has implement more than maximum iterative round number which is pre-assigned or the global optimal fitness unchanged in 10 times of iteration.

3.3   Enhanced Position Assign Mechanism

For the DPSO algorithm discussed above, the most important thing is the initial position assignment and function *updatePosition ( )* of particles. This is because we introduce repeatable features in VNE solution and we want to map virtual nodes of a VNR onto the least number of physical nodes to save more physical link resource, i.e. try to make the best position converge to smallest substrate node set.

The goal of convergence is achieved by improving the *initial_position( )* and *updatePosition ( )*. We aim at allocating repeatable node forwardly in particle initial position and update position operations. For the number of nodes of a VNR may be different from others, we introduce a count $k = (rnd.nextInt(numofbynodes)+2)/2$, which is corresponding to the number of the nodes of VNR. Then for each dimension of the particle, if serial number of this dimension mod $k=0$, this dimension should be allocate a new random number of substrate node; otherwise, the dimension keeps the previous substrate node number. Thus we can limit the position in a smaller set of the search space. The core of the initial position assign function is as follows:

```
programe initialposition(VNR) {
   {numofsnodes: the total number of substrate network;
    numofvnodes: the total number of nodes in VNR}
   k=(rnd.nextInt(numofvnodes)+2)/2
   maph=rnd.nextInt(numofsnodes)
   for(i=0;i<numofvnodes;i++){
      if(i%k==0){
         maph=rnd.nextInt(numofsnodes) }
      position.put(i,maph) }
}
```

The update position function is similar to initial position assignment. Note that in update position phase, not all the dimension will be modified, according to Definition 3 just the dimensions with corresponding speed is 0 need to be changed. Under the premise of fixed iterative times to keep efficiency, this mechanism can bring about more repeatable mapping features.

4 Performance Evaluation

We implemented the algorithm using the CloudSim3.0.1 simulator [8] on a high level PC which has one Intel Core i7-3770 CPU and 20G RAM. Both topologies of substrate network and virtual network are generated randomly by a topology generator write in java. The main parameters of our simulation are listed in Table 1.

Table 1. Parameters in simulation

Topology:	Substrate Network	Virtual Network
Number of Nodes:	60; 80; 100	4
Connectivity:	0.2	0.4
Node Capacity:	100 unit	3-30 unit uniform distribution
Bandwidth Capacity	100 unit	3-30 unit uniform distribution

Each virtual network's living time uniformly distributed between 100 and 1000 time unit. And $\varphi_1 = \varphi_2 = 0.5$ in equation (3). We analyzed the performance of the new algorithm present in this paper with and without enhanced position assign mechanism and compared our algorithm with un-repeatable mapping PSO algorithm VEN-R-PSO which present in [8].

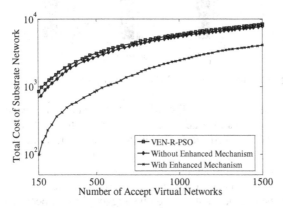

Fig. 1. Total cost of substrate network correspond to accept virtual networks

In the first experiment, we set the number of nodes of the substrate network to be 100 and simulate 2000 VNRs. We tested the condition with and without enhanced position assign mechanism and compared them to the VEN-R-PSO algorithm. Each test run 10 times, every implement take about more than 3 hours. For a VNR the max iterative count of particle swarm is 30. When the substrate network accepts about 120 VNRs it reaches full load condition, then the other VNRs should wait to be implement. Thus we can measure the extreme effect of the three algorithms. Because of the connectivity of VN is bigger than substrate network, which may lead no solution for some VNRs, so we just plot the average result of total cost definite in (1) verified to accessed number of VNRs from 0 to 1500 for the three algorithms as shown in Figure 1. Note that if virtual links are mapped onto substrate links we calculate cost; if they are mapped as a ram switch link the cost is 0.

Fig. 1 shows that the repeatable mapping can reduce cost of substrate network. No matter with and without enhanced position assign mechanism the algorithm present in this paper always produce less cost than VEN-R-PSO at the same time. However the enhanced position assign mechanism can save much more cost because it can make the solution onto as less substrate nodes as possible. The reason is that repeatable mapping can save substrate bandwidth, many virtual nodes from same VN are mapped on to single substrate node and the links between them are just instead by ram switch; in VEN-R-PSO, the bottleneck is the link resource limitation, if the substrate nodes which can satisfy the connectivity of a VNR are all full load, the VNR must wait for completion of running ones.

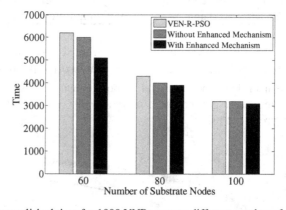

Fig. 2. Accomplished time for 1000 VNRs vary to different number of substrate nodes

In the second experiment, we validated the performance of enhanced position assign mechanism when the substrate nodes number is set to 60, 80 and 100. Each test also runs 10 times. We record the accomplished average time of 1000 VNRs using three algorithms according to three number conditions of substrate network. As shown in Fig. 2, smaller number of substrate nodes takes more time. With the decrease of the number of substrate nodes the differentials between with and without enhanced position assign mechanism is grows bigger. That is because when substrate nodes number decrease, the search space for particles is smaller. If there is no enhanced position assignment, the algorithm will terminate in early time due to the second terminate condition. That means the virtual nodes will be easier to map dispersedly onto the substrate network, thus the repeatable features will be weakened.

5   Conclusions

This paper introduced a new VNE algorithm, the major characteristics of our work is that: Firstly, we considered repeatable embedding by leveraging the advantage of inter ram switch techniques; Secondly, we introduce an enhanced position assign mechanism to accelerate the convergence of the algorithm and make the solution space more narrow so as to bring into play more repeatable feature to save more substrate bandwidth. In our future work, we will continually study the acceleration for the VNE DPSO algorithm and try to design a distribute algorithm for searching the mapping solution for VNE problem.

Acknowledgments. This work was supported by The Central University Fundamental Research Foundation, under Grant. N110323009, N110323007.

References

1. Mosharaf, C., Muntasir, R.R., Raouf, B.: ViNEYard: Virtual Network Embedding Algorithms With Coordinated Node and Link Mapping. J. IEEE/ACM Transactions on Networking 20(1), 206–219 (2011)
2. Wang, J., Wright, K.-L., Gopalan, K.: XenLoop: a ransparent high performance inter-vm network loopback. In: HPDC 2008: 17th IEEE International Symposium on High Performance Distributed Computing, pp. 109–118. IEEE Press, Boston (2008)
3. Ines, H., Wajdi, L., Djamal, Z.: A distributed virtual network mapping algorithm. In: IEEE ICC 2008, pp. 5634–5640. IEEE Press, Beijing (2008)
4. Jens, L., Holger, K.: A virtual network mapping algorithm based on subgraph isomorphism detection. In: VISA 2009, pp. 81–88. ACM, New York (2009)
5. Minlan, Y., Yung, Y., Jennifer, R., Mung, C.: Rethinking virtual network embedding: substrate support for path splitting and migration. J. ACM SIGCOMM CCR 38(2), 17–29 (2008)
6. Yong, Z., Mostafa, A.: Algorithms for assigning substrate network resources to virtual network components. In: 25th IEEE International Conference on Computer Communications, INFOCOM 2006, pp. 1–12. IEEE Press, Barcelona (2006)
7. Xiang, C., Zhong, B.Z., Sen, S., Kai, S., Fang, C.Y.: Virtual Network Embedding Based on Particle Swarm Optimization. J. Acta Electronica Sinica 39(10), 2240–2244 (2011)
8. CloudSim, A.: Framework For Modeling And Simulation Of Cloud Computing Infrastructures And Services, http://www.cloudbus.org/cloudsim/

The Application of Particle Swarm Optimization Arithmetic in Propeller Design

Chao Wang[1,2], Guoliang Wang[1], Wanlong Ren[1], Chunyu Guo[1], and Bin Zhou[3]

[1] College of Shipbuilding Engineering, Harbin Engineering University, Harbin 150001, China
[2] College of Naval Architecture and Marine Power, Naval University of Engineering,
Wuhan 430033, China
[3] China Ship Scientific Research Center, Wuxi 214082, China

Abstract. In order to obtain a propeller with good efficiency and cavitation performance, the propeller sections were optimization designed using particle swarm optimization (PSO) method. An interactive calculation method was used in design process for the circulation distribution of designed propeller was not coincident with optimization circulation. The difference of circulation was defined as correction factor to adjust lift coefficients of sections. PSO method was used to optimize sections to improve the lift-to-drag ratio and pressure distribution. The convergence condition was the circulation distribution fulfilled optimum circulation distribution form. A MAU propeller was optimized using the method. Hydrodynamic performances of propeller and sections' pressure distribution of original propeller were compared with optimized propeller. It indicates from the results that compared with traditional method, the PSO method is simpler in theory and cost less computing time. The open water efficiency of optimized propeller advanced obviously. The min negative pressure is smaller which means the cavitation performance is better.

Keywords: PSO, Propeller, Optimization design, Panel method, Open water efficiency.

With the development of the ship industry, the requirement of propeller performance increases constantly. It's hard to further satisfy the requirement of enhancing propeller's efficiency with the traditional design method. Thus, the propeller's optimization design is of great significance. Particle Swarm Optimization (PSO) [1] developed rapidly in the recent years is one kind of intelligent optimization method, which is widely used in the field of aviation and marine [2-7]. Chang Xin [8] and Xu Weibao[9] who first use the PSO in optimizing hydrofoil section thought that appropriate fitness functions can lead to the increase of lift and the decrease of drag.

This paper brought PSO in the propeller design and used it to design and optimize blade section combined with panel method. In order to validate the feasibility of this method, this paper took MAU propeller as an example to analyze the optimization design.

Y. Tan, Y. Shi, and H. Mo (Eds.): ICSI 2013, Part I, LNCS 7928, pp. 218–224, 2013.

1   Particle Swarm Optimization (PSO) Algorithm

Particle Swarm Optimization (PSO), also known as fine grain swarm optimization, had been developed into an evolutionary computation technique by J. Kennedy and R. C. Ebehtart and so on in 1995, which comes from simulation of a simplified social model [10].

In order to keep the smooth shape of hydrofoil section, the linear superposition of Hicks-Henne functions family can be used for re-expression of hydrofoil surface [11] during the optimization process. Finally, hydrofoil section can be expressed as:

$$y_{up}(x) = y_{low}(x) + \sum_{k=1}^{7} c_k f_k(x)$$

$$y_{oup}(x) = y_{olow}(x) + \sum_{k=1}^{7} c_{k+7} f_k(x)$$

(1)

Where $y_{up}(x)$, $y_{low}(x)$ means the vertical axis of upper and lower surfaces respectively, while $y_{oup}(x)$, $y_{olow}(x)$ are the original hydrofoil. C_k is a variable quantity between -0.0005 and 0.0005 which controls the change of hydrofoil and $f_k(x)$ is a Hicks-Henne function, whose expression is:

$$f_k(x) = \begin{cases} x^{0.25}(1-x)\exp^{-20x}, & k = 1 \\ \sin^3(\pi x^{(\lg 0.5/\lg xk)}), & k > 1, \quad 0 \le x_k \le 1 \end{cases}$$

(2)

The number of particles n is the candidate for Hydrofoil shape, which called as population size. After the specific concepts, we can optimize the hydrofoil section according to Particle Swarm Optimization theory.

2   The Panel Method Design of Propeller

When using traditional panel method, section attack angle should be adjusted to satisfy the optimization circulation distribution. But, there are still some vices during the design process. First, there is a direct relation between section design and the given attack angle. Section type will be changed when attack angle is adjusted. Second, there is a large calculation process which can make the enhancement of the grid density even harder. Third, the distribution type of flat pressure has randomicity.

This paper improves the propeller design method on the basis of the problems above. Specified improving methods are as follows: First, after the design of propeller section by panel method, PSO was used to optimize the design in order to enhance the lift-to-drag ratio. Second, the difference between design circulation and calculation circulation should be taken account into the optimization. According to the fact, attack angle should not be adjusted.

The main design processes in this paper are as follows:

1. Choosing the MAU propeller as model propeller, the main parameter of the design propeller can be determined by design condition.
2. Calculating the optimization circulation distribution G_{opt} and the cavitations number of each blade section.
3. Determining the thickness and chord length of blade section.
4. Designing the geometry shape which can satisfy the requirement of thickness and circulation distribution by panel method.
5. Using PSO to optimize blade section.
6. Forming design propeller, and calculating its hydrodynamics performance, pressure and circulation distribution by panel method.
7. Comparing the circulation distribution of design propeller and optimization circulation distribution, lift coefficient should be adjusted by the function below.

$$cl^{k+1}(i) = cl^k(i) \times ((1 + G_{opt}(i) - G_k(i))/G_{opt}(i) \times f_k), k = 1, 2... \qquad (3)$$

8. Restarting to optimizing blade section, calculating hydrodynamics performance and circulation distribution of propeller with the adjusted lift coefficient. If results are among the allowable precision range, the design propeller can be determined. Otherwise, return to step (5).

3   Calculation and Analysis

3.1   The Optimization of Propeller Section Shape

The aerofoil of Naca66mod is chose to be optimized. This aerofoil has good cavitations performance and large lift-drag ratio and is adopted in the design of propellers widely [11]. There is practical significance to optimize this aerofoil section with PSO. The attack angle is 1.5 degree during optimizing the section. These symbols of Cl, Cd, Tm represent lifting coefficient, drag coefficient, maximum thickness ratio. While calculating, some parameters are set, such as the uniform flow velocity 10m/s, aspect ratio 4, $b1 = b2 = 2$, $\omega_{max} = 1.0$, $\omega_{min} = 0.2$ and the velocity constraints k=0.5 . The particle swarm scale is set as 20 and the particle dimension is set as 14. It is hoped that the surface of optimized hydrofoil can obtain the uniform pressure distribution and this case $(-C_{p\,min} > \sigma)$ doesn't occur. The change of lift-drag ratio in the optimization process is shown in Table1. The change of C_p with eight optimization processes and the Geometry form change of hydrofoil with optimization process are shown in Fig.1 and Fig.2 respectively.

The optimization results of Table 1 shows that lift coefficient of hydrofoil is increased by 1.8% at 1.50 attack angle, lift-drag ratio is increased by 2.3% and the variation of thickness is not more than 1.02% compared the optimized hydrofoil with the original hydrofoil. This shows that PSO can improve the lift-drag ratio of the aerofoil section.

The optimization results of Table 1 shows that lift coefficient of hydrofoil is increased by 1.8% at 1.5^0 attack angle, lift-drag ratio is increased by 2.3% and the variation of thickness is not more than 1.02% compared the optimized hydrofoil with the original hydrofoil. This shows that PSO can improve the lift-drag ratio of the aerofoil section.

Table 1. Optimization process of blade section at 1.5^0 attack angle

	cl	cd	tm	cl/cd
Original aerofoil	0.56956	0.02224	0.07152	25.60971
1st Iteration	0.57062	0.02223	0.07187	25.66892
2nd Iteration	0.57166	0.02223	0.07225	25.7157
3rd Iteration	0.5728	0.02221	0.07194	25.79018
4th Iteration	0.57401	0.02219	0.07172	25.86796
5th Iteration	0.57525	0.02217	0.07157	25.94723
6th Iteration	0.57644	0.02215	0.07158	26.02438
7th Iteration	0.57762	0.02213	0.07151	26.10122
8th Iteration	0.57957	0.02212	0.07127	26.20118

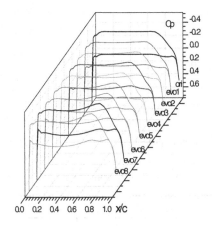

Fig. 1. The change of Cp with the wind optimization process

Fig. 2. The Geometry form change of hydrofoil with optimization process

In Fig.1, it shows that pressure coefficient changes a lot near the rear of hydrofoil with the improvement of lift-drag ratio (increasing lift force and reducing drag force). In Fig.2, it also shows the geometric shape of the hydrofoil began to change accordingly. This method can counteract drag of hydrofoil surface in the flow direction effectively.

3.2   The Optimization Design of MAU

According to the model propeller of charts design, the main parameters are shown in Table 2. It is hoped to get the optimized circulation distribution through adjusting the circulation of the design propeller. The circulation distribution is shown in Fig.3 and the open water performance before and after optimization is shown in Table 3.

Table 2. Main parameters of prototype propeller

Blade number	Diameter (m)	d/D	As/A0	P/D	Trim angle(°)	Design speed(kn)	Design rotate speed(rpm)
4	4.78	0.18	0.545	0.6825	8	15.48	155

The model propeller's circulation was chose as initial circulation of optimization design. In Fig.3, it shows the result that the circulation distribution of design propeller can almost overlap with the optimized circulation distribution. Table 3 illustrates that the hydrodynamics performance of design propeller increased about 8% than model propeller, the absorbing power only cost 99% and the thrust coefficient increased about 10.8%. Thus, the entirety efficiency of propeller can be enhanced.

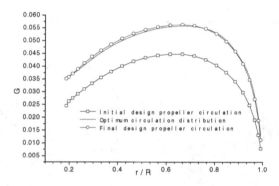

Fig. 3. Circulation distribution of propeller

Table 3. Open water performance of prototype and design propeller

J	Design propeller			Model propeller		
	KT	10KQ	η_0	KT	10KQ	η_0
0.1	0.3398	0.3223	0.16782	0.33277	0.3432	0.15433
0.2	0.29172	0.2892	0.32107	0.28512	0.3022	0.30035
0.3	0.24256	0.2525	0.45869	0.23647	0.2589	0.43616
0.4	0.19234	0.2121	0.5774	0.18683	0.2132	0.55779
0.419157	**0.18259**	**0.2039**	**0.59741**	**0.17721**	**0.2042**	**0.57866**
0.5	0.14104	0.1679	0.66838	0.1362	0.1653	0.65582
0.6	0.08868	0.1201	0.70526	0.08455	0.1149	0.70258
0.7	0.03506	0.0683	0.57233	0.03191	0.0622	0.57111

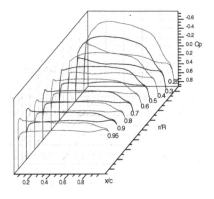

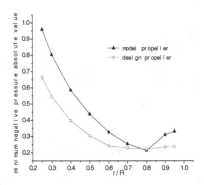

Fig. 4. Pressure coefficients distribution of different sections

Fig. 5. Comparison of min negative pressure at different sections

In Fig. 5, it shows that the pressure coefficient of each section distribute uniformity after optimizing, especially the part from 0.5R to the top which almost reached the flat pressure distribution. At the same time, in Fig. 6, it shows that the minimum negative pressure of the design propeller is much less than model propeller.

4   Conclusions

This paper brings PSO into the theory design of propeller, and improves the traditional method. Through the optimized design of MAU propeller, the comparisons between the design propeller and the model propeller are as follows:

- The difference between design circulation and calculation circulation should be taken account into the optimization. It can be closer to the actual situation that it reduced much calculation process because of avoiding the iteration of adjusting the attack angel.
- The propeller which was optimized can satisfy the optimum circulation distribution. And the pressure coefficient of each section distribute is uniformity which almost reached the flat pressure distribution.
- The minimum negative pressure of the optimized propeller decreased, and the cavitations performance was improved.
- With the same advance speed, the efficiency of the optimized propeller increased about 8% than the model under the design working condition.

Acknowledgement. This work is supported by the China postdoctoral science foundation (grant NO.2012M512133), Ph.D programs foundation of ministry of education of China (grant NO. 20102304120026) and the national natural science foundation of China under grant NO. 41176074.

References

1. Kennedy, J., Ebehtart, R., Shi, Y.: Swarm Intelligence. Morgan Kaufmann, San rnaeiseo (2010)
2. Xu, P., Jiang, C.: Aerodynamic Optimization Design of Airfoil Based on Particle Swarm Optimization. Aircraft Design 128(15), 6–9 (2008)
3. Eberhart, R.C., Shi, Y.H.: Particle swarm optimization: development, applications and resources. In: Seoul, Korea: Proceedings of the 2001 Congress on Evolutionary Computation, pp. 81–86 (2001)
4. Luo, D.L., Yang, Z., Duan, H.B.: Heuristic particle swarm optimization algorithm for air combat decision making on CMTA. Transactions of Nanjing University of Aeronautics & Aeronauts 23(1), 20–27 (2006)
5. Zhang, Y., Song, B.: Aerodynamic optimization design and experimental study of a high-lift airfoil. Flight Dynamics 24(4), 70–72 (2006)
6. The research of optimization design of propeller performance based on genetic algorithm. Dalian University of Technology
7. Wang, X., Gao, Z.: Aerodynamic optimization design of airfoil based on genetic algorithm. Acta Aerodynamic Sinica 21(3), 70–75 (2000)
8. Chang, X., Guo, C.: Design of hydrofoil section based on particle swarm optimization. Ship Engineering 32(5), 1–3 (2010)
9. Xu, W., Wang, C., Huang, S., Yu, K., Zhou, B.: The application of Particle swarm optimization theory in the hydrofoil design. Journal of SHIP Mechanics 15(6), 598–604 (2011)
10. Gao, S.: Theory and Applications of Ant Colony Algorithm and Hybridizing Other Algorithms. In: A Dissertation Submitted to Nanjing University of Science and Technology for the Degree of Doctor of Philosophy, Nanjing (2005)
11. Hicks, R.M., Henne, P.A.: Wing Design by Numerical Optimization. Journal of Aircraft 15(7), 407–412 (1978)

Application of an Improved Particle Swarm Optimization Algorithm in Hydrodynamic Design

Sheng Huang[1], Wanlong Ren[1], Chao Wang[1,2], and Chunyu Guo[1]

[1] College of Shipbuilding Engineering, Harbin Engineering University, Harbin 150001, China
[2] College of Naval Architecture and Marine Power, Naval University of Engineering, Wuhan 430033, China

Abstract. In order to design the hydrofoil section with good lift-drag ratio performance, the airfoil which received by Improved Particle Swarm Optimization algorithm and Particle Swarm Optimization algorithm should be compared to find the best way in accord with the target. Airfoils are represented by analytic functions, and objective function and fitness function are provided by numerical solution of Panel method. In entire optimization process Improved Particle Swarm Optimization algorithm only changed the weight which influences the speed of particles flying, and optimized airfoil that compared to the original airfoil hydrodynamic performance has improved significantly, and has better result than the elementary particle swarm algorithm. Results of Optimization verified the feasibility of the improved Particle Swarm Optimization algorithm in the optimization of airfoil section design, and in the future this algorithm has certain significance.

Keywords: Hydrofoil section optimization, Improved Particle Swarm Optimization algorithm, Panel method, Optimization design.

1   Introduction

Airfoils is not only a component of aircraft performance, but also an important part of propeller design and the design of all kinds of water sports equipment, and has a wide range of applications in marine engineering[1]. Airfoil hydrodynamic performance calculation includes lift by airfoil section and the fluid resistance. The lift-to-drag ratio of propeller blade section is directly related to the propeller efficiency [2].

Particle Swarm Optimization algorithms was a new optimization algorithm which developed in recent years. Optimization of Airfoil section is to get high lift-to-dag ratio airfoils shape under the given constraints [3]. Zhou bin .etc applied linear particle swarm optimization algorithm to the design of 3D airfoil and received higher lift-to-drag ratio than original airfoil by using linear weights to get particle velocity. Due to linear PSO has a lower accuracy, this paper proposes an improved Particle Swarm Optimization algorithm and applies to the airfoil optimization.

In this paper, improved particle swarm optimization algorithm combined with the panel method was used to design the three-dimensional airfoil for the purpose of high

Y. Tan, Y. Shi, and H. Mo (Eds.): ICSI 2013, Part I, LNCS 7928, pp. 225–231, 2013.

lift-to-drag ratio. In order to validate the feasibility of this method, this paper put the optimization results of different situations to compare with the results by linear PSO.

2   Improvement of PSO Algorithm

Particle Swarm Optimization (PSO) is an evolutionary computation technology (evolutionary computation), in 1995 Eberhart and Dr. Dr. kennedy proposed from the study of the predation behavior of bird [4]. Particle swarm algorithm is based on observed cluster activity behavior of animals, used the sharing information of individual of groups so that the movement of the entire group in the problem space changed from disorder to order, so as to obtain the optimal result [5].

After many years of development, the particle swarm algorithm has been widely used in engineering, and has been made considerable progress. Due to Particle Swarm Optimization has some defects that is difficult to overcome, the paper proposes an implementation easy and effective way to improve algorithm, called Improved particle swarm optimization (Modified Particle Swarm Optimization).

The mathematical description is: Assume the search space is D-dimensional, and the total number of particle is n, position of i-particle is expressed as vector $x_i = (x_{i1}, x_{i2}...x_{iD})$, stand for one point of space. The position variation of i-particle is vector $v_i = (v_{i1}, v_{i2}...v_{iD})$. Compare the best or worst position of particle according to the result of objective function, so far the position of i-particle is $pbesi = (P_{i1}, P_{i2}...P_{iD})$ and the optimal location of the whole particle swarm is $gbest = (g_1, g_2...g_N)$. Particle velocity of each dimension and position in the evolutionary process changes as following formula:

$$V_{id}(t+1) = w \cdot V_{id}(t) + c_1 \cdot r_1 \cdot (P_{id}(i) - x_{id}(t)) + c_2 \cdot r_2 \cdot (P_{gd}(t) - x_{id}(t))$$
$$x_{id}(t+1) = x_{id}(t) + k \cdot V_{id}(t+1) \tag{1}$$
$$1 \le i \le n, 1 \le d \le n$$

Where ω is inertia weight which reflects the choices of algorithm between global search and local search; while $c1$ and $c2$ called as cognitive and social parameters are non-negative constant; $r1$ and $r2$ are the random numbers between[0,1]; and k is compressibility factor which limit the speed of particles[2].

2.1   The Nonlinear Decreasing Strategy Adjustment Update Particle Swarm Algorithm

In some adjustable parameter of PSO algorithm, inertia weight ω has a great influence of the algorithm, directly affect the position distance of next generation particle and the present generation particle [6]. Nonlinear ideas inspired by linear adjustment strategy construct a non-linear function on adjustments ω, trying to be more reasonable reflection of the particle swarm search. Specific expressed as follows:

$$\omega = \omega_{\max} - (\omega_{\max} - \omega_{\min})/(Gen-1)\cdot iter \qquad (2)$$

This is a sloping straight line, and it is a linear inertia weight adjustment method, denoted as the linear weight particle swarm optimization.

$$\omega = (\omega_{\max} - \omega_{\min})\cdot (iter/Gen)^3 + (\omega_{\max} - \omega_{\min})\cdot (iter/Gen)^2 + (\omega_{\max} - \omega_{\min})\cdot (iter/Gen) + \omega_{\max} \qquad (3)$$

This is an opening upward nonlinear curve, particle swarm algorithm using this non-linear adjustment recorded as a non-linear weight particle swarm algorithm to represent [6].

This article ω was in the range of [0.2, 1.2], $c1 = c2 = 2$ [7]. For convenience, the combination of the above-described improved method and Particle Swarm Optimization called Nonlinear PSO; the linear decreasing the strategy adjustment method and the particle swarm algorithm was combined with linear PSO[8].

3   Hydrofoil Optimization

3.1   Function of Hydrofoil Section

In order to keep the smooth shape of airfoil section, the linear superposition of Hicks-Henne functions family can be used for re-expression of hydrofoil surface during the optimization process [8]. Seven Hicks-Henne functions are selected here. Finally, hydrofoil section can be expressed as:

$$y_{up}(x) = y_{op}(x) + \sum_{k=1}^{7} c_k f_k(x)$$
$$y_{low}(x) = y_{olow}(x) + \sum_{k=1}^{7} c_{k+7} f_k(x) \qquad (4)$$

Where $y_{up}(x)$, $y_{low}(x)$ means the vertical axis of upper and lower surfaces respectively, while $y_{oup}(x)$, $y_{olow}(x)$ are the original hydrofoil. Ck is a variable quantity between -0.0005 and 0.0005 which controls the change of hydrofoil and $fk(x)$ is a Hicks-Henne function, whose expression is:

$$f_k(x) = \begin{cases} x^{0.25}(1-x)\exp(-20x), & k=1 \\ \sin^3(\pi x^{\frac{\lg 0.5}{\lg x_k}}), & k>1, \ 0 \le x_k \le 1 \end{cases} \qquad (5)$$

Where $k =2$、3、4、5、6、7, $x_k =0.15$、0.3、0.45、0.60、0.75、0.9 respectively [8]. In the application it should be in accordance with the concept of mathematical model of particle swarm optimization.

3.2   Selection of Fitness Function

This paper is based on potential flow theory panel method to calculate the three-dimensional airfoil hydrodynamic performance. Due to its fast development and accurate and relatively mature, so it is widely used in the field of hydrodynamic calculations. Compared with using CFD software to solve N-S function, panel method can not only greatly reduce the computational time, but also get more accurate results with viscous correction, which can compare the advantages and disadvantages of various designs as the optimal design process [9].

Fitness(y) is selected as the lift-drag ratio of hydrofoil section optimization, where y means the control variable quantity Ck . In the optimization process it can not only pursue a high lift-to-drag ratio, but should be stopped when the minimum negative pressure coefficient is close to critical cavitations number [10].

4   Airfoil Optimization Results Analysis

In this paper, nonlinear PSO and linear PSO should be used to airfoil optimization in target of higher lift-drag ratio of section. The airfoil cross-sectional shape and the pressure coefficient also changed in the process of optimization. The paper selected $naca66\,mod$ profile to optimize for the original section. This airfoil was chosen because this airfoil has a good vacuoles performance with high lift-to-drag ratio and widely used in propeller design.

In order to compare the two algorithms objectively, angle of attack and flow speed is respectively changed and the result of two algorithms is compared. The compare ion between two algorithms comes from the convergence and effectiveness.

4.1   The Analysis of Airfoil Optimization Convergence Precision

The Airfoil Optimization Results Analysis in Different Speed Conditions

The airfoil optimal conditions: the angle of attack is 0.5, airfoil of aspect ratio is 4; ω_{max} =1.2 , ω_{min} =0.2, $c1 = c2 = 2$, and the velocity constraints k=0.5. The particle swarm scale is set as 20 and the particle dimension is set as 14. It is hoped that the surface of optimized hydrofoil can obtain the uniform pressure distribution and this case $(-C_{p\,min} > \sigma)$ doesn't occur. The previous two optimization algorithms were used for optimization in the condition that different flow velocity. the curve in Figure 1-3show the process of optimization.

In fig1-3, Cl , Cd , Cl/Cd represents lift coefficient, drag coefficient, lift-to-drag ratio ,and the picture display the results changed in different speeds change with iterations. From the curve of the picture, when the flow velocity was changed, both of two algorithms can get airfoil of higher lift-to-drag ratio, and airfoil of nonlinear PSO optimization has higher lift-to-drag ratio than linear PSO.

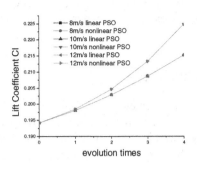

Fig. 1. Cl of different speed

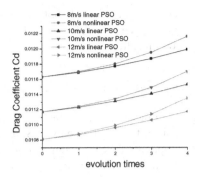

Fig. 2. Cd of different speed

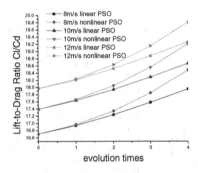

Fig. 3. Cl/Cd of different speed

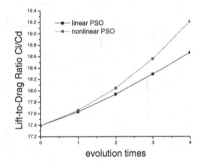

Fig. 4. naca66mod（0.5）Cl/Cd

Analysis of Airfoil Optimization in Different Angle of Attack

Keep the airfoil flow speed as 10m/s constantly, the two PSO algorithms is used to optimize the airfoil in different angle of 0.5、1.5、2.5. The curve in figure 4-6 display that the lift-to-drag ratio changed under different angle of attack.

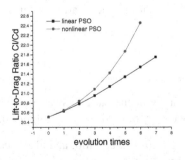

Fig. 5. naca66mod（1.5）Cl/Cd

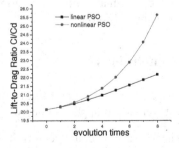

Fig. 6. naca66mod（2.5）Cl/Cd

It can be seen in the fig 4-6, both of two algorithms get airfoil of higher lift-to-drag ratio when changed the angle of attack. The nonlinear PSO do better than linear PSO in airfoil optimization.

4.2   The Analysis of Airfoil Optimization Effectiveness

In Fig.6, nonlinear PSO meet the convergence requirements after 5 iterative, and linear PSO meet the requirements in 8 iterative. In order to compare the convergence rate, the angle of attack is 1.5 degrees, the flow speed is 10m/s, other conditions being equal, the data in iterative process of two algorithms are shown in Table 1.

Table 1. comparison between nonlinear PSO and linear PSO

	Linear PSO				Nonlinear PSO			
	cl	cd	tm	cl/cd	cl	cd	tm	cl/cd
Original aerofoil	0.26175	0.01276	0.04174	20.5181	0.26175	0.01276	0.04174	20.5181
1st Iteration	0.26556	0.01287	0.04177	20.63857	0.26598	0.01288	0.04177	20.65155
2nd Iteration	0.27046	0.01301	0.0418	20.78773	0.27221	0.01306	0.04181	20.83934
3rd Iteration	0.27446	0.0131	0.04174	20.9573	0.27717	0.01314	0.04167	21.08995
4th Iteration	0.2784	0.01317	0.04164	21.14398	0.28195	0.01316	0.04137	21.42482
5th Iteration	0.28232	0.01323	0.04152	21.34232	0.28648	0.0131	0.04088	21.86962
6th Iteration	0.28621	0.01328	0.04139	21.54653	0.29074	0.01295	0.04017	22.45841
7th Iteration	0.29009	0.01334	0.04125	21.75041				

In Table 1, it can be seen that the nonlinear PSO has less iteration times than linear PSO, but the lift-to-drag ratio of nonlinear PSO improved 3.2% than linear PSO. In Particle Swarm Optimization algorithm speed formula (1), weighting factors greatly influenced the speed. When selecting nonlinear dynamic weight, the speed of the particles is modified, then speed of the algorithm optimization is changed. So nonlinear particle swarm algorithm is superior to the linear particle swarm optimization in speed of the optimization.

5   Conclusions

In this paper, to solve the problem of the accuracy of the particle swarm algorithm, nonlinear weight was applied to improve the accuracy of particle swarm algorithm.

1. The nonlinear PSO could get airfoil with high lift-to-drag ratio, and also meet the requirements of lift coefficient in airfoil optimization.

2. When the flow speed of the airfoil was changed, particle swarm optimization algorithm can get satisfactory airfoil; When the angle of attack of the airfoil was changed, it can get similar results above, and the nonlinear PSO to get better results than the linear PSO.
3. During the optimization process of the calculation, the nonlinear PSO is faster than linear PSO. It is of great significance for optimization of complex aerofoil or paddle optimization.

Acknowledgments. This work is supported by the China postdoctoral science foundation (grant No.2012M512133), Ph.D programs foundation of ministry of education of China (grant No. 20102304120026) and the national natural science foundation of China under grant NO. 41176074.

References

1. Wang, X.: Marine wing theory. National Defence Industry Press, Beijing (1988)
2. Zhou, B.: The study of four paddle two rudder of a large ship's propeller panel method designed. Harbin, Harbin Engineering University. master's thesis (2010)
3. Xu, P., Jiang, C.: Aerodynamic Optimization Design of Airfoil Based on Partical Swarm Optimuzation. Aircraft Design 128(15), 6–9 (2008)
4. Gao, S.: Ant colony algorithm theory, application and mixing with other algorithms. Nanjing University of Science and Technology doctoral thesis, NanJing (2005)
5. Kennedy, J., Ebehtart, R., et al.: SwarmIntelligence. Morgan Kuafann publishers, San rnaeiseo (2001)
6. Xu, Q., Liu, S., Liu, Q.: An improved particle swarm algorithm. Hangzhou University of Electronic Technology 28(6), 103–106 (2008)
7. Wang, X., Gao, Z.: Aerodynamic optimization design of airfoil based on genetic algorithm. Acta Aerodynamica Sinica 21(3), 70–75 (2000)
8. Xu, W.-B., Wang, C., Huang, S., Yu, K., Zhou, B.: The application of Particle swarm optimization theory in the hydrofoil design. Journal of SHIP Mechanics 15(6), 598–604 (2011)
9. Chang, X., Guo, C., Meng, X., Zhou, B.: Design of the airfoil section of the particle swarm algorithm. Marine Engineering 32(5), 1–3 (2010)
10. Su, Y., Huang, S.: Ship propeller theory. Harbin Engineering University Press, Harbin (2003)

Modeling of Manufacturing N-phase Multiphase Motor Using Orthogonal Particle Swarm Optimization

Jian-Long Kuo

Dept. of Mechanical and Automation Engineering,
National Kaohsiung First University of Sci. and Tech., Nan-Tze, Kaoh-Siung 811, Taiwan
jlkuo@nkfust.edu.tw

Abstract. This paper intends to propose an energy functional based modeling technique on an n-phase multiphase motor. In motor control area, the multiphase motor is becoming more and more popular recently. The multiphase can be applied in direct-drive electric vehicle. However, the associated mathematical model for energy functional is seldom discussed. This paper will discuss the modeling of the motor system by energy functional optimization. Orthogonal particle swarm optimization (OPSO) is used to derive the optimal solution set for the dynamic system. The Simulation and experimental results shows the validity of the proposed model. It is believed that the developed system model can be used in the energy functional of the multiphase motor.

Keywords: Modeling, Manufacturing, Multiphase Motor, Orthogonal Particle Swarm Optimization.

1 Introduction

Recently, the multiphase motor is becoming a potential machine in industrial applications. Levi [1] investigated tremendous papers in the area of multiphase motor. Multiphase motors have many advantages over the conventional three-phase motors, such as high reliability, higher torque density, and lower torque pulsation. The multiphase driver also gains the benefit of low voltage and current rating for the power devices.

This paper will focus on the axial-flux type multiphase motor. The axial-flux motor has flat-type structure which is convenient for the vehicle application [2]. The power converter can be decomposed into many smaller power modules with lower voltage and current rating to provide the required power for each phase winding [3]. The proposed axial-flux motor has two parallel windings wound around each phase. Different from the full-bridge power converter structure in [3], the proposed motor only requires single transistor structure to provide the required power for each phase winding. Assume the phase number is n, the required power transistors for the conventional converter are two times of the phase number $2n$. However, the proposed motor with two parallel windings only requires only half number of the conventional multiphase converter. The number of power transistors is the same as phase number n.

Y. Tan, Y. Shi, and H. Mo (Eds.): ICSI 2013, Part I, LNCS 7928, pp. 232–240, 2013.

Though this kind of motor has a lot of advantages, it is more complicated to derive simple mathematical model for the associated energy functional due to its inherent multiphase feature. This paper will develop a modeling technique to reduce multiphase model to unified two-phase model for easier energy functional derivation. Eight-phase prototype is illustrated in this paper to verify the experimental results.

Similarity Transformation is a powerful technique to deal with a control system problem. Lee [4] once designed a state observer for a special MIMO nonlinear system such as induction motor. The similarity transformation is used to provide the convergence of the state observer. The energy functional [5] relies on matrix transformation to decouple the complicated relation in a nonlinear system. Park transformation is the most well-known transformation in motor control area. Actually, it is also a similarity transformation in the viewpoint of the linear algebra. D-Q decomposition in [5] is used to decouple the mutual interaction in the current variables for a motor system.

The similarity transformation can also be used for the field-oriented control in the two-phase AC induction motor system [6]. Many transformation methods had ever summarized as in [7] for the nonlinear control of the induction motor. The linear relation between the torque and the current can be obtained under such transformations.

Multiphase motor is actually an n-dimensional system in state-space point of view. Conventionally, three-phase motor is a popular one in many motor applications. Recently, the technology of multi-phase motor grows rapidly. It can be easily applied in direct-drive electric vehicle or inverter-drive home appliances. However, the associated system model that can be used in its energy functional is seldom discussed. Therefore, this paper intends to develop its effective system model that can be used in energy functional. Two transformation methods are used. Singular value decomposition is used to transform the n-dimensional system into a coupled two-dimensional system. Then, similarity transformation is again used to transform the coupled relation into a decoupled relation. Therefore, a decoupled two-dimensional model for its energy functional can be derived. Due to the page limit, the continued paper in the future will discuss more details about its energy functional.

2   System Description of Multiphase Axial-Flux Brushless Motor

Recently, the axial-flux type brushless motor with two parallel windings is widely used in many information products in which the motive force is required. The CPU cooling fan is often designed as the axial-flux type. The disc rotation of the DVD-ROM will also require such motor to provide the required rotation. Actually, it belongs to a permanent magnet synchronous motor with Hall sensors.

Axial-flux type motor is also widely used in many applications such as electric vehicles [8]. This type motor has good feature of controllable field current. Field weakening can be achieved to control this motor. Therefore, the axial-flux motor can operate under very wide speed range. The field weakening can be easily carried out by eliminating the effects of d-axis current injection.

As shown in Fig. 1 and Fig. 2, this kind of motor has winding configuration different from the conventional one discussed in the general three-phase radial-flux motor system. To simplify the driver, the parallel windings with unipolar direction are designed to reduce the complexity of the driver as shown in Fig. 2. The flat-type axial-flux brushless motors are thereby widely used in many information product and industrial applications.

Flat type brushless motors have many different types of structures. The common flat type motor can be possibly driven in many ways. Bipolar full-bridge driver is frequently illustrated to drive the proposed motor. However, the full bridge driver might require expensive cost. Only simple unipolar driver is required to drive the discussed motor in this paper. The parallel windings configuration is designed for the cost-down requirement of the motor driver. Unipolar driver can reduce the cost in many information products.

As shown in Fig. 1(d)-1(e), the two parallel windings are not separated by 90 degree between each other like the radial-flux brushless motor. Actually, the parallel windings for the axial-flux brushless motor is separated by 180 degree between each other in space. The polarity definition for the two windings is reverse in the reference direction. Therefore, the dynamic system model for this kind of motor is different from the conventional model such as Park transformation which only can be used in general radial-flux motor system. To deal with the axial-flux motor, the proposed orthogonal transformation for a system will cover the two topics. One is similarity transformation and the other is singular value decomposition.

This paper intends to provide an alternative system modeling by energy functional approach. Optimization process via orthogonal particle swarm optimization (OPSO) is studied. The developed formulation will be very suitable for the associated dynamic analysis and the practical energy functional applications.

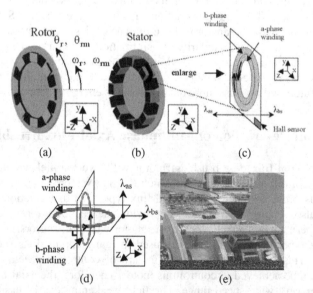

Fig. 1. Manufacturing Structure of (a) rotor, (b) stator, current and flux representation for (c) the studied axial-flux motor and (d) conventional radial-flux motor (e) prototype for the investigated brushless motor.

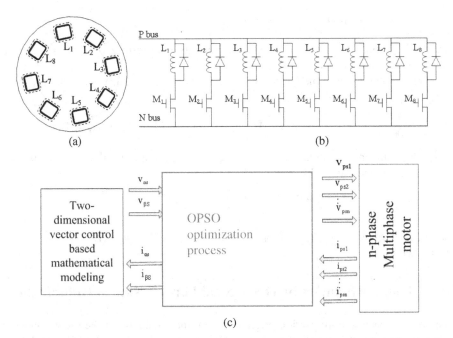

Fig. 2. Modeling technology (a) Winding configuration and (b) simple uni-polar driver of the investigated brushless motor (c) overall representation

3   Multi-to-Two Phase Transformation and Unified Two-Phase Modeling by Energy Functional

The mechanical dynamics can be derived from the Newton's law of motion. The total magnetic field energy of this electromechanical system with respect to energy functional may be expressed as

$$W_f(i_{as}, i_{bs}, \theta_{rm}) = \frac{1}{2} \sum_j \lambda_j i_j \tag{1}$$

The coenergy of the electromechanical system with respect to energy functional can be defined as

$$W_c(i_{as}, i_{bs}, \theta_{rm}) = \sum_j \lambda_j i_j - W_f(i_{as}, i_{bs}, \theta_{rm}) \tag{2}$$

The electromechanical force can be obtained from the derivative of the coenergy.

$$\frac{\partial W_c(i_{as}, i_{bs}, \theta_{rm})}{\partial \theta_{rm}} = \sum_j i_j \frac{\partial \lambda_j(i_{as}, i_{bs}, \theta_{rm})}{\partial \theta_{rm}} - \frac{\partial W_f(i_{as}, i_{bs}, \theta_{rm})}{\partial \theta_{rm}} \tag{3}$$

If the magnetic system is a linear problem, the relation $W_c = W_f$ can hold. By substituting all the related variables into the coenergy function with respect to energy functional in Eq. (6)

$$W_c = (1/2)(\mathbf{I}^T_{sys})_{1 \times n}(\mathbf{L}_{sys})_{n \times n}(\mathbf{I}_{sys})_{n \times 1} + (\lambda^T_{sys\_m})_{1 \times n}(\mathbf{I}_{sys})_{n \times 1} \tag{4}$$

where W_{pm} relates to the offset level of the energy with respective to the permanent magnets, which is constant in this motor.

The differentiation of the coenergy with respect to energy functional can derive the torque equation as below.

$$T_e = \frac{\partial W_c}{\partial \theta_{rm}} = \frac{\partial}{\partial \theta_{rm}}[(1/2)(\mathbf{I}^T_{sys})_{1 \times n}(\mathbf{L}_{sys})_{n \times n}(\mathbf{I}_{sys})_{n \times 1} + (\lambda^T_{sys\_m})_{1 \times n}(\mathbf{I}_{sys})_{n \times 1}] \tag{5}$$

The electromagnetic torque T_e can be equal to the mechanical net force by the Newton's Second Law of Motion.

$$
\begin{aligned}
T_e(i_{as}, i_{bs}, \theta_{rm}) &= J_m \dot{\omega}_m + B_m \omega_{rm} + T_L \\
&= J_m \ddot{\theta}_m + B_m \dot{\theta}_{rm} + T_L
\end{aligned}
\tag{6}
$$

where the moment of inertia is J_m and the damping coefficient is B_m. The load torque is defined as T_L.

4   Optimal Solution by Orthogonal Particle Swarm Optimization

The particle swarm optimization originates from the emulation of the group dynamic behavior of animal. For each particle in a group, it is not only affected by its respective particle, but also affected by the overall group. There are position and velocity vectors for each particle. The searching method combines the experience of the individual particle with the experience of the group. For a particle as a point in a searching space with D-dimensional can be defined as
The i-th duty cycle particle associated with the MPPT controller can be defined as:

$$X_{id} = (x_{i1}, x_{i2}, ..., x_{iD}) \tag{7}$$

where d=1,2,...,D and i=1,2,...,PS, PS is the population size. The respective particle electric power and group electric power associated with each duty cycle X_{id} are defined as

$$P_{pd} = (p_{p1}, p_{p2}, ..., p_{pD}) \tag{8}$$

$$P_{gd} = (p_{g1}, p_{g2}, ..., p_{gD}) \tag{9}$$

The refreshing speed vector can be defined as

$$V_{id} = (v_{i1}, v_{i2}, ..., v_{iD}) \tag{10}$$

The refreshing position and velocity vectors can be expressed as

$$
\begin{aligned}
V^{n+1}_{id} = V^n_{id} + c_1 \times rand() \times \left(P_{pd} - X^n_{id}\right) \\
+ c_2 \times rand() \times \left(P_{gd} - X^n_{id}\right)
\end{aligned}
\tag{11}
$$

$$X^{n+1}_{id} = X^n_{id} + V^n_{id} \tag{12}$$

When the searching begins, the initial solution is set. In the iteration process, the particle is updated by the value coming from group duty cycle and particle duty cycle.

The convergence condition is dependent on the minimum of the mean square error of the particle. Both the experience of the individual particle and the experience of the group are mixed into the searching process.

In the optimization problem, there might be a local minimum problem. The optimal solution might jump into a local trap and can not jump out of the trap. Actually, a local minimum point does not represent a global solution in a wide range. In the group experience, random function is used to jump out of the local interval. An inertia weighting factor is considered in this algorithm to increase the convergence rate. An inertia weighting factor is added in the following expression. The modified formula can be expressed as:

$$V_{id}^{n+1} = W \times V_{id}^{n} + c_1 \times rand() \times \left(P_{pd} - X_{id}^{n} \right)$$
$$+ c_2 \times rand() \times \left(P_{gd} - X_{id}^{n} \right) \tag{13}$$

$$W = W_{max} - \frac{W_{max} - W_{min}}{gen_{max}} \times gen \tag{14}$$

where the c1 and c2 are both constants. W_{max} is The initial weighting value. W_{min} is the final weighting value. gen is the number of current generation. gen_{max} is the number of final generation. However, the above mentioned is actually a kind of linear modification. To make the algorithm suitable for nonlinear searching problem, there is many nonlinear modification methods proposed to refresh the velocity vector. The modified term is defined as the key factor. By setting the learning factors c_1 and c_2 which are larger than 4, the modification for the speed vector can be expressed as:

$$V_{id}^{n+1} = K \times \begin{bmatrix} V_{id}^{n} + c_1 \times rand() \times \left(P_{pd} - X_{id}^{n} \right) \\ + c_2 \times rand() \times \left(P_{gd} - X_{id}^{n} \right) \end{bmatrix} \tag{15}$$

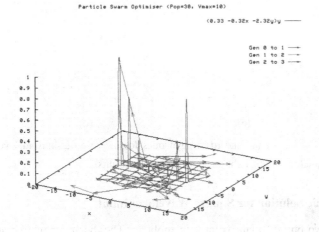

Fig. 3. Illustration of Particle refreshing process in OPSO optimization

$$K = \frac{2}{\left| 2 - \left(c_1 + c_2 - \sqrt{\left(c_1 + c_2 \right)^2 - 4 \times \left(c_1 + c_2 \right)} \right) \right|} \tag{16}$$

A modified PSO method called orthogonal PSO (OPSO) is proposed to solve the update problem effectively. A simple orthogonal array in Taguchi method is used in this algorithm to help the update as shown in Fig. 3.

5   Verification

In order to verify the validity of the optimization process and optimal solution set. Dynamic simulation is performed as follows. Three testing cases are studied in the following.

5.1   Dynamic Solution for Sinusoidal Waveform Input

First, two-phase sinusoidal voltage is applied to simulate the free acceleration of the flat type brushless motor. The experimental and simulation results in Fig. 4(a)~(c) can show the dynamics of the motor with free acceleration up to $\omega_r = 45$ rad/s. Since this is an open-loop simulation, the objective speed keeps constant at the steady state under no-load condition. In addition, the torque approaches to zero at the steady state in Fig. 4(d).

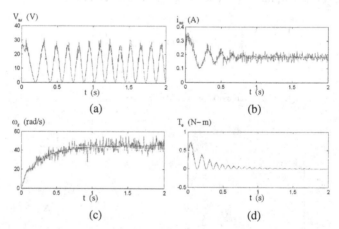

Fig. 4. (a) V_{as}, (b) i_{as}, (c) ω_r and (d) T_e dynamics for the case of sinusoidal voltage input. (deep blue: simulation results, light green: experimental results)

5.2   Dynamic Solution for Square Waveform Input

Second, the two-phase square voltage is applied. The dynamics of free acceleration for this motor are shown in Fig. 5. As compared with the previous sinusoidal study case, the motor has naturally non-smooth harmonic current due to the square waveform. However, the velocity still goes up to $\omega_r = 55$ rad/s eventually. This is also an open-loop case under no-load condition. Therefore, the torque decreases down to zero finally in Fig. 5(d).

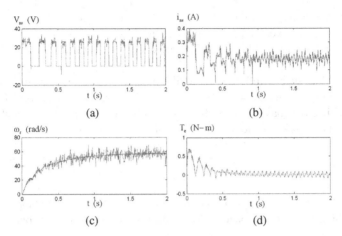

Fig. 5. (a) V_{as}, (b) i_{as}, (c) ω_r and (d) T_e dynamics for the case of the square voltage input. (deep blue: simulation results, light green: experimental results)

5.3   Steady State Comparison with Simulation and Experimental Dynamic Solutions

To verify the validity of the simulation, steady state experimental waveform is provided for comparison with the simulation result as shown in Fig. 6. The current i_{as} can have good match between the simulated and experimental results. The results are coincident with each other. The proposed model is proved to be effective for simulation of the system dynamics. Since the rotor structure is made by octagon arrangement, sinusoidal approximation for the flux terms is used to formulate the model derivation for simplification. The switching transient exhibits a little bit difference in the harmonic component. However, the DC component can match very well.

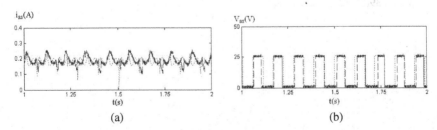

Fig. 6. Steady state comparison between (a) the phase current i_{as} and (b) the voltage variables V_{as} for the simulation (red dotted) and experimental (blue dashed) results.

6   Conclusion

Field oriented energy functional is a popular advanced control technique in motor control area. Unfortunately, the conventional form is usually suitable for the three-phase motor. Recently, the multi-phase motor is becoming more and more important in direct-drive electric vehicle. Multi-phase motor is actually an n-dimensional system in system point of view. However, the associated mathematical model is seldom discussed.

This paper has successfully described the mathematical modeling on the n-phase multiphase motor. This paper proposed the optimization of energy functional. Finally, verifications have shown that the validity of the proposed method. It is valuable for the associated energy functional of the multiphase motor.

References

1. Levi, E.: Recent developments in high performance variable-speed multiphase induction motor drives. In: Sixth International Symposium Nikola Tesla, EM1.2, Belgrade, Serbia (2006)
2. Hou, L., Su, Y., Chen, L.: DSP-based indirect rotor field oriented control for multiphase induction machines. In: IEEE International Electric Machines and Drives Conference, vol. 2, pp. 976–980 (2003)
3. Locment, F., Semail, E., Piriou, F.: Design and study of a multiphase axial-flux machine. IEEE Trans. Magnetics 42(4), 1427–1430 (2006)
4. Lee, S., Kim, S., Park, M.: A state observer for a special class of MIMO nonlinear system and its application to induction motor. In: Proc. IEEE Conference on Decision and Control, vol. 1, pp. 154–159 (2002)
5. Kim, W.J., Trumper, D.L., Lang, J.H.: Modeling and vector control of planar magnetic levitator. IEEE Trans. Industry Applications 34(6), 1254–1262 (1998)
6. Sattler, K., Schafer, U., Gheysens, R.: Field oriented control of an induction motor with field weakening under consideration of saturation and rotor heating. In: International Conference on Power Electronics and Variable-Speed Drives, pp. 286–291 (1991)
7. Sullivan, C.R., Kao, C., Acker, B.M., Sanders, S.R.: Control systems for induction machines with magnetic saturation. IEEE Trans. Industry Electronics 43(1), 142–152 (1996)
8. Liu, C.T., Chiang, T.S., Díaz Zamora, J.F., Lin, S.C.: Field-oriented control evaluations of a single-sided permanent magnet axial-flux motor for an electric vehicle. IEEE Trans. Magnetics 39(5), 3280–3282 (2003)

An Effective Transactional Service Selection Approach with Global Optimization Based on Skyline and Particle Swarm Optimization

Wanchun Yang[1,2] and Chenxi Zhang[3]

[1] School of Electronics and Information Engineering, Tongji University,
201804 Shanghai, China
[2] School of Sciences, Shandong Jiaotong University, 250023 Jinan, China
[3] School of Software Engineering, Tongji University, 201804 Shanghai, China
wcyang.tj@gmail.com, chenxizhang10@126.com

Abstract. With the growing number of alternative services in the open service environment, service selection with global optimization in service composition is a critical issue. In this paper, we propose an approach SPSO-GOTSS (global optimization of transactional service selection based on skyline and particle swarm optimization) to implement transactional service selection with global optimal QoS and semantic matching degree. This approach first adopts skyline operator to prune redundant services, then employs particle swarm optimization to select service from amount of candidates. When computing the final skyline service, we consider both dominance and incompatibility checking. The mutation operation is used to overcome the premature convergence of traditional PSO. The experimental results show that our proposed approach is feasible and effective.

Keywords: Service Selection, Global Optimization, Semantic Matching, QoS, Transaction, Skyline, Particle Swarm Optimization.

1 Introduction

Service-oriented architecture (SOA) is an emerging style of software architectures that reuses and combines loosely coupled services for building and integrating applications in order to improve productivity and cost effectiveness throughout application life cycle. Service selection is a key step in the process of service composition. Existing researches consider service selection from two aspects. On the one hand, there are more and more services which have the similar functions but with different nonfunctional attributes. How to choose services from a large number of candidate services is a challengeable topic. In [1], multidimensional multi-choice knapsack problem has been used to model this problem. Tang and Ai [2] provided a hybrid genetic algorithm for the service selection problem. Alrifai et al. [3] proposed an approach based on the notion of skyline to effectively and efficiently select services for composition. In [4], authors proposed a strategy to implement service selection with global QoS optimization based on PSO. Zhao et al. [5] proposed an improved discrete immune optimization algorithm based on PSO for QoS-driven service composition.

Y. Tan, Y. Shi, and H. Mo (Eds.): ICSI 2013, Part I, LNCS 7928, pp. 241–249, 2013.
© Springer-Verlag Berlin Heidelberg 2013

On the other hand, transactional properties of composite service have been taken a great attention. In order to satisfy QoS and transactional constraints, Haddad et al. [6] proposed a transactional and QoS-aware selection algorithm for automatic service composition. However, it can only ensure QoS local optimum while satisfying atomic consistency. In [7], authors proposed a new model based on 0-1 linear programming for determining a composite service maximizing QoS aggregate measure and satisfying transactional properties. In [8], an ant colony system based service selection algorithm is proposed to guarantee the end-to-end QoS constraints on the premise of ensuring the atomic consistency during service selection.

However, current researches do not take QoS, semantic matching degree and transactional properties into consideration simultaneously. The execution efficiency of service selection approaches still needs to be improved further. In this paper, we propose an effective transactional service selection approach with global optimal QoS and semantic matching degree based on skyline and particle swarm optimization (PSO) algorithm. This approach first adopts skyline operator [9] to improve the efficiency of selection by using the dominance relationship of skyline to prune services, then employs particle swarm optimization to select service from amount of candidates. When computing the final skyline service, we consider both dominance and incompatibility checking. The mutation operation is used to overcome the premature convergence of traditional PSO.

2   Problem Description

In service composition, an ordinary service composition flow model consists of multiple abstract services. Each abstract service (AS) corresponds to a service class (SC). The services in same service class have same or similar function but different QoS and transactional properties. The problem of service selection is to select a concrete service from SC for each abstract service to conduct an executable service process which meets the constraints, and multi-objective functions are maximally optimized, as shown in Fig. 1.

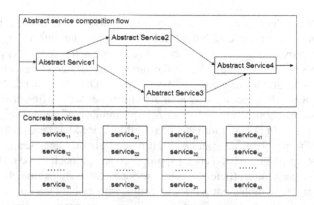

Fig. 1. Abstract services correlated with concrete services

3   The Comprehensive Evaluation Model of Composite Service

3.1   QoS Model and Semantic Matching Degree of Composite Service

When computing the QoS of composite service, we consider quantitative non-functional properties of services. These can include generic QoS attributes like response time, availability, price, reputation etc, as well as domain-specific QoS attributes, for example bandwidth for multimedia services. The QoS values of composite service are determined by the QoS values of its component. Here, we focus on the sequential composition model. Other models may be reduced or transformed to the sequential model. In this paper, we consider three types of aggregation functions: (1) summation, (2) multiplication and (3) minimum.

Based on the semantic description of service, we adopt two steps of semantic matching [10]. One is local semantic matching and the other is global semantic matching. $sim(as_i, s_i)$ represents the semantic similarity of abstract as_i and concrete service s_i. The value of $sim(as_i, s_i)$ comes from results of local semantic matching [10]. The corresponding semantic matching degree and QoS aggregation functions are illustrated in Table 1.

Table 1. Semantic matching degree and QoS aggregation functions

Aggregation type	Properties	Function
Summation	Response time	$\sum_{i=1}^{n} Time(t_i)$
	Price	$\sum_{i=1}^{n} Price(p_i)$
	Reputation	$\frac{1}{n}\sum_{i=1}^{n} Reputation(r_i)$
	Semantic matching degree	$\sum_{i=1}^{n} Sim(as_i, s_i)$
Multiplication	Availability	$\prod_{i=1}^{n} Availability(a_i)$
	Reliability	$\prod_{i=1}^{n} Reliability(r_i)$
Minimum	Throughput	$\min_{i=1}^{n} Throughput(t_i)$

3.2   Transactional Property of Composite Service

Transactional property presents the facility of service when fault is occurred. In this paper, we use the more common transactional properties for a service: pivot, retriable and compensatable.

- Pivot (P). A service with pivot property will roll back if it has any fault happened during the execution time. However, it is unable to recovery its effect semantically once the service is completed successfully.

- Compensatable (C). A service has compensatable property can be undo in certain manner even the service is completed successfully.
- Retriable(R). A service can be repeatedly invoked until it is successfully completed. This property is always combined with the two previous one, defining pivot retriable (PR) or compensatable retriable (CR) service.

The authors of [6] proposed transactional rules defining the possible combinations of component service to obtain a transactional composite service.

3.3   The Comprehensive Evaluation Model of Composite Service

The QoS and semantic matching degree must be maximized and using aforementioned constraints we obtain the comprehensive evaluation model of composite service:

$$Maximum \quad F(CS) = \sum_{i=1}^{m} W_i Q_i'(CS) + W_s S'(CS) \quad (1)$$

s.t (1) $Q_i(CS) < C^i$ if Q_i is negative parameter

or $Q_i(CS) > C^i$ if Q_i is positive parameter

(2) $S(CS) > C^s$

(3) $TP(CS) \in \{p, c, r, cr\}$

(4) $W_i > 0, \ W_s > 0, \ \sum_{i=1}^{m} W_i + W_s = 1$

$F(CS)$ is the objective function and CS represents the composite service. W_i and W_s represent weights assigned by users to each parameter. C^i and C^s represent the QoS and semantic matching degree constraints. $Q_i(CS)$ and $S(CS)$ represent the values of the ith QoS and semantic matching degree of CS. $Q_i'(CS)$ and $S'(CS)$ are the standardized values of $Q_i(CS)$ and $S(CS)$. $TP(CS)$ represents the transactional property of CS.

4   SPSO-GOTSS Algorithm

4.1   Skyline Service

Skyline computation has received considerable attention in database community. For a d-dimensional data set, the skyline consists of a set of points which are not dominated by any other points. The basic idea of our approach is to perform skyline operator on all composite services of each class to distinguish between those services that are potential candidates, and those that cannot possibly be part of the final solution.

Inspired by [11], we consider incomparability when computing the final skyline services. We design an improved branch and bound skyline algorithm using hybrid Index (BBS-HI) which optimizes both dominance and incompatibility checking to achieve good performance. The index based on R-tree and dynamic index based on partition tree [11] are integrated into the skyline service computation.

The corresponding BBS-HI algorithm description is as follows:

Step1: Insert all entries of root R-tree in the heap, an intermediate entry corresponds to minimum bounding rectangle (MBR), while a leaf entry corresponds to a data points; the partition tree is initialized when the first skyline service is identified.

Step2: The entries output from the heap are evaluated using the partition tree. If the entry is dominated it will be pruned. Otherwise, if the entry is an intermediate node, its child entries will be evaluated using the partition tree before inserting into the heap; if the entry is a leaf node, it will be inserted into the partition tree. Repeat Step2 until the heap is empty.

4.2   Particle Swarm Optimization (PSO)

Particle swarm optimization (PSO), introduced by Kennedy and Eberhart in 1995[12], was inspired by the social behavior of bird flocking and fish schooling. A particle is attracted by its personal best position p_{best} and the best position of all particles in its neighborhood p_{gbest}. By randomly changing the magnitude of these attractions, particles can search for better position in the regions around p_{best} and p_{gbest}. We assume that the search space is d-dimensional and the particle population is N. At the beginning, the N particles are initialized with a random position. The fitness of all initial position is evaluated by the fitness function, leading to an initial p_{gbest}. During every iteration step of PSO, the new position x_{id}^{t+1} and new velocity v_{id}^{t+1} of each particle are calculated based on the following equations:

$$v_{id}^{t+1} = w \cdot v_{id}^{t} + c_1 \cdot r_1 \cdot (p_{best}(t) - x_{id}^{t}) + c_2 \cdot r_2 \cdot (p_{gbest}(t) - x_{id}^{t})$$
$$x_{id}^{t+1} = x_{id}^{t} + v_{id}^{t+1} \tag{2}$$

where the parameter w is called the inertia weigh, which is a measure for the sensitivity to influences of p_{gbest} and p_{best}, and controls the exploration behavior of the swarm, the parameters c_1 and c_2 are the cognitive ratio and the social ratio which are used to control the influence of p_{gbest} and p_{best} on a particle's new velocity, the random variables r_1 and r_2 are uniformly distributed $U(0,1)$.

In order to achieve an effective balance between global and local searches, we adopt mutation strategy to further enhance the global search ability of the PSO. We set a threshold value denoted as ϕ_d to indicate whether the current particle can use a mutation operator on dimension d. If the absolute value of difference between x_{id}^{t+1} and x_{id}^{t} is smaller than ϕ_d, the mutation operator should be used. ϕ_d is calculated using the following equation:

$$\text{If } (f_d = \theta) \text{ then } f_d = 0 \text{ and } \phi_d = \phi_d / \mu. \tag{3}$$

Parameter θ controls the mutation frequency of particles, μ controls the decline rate, f_d denotes the number of particles that has used the random mutation operator on dimension d, the initial values of f_d is equal to 0.

4.3   Algorithm Design of SPSO-GOTSS

The corresponding algorithm description is as follows:

Step1: For each abstract service A_i, S_i (the set of concrete services) is partitioned into four subsets. $S_i = SP_i \cup SR_i \cup SC_i \cup SCR_i$ with SP_i being the set of pivot services which can perform A_i, SR_i being the set of retriable services, SC_i being the set of compensatable services and SCR_i being the set of compensatable retriable services. Our algorithm uses the dominance relationship of skyline to prune services.

Step2: Initialize N particles and set the corresponding parameters; code the position of the particle into a candidate of composite services. Check whether the candidate satisfies the constraint conditions. If it is true, compute the value of the particle with equation (1).

Step3: Iteration number $t=1$.

Step4: Check whether the algorithm meets the condition of termination (iteration number), If it is not the case, update the velocity and position of the particle with equation (2), meanwhile the random mutation on dimension d is adopted if the absolute value of difference between x_{id}^{t+1} and x_{id}^{t} is smaller than ϕ_d; otherwise, go to step 7.

Step5: Compute the value of each particle which satisfies the constraint conditions. For the current position x_i of the particle, compare its value with that of best position that it experienced before. If the current fitness value is better, then let x_i be the best position that it experienced; for the best position p_{best} of the particle, compare its fitness value with that of global best position p_{gbest} that all particles experienced before. If the current fitness value is better, then let p_{best} be the global best position p_{gbest} that all particles experienced.

Step6: $t = t + 1$; return to Step4.

Step7: Get the best solution and stop the algorithm.

5   Experimental Analysis

In this section, we conduct a set of experiments to assess the effectiveness of the proposed algorithms. The experiment is running on a PC machine with Pentium 2.0GHz processor, 4.0GB of RAM and Windows XP SP3. Nine kinds of QoS attributes including price, response time, availability and reputation are considered [3]. The QoS attributes and semantic matching degree of each service are randomly generated within certain ranges. The transactional property of each service is generated among the set $\{p, c, r, cr\}$ randomly.

5.1   Performance of BBS-HI

Since there is not any sizable service test case that is in the public domain and that can be used for experimentation purposes, we focus on evaluating the performance of BBS-HI by using synthetic services. We used a publicly available synthetic generator to obtain an independent dataset, in which the QoS attributes and semantic matching degree are randomly set. The dataset comprises 100K vectors, and each vector represents the 9 QoS attributes and semantic matching degree of one service.

Fig.2 shows the computation time of BBS-HI is fewer than that of OSP. The performance of BBS-HI and OSP[11] tend to be similar with the increase of dimensions. As more services become incomparable to each other, the number of services that can be pruned by the R-tree will be significantly reduced, then the performance of BBS-HI is similar with OSP.

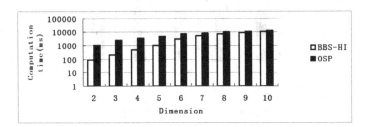

Fig. 2. Comparison on computation time

5.2   Performance of SPSO-GOTSS

This part will compare SPSO-GOTSS with the MIP approach used in [3] on the computation time and optimality.

1)   Comparison on computation time
We investigate the time cost of SPSO-GOTSS and MIP using the sequential composite service which contains 5 abstract services. The service class of each abstract service is partitioned into four subsets according to transactional properties. The number of concrete service candidates in each subset increases from 100 to 300.

Fig. 3 shows the computation time of SPSO-GOTSS is much fewer than that of MIP. By using the dominance relationship of skyline to prune services, the search space is reduced. By using PSO, it can make good use of the information sharing mechanism and the parallel global search to converge quickly to a reasonably good result.

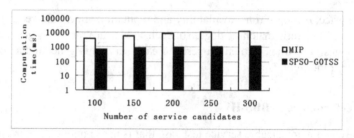

Fig. 3. Comparison on computation time

2)   Comparison on optimality

In this comparison, we evaluate the optimality of the results of the two approaches. The optimality can be calculated by

$$optimality = SPSO - GOTSS / MIP \qquad (4)$$

Fig. 4 shows the optimality of SPSO-GOTSS is in all cases above 95%, which indicates the ability of SPSO-GOTSS to achieve close-to-optimal results.

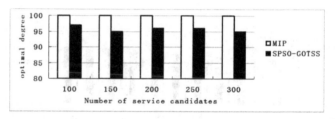

Fig. 4. Comparison on optimality

6   Conclusion

This paper proposes an effective transactional service selection approach with global optimal QoS and semantic matching degree based on skyline and particle swarm optimization (PSO) algorithm. This approach first adopts skyline operator to improve the efficiency of selection by using the dominance relationship of skyline to prune services, then employs particle swarm optimization to select service from amount of candidates.

In our future work, we will consider the recovery cost of the failed services in service selection. The strategies which can guarantee the balance between coarse-grained exploration and fine-grained exploitation search are also left for future research.

Acknowledgments. This work is supported by National Key Basic Research program of China under Grants No. 2010CB328106.

References

1. Yu, T., Zhang, Y., Lin, K.J.: Efficient Algorithms for Web Services Selection with End-to-End QoS Constraints. ACM Transactions on the Web 1(1), 1–26 (2007)
2. Tang, M.L., Ai, L.F.: A Hybrid Genetic Algorithm for the Optimal Constrained Web Service Selection Problem in Web Service Composition. In: Proceeding of the 2010 World Congress on Computational Intelligence, pp. 1–8. IEEE Press, Barcelona (2010)
3. Alrifai, M., Skoutas, D., Risse, T.: Selecting Skyline Services for QoS-based Web Service Composition. In: 19th International Conference on World Wide Web, pp. 11–20. ACM, Raleigh NC (2010)
4. Kang, G.S., Liu, J.X., Tang, M.D., Xu, Y.: An Effective Dynamic Web Service Selection Strategy with Global Optimal QoS Based on Particle Swarm Optimization Algorithm. In: 26th International Parallel and Distributed Processing Symposium Workshops & PhD Forum, pp. 2280–2285. IEEE Press, Shanghai (2012)
5. Zhao, X.C., Song, B.Q., Huang, P.Y., et al.: An Improved Discrete Immune Optimization Algorithm based on PSO for QoS-driven Web Service Composition. Applied Soft Computing 12(8), 2208–2216 (2012)
6. Haddad, J.E., Ramirez, G., Manouvrier, M., Rukoz, M.: TQoS: Transactional and QoS-aware Selection Algorithm for Automatic Web Service Composition. IEEE Transactions on Services Computing 3(1), 73–85 (2010)
7. Gabrel, V., Manouvrier, M., et al.: A New 0–1 Linear Program for QoS and Transactional-aware Web Service Composition. In: IEEE Symposium on Computers and Communications, pp. 845–850. IEEE Press, Cappadocia (2012)
8. Cao, J.X., Zhu, G.R., et al.: TASS: Transaction Assurance in Service Selection. In: IEEE International Conference on Web Services, pp. 472–479. IEEE Press, Honolulu (2012)
9. Borzsony, S., Kossmann, D., Stocker, K.: The Skyline Operator. In: 17th International Conference on Data Engineering, pp. 421–430. IEEE Press, Heidelberg (2001)
10. Zhang, P.Y., Huang, B., Sun, Y.M.: Web Services Composition based on Global Semantics and QoS-aware Aspect. Journal of Southeast University 24(3), 296–299 (2008)
11. Zhang, S.M., Mamoulis, N., Cheung, D.W.: Scalable Skyline Computation using Object-based Space Partitioning. In: 35th SIGMOD International Conference on Management of Data, pp. 483–494. ACM, Rhode Island (2009)
12. Kennedy, J., Eberhart, R.: Particle Swarm Optimization. In: IEEE International Joint Conference on Neural Networks, pp. 1942–1948 (1995)

Discrete Particle Swarm Optimization Algorithm for Virtual Network Reconfiguration

Ying Yuan[1], Cuirong Wang[2], Cong Wang[2], Shiming Zhu[2], and Siwei Zhao[3]

[1] School of Information Science and Engineering, Northeastern University,
Shenyang, 11004, China
[2] School of Northeastern University at Qinhuangdao, Qinhuangdao, 066004, China
[3] School of Electronic and Information Engineering, Beijing Jiaotong University,
Beijing, 100044, China
{yuanying1121,Congw1981}@gmail.com, wangcr@mail.neuq.edu.cn

Abstract. Network virtualization allows multiple virtual networks (VNs) to coexist on a shared physical substrate infrastructure. Efficient network resource utilization is crucial for such problem. Most of the current researches focus on algorithms to allocate resources to VNs in mapping. However, reconfiguration problem of running VNs is relatively less explored. Aiming at dynamic scheduling of running VNs, this paper introduces a virtual network reconfiguration model to achieve more substrate network resource utilization. We formulate the virtual network reconfiguration problem as a multi object optimal problem and use discrete particle swarm optimization (DPSO) algorithm to search optimal solution. Experimental results show that by rescheduling the running VNs on substrate network according to the optimal reconfiguration solution our approach can observably reduce the biggest load in both physical node and link load, balance average load and avoid bottlenecks in substrate network so as to gain high VNs accept ratio.

Keywords: network virtualization, reconfiguration algorithm, load balancing, discrete particle swarm optimization.

1 Introduction

With the rapid development of Internet, it has been difficult for existing network architecture to meet the development of new applications, and ossification has grown to some extent. Network virtualization is considered to be the main means of solving the ossification problem [1]. Network virtualization can provide more flexiblity by separating the network provider from infrastructure provider. InPs manage the physical Infrastructure while multiple SPs will be able to create heterogeneous VNs to offer customized end-to-end services to the users by leasing shared resources from one or more InPs [2]. The main of challenges in network virtualization is the efficient allocation of substrate resources to the incoming VNs, a problem known as virtual network mapping (VNM) [3].

Virtual network mapping problem is known to be NP-hard even in the offline case [4]. Even if all the virtual nodes are mapped, it is still NP-hard to mapped virtual

Y. Tan, Y. Shi, and H. Mo (Eds.): ICSI 2013, Part I, LNCS 7928, pp. 250–257, 2013.

links [5]. Recently, a number of heuristic-based algorithms or customized algorithms have appeared in the relevant literature [3-10].The problem of reconfiguration, however, is relatively less explored.

Each virtual network is a collection of virtual nodes and virtual links that connect a subset of the underlying physical network resources. Substrate networks have finite resources and utilizing them efficiently is an important objective for a VNM method. Excessive using of scarce resources will lead to accept rate of virtual networks decrease. Through reconfiguring the VN requests, it is possible to optimize the allocation of resources to VNs.

In this paper, we will propose a new virtual network reconfiguration algorithm based on discrete particle swarm optimization, denoted by VNRC-DPSO. We attempt to reconfigure the VNs that reducing the peak load as well as the number of overloaded nodes and links in the substrate network. VNRC-DPSO algorithm reconfigures the VNs via virtual node and link migrations that improving the substrate's capability to admit more requests in the future.

The remainder of the paper is organized as follows. In the next Section we will summarize the related approaches to solving the VN reconfiguration problem. In section 3, we give the detailed description of virtual network reconfiguration and its general model. In section 4, firstly, we briefly introduce DPSO, the parameters and operations of the DPSO are redefined. Then we discuss how to deal with virtual network reconfiguration problem with DPSO. The simulation environment and results are given in Section 5. Section 6 gives the conclusion.

2   Related Work

Virtual network reconfiguration is a challenging problem that has only been addressed by a few research papers. Zhu and Ammar [6] have provided a reconfiguration algorithm VNA-II. The authors developed a selective VN reconfiguration scheme that prioritizes the reconfiguration for the most critical VNs. The actual reconfiguration happens during the each virtual network reconfiguration phase.

Yu et al. [7] employ path migration to periodically re-optimize the utilization of the substrate network to help accommodate new requests. That algorithm only looks at reconfiguration of virtual links and does not consider migration of virtual nodes.

Butt et al. [8] propose a VN reconfiguration algorithm. When a VN request is rejected by the VN mapping strategy the algorithm is run only. To achieve this, the proposed algorithm first detects the unmapped virtual nodes and links causing the VN request rejection. The authors migrates these unmapped virtual nodes to one node among the potential candidate nodes. The algorithm also finds the bottlenecked substrate links causing the mapping to be blocked. Then, the authors reassign the overloaded substrate link. Note that the authors assume that candidate substrate nodes are predefined for each virtual node, which is not realistic.

Marquezan et al. [9] propose a distributed self-organizing algorithm to manage the substrate network resources. The main idea is to shorten the physical path embedding a virtual link that overloads at least one substrate link according to its traffic. The virtual node is moved in order to shunt the overloaded substrate link.

Masti and Raghavan [3] propose a simulated annealing algorithm for reconfiguring the VNs in order to balance the load across the substrate network, thereby reducing the peak node and link load on the substrate network.

Fajjari et al. [10] propose a new greedy virtual network reconfiguration algorithm VNR. In order to minimize the number of overloaded substrate links, while also reducing the cost of reconfiguration. The main idea is to re-assign canonical star virtual topologies hosted in the overloaded substrate nodes and links.

Unlike [6], [7], [8], VNRC-DPSO is not interrelated with the embedding strategy. Unlike [8], the algorithm does not need any details about the virtual network request rejected. Unlike [10], the algorithm is not periodically executed. Our proposal is not based on path splitting as in [7] and be used to reconfigure the VNs such that there is a balanced distribution of load across the substrate network.

3   Network Model and Problem Description

In this section, we first provide the network model, including substrate network, virtual network request. The proposed solution represents virtual as well as substrate networks as undirected graphs. We then introduce the virtual network reconfiguration problem, including basic definitions and formulations. In this paper, we consider central processing unit (CPU) as a resource for substrate nodes and virtual node, and bandwidth to be the resource for substrate paths and virtual link.

3.1   Substrate Network

The substrate network is represented by $S = (\hbar^s, \xi_h^s, \ell^s, \xi_\ell^s)$, where $\hbar^s$ is the set of substrate nodes and ℓ^s is the set of substrate link. We denote the set of loop-free substrate paths by $\wp^s$. The notations ξ_h^s and ξ_ℓ^s denote the attributes of the substrate nodes and link, respectively.

3.2   Virtual Network Request

Let $VN = \{vn_1, vn_2, \cdots, vn_\tau\}$ is the set of VN requests. The VN is shown by $V = (\hbar^V, \zeta_h^V, \ell^V, \zeta_\ell^V)$, where $\hbar^v$ and ℓ^v denote the set of virtual nodes and virtual link. The notations ζ_h^v and ζ_ℓ^v denote the constraints of virtual nodes and edges.

3.3   Virtual Network Reconfiguration

The virtual network mapping problem is defined by a mapping $\nabla : V \mapsto (\hbar^{s^*}, \psi_h, \ell^{s^*}, \psi_\ell)$, from V to S, where $\hbar^{s^*} \subseteq \hbar^s$, $\ell^{s^*} \subseteq \ell^s$.

The $VN = \{vn_1, vn_2, \cdots, vn_\tau\}$ have been allocated resources on the substrate network. Give a mapping ς of a set of virtual networks on the substrate network, reconfigure ς to obtain a new mapping ς^* such that $\Im(\varsigma^*) \leq \Im(\varsigma)$.

3.4   System Object

The main objective of reconfiguration algorithm is to reconfigure the existing mapping ς to get a new mapping ς^* that is better than ς with respect to a cost.

Similar to the early work in [3], Firstly, we define the cost as

$$\Im(S,\varsigma) = \varpi_h + \sigma_\ell + \vartheta_\dagger \tag{1}$$

$$\varpi_h = \alpha_h \cdot \Upsilon_h(S,\varsigma) + \beta_h \cdot T_h(S,\varsigma) \tag{2}$$

$$\sigma_\ell = \alpha_\ell \cdot \Upsilon_\ell(S,\varsigma) + \beta_\ell \cdot T_\ell(S,\varsigma) \tag{3}$$

$$\vartheta_\dagger = \delta_h \cdot \vartheta_h + \delta_\ell \cdot \vartheta_\ell \tag{4}$$

Where ϖ_h, σ_ℓ and $\vartheta_\dagger$ is node cost, link cost and migration costs, respectively, $\Upsilon_h(S,\varsigma)$ and $\Upsilon_\ell(S,\varsigma)$ is maximum node and link load, $T_h(S,\varsigma)$ and $T_\ell(S,\varsigma)$ is average node and link load, ϑ_h and ϑ_ℓ is node and link migration costs. $\alpha_h, \alpha_\ell, \beta_h, \beta_\ell, \delta_h$ and δ_ℓ are weighting coefficients.

4   VNRC-DPSO Algorithms

4.1   Discrete PSO for Virtual Reconfiguration

Particle swarm optimization (PSO) is a population based stochastic optimization technique developed by Dr. Eberhart and Dr. Kennedy in 1995, inspired by social behavior of bird flocking or fish schooling.

Standard PSO is not directly applicable to the optimal reconfiguration problem, so we used variants of PSO for discrete optimization problems to solve the optimal reconfiguration problem.

Redefine the position parameters, the velocity parameters and the operations for discrete PSO as follows:

Definition 1. Position Λ Let $\Lambda^q = \{\Lambda_1^q, \Lambda_2^q, \cdots, \Lambda_\chi^q\}$ is a position matrix of a particle, which denotes a possible virtual network reconfiguration solution. χ is the number of virtual networks running on the substrate network. For each column vector $\Lambda_i^q \in \Lambda^q$, Λ_i^q is a vector, denotes a possible VNM solution for ith virtual networks according to the substrate network.

Definition 2. Velocity Ψ Let $\Psi^q = \{\Psi_1^\theta, \Psi_2^q, \cdots, \Psi_\chi^q\}$ is a velocity matrix, denote which position of each virtual network mapping solution should be adjust.

Definition 3. $\odot$ $\Lambda^q \odot \Psi^q$ a new position matrix that corresponds to a new virtual network reconfiguration solution. If all the values of ith column vector of Ψ^q is equals to 1, the VNM solution for ith virtual networks Λ_i^q will be kept; otherwise,

Λ_i^q should be adjust. For example, $\begin{pmatrix} 2 & 3 & 5 & 6 \\ 1 & 2 & 9 & \\ 4 & & 7 & \end{pmatrix} \odot \begin{pmatrix} 1 & 1 & 0 & 1 \\ 1 & 1 & 1 \\ 1 & & 0 & \end{pmatrix}$ denotes Λ_3^q be

adjusted, where the first and third virtual node of the virtual network will be moved.

Definition 4. Θ $\Lambda^q \Theta \Lambda^p$ a velocity matrix. It indicates the differences of the two virtual network mapping solutions. If Λ_i^q and Λ_i^p have the same values at the same dimension, the resulted value is 1, otherwise, the value is 0. For example,

$$\begin{pmatrix} 2 & 3 & 5 & 6 \\ 1 & 2 & 9 & \\ 4 & & 7 & \end{pmatrix} \Theta \begin{pmatrix} 2 & 3 & 7 & 1 \\ 5 & 2 & 3 & \\ 1 & & 9 & \end{pmatrix} = \begin{pmatrix} 1 & 1 & 0 & 0 \\ 0 & 1 & 0 & \\ 0 & & 0 & \end{pmatrix}$$

Definition 5. $\Diamond$ $\aleph_1 \Psi^q \Diamond \aleph_2 \Psi^p$ If Ψ_i^q and Ψ_i^p have the same values at the same dimension, the resulted value of the corresponding dimension will be kept; otherwise, keep Ψ_i^q with probability $\aleph_1$ and keep Ψ_i^p with probability. For exam-

ple, $0.3 \begin{pmatrix} 1 & 0 \\ 0 & 1 \\ 0 & 1 \end{pmatrix} + 0.7 \begin{pmatrix} 1 & 0 \\ 1 & 1 \\ 0 & 1 \end{pmatrix} = \begin{pmatrix} 1 & 0 \\ \Xi & 1 \\ 0 & 1 \end{pmatrix}$ where Ξ denotes the probability of being

0 or 1. In this example, the Ξ is equal to 0 with probability 0.3 and equal to 1 with probability 0.7.

Because of the specificity of discrete quantity operation, we modify the particle motion equation and cancel the original inertia item. The position and velocity of particle q are determined according to the following velocity and position update recurrence relations:

$$\Psi^q \leftarrow \aleph_1 (pB^q \Theta \Lambda^q) \Diamond \aleph_2 (gB^p \Theta \Lambda^q) \tag{5}$$

$$\Lambda^q \leftarrow \Lambda^q \odot \Psi^q \tag{6}$$

Where $\aleph_1$ and $\aleph_2$ are set to constant values that satisfy the inequality $\aleph_1 + \aleph_2 = 1$; pB^q denote the position with the best fitness found so far for the qth particle; gB^q denote the best global position in the swarm.

4.2   VNRC-DPSO Algorithm Description

In this section, we present the VNRC-DPSO algorithm for VN reconfiguration. Reconfiguration of VN involves migration of virtual nodes and links. The algorithm takes as input a mapping ς of a set of VNs and tries to look for another mapping ς^* that has the least cost among all possible mappings. The solution space is the set of all feasible solutions. The cost function gives the cost associated with a solution belonging to the solution space. The algorithm takes the objective function (1) as fitness function. The details of the VNRC-DPSO algorithm are listed in Table 1.

Table 1. VNRC-DPSO algorithm

Algorithm 1 VNRC-DPSO algorithm

Input: Substrate network $S = (\hbar^s, \xi_\hbar^s, \ell^s, \xi_\ell^s)$

      Mapping solution ς of current VNs running on substrate network

Output: reconfiguration mapping solution ς^*

1. Initialize particle count N and max iterative count M;
2. Calculate fitness $G_{fitness}$ of ς and set the initial gB^q as ς;
3. For each particle randomly set it position matrix Λ^q;
4. Do while iterative count $< M$ & gB^q changes in every 5 rounds;
5.   Update velocity matrix according to gB^q and pB^q, using equation (5);
6.   Update position using equation (6), i.e. for each column vector of Ψ^q randomly set the mapping target in substrate nodes according to Definition 3;
7.   Calculate fitness of current position using equation (1);
8.   If the new fitness is smaller than $G_{fitness}$, update gB^q;
9.   If the new fitness is smaller than $P_{fitness}$, update pB^q;
10. End while and set gB^q as output solution ς^*.

5   Simulation

We implemented the VNRC-DPSO algorithm using the CloudSim3.0.1 on a PC which has one Intel Core i7-3770 CPU and 20G DDR3 1600 RAM. We write a random topological generator in java to generate topologies for the underlying substrate networks and virtual networks in CloudSim. Substrate networks in our experiments have 60 nodes, each node connect to other nodes with probability 0.2, so there are about 300 links in the networks. The substrate nodes and links were assigned resources by generating a uniform random number between 50 and 100. We assume that VN requests arrive in a Poisson process with an average rate of 4 VNs per 100 time units. For each VN request, the number of virtual nodes was randomly determined by a uniform distribution between 4 and 10, with the probability of a virtual link between any two virtual nodes set to 0.5. The CPU and bandwidth requirements of virtual nodes and links are real numbers uniformly distributed between 3 and 30 units. Each virtual network's living time uniformly distributed between 100 and 1000 time unit. The parameters were set as following $\alpha_\hbar = \alpha_\ell = \delta_\hbar = \delta_\ell = 0.5$, $\beta_\hbar = \beta_\ell = 0.25$.

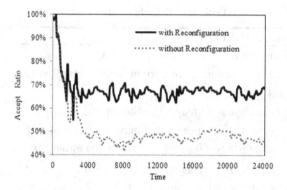

Fig. 1. Comparison of accept ratio with reconfiguration algorithm or without

In first experiment, the simulation was fixed at one reconfiguration every 30 VN requests. We simulated 1000 virtual network request for each algorithm, each test run 20 times. All the results presented for an experiment are an average of 20 runs of simulation. We ran the reconfiguration algorithm in VNM solution and compered the acceptance ratio to without any reconfigurations. As shown in Fig.1, we can easily see that with periodic reconfiguration the VNs accept ratio can be increased about 20% in major running time.

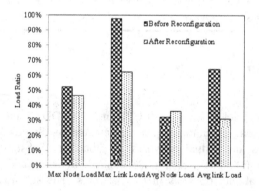

Fig. 2. Comparison of Load ratio for before and after reconfigure

The second experiment verifies load balance of the substrate network with and without reconfiguration. We simulated 1000 virtual network request and run the algorithm present in this paper every 30 VN requests are mapped into the substrate network. Fig. 2 plots the average result of 300 rounds of reconfigurations. Max node and link load are sampled individual biggest load node and link, Average node and link load are the average load ratio of all the substrate nodes and links. From the figure we can see that the reconfiguration algorithm significantly reduce max Link load and average link load, it also cut down the max node load. In average node load because the reconfigure algorithm present in this paper can balance average load and avoid bottlenecks in substrate network, so it to can achieve higher VNs access ratio. For this reason, after every reconfigure the substrate can accept more VNs, i.e. there are more VNs running on the substrate network at the same time with reconfigure algorithm.

6   Conclusions

This paper introduced a new virtual network reconfiguration algorithm base on discrete particle swarm optimization. The algorithm reconfigures the VNs by migrations and optimizes the allocation of resources to VNs such that the load is balanced across the substrate network, thereby reducing the peak node and link load on the substrate network. Simulation results show that the biggest load in both physical node and link load are significantly reduced and can improve load balance of substrate network so as to achieve higher VNs accept ratio.

Acknowledgments. This work was supported by The Central University Fundamental Research Foundation, under Grant. N110323009.

References

1. Turner, J., Taylor, D.: Diversifying the Internet. In: GLOBECOM 2005, pp. 755–760. IEEE Press, St. Louis (2005)
2. Nick, F., Li, X.G., Jennifer, R.: How to lease the Internet in your spare time. J. ACM SIGCOMM CCF 37(1), 61–64 (2007)
3. Sarang, B.M., Raghavan, S.V.: Simulated Annealing Algorithm for Virtual Network Reconfiguration. In: 2012 8th EURO-NGI, pp. 95–102. IEEE Press, Karlskrona (2012)
4. Mosharaf, C., Muntasir, R.R., Raouf, B.: ViNEYard:Virtual Network Embedding Algorithms With Coordinated Node and Link Mapping. J. IEEE/ACM Transactions on Networking 20(1), 206–219 (2011)
5. Xiang, C., Sen, S., Zhong, B.Z., Kai, S., Fang, C.Y.: Virtual network embedding Through Topology-Aware Node Ranking. J. ACM SIGCOMM CCR 4(2), 39–47 (2011)
6. Yong, Z., Mostafa, A.: Algorithms for assigning substrate network resources to virtual network components. In: 25th IEEE International Conference on Computer Communications, INFOCOM 2006, pp. 1–12. IEEE Press, Barcelona (2006)
7. Minlan, Y., Yung, Y., Jennifer, R., Mung, C.: Rethinking virtual network embedding: substrate support for path splitting and migration. J. ACM SIGCOMM CCR 38(2), 17–29 (2008)
8. Farooq Butt, N., Chowdhury, M., Boutaba, R.: Topology-awareness and reoptimization mechanism for virtual network embedding. In: Crovella, M., Feeney, L.M., Rubenstein, D., Raghavan, S.V. (eds.) NETWORKING 2010. LNCS, vol. 6091, pp. 27–39. Springer, Heidelberg (2010)
9. Clarissa, C.M., Lisandro, Z.G., Giorgio, N., Marcus, B.: Distributed autonomic resource management for network virtualization. In: NOMS 2010, pp. 463–470. IEEE Press, Osaka (2010)
10. Ilhem, F., Nadjib, A., Guy, P., Hubert, Z.: VNR Algorithm: A Greedy Approach For Virtual Networks Reconfigurations. In: GLOBECOM 2011, pp. 1–6. IEEE Press, Houston (2011)

Power Distribution Network Planning Application Based on Multi-Objective Binary Particle Swarm Optimization Algorithm

José Roberto Bezerra[1], Giovanni Cordeiro Barroso[2],
Ruth Pastôra Saraiva Leão[2], Raimundo Furtado[2],
and Eudes Barbosa de Medeiros[3]

[1] Instituto Federal do Ceará, Brazil
[2] Universidade Federal do Ceará, Brazil
[3] Companhia Energética do Ceará, Brazil

Abstract. Power distribution networks are the most susceptible sector of the whole electric grid in terms of reliability. Failures along the lines cause the disconnection of a great number of customers what have an immediate impact on quality and security indices. Innovations capable to mitigate impacts or improve reliability are ever pursued by the electric utilities. In view of that, the planning of the modern distribution networks must consider the installation of switches along the network as an important procedure to isolate failures reducing the impact and the number of customers not supplied. However, the complexity and the dimension of the current distribution networks, makes the task of proper allocation of switches strongly dependent on the expertise of engineers. This paper proposes an application based on a Multi-Objective Particle Swarm Optimization algorithm that determines the suitable placement and a feasible number of switches on the power distribution networks in order to minimize the number of customers affected by faults. Detailed information about the algorithm and its application in a test distribution system is presented. The effectiveness of the algorithm is presented in a case study applied to the IEEE 123-Node Test Feeder.

1 Introduction

Smart grid is an important and novel concept that aims to add intelligence to the whole power network from generation to transmission and distribution network up to the end user premises. This effort intends to improve the efficiency, availability and quality of the electricity supplied. The distribution system and the end users although are the entities that shall take full advantage of the smart technologies.

The distribution networks are much susceptible to failures [1], and its availability is closely related to the reliability indices. Improvements in the System Average Interruption Duration (SAIDI) index or the number of Customers Not Supplied (CNS) during outages, among others, are continuously pursued by the electric power distribution companies to avoid penalties by regulatory organizations. Furthermore, innovations are evaluated to be effectively applied by utilities in their networks, devices and systems of information and control.

Y. Tan, Y. Shi, and H. Mo (Eds.): ICSI 2013, Part I, LNCS 7928, pp. 258–267, 2013.
© Springer-Verlag Berlin Heidelberg 2013

Heuristic techniques play an important role in the planning of smart distribution networks and several papers propose solutions based on evolutionary algorithms like Simulated Annealing (SA), Genetic Algorithms (GA), Ant Colony Optimization (ACO) and Particle Swarm Optimization (PSO).

PSO algorithm was originally proposed by [2] and is commonly applied to power systems due to its characteristics of flexibility, robustness and easy implementation [3]. A solution to reactive power optimization is proposed in [4] using an improved PSO algorithm. The focus of the improvement is to avoid the convergence to a local minimum which is commonly appointed as a disadvantage in PSO algorithms. Features such as inertia weight, shrinkage factor, crossover and mutation are used to improve the algorithm efficiency.

Many extensions to the original PSO algorithm have been proposed to broaden its capabilities and applications. The Discrete Binary PSO (DBPSO), proposed in [5], is one of the most important extensions because it operates on discrete binary variables, what has drastically increased the universe of applicable problems. An optimal switch placement algorithm for power systems is proposed in [6] that uses a modified version of DBPSO. The authors deals not only with discrete binary values $\{0, 1\}$, but instead a Trinary PSO (TPSO) algorithm is proposed which, in fact, is a multistate version of DBPSO.

More recently, multi-objective optimization problems have been solved by PSO algorithms [7]. However, few of them can deal with discrete problems. A modified Multi-Objective Binary PSO (MOBPSO) algorithm is proposed in [8] for improving the multi-objective optimization performance. Mutation and dissipation operators are introduced to improve the search without impair the diversity.

The optimal switch placement is an useful procedure to improve the reliability of power distribution networks. The use of switches reduces the impact of outages caused by faults as detailed in [9]. Albeit, the reliability improvement with minimum costs are conflicting objectives. The high cost of switches makes its use to be carefully assessed by utilities. A feasible number of switches to achieve an affordable reliability level must be evaluated. Despite the high costs involved, the penalties that the utility may suffer from regulatory agencies due to infringements in reliability indices, is the main reason to invest in switch placement. Accordingly, a trade-off between reliability and cost has to be stablished.

This paper proposes a multi-objective approach based on MOBPSO algorithm that aids the decision making in switch placement in electric power distribution networks. The algorithm aims to determine an affordable result for application in the distribution network of a power utility. Tests results when applied to the IEEE 123-Node Test System [10] are presented. The results found are compared to the ones presented by [6]. Better results are presented considering the economic cost of switches and improved reliability.

The article is organized as follows. In section 2 the switch placement problem is explained. The mathematical representation of the distribution network topology is introduced in section 3. The complete description of the algorithm is given in section 4. In section 5 the case study is presented. Finally, the conclusions are presented in section 6.

2 Switch Placement in the Distribution Networks

The electric distribution networks are of the power system as a whole the closest part to the end users and the most susceptible to failures. The distribution networks are very complex and that complexity comes from: the various options for substation, transformer and switch location; several alternatives for cable or line sizes and routes; multistage investment decisions; complex objectives; and uncertainty about demand variation and location, equipment availability, and faults [11]. Despite these challenges, the electric distribution companies are responsible to deliver electricity to customers within criteria of reliability, security and affordable cost. Thus, the reliability of the distribution networks is a very important issue and it is evaluated by many indices like SAIDI (System Average Interruption Duration Index), SAIFI (System Average Interruption Frequency Index), ECOST (Expected Outage Cost), number of Customers Not Supplied (CNS), among others. Regulatory agencies in many countries can apply heavy penalties to the utilities based on the reliability indices infringement. Therefore, the reliability improvement is fundamental and it is always pursued by electric distribution companies.

A representation of a simplified distribution network is presented in Fig.1a. It comprises a bus that connects two feeders ($F1$ and $F2$). Each feeder has two load points. $LP1$ and $LP2$ connected to $F1$. $LP3$ and $LP4$ connected to $F2$. There is also one circuit breaker at the input of each feeder, $CB1$ and $CB2$. The circuit breakers are the primary protection of the feeders. They are responsible to open the circuit when short circuits are detected on each feeder what is called a fault. When a fault occurs on a feeder, all the customers connected to that feeder are switched off.

An alternative configuration to the distribution network showed in Fig.1a is presented in Fig.1b. Three new elements were added to the network, the sectionalizer switches $nc1$, $nc2$ and no. The first two switches are Normally Close (NC) and the latter is Normally Open (NO). The presence of these switches

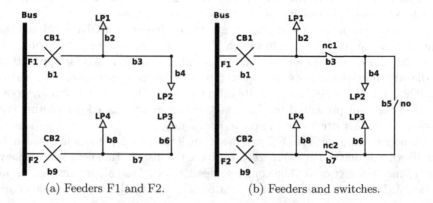

(a) Feeders F1 and F2.   (b) Feeders and switches.

Fig. 1. Simplified distribution network

reduces the fault impacts isolating the branch (segment of distribution network) in fault. Suppose that a short circuit occurs at $LP2$. The primary protection $CB1$ opens and all the customers connected to $F1$ are affected. But, $nc1$ can be opened isolating the faulty section ($b4$). After that, the energy supply can be re-established by closing $CB1$. As a result, only $LP2$ is switched off and $LP1$ is normally supplied by the bus.

Now, if a short circuit occurs at branch $b1$, the primary protection $CB1$ automatically detects and eliminates the fault by opening its contacts. Load points $LP1$ and $LP2$ are then de-energized. Next, the isolation of the faulty branch, $b1$, can be made by opening the switch $nc1$. $LP2$ can be then supplied by feeder $F2$ by closing the no switch. The goal of the network reconfiguration is to keep the energy supply to a greater number of consumers that are connected to the feeder.

As it has been shown, the installation of sectionalizer switches or reclosers along the distribution networks reduces the impact caused by faults by reducing the number of customers not supplied. For the simple distribution network presented in the example, it is easy to determine at which points the switches should be installed. However, the complexity and dimension of the actual distribution networks makes the switch placement a hard task hard to be performed. It is not commonplace the optimal switch placement in a distribution network with 3 or more feeders with thousands of customers connected to each one. The number of switches is also an important issue due to the necessary investments in assets besides the maintenance cost to the utilities.

3   Mathematical Modeling of the Distribution Networks

The topology of the distribution network initially has to be modeled mathematically in order to run the MOBPSO algorithm. Graph theory concepts are used for modeling the network. A graph G is a triple consisting of a vertex set $V(G)$, an edge set $E(G)$ and a relation that is associated to each edge. Adjacency matrix is used to describe the network relating the connectivity between the nodes or the available paths between them. Integer numbers are used to name the nodes. The adjacency matrix (I) is defined as

$$I_{ij} = \begin{cases} 1 \text{ if } (i,j) \in E; \\ 0 \text{ otherwise} \end{cases} \tag{1}$$

where i and j are row and column numbers, respectively.

The node numbering for the simple distribution network presented in Fig.1a is showed in Fig.2a in parenthesis and its adjacency matrix is presented in (3). The element I_{24} in (3) is 1, which indicates that there is a branch between nodes 2 and 4. I_{46} in turn is 0 because there is not any path between nodes 4 and 6.

When the switches no, $nc1$ and $nc2$ are added to the network as represented in Fig.2b, the adjacency matrix is updated to consider the presence of the switches at each network branch as defined in (2).

$$
I_{ij} = \begin{cases}
2 & \text{if it has a circuit breaker between nodes } i \text{ and } j, \\
1 & \text{if it has a line between nodes } i \text{ and } j, \\
-1 & \text{if it has a NO switch between nodes } i \text{ and } j, \\
0 & \text{if it has not any element between nodes } i \text{ and } j.
\end{cases} \tag{2}
$$

For example, in (4) the element I_{24} still have 1, but the element I_{46} is now -1, instead of 0, indicating the presence of a NO switch between nodes 4 and 6. Similarly, the elements I_{12} and I_{18} both are equal to 2 indicating a circuit breaker.

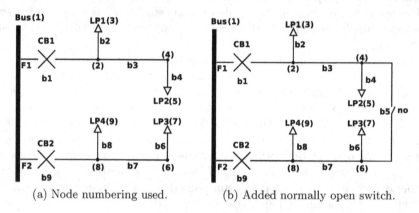

(a) Node numbering used.   (b) Added normally open switch.

Fig. 2. Node numbering and the switch.

	1 2 3 4 5 6 7 8 9
1	0 1 0 0 0 0 0 1 0
2	0 0 1 1 0 0 0 0 0
3	0 0 0 0 0 0 0 0 0
4	0 0 0 0 1 0 0 0 0
5	0 0 0 0 0 0 0 0 0
6	0 0 0 0 0 0 1 0 0
7	0 0 0 0 0 0 0 0 0
8	0 0 0 0 0 1 0 0 1
9	0 0 0 0 0 0 0 0 0

(3)

	1 2 3 4 5 6 7 8 9
1	0 2 0 0 0 0 0 2 0
2	0 0 1 1 0 0 0 0 0
3	0 0 0 1 0 0 0 0 0
4	0 0 0 0 1 -1 0 0 0
5	0 0 0 0 0 0 0 0 0
6	0 0 0 0 0 0 1 1 0
7	0 0 0 0 0 0 0 0 0
8	0 0 0 0 0 1 0 0 1
9	0 0 0 0 0 1 0 0 0

(4)

4   Algorithm Description

A Multi-Objective Binary Particle Swarm Optimization algorithm (MOBPSO) is developed for application of optimal placement of switches in power distribution networks. In this study, the array x represents the solution (a set of switches and its locations) and it is randomly initialized at the beginning of the algorithm.

The dimension, d, of x is given by (5), where NC is the number of network branches, NO is the number of normally open switches and CB is the number of feeder circuit breakers. Each element of x contains binary values that indicate the presence (1) or the absence (0) of a switch in the correspondent branch.

$$d = NC + NO - CB \tag{5}$$

The main objective of the algorithm is to achieve a set of switches and its placements in the network that minimizes the impacts of the faults. It can be noticed that minimum cost solutions are usually not maximally reliable and maximally reliable solutions are not often the cheapest. In such a scenario, none of these two extreme solutions (the cheapest and the most reliable solutions) can be declared as an absolute optimum solution complying to both objectives of design [12]. The most widely adopted approach to multi-objective problems like this is to convert it into a single objective optimization problem, easier to solve, using the Weighted Sum Method(WSM) [13]. Using WSM, the composite objective function, given in (6), is evaluated at each iteration i of the algorithm to obtain a solution x_i that corresponds to a set of switches. The array w, given in (7), contains the normalized weights that represent the importance of each particular objective, reliability (w_r) or cost (w_c), for the problem.

$$F(x) = w \left[q_r(x) \quad q_c(x) \right]^T \tag{6}$$

$$w = \left[w_r \quad w_c \right] \tag{7}$$

$$w_r + w_c = 1 \tag{8}$$

For the switch placement problem, the choice of the weights is based on the utility preferences or needs. For economic valuable areas that are priority for the utility is common sense that better reliability indices are desirable. On the other hand, when the economic resources available are scarce the reliability is deprecated.

Simply apply the weights chosen as WSM indicates achieves good results in relative few iterations. However, a different procedure is adopted to improve the quality of the solutions. The algorithm is divided to run in two stages. At first stage, the objective function (6) is compounded only by the reliability term, ($w_c = 0$). The goal is to deal each objective individually and obtain the optimum placement of switches neglecting the number of switches needed. At the end of first stage, the best configuration of switches, considering only the minor number of CNS, is determined.

The goal in the second stage is to reduce the number of switches needed, reducing the finance cost of the solution, making it feasible. Assuming a nonzero value for w_c, accordingly to (8), an affordable solution is determined.

The steps of the algorithm are detailed next.

Step 1. The topology of the distribution network is loaded and the dimension of the solution is determined. The size of swarm (M), the learning factors (c_1 and c_2) and the number of iterations (N_{itr}) are stated.

Step 2. The arrays x and v (velocity) are randomly initialized. The weights w_r and w_c assume its values according to the current stage.

Step 3. The solution fitness is evaluated by (6). q_r is given by (9), where q_j is the number of Customers Not Supplied by the j contingency. q_c is given by (10), where NS_i is the number of switches of the solution x_i.

$$q_r = \sum_{j=1}^{b} q_j \tag{9}$$

$$q_c = \frac{NS_i}{NC} \tag{10}$$

Step 4. The fitness of the partial solution, x_i, is compared to *pbest* and *gbest* and the best one of them is adopted as the current solution.

Step 5. The velocity, v, and x_i are updated to a new iteration.

The steps 3 to 5 are repeated until the number of iterations, N_{itr}, is achieved for the first stage. The step 2 is repeated to update the weights (w_r and w_c), the velocity(v) and a new solution to test (x). Steps 3 to 5 are repeated N_{itr} times again for the second stage and the solution is found. An important remark is that the first proposed solution (x) in the second stage is the *gbest* solution found at the end of the first stage. This act as a reference point to the algorithm leading it to continue the search and do not jump to another area of search space when second stage begins.

5   Case Study

To demonstrate the effectiveness of the aforementioned algorithm it was applied to the IEEE 123-Node Test System [10] which is widely applied as a reference distribution network by power systems researches. The IEEE-123 Node Test System is presented in Fig.3. It is a single radial distribution feeder with 114 load points and estimated $5,410$ customers connected.

The number of iterations ($N_{itr} = 100$) and the learning factors ($c_1 = c_2 = 2$) are empirically determined by tests. Greater test systems may require more iterations to converge. The dimension given by (5), $d = 116$. The number of agents assume the same value of dimension. The weights w_r and w_c assume the values 1.0 and 0 at first stage, respectively. For the second stage, w_r and w_c assume 0.7 and 0.3, respectively.

The simulation results are presented in Fig.4 and Table 2. From Fig.4 can be seen that the convergence is achieved on both stages. For the first stage the CNS index is $1,304$ with 42 switches. At second stage, the CNS index is $1,405$, but with only 13 switches. In other words, installing the number of switches indicated by the algorithm, $1,405$ customers are switched off, in worst case, instead of all the customers connected to the feeder. The final configuration of switches in the distribution network is showed in Table 2.

Comparing the proposed algorithm to the TPSO algorithm presented in [6], a reduction from 26 to 13 sectionalizer switches is achieved. Additionally, the

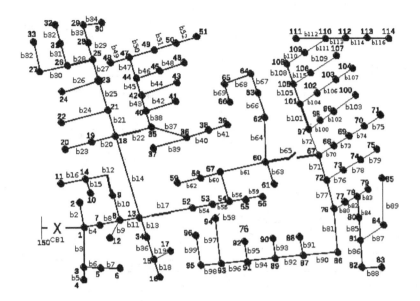

Fig. 3. IEEE 123-Node Test System

Table 1. Comparison against TPSO and MOBPSO

Algorithm	CNS	Number of Switches
TPSO	2, 034	26
MOBPSO	1, 405	13

CNS index of the TPSO resulted in 2, 034 customers. On the other hand, the proposed MOBPSO algorithm has evaluated a CNS index of 1, 405 customers. Such values are condensed in Table 1.

Failure rates of equipment of the distribution networks are not usually available and often they are unknown by the utilities. This sort of information is not demanded by the proposed algorithm while they are necessary for the algorithm in [6]. This is an important result that clearly pose the proposed algorithm as a valuable tool for planning reliable distribution networks in a smart grid context.

The development of the MOBPSO has been made with Matlab software. The implementation of the proposed algorithm in a software is under development to be applied in effective distribution networks of Energetic Company of Ceará (COELCE). The COELCE is an important Brazilian utility that supplies next to 1 million customers in the Ceará state. The software intends to be a tool to help the decision making on switch placement for the utilities.

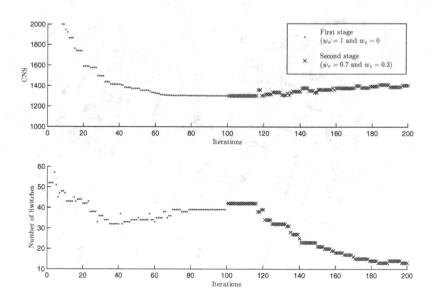

Fig. 4. Convergence for the IEEE123-Node Test Feeder

Table 2. Switch placement in the IEEE123-Node Test Feeder

Sectionalizer Locations	b3 b10 b14 b13 b22 b47 b17 b65 b79 b77 b96 b103 b109

6   Conclusions

An effective tool for power distribution systems planning has been presented. This algorithm is a valuable tool to help the decision making in electric energy distribution companies to place switches along radial distribution networks. The improvements in the reliability indices, particularly in the number of customers not supplied, are achieved.

The effectiveness of the algorithm has been demonstrated by applying it on IEEE 123-Node Test Feeder. The results found are also feasible in terms of the amount of investment costs. The application of the proposed algorithm to a Brazilian distribution company, COELCE, is under development to be the fundamental decision making tool for switch placement in its distribution networks.

References

1. Moradi, A., Fotuhi-Firuzabad, M., Rashidi-Nejad, M.: A reliability cost/worth approach to determine optimum switching placement in distribution systems. In: Transmission and Distribution Conference and Exhibition: Asia and Pacific. IEEE/PES, pp. 1–5 (2005)
2. Kennedy, J., Eberhart, R.: Particle swarm optimization. In: Proceedings of the IEEE International Conference on Neural Networks, vol. 4, pp. 1942–1948 (November- December 1995)
3. Christian Blum, X.L.: Swarm intelligence introduction and applications. Springer, Berlin (2008)
4. Wang, H., Jiang, H., Xu, K., Li, G.: Reactive power optimization of power system based on improved particle swarm optimization. In: 2011 4th International Conference on Electric Utility Deregulation and Restructuring and Power Technologies, DRPT, pp. 606–609 (July 2011)
5. Kennedy, J., Eberhart, R.: A discrete binary version of the particle swarm algorithm. In: 1997 IEEE International Conference on Computational Cybernetics and Simulation, vol. 5 (October 1997)
6. Moradi, A., Fotuhi-Firuzabad, M.: Optimal switch placement in distribution systems using trinary particle swarm optimization algorithm. IEEE Transactions on Power Delivery 23(1), 271–279 (2008)
7. Abdul Latiff, N., Tsimenidis, C., Sharif, B., Ladha, C.: Dynamic clustering using binary multi-objective particle swarm optimization for wireless sensor networks. In: IEEE 19th International Symposium on Personal, Indoor and Mobile Radio Communications, PIMRC 2008, pp. 1–5 (September 2008)
8. Wang, L., Ye, W., Fu, X., Menhas, M.: A modified multi-objective binary particle swarm optimization algorithm. In: Tan, Y., Shi, Y., Chai, Y., Wang, G. (eds.) ICSI 2011, Part II. LNCS, vol. 6729, pp. 41–48. Springer, Heidelberg (2011)
9. Bezerra, J.R., Barroso, G.C., Leao, R.P.: Switch placement algorithm for reducing customers outage impacts on radial distribution networks. In: TENCON 2012 - 2012 IEEE Region 10 Conference, pp. 1–6 (November 2012)
10. Kersting, W.: Radial distribution test feeders. IEEE Power Engineering Society Winter Meeting 2, 908–912 (2001)
11. Singh, K.: Electricity distribution network expansion planning (March 2013), http://www.epoc.org.nz/papers/KaviORSNZPaper.pdf
12. Deb, K.: Multi-objective evolutionary algorithms: Introducing bias among pareto-optimal solutions. In: Ghosh, A., Tsutsui, S. (eds.) Advances in Evolutionary Computing. Natural Computing Series, pp. 263–292. Springer, Heidelberg (2003)
13. Deb, K.: Multi-Objective Optimization Using Evolutionary Algorithms. Wiley Paperback. Wiley (2009)

Improving Chaotic Ant Swarm Performance with Three Strategies

Yu-Ying Li[1], Li-Xiang Li[2], and Hai-Peng Peng[2]

[1] Basis Department,
Institute of Chemical Defense of the Chinese People's Liberation Army,
Beijing, 102205, China
yuying_20042009@yahoo.com.cn
[2] Information Security Center,
Beijing University of Posts and Telecommunications,
Beijing, 100876, China

Abstract. This paper presents an improved chaotic ant swarm (ICAS) by introducing three strategies, which are comprehensive learning strategy, search bound strategy and refinement search strategy, into chaotic ant swarm (CAS) for solving optimization problems. The first two strategies are employed to update ants' positions, which preserve the diversity of the swarm so that the ICAS discourages premature convergence. In addition, the refinement search strategy is adopted to increase the solution quality in the ICAS. Simulations show that the ICAS significantly enhances solution accuracy and convergence stability of the CAS.

1 Introduction

Chaos search has been a novel and potential tool of optimization because the search strategies based on chaos properties have been found to obtain nice capabilities of hill-climbing and escaping from local optima, and to be more effective than random search [1]. Especially, as a special mechanism to avoid being trapped in local optimum, chaos ergodicity has been viewed as an effective search strategy. In recent years, the research based on chaos search has received particular attention and has obtained satisfactory results by combining it with other heuristic algorithms, such as chaos search [2], chaotic neural network [3], chaotic simulated annealing [4], chaotic particle swarm optimization [5] and chaotic ant swarm [6].

Chaotic ant swarm (CAS) is a chaos optimization algorithm inspired by the chaotic and self-organizing behavior of ants in nature. In the CAS, ants use chaotic principles to search for food. Each ant performs the chaotic exploration of its hunting sites and interacts with its neighbors. They search chaotically until they have been organized via pheromone trails, and move to the most successful site among the previously met hunting sites. The CAS uses the principles to implement a meta-heuristic for the search of a global optimum or near-optimum of a function in a search space.

Y. Tan, Y. Shi, and H. Mo (Eds.): ICSI 2013, Part I, LNCS 7928, pp. 268–277, 2013.

The CAS has been successfully used to solve some problems since it was proposed in 2006. However, as a newly emerged optimal algorithm, the CAS still has some problems which require us to do further research such as premature convergence, solution accuracy, and so on. In the past seven years, many adjustments and improvements have been made to the CAS to deal with these problems. As the parameters used in CAS are believed to have great influence on the performance of the algorithm, many researches[7] have been done to adaptively adjust the parameters for different problems. CAS is also hybridized with other methods [8,9] to enhance its ability on a large number of applications. Motivated and inspired by the research work mentioned above, based on the analysis of the CAS, this paper presents an improved chaotic ant swarm (ICAS) by utilizing three strategies, which are comprehensive learning strategy, search bound strategy and refinement search strategy, to increase the performance of the CAS. The first two strategies enable the diversity of the swarm to be preserved to discourage premature convergence. The last strategy improves solution accuracy. Simulations show that the ICAS enhances the global optimization ability greatly.

The rest of the paper is organized as follows. Section 2 introduces the CAS. Section 3 describes the ICAS. Section 4 presents and discusses experimental results that have been obtained on five benchmark functions. Finally, conclusion is given in Section 5.

2   Chaotic Ant Swarm Algorithm

In the CAS, the chaotic system $x(k) = x(k - 1) \times e^{\mu(1-x(k-1))}$ [10] was introduced into the heuristic equation of the CAS for obtaining the chaotic search initially. The adjustment of the chaotic behavior of individual ant is achieved by the introduction of a successively decrement of organization variable y_i and eventually leads the individual to move to the new site acquired with the best fitness value. In order to achieve the information exchange of individuals and the movements to new site taken on the best fitness value, the CAS introduced $(pbest_{id} - x_{id})$. $pbest_{id}$ is selected based on the fitness theory which is very widely developed in optimization theory such as genetic algorithm, tabu search, and so on. x_{id} is the state of the dth dimension of ant i. Considering a D-dimensional optimal problem, the iteration evolution of a swarm of L ants can be represented by those of a position vector, $x_i = (x_{i1}, ..., x_{iD})$, and an organization variable, y_i. At iteration step k, the position and organization variable of ant i are updated by using [6]

$$
\begin{cases}
y_i(k) = y_i(k - 1)^{(1+r_i)} \\
x_{id}(k) = (x_{id}(k - 1) + \frac{7.5}{\psi_d} \times v_i) \times e^{(1-e^{-ay_i(k)})(3-\psi_d(x_{id}(k-1)+\frac{7.5}{\psi_d} \times v_i))} - \frac{7.5}{\psi_d} \times v_i \\
\qquad + (pbest_{id}(k - 1) - x_{id}(k - 1))e^{-2ay_i(k)+b}
\end{cases}
$$

$$(1)$$

where k means the current iteration step, and $k - 1$ is the previous iteration step; $y_i(k)$ is the ith ant's organization variable of the current iteration step,

$y_i(0) = 0.999$; $x_{id}(k)$ is the current state of the dth dimension of ant i, $x_{id}(0) = \frac{7.5}{\psi_d} \times (1 - v_i) \times rand(1)$, where $rand(1)$ is a uniformly distributed random number in $(0, 1)$; $pbest_{id}(k-1)$ is the best position found by the ith ant and its neighbors within $k - 1$ steps; $v_i(0 < v_i < 1)$ determines the search region of ant i; a is a sufficiently large positive constant and can be selected as $a = 200$; $b(0 \leq b \leq 2/3)$ is a constant.

r_i and ψ_d are two important parameters. r_i is the organization factor of ant i, which directly affects the convergence speed of the CAS. If r_i is very large, the iteration step of chaotic search is small, then the system converges quickly and the desired optima or near-optima can not be achieved. If r_i is very small, the iteration step of chaotic search is large, then the system converges slowly and the runtime will be longer. Since small changes are desired as iteration step evolves, the value of r_i is chosen typically as $0 < r_i \leq 0.5$. The format of r_i can be designed according to concrete problems and runtime. Each ant could have different r_i, such as $r_i = 0.1 + 0.2 \times rand(1)$. ψ_d affects the search ranges of the CAS. If the interval of the search is $[-\frac{\omega_d}{2}, \frac{\omega_d}{2}]$, we can obtain an approximate formula $\omega_d \approx \frac{7.5}{\psi_d}$.

Eq. (1) describes the search process of the CAS. The organization variable y_i is used to control the chaotic process of ant moving, and its influence on the ant's behavior is very weak initially. That is, initially the organization capabilities of the ants are very weak so that a non-coordinated process occurs which is characterized by the chaotic walking of ants. This phase lasts until the influence of organization on the individual behavior is sufficiently large. Then, the chaotic behavior of the individual ant disappears and a coordination phase starts. That is, ants do some further searches and move to the best position which they have ever found in search space. Throughout the whole process, these ants exchange information with other ants, then compare and memorize the information.

3   The Improved Chaotic Ant Swarm Algorithm

Based on the understanding of the CAS, we propose an improved chaotic ant swarm (ICAS) by employing three strategies to increase the performance of CAS in this section. We describe the three strategies in detail.

3.1   Comprehensive Learning Strategy

As the fitness value $f(x_i) = f([x_{i1}, x_{i2}, ..., x_{iD}])$ of an ant is possibly determined by values of all D parameters, an ant that has discovered the solutions corresponding to the global optimum in some dimensions may have a high fitness value due to the poor solutions in the other dimensions. This paper defines the fitness value the smaller the better. This good genotype may be lost in this situation because all dimensions of an ant learn from just its historical best information in the CAS. In order to retain this good genotype, this paper introduces a comprehensive learning strategy [11], which makes each ant learn from different ants for different dimensions, to update ants' positions. The updating formulas of position and organization variable in the comprehensive learning strategy are

$$
\begin{cases}
y_i(k) = y_i(k-1)^{(1+r_i)} \\
x_{id}(k) = (x_{id}(k-1) + \frac{7.5}{\psi_d} \times v_i) \times e^{(1-e^{-ay_i(k)})(3-\psi_d(x_{id}(k-1)+\frac{7.5}{\psi_d} \times v_i))} - \frac{7.5}{\psi_d} \times v_i \\
\quad + (pbest_{g_i^d d}(k-1) - x_{id}(k-1))e^{-2ay_i(k)+b}
\end{cases}
$$

$$(2)$$

where $g_i = [g_i^1, g_i^2, ..., g_i^D]$ defines which ants' *pbest* the ant i should follow. $pbest_{g_i^d d}$ can be the corresponding dimension of any ant's *pbest* including its own *pbest*, and the decision depends on probability P, referred to as the learning probability. For each dimension of ant i, we generate a random number within (0,1). If the random number is larger than P_i, the corresponding dimension will learn from its own *pbest*, otherwise it will learn from another ant' *pbest*. When the ant's dimension learns from another ant's *pbest*, we employ the following selection procedure: first randomly choose two ants out of the swarm which exclude the ant whose position is updated, then compare the fitness values of these two ants' *pbests* and select the better one, final use the winner's *pbests* as the exemplar to learn from for that dimension. The details of choosing g_i are given in Fig. 1.

From Eq. (2), each dimension of ant i may learn from the corresponding dimension of a different ant's *pbest* to make the ant generate new positions and preserve the diversity of the swarm. To ensure that an ant learns from good exemplars and to intensify the search of the positions, we allow the ant to

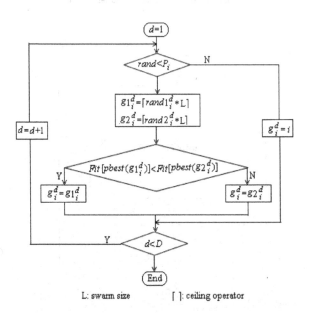

L: swarm size $\lceil\ \rceil$: ceiling operator

Fig. 1. Selection of exemplar dimensions for ant i

learn from the exemplars until the ant stops improving for a certain number of generations called the learning time m, and then we reassign g_i for the ant.

3.2 Search Bound Strategy

From Eq. (1) and $\psi_d \approx \frac{7.5}{w_d}$, the value of x_{id} may move out of the search range $[-\frac{w_d}{2}, \frac{w_d}{2}]$ in the search process. In order to prevent ants from moving out of the search bounds, we use the equation $x_{id} = min(\frac{w_d}{2}, max(-\frac{w_d}{2}, x_{id}))$ to restrain an ant on the border in the ICAS.

3.3 Refinement Search Strategy

In order to improve solution quality further, the refinement search is carried out around the ant with the best fitness value after the last iteration. In this way, the ant can learn from all other ants' historical best information from the first dimension to the final dimension. The flowchart of the refinement search is given in Fig. 2.

The ICAS introduces three strategies, which are comprehensive learning strategy, search bound strategy and refinement search strategy, to increase the performance of the CAS. From Sections 2 and 3, we can observe main differences between the ICAS and the CAS. Firstly, instead of learning from the same ant

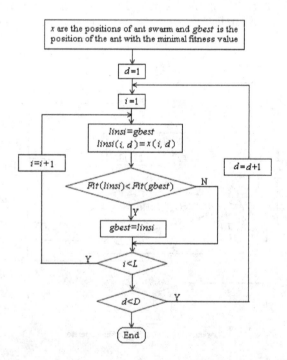

Fig. 2. Flowchart of the refinement searching strategy

for all dimensions, each dimension of an ant may learn from the corresponding dimension of a different ant's *pbest*; Secondly, instead of using an ant's *pbest* as the exemplar, all ants' *pbests* can potentially be used as the exemplars to guide an ant's move direction; Thirdly, instead of all dimensions learning from one exemplar in every generation, each dimension of an ant can learn from one exemplar for a few generations. Fourthly, instead of moving out of search ranges possibly, the variables are restricted in search ranges. Finally, instead of stopping search after the last iteration, the ant with the best fitness value has refinement search after the last iteration.

4   Simulation Experiment

4.1   Benchmark Functions

In our experiments, we choose the five benchmark functions with the global minimum fitness value 0 as follows. All the five functions are widely known for testing the performance of different heuristic optimization algorithms such as evolutionary programming, simulated annealing, genetic algorithms and particle swarm optimization. The five test functions are:

Sphere function:

$$f_1(x) = \sum_{i=1}^{D} x_i^2, \quad s.t. \ x_i \in [-50, 50] \tag{3}$$

DeJongF4 function:

$$f_2(x) = \sum_{i=1}^{D} i x_i^2, \quad s.t. \ x_i \in [-20, 20] \tag{4}$$

Rosenbrock function:

$$f_3(x) = \sum_{i=1}^{D-1} (100(x_{i+1} - x_i^2)^2 + (x_i - 1)^2), s.t. \ x_i \in [-100, 100] \tag{5}$$

Griewank function:

$$f_4(x) = 1 + \sum_{i=1}^{D} (\frac{x_i^2}{4000}) - \prod_{i=1}^{D} \cos(\frac{x_i}{\sqrt[2]{i}}), s.t. x_i \in [-600, 600] \tag{6}$$

Rastrigin function:

$$f_5(x) = \sum_{i=1}^{D} (10 + x_i^2 - 10\cos(2\pi x_i)), s.t. x_i \in [-5.12, 5.12] \tag{7}$$

4.2   Parameters Setting

In the ICAS, we use these parameter settings: $L = 20, a = 200, b = \frac{1}{2}, y(0) = 0.999, r_i = 0.01 + 0.00001 \times rand(1), v_i = rand(1)$. This kind of dynamical neighbors is selected. At first step, the number of neighbors is two. The number of neighbors will increase one every two iterative steps. The max number of neighbors is 19. The value of parameter ψ_d can be selected according to the ranges of intervals. The max number of iterations is 1000. We analyze the effects of learning probability P and learning time m on the search results of the ICAS in Sections 4.3 and 4.4, respectively. We compare the ICAS with the CAS in Section 4.5.

4.3   Learning Probability P

P is the learning probability, which decides how many dimensions are chosen to learn from other ant's *pbest*. In order to show the effects of the learning probability P on the search results of the ICAS, experiments are conducted on five benchmark functions with ten dimensions as defined in Section 4.1. P is set at 0, 0.05, 0.1, 0.2, 0.3, 0.4, 0.5, 0.6, 0.7, 0.8, 0.9 and 1.0, respectively. The ICAS runs 20 times for each P. Table 1 presents the results of the ICAS under different P while the best results among different P are shown in bold (where 1.82e-04 is defined 1.82×10^{-4}). As seen in Table 1, mean values and variances are better when $0.1 \leq P \leq 0.5$. Therefore, there is a problem, that is how to select suitable P to make search results of the ICAS better. Considering this problem, this paper sets different values of P for each ant so that ants achieve different levels of exploration and exploitation ability in the swarm and are able to solve diverse problems. This paper uses the following expression [11] to set a P_i value for each ant:

$$P_i = 0.05 + 0.45 \times \frac{e^{\frac{10(i-1)}{(L-1)}} - 1}{e^{10} - 1}.$$

4.4   Learning Time m

The learning time m ensures that an ant learns from good exemplars and strengthens the search of the positions. If m is too large, the search time at the positions of good exemplars is too long. Because the iteration number is limited, the search process is easy to trap in the local optimization. Considering of this, m should be tuned. In this section, five test functions with 10 dimensions as defined in Section 4.1 are used to investigate the impact of this parameter m on the results of the ICAS from $m = 0$ to $m = 15$. The ICAS runs 20 times on each of these functions. Mean fitness values of the final results are plotted in Fig. 3. As it shows, different learning times can produce different results. For the five test functions, better results are obtained when m is around three. Hence, in our experiments, the learning time m is set at three for all test functions.

Table 1. Mean values and variances of 20 runs on these test functions

P		0	0.05	0.1	0.2	0.3	0.4
Sphere	mean	1.82e-04	2.49e-04	6.44e-04	2.00e-03	1.31e-03	**7.90e-05**
	variances	9.67e-07	2.97e-07	3.43e-06	3.64e-05	2.38e-05	**6.82e-08**
DeJongF4	mean	4.23e-05	2.46e-06	2.29e-04	1.21e-04	1.43e-08	1.99e-06
	variances	2.31e-08	1.18e-10	9.80e-07	2.93e-07	2.29e-15	2.35e-11
Griewank	mean	5.98e-02	5.53e-02	7.19e-02	**8.90e-03**	1.64e-02	9.21e-02
	variances	2.48e-02	2.70e-02	2.61e-02	**5.85e-04**	2.61e-03	5.96e-02
Rosenbrock	mean	7.04	1.41e+01	**4.20**	6.83	1.91e+01	8.32
	variances	2.91e+02	2.10e+02	**5.23e+01**	1.29e+02	2.98e+03	1.03e+02
Rastrigin	mean	3.20e-04	2.64e-04	4.61e-04	2.13e-03	**2.06e-04**	4.85e-04
	variances	2.02e-06	1.05e-06	1.27e-06	4.60e-05	**3.99e-07**	3.29e-06
P		0.5	0.6	0.7	0.8	0.9	1.0
Sphere	mean	3.08e-04	1.80e-03	4.07e-04	1.12e-03	9.69e-04	4.91e-04
	variances	9.66e-07	6.35e-05	2.80e-06	2.28e-05	7.06e-06	2.31e-06
DeJongF4	mean	**6.45e-09**	2.30e-06	7.50e-08	5.66e-07	4.28e-06	1.78e-08
	variances	**6.91e-16**	9.69e-11	6.05e-14	4.95e-12	3.66e-10	2.65e-15
Rosenbrock	mean	1.12e+01	1.26e+01	5.68	5.42	1.16e+01	1.07e+01
	variances	4.14e+02	4.38e+02	9.30e+01	8.75e+01	7.54e+02	6.02e+02
Griewank	mean	5.06e-02	1.49e-01	4.43e-02	9.40e-02	1.51e-02	1.35e-01
	variances	4.56e-02	9.29e-02	1.16e-02	5.72e-02	1.70e-03	7.45e-02
Rastrigin	mean	4.60e-03	3.01e-03	4.12e-04	3.78e-04	6.12e-03	6.61e-03
	variances	4.03e-04	1.91e-04	1.24e-06	1.33e-06	7.15e-04	6.41e-04

4.5 Comparative Study

To evaluate the performance of the ICAS, we have comparisons with the CAS. Each experiment runs 50 times. Experiments are conducted on the five test functions with 30 dimensions. The mean values for 50 times along with the variance are presented in Table 2. The results of the CAS are from Ref. [6]. The experiments were performed on a computer with 2.93 GHz Intel(R) Pentium (R) 4 processor and 512 MB of RAM using Matlab 7.6.

From Table 2, we observe that the ICAS achieves better results in the mean and variance than the CAS on all five test functions because of employing three strategies in the ICAS. Especially, three orders of magnitude are improved in

Table 2. Comparison between the ICAS and the CAS

Function	mean		variance		mean runtime (s)	
	ICAS	CAS	ICAS	CAS	ICAS	CAS
Sphere	2.93E-03	3.81E-01	6.85E-05	5.33E-02	83.26	80.15
DeJongF4	2.84E-05	1.61E-02	1.04E-08	1.62E-03	90.98	87.30
Rosenbrock	2.12E+01	2.34E+01	7.21E+02	1.37E+04	86.43	83.26
Griewank	7.61E-02	4.66E-01	4.56E-02	1.82E-01	92.71	89.78
Rastrigin	1.11E-02	2.26E+01	5.02E-03	1,10E+03	88.69	85.45

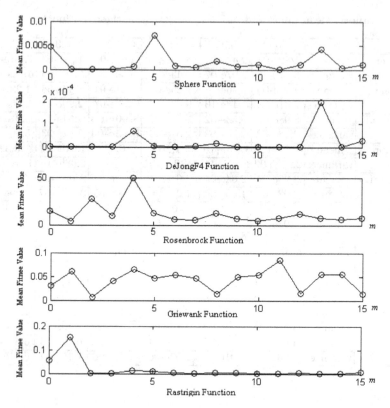

Fig. 3. The results of the ICAS on five test functions with different learning time m

the mean while six orders of magnitude in the variance for Rastrigin function. Although Rosenbrock function has not change the order of magnitude in the mean, it improves two orders of magnitude in the variance. That is, the stability of Rosenbrock function is enhanced in the ICAS. For others functions, the mean and variance both are improved. Such as, Sphere function increases two orders of magnitude in the mean while three orders of magnitude in the variance. DeJongF4 function increases three orders of magnitude in the mean while five orders of magnitude in the variance. Griewank function both improves one order of magnitude in the mean and variance. According to the mean, the ICAS is more efficient to improve solution search ability than the CAS. Such as, from the results of Sphere function, the ICAS increases the local search ability of the CAS; as seen in Griewank and Rastrigin function, the ICAS enhances the global search ability of the CAS. Depending on the variance, solution stability is enhanced in the ICAS compared with the CAS. Thus, the ICAS improves the performance of the CAS in solution search ability and solution stability.

From Table 2, note that the ICAS takes a little higher mean runtime than the CAS because of using three strategies in the ICAS. However, from experimental results (see Table 2), we could see that the ICAS outperforms the CAS on all five benchmark functions. So we could say the ICAS is efficient. Moreover, with the rapid development of computer, the tradeoff between high-quality solutions and computational time tends to the former. So the quality of solutions preponderates when problems could be solved by algorithms in rational time.

5 Conclusion

Based on the mechanism analysis of chaotic ant swarm (CAS), this paper presents an improved chaotic ant swarm (ICAS) by introducing three strategies, which are comprehensive learning strategy, search bound strategy and refinement search strategy, to solve optimization problems. The three strategies enable the ICAS to make use of the information in swarm more effectively and generate better solution quality frequently when compared to the CAS. From the testing results of the two algorithms, we can conclude that ICAS significantly improves solution accuracy and convergence stability of the CAS.

References

1. Hayakawa, Y., Marumoto, A., Sawada, Y.: Effects of the Chaotic Noise on the Performance of a Neural Network Model for Optimization Problems. Phys. Rev. E 51, 2693–2696 (1995)
2. Li, B., Jiang, W.: Optimizing Complex Functions by Chaos Search. Int. J. Cybernet. Syst. 29, 409–419 (1998)
3. Kwok, T., Smith, K.: Experimental Analysis of Chaotic Neural Network Models for Combinatorial Optimization under a Unifying Framework. Neural Networks 13, 731–744 (2000)
4. Ji, M., Tang, H.: Application of Chaos in Simulated Annealing. Chaos Solitons and Fractals 21, 933–941 (2004)
5. Liu, B., Wang, L., Jin, Y., Tang, F., Huang, D.: Improved Particle Swarm Optimization Combined with Chaos. Chaos Solitions and Fractals 25, 1261–1271 (2005)
6. Li, L., Yang, Y., Peng, H., Wang, X.: An Optimization Method Inspired by Chaotic Ant Behavior. International Journal of Bifurcation and Chaos 16, 2351–2364 (2006)
7. Cai, J., li, Q., li, L., Peng, H., Yang, Y.: A Fuzzy Adaptive Chaotic Ant Swarm Optimization for economic dispatch. Electrical Power and Energy Systems 34, 154–160 (2012)
8. Li, Y., Wen, Q., Li, L., Peng, H.: Hybrid Chotic Ant Swarm Optimization. Chaos Solitons and Fractals 42, 880–889 (2009)
9. Li, Y., Wen, Q., Zhang, B.: Chotic Ant Swarm Optimization with Passive Congregation. Nonlinear Dynamics 68, 129–136 (2012)
10. Solé, R.V., Miramontes, O., Goodwill, B.C.: Oscillations and chaos in ant societies. Journal of Theoretical Biology 161, 343–357 (1993)
11. Liang, J., Qin, A.: Comprehensive Learning Particle Swarm Optimizer for Global Optimization of Multimodal Functions. IEEE Transactions on Evolutionary Computation 10, 281–295 (2006)

Improved Ant Colony Classification Algorithm Applied to Membership Classification

Hongxing Wu and Kai Sun

The School of Computer and Information,
Hefei University of Technology, Hefei, China
{hongxingwu,sunkai19870613}@163.com

Abstract. Membership classification belongs to target customer analysis in the customer relationship management. For membership classification, an improved ant colony classification algorithm named mAnt-Miner+ is proposed. This algorithm on the basis of Ant-Miner, draws on the idea of mAnt-Miner (Ant-Miner that uses a population of many ants), and introduces a new heuristic strategy. Experimental results show that, in terms of prediction accuracy, mAnt-Miner+ is competitive with Ant-Miner and higher than mAnt-Miner; in terms of running efficiency, mAnt-Miner+ is more efficient than mAnt-Miner and Ant-Miner.

Keywords: Ant-Miner, Heuristic, Membership Classification, Customer Relationship Management.

1   Introduction

Customer Relationship Management (CRM) includes a set of systems of the business strategy which make enterprises establish long-term cooperative relations with specific customers [1]. Members are the basic customers of retail companies, and many retail companies have stored a large number of member data that contains the basic information and consumption records. Membership classification will benefit retail enterprises mining potential customers. The member data of some retail companies are large, which require improving the operational efficiency of data mining algorithms. For membership classification, this article uses the ant colony classification algorithm, and discusses the method to improve the performance of the algorithm.

Ant-Miner [2] is the first algorithm which implements Ant Colony Optimization (ACO) [3] algorithm for classification task of data mining. However, Ant-Miner works with a single ant in its each iteration, different from the standard definition of ACO which uses a set of ants. GUI Ant-Miner [4] and ACO-Miner [5] have made use of ant populations to improve Ant-Miner. Multi-population parallel strategy is proposed in ACO-Miner. In this paper, we define the algorithm, like GUI Ant-Miner, which uses a population of many ants in Ant-Miner as mAnt-Miner. But mAnt-Miner still applies the heuristic strategy of Ant-Miner that contains the local information and

Y. Tan, Y. Shi, and H. Mo (Eds.): ICSI 2013, Part I, LNCS 7928, pp. 278–287, 2013.

is modified in the running of the algorithm. It makes mAnt-Miner easy to trap in the local optimal solution, and reduces the stability of the algorithm. Therefore, we propose mAnt-Miner+ that is based on mAnt-Miner and uses a new heuristic strategy.

The remaining paper is organized as follows. Section 2 presents an overview of Ant-Miner. Section 3 describes our proposed algorithm mAnt-Miner+. Section 4 reports the experimental results. Finally, conclusions are given in Section 5.

2   Ant-Miner

The discovery of classification rules is the target problem of Ant-Miner. The form of each classification rule is IF <conditions> THEN <class>, where the <conditions> part contains a logical combination of predictor attributes, in the form: term₁ AND term₂ AND ... AND termₙ. Each term is a triple <attribute, operator, value>, where operator represents a relational operator and value belongs to the domain of attribute. Ant-Miner can only cope with categorical attributes, so that operator is always "=". Continuous attributes are discretized in the preprocessing step. The <class> part of the rule specifies the class predicted for cases whose predictor attributes satisfy the <conditions> part of the rule. Ant-Miner consists of three main steps, including rule construction, rule pruning, and pheromone updating.

2.1   Rule Construction

Each ant begins with an empty rule and keeps adding $term_{ij}$ to its current partial rule until meeting the stopping criterion. The $term_{ij}$ is a rule condition, in the form: $A_i = V_{ij}$, where V_{ij} is the jth value in the domain of A_i. The probability that $term_{ij}$ is chosen to be added to the current partial rule is

$$P_{ij}(t) = \frac{\eta_{ij} \cdot \tau_{ij}(t)}{\sum_{i}^{a}\sum_{j}^{b_i}(\eta_{ij} \cdot \tau_{ij}(t)), \forall i \in I} \tag{1}$$

where η_{ij} is the value of the problem-dependent heuristic function for $term_{ij}$, $\tau_{ij}(t)$ is the amount of pheromone associated with $term_{ij}$ at iteration t, corresponding to the amount of pheromone currently available in the position i, j of the trail being followed by the current ant, a is the total number of attributes, b_i is the total number of values in the domain of A_i, and I are the attributes that are not yet used by the ant.

The heuristic function η_{ij} is an estimate of the quality of $term_{ij}$, with respect to its ability to improve the accuracy of the rule. In Ant-Miner, the heuristic function η_{ij} is based on information theory [6] and is defined by

$$\eta_{ij} = \frac{\log_2 k - InfoT_{ij}}{\sum_{i}^{a}\sum_{j}^{b_i}(\log_2 k - InfoT_{ij}), \forall i \in I} \tag{2}$$

$$InfoT_{ij} = -\sum_{w=1}^{k} \left(\frac{FreqT_{ij}^{w}}{|T_{ij}|} \right) \times \log_2 \left(\frac{FreqT_{ij}^{w}}{|T_{ij}|} \right) \tag{3}$$

where k is the number of classes, $|T_{ij}|$ is the total number of cases in partition T_{ij} (partition containing the cases whose attribute A_i has value V_{ij}), $FreqT_{ij}^{w}$ is the number of cases in partition T_{ij} with class w, and a, I have the same meaning as in (1). However, higher the value of $InfoT_{ij}$, lower the predictive power of $term_{ij}$.

2.2   Rule Pruning

The rule pruning procedure is performed for each ant as soon as the ant completes the construction of its rule. The main goal of rule pruning is to improve the rule quality by removing irrelevant terms which might have been added during the rule construction. The rule pruning iteratively removes the term until the rule has just one term or until there is no term whose removal will improve the quality of the rule. The quality of a rule is measured by

$$Q = \frac{TP}{TP+FN} \cdot \frac{TN}{FP+TN} \tag{4}$$

where TP is the number of cases covered by the rule and whose classes are predicted correctly by the rule, FP is the number of cases covered by the rule and whose classes are predicted falsely by the rule, FN is the number of cases that are not covered by the rule while having the class predicted by the rule, and TN is the number of cases that are not covered by the rule and whose classes are predicted falsely by the rule.

2.3   Pheromone Updating

After each ant completes the rule construction and the rule pruning, the amount of pheromone in all segments of all paths must be updated. The pheromone updating has two basic ideas. First, the amount of pheromone associated with each $term_{ij}$ that occurs in the constructed rule is increased. Second, the amount of pheromone associated with each $term_{ij}$ which does not occur in the constructed rule is decreased, simulating pheromone evaporation in real Ant Colony Systems.

The pheromone updating formula of $term_{ij}$ that occurs in the rule is presented as

$$\tau_{ij}(t+1) = \tau_{ij}(t) + \tau_{ij}(t) \cdot Q, \forall i, j \in R \tag{5}$$

where R is the set of terms that occur in the rule constructed by the ant at iteration t and Q has been defined as (4). Q varies within the range of (0, 1) and the larger the value of Q, the higher the quality of the rule.

3 *m*Ant-Miner+

In Ant-Miner, each classification rule is constructed by a single ant. In order to make use of ant populations in ACO concepts, we apply the method of *m*Ant-Miner. However, we find that the heuristic strategy of *m*Ant-Miner contains the local information and is modified during the algorithm running, so that it makes *m*Ant-Miner easy to trap in the local optimal solution and decreases the stability of *m*Ant-Miner. Therefore, we propose *m*Ant-Miner+ using a new heuristic strategy.

Algorithm 1. High level pseudo-code of *m*Ant-Miner+

```
TrainingSet = {all training cases};
RuleList = [];
Calculate heuristic function η;
WHILE (TrainingSet > Max_uncovered_cases)
  t = 1; //iteration index
  j = 1; //convergence test index
  Initialize all trails with
    the same amount of pheromone;
  REPEAT
    k = 1; //ant index
    FOR (k < No_of_ants)
      Antₖ incrementally constructs
        a classification rule;
      Prune the just-constructed rule;
      k = k + 1;
    END FOR
    Select the best ant in k ants as Ant_best,
      and its rule as R_t;
    Update the pheromone of the trail
      followed by Ant_best;
    IF (R_t is equal to R_t-1)
      THEN j = j + 1;
      ELSE j = 1;
    END IF
    t = t + 1;
  UNTIL (t >= No_of_iterations)
    OR (j >= No_rules_converg)
  Choose the best rule R_best among all rules R_t;
  Add rule R_best to RuleList;
  TrainingSet = TrainingSet -
    {cases correctly covered by R_best};
END WHILE
```

A high level pseudo-code of *m*Ant-Miner+ is shown in Algorithm 1. In summary, *m*Ant-Miner+ works as follow. The algorithm starts with an empty rule list and the training set that consists of all the training cases. The values of the heuristic function are computed before the WHILE loop. It is different from Ant-Miner, because the values of this heuristic function are invariable during the running of the algorithm. Each iteration of the WHILE loop including a number of REPEAT-UNTIL loop discovers one classification rule. This rule is added to the rule list and the training cases that are covered correctly by this rule are removed from the training set. Each iteration of the REPEAT-UNTIL loop including a number of FOR loop finds the best ant in ant populations and the rule constructed by this ant. The amount of pheromone in segments of the trail followed by the best ant is updated. Each iteration of the FOR loop, that makes use of ant populations, consists of two steps, rule construction and rule pruning. The processes of rule construction and rule pruning are same as Ant-Miner.

3.1   Ant Populations

In the standard definition of ACO [7], a population is defined as the set of ants that construct solutions between two pheromone updates. But Ant-Miner performs with a population of only one ant, since pheromone is updated until a rule is constructed by a single ant [2].

GUI Ant-Miner [4] is an updated version of Ant-Miner and differs from Ant-Miner as that it makes possible the use of ant populations within the ACO concept. ACO-Miner [5] is an improved algorithm of Ant-Miner. ACO-Miner also makes use of ant populations and uses multi-population parallel strategy that divides the ant colony into several populations. These populations are parallel, run separately and have the same amount of ants.

GUI Ant-Miner is different from ACO-Miner, since GUI Ant-Miner uses the best ant to update pheromone after the best rule constructed by this ant is selected from the rules constructed by a population of many ants. However, ACO-Miner uses each ant to perform the pheromone updating after a rule is constructed by each ant of several ant populations.

Although GUI Ant-Miner introduce the concept of ant populations, it is different from ACO, because it just uses one ant (the best one of ant populations) to update pheromone. However, its cost of computation is smaller than that uses each ant to update pheromone.

This article applies the idea of GUI Ant-Miner, and defines this algorithm as mAnt-Miner. The ant populations used by mAnt-Miner can avoid dependence on initial term due to the random select of term at the beginning of iteration.

3.2   Heuristic Strategy

Ant-Miner2 [8] uses a simpler, though less accurate, density estimation as the heuristic value with the assumption that the idea of pheromone should compensate the small potential errors. This makes Ant-Miner2 computationally less expensive without a significant degradation of the performance. However, in mAnt-Miner+, we believe that not only the idea of the pheromone but also the use of ant populations should compensate the small induced errors in the simple heuristic function.

We also analyze the heuristic strategy in Ant-Miner. The value of η_{ij} contains the local information and is modified during the running of Ant-Miner, since the denominator of η_{ij} varies in the running. The heuristic strategy of Ant-Miner has two minor caveats [2]. First, if the value V_{ij} of attribute A_i does not occur in the training set, then $InfoT_{ij}$ is set to its maximum value of $\log_2 k$. This corresponds to assigning to $term_{ij}$ the lowest possible predictive power. Second, if all the cases in the partition T_{ij} belong to the same class then $InfoT_{ij}$ is set to zero. This corresponds to assigning to $term_{ij}$ the highest possible predictive power. This will also make the value of η_{ij} contain the local information and modified during the running of the algorithm, because the cases of training set is decreased in the running.

The heuristic strategy of Ant-Miner makes mAnt-Miner easy to trap in the local optimal solution, since it contains the local information. However, the heuristic strategy of Ant-Miner is different from ACO. In ACO, the value of η_{ij} is not modified during the running of the algorithm [9]. So we also believe that the heuristic function η_{ij} does not need to vary in the running.

In order to reduce the local information in the value of η_{ij} and make it invariable during the running of Ant-Miner, we remove the denominator of the original heuristic function and delete the two caveats of the original heuristic strategy. Finally, the new heuristic function is defined as

$$\eta_{ij} = \log_2 k - InfoTS_{ij} \tag{6}$$

where TS is different from T in (2), because TS is all the training cases, not the partition. In essence, η_{ij} in (6) is the information gain of $term_{ij}$ based on the total training set. After the new heuristic function finishes the initial computing, it is invariable during the running of the algorithm.

4   Experimental Results

The experiment is divided into two parts, namely: 1) In order to evaluate the performance of the improved algorithm, we selected six standard data sets from the UCI Irvine machine learning repository [10]; 2) In order to apply the improved algorithm in the member classification, we used the actual member data set. The data sets selected from UCI are discrete data; while the actual member data set is continuous data, needing the discrete processing.

All the results of the experiment were obtained by using a Pentium P6000 PC with clock rate of 1.87GHz and 2GB of main memory. The results of Ant-Miner and mAnt-Miner were computed by GUI Ant-Miner. mAnt-Miner+ was developed in Java language and based on GUI Ant-Miner.

4.1   Parameter Setting

Five parameters need to be set in algorithms, described as follow:

1) Number of ants in each ant generation (*No_of_ants*);
2) Minimum number of cases covered by per rule (*Min_cases_per_rule*);

3) Maximum number of uncovered cases in the training set (*Max_uncovered_cases*);
4) Number of rules used to test convergence of ants (*No_rules_converg*);
5) Number of iterations of each ant generation (*No_of_iterations*).

Note that *No_of_ants* is different from the parameter that has the same name in original Ant-Miner, since it represents the number of ants in each ant generation. However, *No_of_ants* in original Ant-Miner has the same meaning of *No_of_iterations*. The roles of *Min_cases_per_rule*, *Max_uncovered_cases* and *No_rules_converg* are same as in original Ant-Miner. Table 1 shows the parameter setting used when testing each algorithm based on Ant-Miner.

Table 1. Parameter setting in experiments

Parameter	Ant-Miner	*m*Ant-Miner	*m*Ant-Miner+
No_of_ants	1	5	5
Min_cases_per_rule	5	5	5
Max_uncovered_cases	10	10	10
No_rules_converg	10	10	10
No_of_iterations	1000	1000	1000

4.2   Data Preparation

Firstly, introduce the UCI standard data sets. In order to reduce the influence of data discretization, the selected data sets have only categorical attributes. The main characteristics of the selected data sets are summarized in Table 2.

Table 2. Summary of the data sets

Data set	Instances	Attributes	Classes
breast-cancer	286	10	2
car	1728	7	4
soybean	683	36	19
sponge	76	46	3
tic-tac-toe	958	10	2
vote	435	17	2

Secondly, introduce the actual member data set. We get two kinds of data from the database of a retail company, namely:

1) The information of 5000 membership card samples, defined as Data 1.
2) The detail records of member consumption in 2011, defined as Data 2.

The attributes of data in Data 1 are the id of membership cards, the kind of membership cards, the time of issuing membership cards, expiration time, and validity. The attributes of data in Data 2 are the serial number, the id of membership cards, transaction time, and transaction amount. The kind of membership cards is the basic of member classification. There are four kinds of membership cards, such as score cards, silver cards, gold cards, and diamond cards.

The Data 2 are the detail records of member consumption in 2011, so the data leaved in Data 1 should meet that the time of issuing membership cards is before 2011, the expiration time is after 2011, and the value of validity is "y". After processing, the number of the remaining data in Data 1 is 1159, including: 892 score cards, 83 silver cards, 137 gold cards and 47 diamond cards. The data in Data 2 are sorted by the transaction time. In order to reduce data density of Data 2 in time line, we merge the data by month, so that we get the monthly consumption of each member in 2011.

Cycle search the data in Data 1 and Data 2, and obtain the data whose id of membership cards are existed in both Data 1 and Data 2. We save the kind of membership cards and the monthly consumption of the obtained data in Data 3. Data 3 contains 13 attributes, namely: the kind of membership cards and the consumption of each month. We delete the data whose some month consumption is negative or total annual consumption is less than or equal to zero. As a result, the number of data reduces to 880 from 1159, including: 640 score cards, 75 silver cards, 124 gold cards and 41 diamond cards. Finally, we define the kind of membership cards as the class attribute, the monthly consumption as the general attributes.

The data need to be discretized, since Ant-Miner can only deal with the discrete data. This article uses the unsupervised discretization to process data. We divide the value of the monthly consumption into 10 equal-width intervals, and obtain the member data sets used in experiment.

4.3   Comparison of Results

The comparison is carried out across two criteria: the predictive accuracy and the processing time. Ten-fold cross validation is performed on each of data sets. We compare the experimental results of Ant-Miner, mAnt-Miner and mAnt-Miner+ on the two kinds of data sets that are introduced in section 4.2.

Table 3 shows a comparison of the predictive accuracy for each algorithm on UCI data sets. The numbers after the "±" symbol denote standard deviations. As shown in the table, both mAnt-Miner+ and Ant-Miner discover rules with a better predictive accuracy than mAnt-Miner in all six data sets. However, mAnt-Miner+ discovers rules with a better predictive accuracy than Ant-Miner in four data sets. In other two data sets, Ant-Miner is more accurate than mAnt-Miner+.

Table 3. Predictive accuracy (%) of each algorithm on UCI data sets

Data set	Ant-Miner	mAnt-Miner	mAnt-Miner+
breast-cancer	74.37 ±2.55	73.85 ±2.19	74.79 ±2.19
car	84.90 ±0.90	83.80 ±1.38	84.84 ±1.49
soybean	87.91 ±1.80	87.58 ±1.12	88.17 ±1.13
sponge	91.67 ±3.88	90.48 ±2.82	91.25 ±2.67
tic-tac-toe	72.30 ±2.82	71.28 ±2.37	75.57 ±1.47
vote	94.95 ±1.11	94.47 ±0.78	95.65 ±0.92

Table 4 shows the processing time of each algorithm on UCI data sets. From Table 4, we can conclude that *m*Ant-Miner+ is faster than *m*Ant-Miner and Ant-Miner in all six data sets. *m*Ant-Miner is faster than Ant-Miner in four data sets. In other two data sets, the processing time of Ant-Miner and *m*Ant-Miner are the same.

Table 4. Processing time (s) of each algorithm on UCI data sets

Data set	Ant-Miner	*m*Ant-Miner	*m*Ant-Miner+
breast-cancer	4	3	1
car	12	8	5
soybean	1011	450	282
sponge	123	58	6
tic-tac-toe	8	8	2
vote	4	4	3

We can analysis from the results of Table 3 and Table 4 that *m*Ant-Miner is easy to trap in the local optimal solution, since the predictive accuracy of *m*Ant-Miner is lower than Ant-Miner. *m*Ant-Miner exists the problem of premature convergence, so *m*Ant-Miner is faster than Ant-Miner in two thirds of the data sets. However, the predictive accuracy of *m*Ant-Miner+ is competitive with Ant-Miner and better than *m*Ant-Miner, and *m*Ant-Miner+ is the fastest one in three algorithms. So *m*Ant-Miner+ avoids trapping in the local optimal solution, and improves the efficiency of the algorithm.

Table 5 shows the predictive accuracy and the processing time of each algorithm on the actual member data sets. As shown in the table, the predictive accuracy of *m*Ant-Miner+ and Ant-Miner are approximate, but the processing time of *m*Ant-Miner+ is fastest. The predictive accuracy of *m*Ant-Miner is still lowest, namely, *m*Ant-Miner traps in the local optimal solution again. However, *m*Ant-Miner does not premature converge this time, and the processing time of *m*Ant-Miner is the longest. It illustrates that *m*Ant-Miner is not stable in experiment on different type of data sets.

Table 5. Predictive accuracy (%) and processing time (s) of each algorithm on member data sets

Data set	Predictive accuracy	Processing time
Ant-Miner	71.59 ±1.80	54
*m*Ant-Miner	70.23 ±3.03	86
*m*Ant-Miner+	71.34 ±1.10	34

5   Conclusion

An improved ant colony classification algorithm named *m*Ant-Miner+ is proposed in this article, and is applied to membership classification. *m*Ant-Miner+ is based on *m*Ant-Miner and uses a new heuristic strategy that reduces the local information. In order to be close to the operating mechanism of ACO, we make the value of new heuristic strategy invariable in the running. The results of experiments show that, mAnt-Miner is easy to trap in the local optimal solution, and is not stable in experiment on different types of data sets. However, mAnt-Miner+ successes in avoiding the weakness of mAnt-Miner and is competitive with Ant-Miner and higher than mAnt-Miner with respect to predictive accuracy; the processing time of mAnt-Miner+ is faster than mAnt-Miner and Ant-Miner.

Acknowledgment. This paper is supported by the National Natural Science Foundation of China under Grant No. 61070131.

References

1. Ngai, E.W.T., Xiu, L., Chau, D.C.K.: Application of data mining techniques in customer relationship management: A literature review and classification. Expert Systems with Applications 36(2), 2592–2602 (2009)
2. Parpinelli, R.S., Lopes, H.S., Freitas, A.A.: Data mining with an ant colony optimization algorithm. IEEE Trans on Evolutionary Computation 6(4), 321–322 (2002)
3. Colorni, A., Dorigo, M., Maniezzo, V.: Distributed Optimization by Ant Colonies. In: 1st European Conference on Artificial Life, Paris, France, pp. 134–142 (1991)
4. ACO: Public Software,
   http://www.aco-metaheuristic.org/aco-code/public-software.html
5. Jin, P., Zhu, Y., Hu, K., Li, S.: Classification Rule Mining Based on Ant Colony Optimization Algorithm. In: Huang, D.-S., Li, K., Irwin, G.W. (eds.) ICIC 2006. LNCIS, vol. 344, pp. 654–663. Springer, Heidelberg (2006)
6. Cover, T.M., Thomas, J.A.: Elements of Information Theory, pp. 12–49. Wiley (1991)
7. Dorigo, M., Caro, G.D.: The ant colony optimization meta-heuristic. In: Corne, D., Dorigo, M., Glover, F. (eds.) New Ideas in Optimization, pp. 11–32. McGraw-Hill, New York (1999)
8. Liu, B., Abbass, H.A., McKay, B.: Density-based heuristic for rule discovery with ant-miner. In: 6th Australasia-Japan Joint Workshop on Intelligent and Evolutionary System, Canberra, Australia, pp. 180–184 (2002)
9. Dorigo, M., Maniezzo, V., Colorni, A.: Ant System: Optimization by a Colony of Cooperating Agents. IEEE Trans on Systems, Man, and Cybernetics B 26(1), 29–41 (1996)
10. Asuncion, A., Newman, D.: UCI machine learning repository (2007),
    http://www.ics.uci.edu/~mlearn/MLRepository.html

Two Ant Decision Levels and Its Application
to Multi-Cell Tracking

Benlian Xu[1], Qinglan Chen[2], Mingli Lu[1], and Peiyi Zhu[1]

[1] School of Mechanical Engineering,
Changshu Institute of Technology, 215500 Changshu, China
[2] School of Electrical & Automatic Engineering,
Changshu Institute of Technology, 215500 Changshu, China
xu_benlian@yahoo.com.cn, chenqinglan@yahoo.com, zpy2000@126.com

Abstract. Inspired by ant's stochastic behavior in searching of multiple food sources, a novel ant system with two ant decision levels are proposed to track multiple cells in biological field. In the ant individual level, ants within the same colony perform independently, and ant decision is determined in probability by both its intended motion model and likelihood function. In the ant cooperation level, each ant adjusts individual state within its influence region, while the global best template at current iteration is found among all ant colonies and further utilized to update ant model probability, influence region, and the probability of fulfilling task. Simulation results demonstrate that our algorithm could automatically track numerous cells and its performance is compared with the multi-Bernoulli filtering method.

Keywords: Ant Colony, Multi-Cell Tracking, Parameter Estimate.

1   Introduction

The study of cellular behavior analysis involves many challenges, such as model uncertainty, morphological variance, overlapping and colliding between cells, and conventional or manual analysis is definitely a tedious and time consuming process even for experts in this field. As such, the automated analysis of cellular behavior is demanded eagerly for large number of cell image data, and related promising reports in this field have been reported over the past decades [1-5]. Among these reports, three categories are summarized, i.e., model propagation based method, detection based method, and multi-object Bayesian probabilistic method. In terms of the model propagation based methods, both snakes [6] and Level set [7] require cells to be partially overlapping in adjacent frames; Mean-shift algorithms [8] give a fast solution for object tracking in video sequences, but usually do not give object contours. In the category of detection based method, the typical advantage is computational efficiency with respect to segmentation, but the algorithms encounter problems during the temporal data association stage [9]. The last category of cell tracking algorithm is the multi-object Bayesian probabilistic method, the Random Finite Sets (RFS) based methods [10,11] have demonstrated that they could be extended to successfully track multiple cells such as E. coli bacteria and T cell.

Y. Tan, Y. Shi, and H. Mo (Eds.): ICSI 2013, Part I, LNCS 7928, pp. 288–296, 2013.

Ant colony optimization (ACO), a population-based and meta-heuristic approach, is to mainly solve various optimization problems [12]. In this paper, however, we present a novel ant system algorithm to track multiple cells, which consists of two levels, namely, the ant individual level and the ant cooperation level. The ant individual level encourages each ant to move closely to the location of object, while in the ant cooperation level the cooperation mechanism among colonies is utilized to update and regulate the behavior of ant for moving towards regions where cells occurs.

2   Bayesian Multi-Object Tracking Problem Description

With the assumption that each object follows Markov process, the Bayesian filtering algorithm offers a concise way to describe the multi-object tracking problem. For tracking n objects, we denote the multi-object state by $X(k) = \{x_1(k), x_2(k), ..., x_n(k)\}$ at tim k, where $x_i(k)$ is the state vector of i-th object. Let $z(k) = [z_1, z_2, ..., z_{\bar{m}}]$ denote the image observation comprising an array of $\bar{m}$ pixel values, and $z(1:k)$ is defined as a cumulative image sequences up to time k. Therefore, if the posterior density of multi-object state is denoted by $\pi(\bullet \mid z(1:k))$, the theoretic optimal approach to multi-object state estimation problem can be formulated as

$$\pi(X(k) \mid z(1:k-1)) = \int f(X(k) \mid X)\pi(X \mid z(1:k-1))\delta X$$

$$\pi(X(k) \mid z(1:k)) = \frac{h(z(k) \mid X(k))\pi(X(k) \mid z(1:k-1))}{\int h(z(k) \mid X)\pi(X \mid z(1:k-1))\delta X} \qquad (1)$$

where $f(\bullet)$ is the multi-object transition density function, $h(\bullet)$ is the observation likelihood function, and δ is an appropriate reference measure on some state space. In the next section, we will use proposed ant system to approximately estimate the posterior density of multi-object state.

3   Ant System with Two Ant Decision Levels

3.1   Ant Individual Decision Level

Considering the case of tracking numerous cells each represented approximately by the distribution of N ants, we use m ant colonies to cooperatively estimate the state of each object. *In the ant individual decision level*, with the assumption of each colony working independently, multi-object tracking system can be treated as a single object tracking problem in which only one ant colony is employed. Next, without loss of generality, we only describe the decision behavior of ant i ($i \in \{1, ..., N\}$) in colony s ($s \in \{1, ...m\}$) when searching for potential position where cell probably appears. At time k, the decision of ant i at next step can be defined with a probability as

$$p_s^{(i,j)}(k+1\mid k)$$

$$= \frac{\left[\pi_s^{(i),j}(k+1\mid k)f^{(j)}(x_s^{(j)}(k+1\mid k)\mid x_s^{(i)}(k))\right]^\alpha [h(z(k+1)\mid x_s^{(j)}(k+1\mid k))]^\beta}{\sum_{l\in\Omega_s^{(i)}}\left[\pi_s^{(i),l}(k+1\mid k)f^{(l)}(x_s^{(l)}(k+1\mid k)\mid x_s^{(i)}(k))\right]^\alpha [h(z(k+1)\mid x_s^{(l)}(k+1\mid k))]^\beta} \tag{2}$$

where model probability $\pi_s^{(i),j}(k)$ denotes the j-th element in $\pi_s^{(i)}(k)=[\pi_s^{(i),1}(k),\pi_s^{(i),2}(k),\pi_s^{(i),3}(k)]^T$ with the assumption that each ant has three modes, namely, *moving forward, turning right, and turning left* ; $x_s^{(j)}(k+1\mid k)$ is the one-step prediction of state $x_s^{(i)}(k)$ following a Markov process with a transition density $f^{(j)}(\bullet)$; Observation image $z(\bullet)$ is usually modeled by the likelihood function $h(z(\bullet)\mid x_s^{(j)}(\bullet))$ conditioned on the current state $x_s^{(j)}(\bullet)$; j corresponds to one of positions obtained from three modes, α and β are the importance adjustment parameter, and $\Omega_{(s)}^i$ denotes the set of three predicted positions of ant i .

As implicated in Eq. (2), $\pi_s^{(i),j}(k)$ determines the inclination degree to which ant i move towards j . State transition density $f^{(j)}(\bullet)$ encompasses the information of object moving various modes. Measurement likelihood function $h(\bullet)$ is usually selected according to empirical analysis in computer vision. Our algorithm employs a novel likelihood function to discriminate interested regions easily, and it operates directly on RGB space but requires a small number of templates which is established in a dynamic way. Therefore, for a given histogram u_k of an RGB image, the corresponding likelihood score is computed as

$$h(u_k) = \mu\left(e^{-\rho(1-g_k)^\gamma}\right)^\xi \tag{3}$$

where μ,ρ,γ *and* ξ are the adjustment coefficients designed for achieving better likelihood difference comparison, and g_k is defined as below

$$g_k = \frac{1}{\mid T\{k\}\mid}\sum_{i=1}^{\mid T\{k\}\mid}\sum_{j=1}^{p}\min(u_k(j),\tilde{u}_i(j)) \tag{4}$$

where $\tilde{u}_i(j)$ denotes the value of j -th element of $\tilde{u}_i$ in template pool, p is the total number of elements in a histogram and takes the value of 256×3 for RGB image, and $\mid T\{k\}\mid$ is the number of samples in histogram template pool $T\{k\}$.

3.2   Ant Cooperation Decision Level

We assume that multiple ant colonies are utilized to track multiple cells and each colony in principle when implementation corresponds to one cell, thus information exchange happens both within each colony and among colonies. For ant i

$(i \in \{1,...,N\})$ in colony s ($s \in \{1,...m\}$) at time k, we first use a $c1 \times c2$ rectangle region template whose center is closest to position $(x_s^{(i)}(k), y_s^{(i)}(k))$, then we calculate its corresponding histogram likelihood score according to Eq. (3), and the best template with the highest likelihood score is finally found and denoted by $T_s^{best}(k)$. In this way, the best template $T^{best}(k)$ at current iteration can be further obtained through a set of $\{T_1^{best}(k),...,T_m^{best}(k)\}$. If $h(T^{best}(k)) \geq t_0$, then new augmented template pool is obtained. When $|T\{k+1\}| = N^{max}$ holds, a complete template pool is built and then utilized to track multiple objects at the following iterations.

Updating procedure using the global best information is requisite to further regulate ant behavior in the following iterations. Since this step constitutes the positive feedback mechanism, a well-established update formula guarantees the quality of solution, and accelerates convergence speed as well.

In the ant individual decision level, once the searching behavior of each ant in a given colony is finished, the likelihood score corresponding to each new position is required to calculate for re-evaluating the importance of each ant. Therefore, we have

$$\omega_{s\_temp}^{(i)}(k+1) = \frac{h(z(k+1) \mid x_s^{(i)}(k+1 \mid k))\omega_s^{(i)}(k+1 \mid k)}{\sum_{j=1}^{N} h(z(k+1) \mid x_s^{(j)}(k+1 \mid k))\omega_s^{(j)}(k+1 \mid k)} \tag{5}$$

where the prediction weight $\omega_s^{(i)}(k+1 \mid k)$ is equal to the previous $\omega_s^{(i)}(k)$.

Upon the updating mode probability, we use three different mode candidates as a basis to normalize weights

$$\pi_s^{(i)}(k+1) = \frac{[\pi_s^{(i),1}(k+1), \pi_s^{(i),2}(k+1), \pi_s^{(i),3}(k+1)]^T}{\sum_{j=1}^{3} \pi_s^{(i),j}(k+1)}$$

$$\pi_s^{(i),j}(k+1) = \frac{h(z(k+1) \mid x_s^{(j)}(k+1 \mid k))}{\sum_{l=1}^{3} h(z(k+1) \mid x_s^{(l)}(k+1 \mid k))}\pi_s^{(i),j}(k+1 \mid k) \tag{6}$$

where $\pi_s^{(i)}(k+1 \mid k) = P_M \times \pi_s^{(i)}(k)$ with model jump matrix P_M.

In order to fully characterize the individual difference of each ant, two parameters, i.e., ant state and its influence region denoted by deviation $\sigma_s^{(i)}(k)$, are jointly updated to achieve better tracking performance. In terms of the state update, we follow the same rule as in [13] but velocity and influence region restrictions are introduced to make ants move towards area of interest step by step. For instance, $x_s^{(i)}(k)$ and $x_s^{(j)}(k+1 \mid k)$, respectively, represent the previous state of ant i and the state to be selected by ant i, and the following rule is applied to ant i

$$velo^{(i)}(k) = \begin{cases} velo_{max} & \text{if } \left(x_s^{(j)}(k+1|k) - x_s^{(i)}(k)\right)/T > velo_{max} \\ |x_s^{(j)}(k+1|k) - x_s^{(i)}(k)|/T & \text{if } |x_s^{(j)}(k+1|k) - x_s^{(i)}(k)|/T \le velo_{max} \\ -velo_{max} & \text{if } \left(x_s^{(j)}(k+1|k) - x_s^{(i)}(k)\right)/T < -velo_{max} \end{cases} \quad (7)$$

where $velo_{max}$ denotes the ant's maximum speed, and T is the sampling interval. Meanwhile, ant's one step moving distance d is also within the range of $[\sigma_{min}, \sigma_s^{(i)}(k)]$, where $\sigma_s^{(i)}(k) \le \sigma_{max}$.

Once all ants finished state update, we further use the above information to update the influence region represented by

$$\sigma_s^{(i)}(k+1) = \frac{K \cdot \bar{\omega}_s(k+1)}{\omega_s^{(i)}(k+1)} \sigma_s^{(i)}(k) \quad (8)$$

where $\omega_s^{(i)}(k+1)$ is the update likelihood score using the same form as Eq.(5) with $x_s^{(i)}(k+1)$, $\bar{\omega}_s(k+1)$ is the average likelihood score over all ants in colony s, and K is the adjustment coefficient. Eq. (8) indicates that the bigger the likelihood score the smaller the influence region.

Since we introduce the concept of ant colony of different tasks, and our focus, therefore, is how to describe the degree to which ant colony s has found the corresponding object. the predict and update term can be formulated as

$$q_s(k+1|k) = q_s(k)$$

$$q_s(k+1) = \frac{q_s(k+1|k)\langle f(x_s(k+1)), h(z(k+1|x_s(k+1)))\rangle}{1 - q_s(k+1|k) + q_s(k+1|k)\langle f(x_s(k+1)), h(z(k+1|x_s(k+1)))\rangle} \quad (9)$$

where $\langle f, h \rangle$ denotes the inner product $\int f(x)h(x)dx$, and $1 - q_s(k+1|k)$ describes the degree to which ant colony s hasn't found the corresponding cell.

3.3   Framework of Our Algorithm and Its Implementation

To visualize our proposed algorithm in a full view, we summarize the pseudo-code of main blocks in Table 1. In terms of the issues on algorithm implementation, ant colony at each step consists of two parts, i.e., the remaining ant colonies (except for the first iteration) and a fix number of ant colonies added at each step. We also assume that the total number of ant colonies is larger than that of objects. In addition, both the merge and prune processes are considered at the stage of state estimate for the sake of computation burn and prior constraints.

Table 1. The pseudo-code of main block of our algorithm

---

0.   Initialization
1.   In the Ant individual decision level
For ant colony $s = 1 : m$
    For each ant i in colony s
        Select potential according to Eq. (2);
    End
    Find the highest likelihood score in colony s according to Eq.(3) , and T_s^{best} as well;
End
2.   In the Ant cooperation decision level
Find the global highest likelihood score among above m colonies, and the corresponding template T^{best} is obtained to update the template sample pool T ;
    In each ant colony, ant state update is performed using the same rule as in [13], and Eq. (7);
    Update ant's mode probability $\pi_s^{(i)}(k+1)$, influence region $\sigma_s^{(i)}$, and importance weight $\omega_s^{(i)}$ according to Eqs. (6) , (8) and (5);
    Update the degree to which each ant colony has found the corresponding object, i.e., q_s , according to Eq. (9).
    If $q_s > q_0$ (threshold), return ant colony s to step 1, otherwise delete it.
3.   Output
    Estimate the individual state of each remaining ant colony.

---

4   Experiments

In this section, we will test the performance of our algorithm on low-contrast cell image sequences, which include various scenarios, such as different cell dynamics, cell morphology (shape) variation, and varying number of cells in different frames, etc. As shown in Fig.1, a new cell enters first at the lower rim of image and our algorithm captures the cell instantly when it is fully in the observation area in frame 41. Afterwards, the new cell, as well as the original one, is kept on being tracked with our algorithm in the following frames. In case 2, as shown in Fig. 2, this cell keeps on moving left, and partially leaves observation region in frame 47 and fully in frame 48. Note that our algorithm could track all cells with varying number of cells.

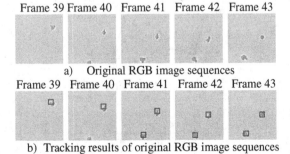

Frame 39 Frame 40 Frame 41 Frame 42 Frame 43
a)   Original RGB image sequences
Frame 39 Frame 40 Frame 41 Frame 42 Frame 43
b) Tracking results of original RGB image sequences

Fig. 1. Cells tracking when new cell enters images ($100\,pix \times 100\,pix$)

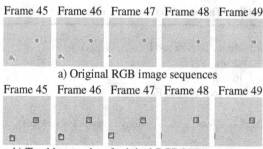

a) Original RGB image sequences

b) Tracking results of original RGB image sequences

Fig. 2. Cells tracking when new cell leaves images ($100\,pix \times 100\,pix$)

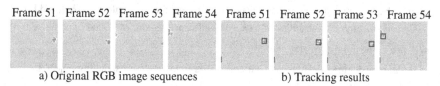

a) Original RGB image sequences   b) Tracking results

Fig. 3. Cells tracking when cell both leaves and enters images ($100\,pix \times 100\,pix$)

In case 3, as shown in Fig. 3, one cell moves right, partially leaves the image in frames 52 and 53, and fully leaves the image in frame 54. Meanwhile, the other cell partially enters from left upper part of image in frames 53 and 54. It can be observed that our algorithm can track existing cell until it disappears in frame 54, and capture new entering one instantly in frame 54. Fig.4 illustrates the false negative reports and false alarm reports, and the averaged FNR and FAR are *10.20%* and *0.20%*, respectively, using our algorithm, and *20.00%* and *0.40%*, respectively, using the method in [11]. It can be seen that our algorithm shows better performance than the method in [11], especially between frame 42 and frame 48. In frame 53, the two algorithms record relatively higher false negative reports, and this is because both are insensitive to the cell that only a small part appears in the left upper region. Fig.5 illustrates the corresponding ant colony distributions of each extracted cell state at selected frames discussed above. It can be observed that each ant colony moves around their individual interested cell, and each is bounded within the image.

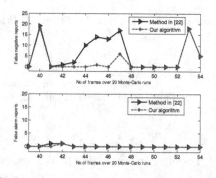

Fig. 4. Performance comparisons in terms of FNR and FAR

| Frame 39 | Frame 41 | Frame 43 | Frame 45 | Frame 47 | Frame 49 | Frame 51 |

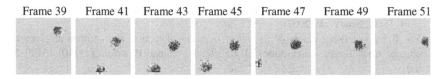

Fig. 5. Ant colony distributions of selected frames

5   Conclusions

In this work, we propose a novel ant system with two ant decision levels to track multiple cells in various challenging scenario. Through introducing two different levels, each ant individual motion is well regulated and the best information is exchanged both within colony itself and among ant colonies. Experiments show that our algorithm could track simultaneously multiple cells of entering and/or leaving image, cells of undergoing drastic dynamic changes, etc. Also, according to statistic results, our algorithm demonstrates a robust tracking performance in terms of the measures of FNR and FAR when comparing with existing methods.

Acknowledgments. This work is supported by national natural science foundation of China (No.61273312), national natural science foundation of Jiangsu (No.BK2010261).

References

1. Meijering, E., Dzyubachyk, O., Smal, I.: Tracking in cell and developmental biology. Seminars in Cell & Developmental Biology 20(8), 894–902 (2009)
2. Nguyen, N.H., Keller, S., Norris, E.: Tracking colliding cells in vivo microscopy. IEEE Trans. Biomed. Eng. 58(8), 2391–2400 (2011)
3. Dewan, M.A.A., Ahmad, M.O., Swamy, M.N.S.: Tracking biological cells in time-lapse microscopy: an adaptive technique combining motion and topological features. IEEE Trans. Biomed. Eng. 58(6), 1637–1647 (2011)
4. Chen, X., Zhou, X., Wong, S.T.C.: Automated segmentation, classification, and tracking of cancer cell nuclei in time-lapse microscopy. IEEE Trans. Biomed. Eng. 53(4), 762–766 (2006)
5. Yang, X., Li, H., Zhou, X.: Nuclei segmentation using marker-controlled watershed, tracking using mean-shift, and Kalman filter in time lapse microscopy. IEEE Trans. Circuits Syst. I 53(11), 2405–2414 (2006)
6. Zimmer, C., Olivo-Marin, J.C.: Coupled parametric active contours. IEEE Trans. Pattern Anal. Mach. Intell. 27(11), 1838–1842 (2005)
7. Mukherjee, D.P., Ray, N., Acton, S.T.: Level set analysis for leukocyte detection and tracking. IEEE Trans. Image Process. 13(4), 562–572 (2004)
8. Debeir, O., Van Ham, P., Kiss, R., Decaestecker, C.: Tracking of migrating cells under phase-contrast video microscopy with combined mean-shift processes. IEEE Trans. Med. Imag. 24(6), 697–711 (2005)

9. Wen, Q., Gao, J., Luby-Phelps, K.: Multiple interacting subcellular structure tracking by sequential Monte Carlo method. In: Proc. IEEE Int. Conf. Bioinf. Biomed., pp. 437–442. IEEE Press, NewYork (2007)

10. Vo, B.T., Vo, B.N., Pham, N.T., Suter, D.: Joint detection and estimation of multiple objects from image observations. IEEE Trans. Signal Processing 58(10), 5129–5141 (2010)

11. Hoseinnezhad, R., Vo, B.N., Vo, B.T., Suter, D.: Visual tracking of numerous targets via multi-Bernoulli filtering of image data. Pattern Recognition 45(10), 3625–3635 (2012)

12. Dorigo, M., Gambardella, L.M.: Ant colony system: A cooperative learning approach to the traveling salesman problem. IEEE Trans. on Evolutional Computation 1(1), 53–66 (1997)

13. Xu, B., Chen, Q., Wang, Z.: A novel estimator with moving ants. Simulation Modeling Practice and Theory 17(10), 1663–1677 (2009)

An Ant Colony System
Based on the *Physarum* Network

Tao Qian[1], Zili Zhang[1,2,*], Chao Gao[1], Yuheng Wu[1], and Yuxin Liu[1]

[1] School of Computer and Information Science
Southwest University, Chongqing 400715, China
[2] School of Information Technology, Deakin University, VIC 3217, Australia
zhangzl@swu.edu.cn

Abstract. The *Physarum* Network model exhibits the feature of important pipelines being reserved with the evolution of network during the process of solving a maze problem. Drawing on this feature, an Ant Colony System (ACS), denoted as PNACS, is proposed based on the *Physarum* Network (PN). When updating pheromone matrix, we should update both pheromone trails released by ants and the pheromones flowing in a network. This hybrid algorithm can overcome the low convergence rate and local optimal solution of ACS when solving the Traveling Salesman Problem (TSP). Some experiments in synthetic and benchmark networks show that the efficiency of PNACS is higher than that of ACS. More important, PNACS has strong robustness that is very useful for solving a higher dimension TSP.

Keywords: *Physarum* Network, Ant Colony System, TSP.

1 Introduction

There are lots of biological phenomena in the nature, which are very complicated and cannot be explained easily. More and more scientists take their life to explore and discover the essential rules hidden in the observed biological phenomenons. For example, Toshiyuki Nakagaki et al. have found a kind of slime mold called *Physarum ploycephalum*, which extends its pseudopodia to form divers tubular networks[1]. Specially, such *Physarum* can be used to solve a maze problem. Tero et al. have proposed a mathematical model to describe this intelligent behaviors for maze solving[2,3]. Meanwhile, Tero et al. have reported that *Physarum* is cultivated to simulate the Tokyo rail system on a solid medium[4]. They deploy some *Physarum*'s foods (rolled oats) on a wet surface according to the positions of cities around Tokyo, and cultivate *Physarum* on foods in an appropriate environment. After 26 hours, they find the network emerged by *Physarum* is very similar to the real-world Tokyo rail system. What's more, statistical analyses have demonstrated that such biological network has strong robustness, great fault tolerance and high transport efficiency on the transportation[5].

---

* Corresponding author.

Y. Tan, Y. Shi, and H. Mo (Eds.): ICSI 2013, Part I, LNCS 7928, pp. 297–305, 2013.

The current *Physarum* Network (PN) model, proposed by Tero at al., is mainly focused on the network design and optimization. There is few reports about how to use the PN model to solve combinatorial optimization problems. This paper designs an ACS based on the *Physarum* Network (PNACS) for solving the TSP. In PNACS, we update pheromone trails both released by ants and flowing in a network. We have found that the PN model can exhibit the feature of important pipelines being reserved during the process of building efficient transportation network[2]. Taking advantage of this feature, we can overcome some shortcomings of ACS, such as slow convergence rate and local optimal solution.

2   The Definition of Traveling Salesman Problem

TSP is a *NP*-hard problem. There are n cities, defined as $V = \{i|i = 1, 2, ..., n\}$. A salesman leaves from city i and visits all of other cities. The distance between city i and city j is defined as d_{ij}, where $i, j \in V$. The salesman goes back to the first city where he left after he has travelled all of other cities. The sequence of visited cities that the salesman has traveled is defined as $T = (t_1, t_2, ..., t_n)$, where $t_n \in V$. Therefore, the shortest-path $S_{\min}$ is defined as $\min_{T \in \Omega} \sum_{i=1}^{n} d_{t_i t_{i+1}}$, where Ω represents all of the directed Hamiltonian circuit sequences of n cities.

The current solutions of higher dimension TSP are mostly approximate results obtained by intelligence algorithms such as ant colony optimization algorithm and genetic algorithm. In order to obtain better results, a dataset is usually calculated iteratively and computed several times repeatedly. This paper will use some benchmarks as follows to compare ACS with PNACS.

1. *Steps* stands for the number of iterative steps. If an algorithm obtains the same optimal solution as other algorithms in a few iterative steps, it means the algorithm has a faster convergence rate.
2. $S_{average}$, $S_{midvale}$, *Variance* stand for the average value, mid-value and variance of results respectively. They are obtained after C times computation repeatedly in each iterative step, such as $S_{average}$ is calculated as $\sum_{i=1}^{C} S_{i,steps(k)} \big/ C$, where $S_{i,step(k)}$ represents a best result of a TSP in the kth step for the ith time. Through comparing these values, we further distinguish the advantages, convergence rates and variances of different algorithms.

3   An Ant Colony System Based on the *Physarum* Network

3.1   Ant Colony System

Dorigo et al. have proposed an ACS that has been used for solving the TSP[6]. In a TSP, there are n cities and m ants. And ants leave from different cities respectively and the cities they have visited won't be selected again. At time t, an ant k in a city i selects an unvisited city j from its neighbors as its next visiting target based on a certain probability P_{ij}. The definition of P_{ij} is shown in Eq. (1), where $\tau_{ij}(t)$ represents the value of pheromone trails in the path connecting the city i and j. And u stands for an unvisited city. The local heuristic value $\eta_{ij}(t) = 1/d_{ij}$ represents the expectation that the ant k moves from the city i to j, which shows that the closest cities are more likely to be selected. N_i is the unvisited city set of neighbors of node i.

$$P_{ij}^k = \begin{cases} \frac{\tau_{ij}^\alpha(t) \times \eta_{ij}^\beta(t)}{\sum \tau_{iu}^\alpha(t) \times \eta_{iu}^\beta} & j \in N_i \\ 0 & j \notin N_i \end{cases} \tag{1}$$

During visiting cities, an ant k releases pheromone trails on roads. Meanwhile, the trails evaporate at a speed ρ $(0 < \rho < 1)$. After the ant k has visited all of cities, the trails are updated based on Eq. (2), where the constant parameter F represents the value of pheromone trails released by the ant k. S_k is defined as the total length of the route that the ant k has travelled.

$$\tau_{ij}(t+1) = (1 - \rho)\tau_{ij}(t) + \rho(\sum_{k=1}^{m} \frac{F}{|S_k|}) \tag{2}$$

However, the final results are mainly affected by the initial formed pathes by a few ants, and the local optimization solutions are obtained at the later iteration in ACS[7,8].

3.2   The *Physarum* Network Model

The PN model also can be called the maze-solving model or the PN model with one pair of inlet/outlet nodes[3][5]. The main idea of PN model comes from the maze-solving experiment made by Nakagaki et al[1]. They deploy plasmodial pieces of one *Physarum* in a maze. Then, they place foods at the start and end points. During *Physarum* foraging, the tubular pseudopodia that cannot get foods will shrink and disappear. In contrast, other pseudopodia that can obtain energy from foods will become thick. At the end, only pipelines in the shortest-path are reserved.

We assume that the edges of a maze network are pipelines with water inside, as shown in Fig. 1(a), where N_{in} represents the inlet node of the maze and N_{out} represents the outlet node. D_{ij} is defined as a measure of the conductivity of the pipeline connecting nodes i and j, which related to the thickness of the pipeline. When a pipeline becomes thicker, the conductivity will be enhanced. And at the same time, the flux Q_{ij}, which represents the flux of the pipeline connecting nodes i and j, will be enhanced too.

In each time step, I_0 is defined as the fixed flow of N_{in} and N_{out}. According to the Kirchhoffs law, it is known that the flux input at a node is equal to the flux output. This process can be denoted in Eq. (3), where $Q_{ij} = D_{ij} (p_i - p_j)/d_{ij}$,

p_i and p_j stand for the pressures at nodes i and j respectively. Through Eq. (3), the pressure of each node in a network can be calculated.

$$\begin{cases} \sum_i Q_{i,in} = -I_0 \\ \sum_i Q_{i,out} = I_0 \\ \sum_i Q_{i,j} = 0 \end{cases} \tag{3}$$

Then, the flux Q_{ij} is computable. With water flowing into the pipelines of the network, the conductivity of pipeline in the next time step will grow as the flux increases, as shown in Eq. (4). To sum up, one iterative step is completed.

$$\frac{dD_{ij}}{dt} = \frac{|Q_{ij}|}{1 + |Q_{ij}|} - D_{ij} \tag{4}$$

Finally, the conductivities at time $t+1$ will be fed back to Eq. (3), and the above process will continue loop iteration. The iteration does not terminate until the constraint of $\left| D_{ij}^{t+1} - D_{ij}^t \right| \leq 10^{-6}$ is satisfied. D_{ij}^{t+1} stands for the conductivity of the pipeline connecting nodes i and j at time $t+1$. The final network is shown in Fig. 1(b).

We regard the iterative process as a process that "develop" pipelines becoming thicker. Some pipelines developed deeply become thicker with higher conductivities. In this paper, we call this kind of feature as deep-developing. And the pipelines remained in the shortest-path are called as "important pipelines", as the solid edges shown in Fig. 1(a). Therefore, with the increment of fluxes, these important pipelines emerge through consecutive iteration.

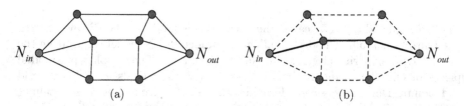

Fig. 1. The PN model with one pair of inlet/outlet nodes: (a) the initial network, (b) the final network. N_{in}, the leftmost node, represents the inlet node and N_{out}, the rightmost node, represents the outlet node.

3.3 PNACS

Taking the advantage of PN model, i.e., the important pipelines being deeply developed during the consecutive iteration, this paper proposes a new algorithm, named as PNACS, to solve the TSP. In PNACS, pheromone trails are filled in the whole tubular network. In this hybrid algorithm, when updating pheromone in the network, we need to consider both the pheromone released by ants and

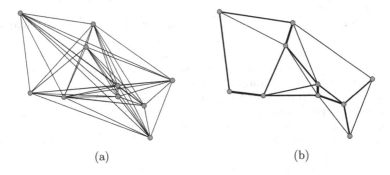

Fig. 2. The PN model with multi-pairs of inlet/outlet nodes: (a) the initial network, (b) the final network.

the pheromone tails flowing in pipelines. However, we find that one pair of inlet/outlet nodes cannot exhibit the flux of pheromone tails in the whole network. Therefore, the PN model with multi-pairs of inlet/outlet nodes is proposed.

At time t, every pair of two nodes in each link of a network is selected as inlet/outlet nodes and the pressures of nodes are calculated by Eq. (3). The initial flow I_0 is set as I_0/M, where M represents the number of pipelines in the network. Then, the flux Q_{ij} in the network is substituted with an average flux $\overline{Q_{ij}}$, as shown in Eq. (5). According to $\overline{Q_{ij}}$, we can calculate the conductivities at time $t+1$ using Eq. (4). The above steps are repeated until there're no changes about all of the conductivities.

$$\overline{Q_{ij}} = \frac{1}{M}\sum_{k=1}^{M}\left|Q_{ij}^{(k)}\right| \tag{5}$$

The initial network is built in Fig. 2(a). After calculating based on the PN model with multi-pairs of inlet/outlet nodes, the final network is shown in Fig. 2(b). We find that some pipelines become thicker and others become thinner (some thinnest pipelines aren't shown in Fig. 2(b) and they are regarded as disappeared). All of these reserved pipelines are also called as important pipelines.

In order to synthesize the influence of the pheromone fluxes in pipelines, Eq. (2) is substituted by Eq. (6), where ε is defined as an impact factor to measure the effect of pheromone trails in pipelines on the total pheromone trails in a network. I_0 is set as F/M. In Eq. (7), $totalsteps$ stands for the total steps of iteration and $\lambda \in (1, 1.2)$. The description of PNACS is shown in Algorithm 1.

$$\tau_{ij}(t+1) = (1-\rho)\tau_{ij}(t) + \rho(\sum_{k=1}^{m}\frac{F}{|S_k|} + \varepsilon\frac{Q_{ij}(t+1)}{I_0}) \tag{6}$$

$$\varepsilon = 1 - \frac{1}{1 + \lambda^{totalsteps/2-(t+1)}} \tag{7}$$

Algorithm 1. PNACS

Step 1: Initializing a pheromone matrix and a conductivity matrix as all 1-matrixes. And setting the iteration counter $N := 0$.

Step 2: Deploying m ants on the locations of cities respectively. Ants select the next unvisited cities with the probabilities calculated using Eq. (1). At last, all of the ants return to the start city after visiting all of other cities.

Step 3: Recording tracks that ants have travlled and finding the shortest path S_{min}.

Step 4: Calculating pheromone fluxes in the pipelines using Eq. (3) and Eq. (5). Then, computing the conductivities of the next time step by Eq. (4).

Step 5: Updating the pheromone matrix of the network, which is calculated using Eq. (6) and Eq. (7). Meanwhile, setting $N := N + 1$.

Step 6: If $N < totalsteps$, going to **Step 2**.

Step 7: Outputting the final optimization solution S_{min}.

---

4   Simulation Experiments

4.1   Dataset

Two types of datasets are used in this paper. One is a synthetic dataset. we randomly generate 30 cities, whose coordinates are shown in Fig. 3(b). The other are two benchmark datasets downloaded from the website TSPLIB[1], gr17 and bays29 (Bavaria). Based on these data, we build undirected weighted networks that are fully connected. The weight of edge is a straight-line distance in a synthetic network, while the weight of edge in the benchmark network is an actual mileage.

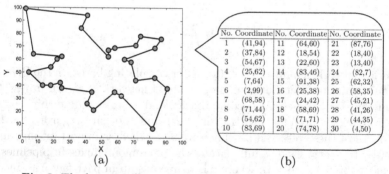

No.	Coordinate	No.	Coordinate	No.	Coordinate
1	(41,94)	11	(64,60)	21	(87,76)
2	(37,84)	12	(18,54)	22	(18,40)
3	(54,67)	13	(22,60)	23	(13,40)
4	(25,62)	14	(83,46)	24	(82,7)
5	(7,64)	15	(91,38)	25	(62,32)
6	(2,99)	16	(25,38)	26	(58,35)
7	(68,58)	17	(24,42)	27	(45,21)
8	(71,44)	18	(58,69)	28	(41,26)
9	(54,62)	19	(71,71)	29	(44,35)
10	(83,69)	20	(74,78)	30	(4,50)

(a)   (b)

Fig. 3. The location illustration of 30 cities and their coordinates

4.2   Experiment Analysis

All of the experiments have been undertaken in one computer and the parameters of ACS and PNACS are set as $\alpha = 1$, $\beta = 2$, $\rho = 0.7$, $F = 100$ and $totalsteps = 300$. These experiments are repeated 50 times.

---

[1] http://www.iwr.uni-heidelberg.de/groups/comopt/software/TSPLIB95/

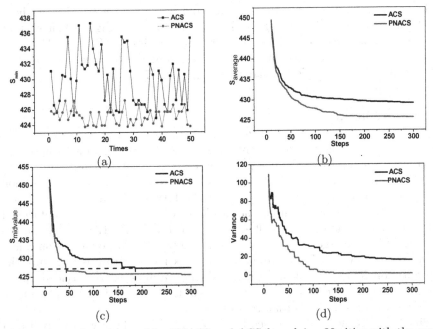

Fig. 4. The results calculated by PNACS and ACS for solving 30 cities with the synthetic dataset

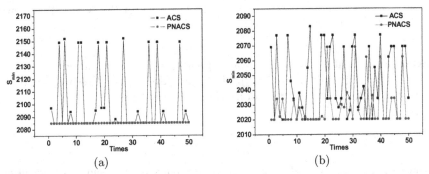

Fig. 5. The shortest-path S_{min} by PNACS and ACS with the benchmark datasets of (a) gr17 and (b) bays29

In Fig. 4(a), the best optimized result of PNACS is 423.7406, which is better than 424.8439 of ACS. The average value of PNACS decreases more obviously than that of ACS with iteration in Fig. 4(b). As shown in Fig. 4(c), in order to obtain $S_{midvalue} = 427$, PNACS needs only 43 steps while there are 160 steps by ACS. So PNACS has a higher convergence rate than that of ACS. Fig. 4(d) shows that PNACS has a lower variance than that of ACS with iterative step growing, which means that PNACS has a stronger robustness than that of ACS.

In order to further verify the accuracy and robustness of PNACS, we have used two benchmark datasets, gr17 and bays29 (Bavaria). Fig. 5(a) shows that PNACS can observe the best solution 100% for 17 cities TSP, while ACS only has 62%. Fig. 5(b) also exhibits PNACS has stronger robustness than that of ACS through comparing the dynamic changes of variance.

5   Conclusion

In order to overcome the slow convergence rate and local optimal solution of ACS when solving the TSP, this paper proposes a PNACS algorithm drawing on the feature of the important pipelines being reserved during the evolution of *Physarum* network. Some experiments in both synthetic and benchmark networks show that PNACS is more efficient than ACS in the field of computational efficiency and robustness in small and medium-scale TSPs. However, when the scale of TSP is enlarged, PNASC will spend more computational cost on calculating the PN model. Therefore, our main work in the future is to reduce the computational cost. Meanwhile, the initial trails in pipelines are decided by the total trails released by an ant, so the value of trails released by the ant affects the final results of PNACS than that of ACS.

Since the PN model is proposed by Tero et al., it has been widely applied in the field of the transportation and complex network[9]. It is innovative for this study to incorporate the PN into ACS for solving the combinatorial optimization problem. And it is also a new application of the PN model. In our future work, we will do more experiments to improve the efficiency of PNACS and compare PNACS with other heuristic algorithms.

Acknowledgment. This project was supported by the Natural Science Foundation of Chongqing (No. cstc2012jjA40013) and Specialized Research Fund for the Doctoral Program of Higher Education (No. 20120182120016).

References

1. Nakagaki, T., Yamada, H., Toth, A.: Maze-Solving by an Amoeboid Organism. Nature 407(6803), 470 (2000)
2. Miyaji, T., Ohnishi, I.: Mathematical Analysis to an Adaptive Network of the Plasmodium System. Hokkaido Mathematical Journal 36(2), 445–465 (2007)
3. Tero, A., Kobayashi, R., Nakagaki, T.: A Mathematical Model for Adaptive Transport Network in Path Finding by True Slime Mold. Journal of Theoretical Biology 244(4), 553–564 (2007)
4. Tero, A., Takagi, S., Saigusa, T., Ito, K., Bebber, D.P., Fricker, M.D., Yumiki, K., Kobayashi, R., Nakagaki, T.: Rules for Biologically Inspired Adaptive Network Design. Science Signalling 327(5964), 439 (2010)
5. Watanabe, S., Tero, A., Takamatsu, A., Nakagaki, T.: Traffic Optimization in Railroad Networks Using an Algorithm Mimicking an Amoeba-like Organism, *Physarum* Plasmodium. Biosystems 105(3), 225–232 (2011)

6. Dorigo, M., Gambardella, L.M.: Ant Colony System: a Cooperative Learning Approach to the Traveling Salesman Problem. IEEE Transactions on Evolutionary Computation 1(1), 53–66 (1997)
7. Zhao, N., Wu, Z., Zhao, Y., Quan, T.: Ant Colony Optimization Algorithm with Mutation Mechanism and Its Applications. Expert Systems with Applications 37(7), 4805–4810 (2010)
8. Blum, C.: Ant Colony Optimization: Introduction and Recent Trends. Physics of Life Reviews 2(4), 353–373 (2005)
9. Zhang, Y., Zhang, Z., Wei, D., Deng, Y.: Centrality Measure in Weighted Networks Based on an Amoeboid Algorithm. Journal of Information and Computational Science 9(2), 369–376 (2012)

Solving the Minimum Common String Partition
Problem with the Help of Ants

S.M. Ferdous and M. Sohel Rahman[*]

AℓEDA Group, Department of CSE, BUET, Dhaka-1000, Bangladesh

Abstract. In this paper, we consider the problem of finding minimum common partition of two strings (MCSP). The problem has its application in genome comparison. As it is an NP-hard, discrete combinatorial optimization problem, we employ a metaheuristic technique, namely, MAX-MIN ant system to solve this. The preliminary experimental results are found to be promising.

Keywords: Ant Colony Optimization, Stringology, Genome sequencing, Combinatorial Optimization, Swarm Intelligence, String partition.

1 Introduction

String comparison is one of the important problems in Computer Science with diverse applications in different areas including genome sequencing, text processing and compressions. In this paper, we address the problem of finding minimum common partition (MCSP) of two strings. MCSP is closely related to genome arrangement which is an important field in computational biology. More detailed study of the application of MCSP can be found at [5], [6] and [8].

In MCSP problem, we are given two *related* strings (X, Y). Two strings are related if every letter appears the same number of times in each of them. Clearly, two strings have a common partition if and only if they are related. So, the length of the two strings are also the same (say, n). A partition of a string X is a sequence $P = (B_1, B_2, \cdots, B_c)$ of strings whose concatenation is equal to X, that is $B_1 B_2 \cdots B_c = X$. The strings B_i are called the blocks of P. Given a partition P of a string X and a partition Q of a string Y, we say that the pair $\pi = < P, Q >$ is a common partition of X and Y if Q is a permutation of P. The minimum common string partition problem is to find a common partition of X, Y with the minimum number of blocks, that is to minimize c. For example, if $(X, Y) = \{$"ababcab","abcabab"$\}$, then one of the minimum common partition sets is $\pi = \{$"ab","abc","ab"$\}$ and the minimum common partition size is 3. The restricted version of MCSP where each letter occurs at most k times in each input string, is denoted by k-MCSP.

1.1 Related Works

In [9], the authors investigated k-MCSP along with two other variants: $MCSP^c$, where the alphabet size is at most c; and x-balanced MCSP, which requires that the length of

---

[*] Partially supported by a Commonwealth Fellowship and an ACU Titular Award.

Y. Tan, Y. Shi, and H. Mo (Eds.): ICSI 2013, Part I, LNCS 7928, pp. 306–313, 2013.
© Springer-Verlag Berlin Heidelberg 2013

the blocks must be witnin the range $(n/d - x, n/d + x)$, where d is the number of blocks in the optimal common partition and x is a constant integer. They showed that $MCSP^c$ is NP-hard when $c \geq 2$. As for k-MCSP, they presented an FPT algorithm which runs in $O^*((d!)^{2k})$ time.

Chrobak et al. [8] analyzed a natural greedy heuristic for MCSP: iteratively, at each step, it extracts a longest common substring from the input strings. They showed that for 2-MCSP, the approximation ratio (for the greedy heuristic) is exactly 3, for 4-MCSP, $\log n$ and for the general MCSP, between $\Omega(n^{0.43})$ and $O(n^{0.67})$.

In this paper, we apply an Ant Colony Optimization (ACO) algorithm to solve the MCSP problem. We conduct experiments on both random and real data to compare our algorithm with the state of the art algorithm in the literature and achieve promising results.

2 Preliminaries

In this section, we present some defitnitions and notations that are used throughout the paper. Two strings (X, Y), each of length n, over an alphabet Σ are called *related* if every letter appears the same number of times in each of them. A block $B = ([id, i, j])$, $0 \leq i \leq j < n$, of a string S is a data structure having three fields: id is an identifier of S and the starting and ending positions of the block in S are represented by i and j, respectively. Naturally, the *length* of a block $[id, i, j]$ is $(j - i + 1)$. We use $substring([id, i, j])$ to denote a substring of S induced by the block $[id, i, j]$. Throughout the paper we will use 0 and 1 as the identifiers of X and Y respectively. We use $[]$ to denote the empty block.

For example, if we have two strings $(X, Y) = \{$"abcdab","bcdaba"$\}$, then $[0, 0, 1]$ and $[0, 4, 5]$ both represent the substring "ab" of X. In other words, $substring([0, 0, 1]) = substring([0, 4, 5]) = $ "ab".

Two blocks can be intersected or unioned. The intersection of two blocks is a block that contains the common portion of the two. Formally, the intersection operation of $B_1=[id, i, j]$ and $B_2=[id, i', j']$ is defined as follows:

$$B_1 \cap B_2 = \begin{cases} [] & \text{if } i' > j \text{ or } i > j' \\ [id, i', j] & \text{if } i' \leq j \\ [id, i, j'] & \text{else} \end{cases} \tag{1}$$

Union of two blocks is either another block or an ordered (based on the starting position) set of blocks. Without the loss of generality we suppose that, $i <= i'$ for $B_1=[id, i, j]$ and $B_2=[id, i', j']$. Then, formally the union operation of B_1 and B_2 is defined as follows:

$$B_1 \cup B_2 = \begin{cases} [id, i, j] & \text{if } j' <= j \\ [id, i, j'] & \text{if } j' > j \text{ or } i' == j + 1 \\ \{B_1, B_2\} & \text{else} \end{cases} \tag{2}$$

The union rule with an ordered set of blocks, B_{lst} and a block, B' can be defined as follows. We have to find the position where B' can be placed in B_{lst}, i.e., we have to find $B_k \in B_{lst}$ after which B' can be placed. Then, we have to replace the ordered subset $\{B_k, B_{k+1}\}$ with $B_k \cup B' \cup B_{k+1}$. As an example, suppose we have three blocks, namely, $B_1 = [0,5,7], B_2 = [0,11,12]$ and $B_3 = [0,8,10]$. Then $B_1 \cup B_2 = B'_{lst} = \{[0,5,7],$ $[0,11,12]\}$. On the other hand, $B'_{lst} \cup B_3 = [0,5,12]$, which is basically identical to $B_1 \cup B_2 \cup B_3$.

Two blocks B_1 and B_2 (in the same string or in two different strings) matches if $substring(B_1) = substring(B_2)$. If the two matched blocks are in two different strings then the the matched substring is called a common substring of the two strings denoted by $cstring(B_1, B_2)$.

The span of a block, $B = [id, i, j]$, denoted by, $span(B)$ is the length of the maximum block that contains B. More formally, $span(B) = \max\{\ell \mid \ell = length(B'), B \subseteq B', \forall B'\}$. For example, if three blocks B_1, B_2 and B_3 are respectively $[0,0,0]$, $[0,0,1]$ and $[0,0,2]$, then $span(B_1) = span(B_2) = span(B_3) = 2$.

3 Our Approach: Max Min Ant System on the Common Substring Graph

3.1 Formulation of *Common Substring Graph*

We define a common substring graph, $G_{cs}(V, E, id(X))$ of two strings (X, Y) as follows. Here V is the vertex set of the graph and E is the edge set. Vertices are the positions of string X, i.e., for each $v \in V$, $v \in [0, |X| - 1]$. Two vertices $v_i \leq v_j$ are connected with and edge, i.e, $(v_i, v_j) \in E$, if the substring induced by the block $[id(X), v_i, v_j]$ matches some substring of Y. More formally, we have:

$$(v_i, v_j) \in E \Leftrightarrow cstring([id(X), v_i, v_j], B') \text{ is not empty } \exists B' \in Y$$

In other words, each edge in the edge set corresponds to a *block* satisfying the above condition. For convenience, we will denote the edges as *edge blocks* and use the list of edge blocks (instead of edges) to define the edgeset E. Notably, each *edge block* on the edge set of $G_{cs}(V, E, id(X))$ of string (X, Y) may match with more than one blocks of Y. For each *edge block* B a list is maintained containing all the matched blocks of string Y to that *edge block*. This list is called the *matchList(B)*.

For example, suppose $(X, Y) = \{$"abcdba","abcdab"$\}$. Now consider the corresponding common substring graph. Then, we have vertex set, $V = \{0,1,2,3,4,5\}$ and edge set, $E = \{[0,0,0], [0,0,1], [0,1,1], [0,2,2], [0,2,3], [0,3,3], [0,4,4], [0,5,5]\}$. The *matchList* of the second *edge block*, i.e., $matchList([0,0,1]) = \{[1,0,1], [1,4,5]\}$.

To find a common partition of two strings (X, Y) we first construct the common substring graph of (X, Y). Then from a vertex v_i on the graph we take an edge block $[id(X), v_i, v_j]$. Suppose M_i is the *matchList* of this block. We take a block B'_i from M_i. Then we advance to the next vertex that is $v_j + 1 \ MOD \ |X|$ and choose another corresponding edge block as before. We continue this until we come back to the starting vertex. Let *partitionList* and *mappedList* are two lists, each of length c, containing the traversed edge blocks and the corresponding matched blocks. Now we have the following lemma.

Lemma 1. *partitionList is a common partition of length c if the blocks of mappedList obeys,*

$$B_i \cap B_j = [] \ \forall B_i, B_j \in mappedList, i \neq j \tag{3}$$

and

$$B_1 \cup B_2 \cup \cdots \cup B_c = [id(Y), 0, |Y| - 1] \tag{4}$$

3.2   Heuristics

Heuristics (η) contain the problem specific information. We propose two different (types of) heuristics for MCSP. Firstly, we propose a static heuristic that does not change during the runs of algorithm. The other heuristic we propose is dynamic in the sense that it changes between the runs.

The Static Heuristic for MCSP. We employ a very naive and intuitive idea. It is obvious that the larger is the size of the blocks the smaller is the partition set. To capture this phenomenon, we assign on each edge of the common substring graph a numerical value that is proportional to the length of the substring corresponding to the edge block. Formally, the static heuristic (η_s) of an edge block $[id, i, j]$ is defined as follows:

$$\eta_s([id, i, j]) \propto length([id, i, j]) \tag{5}$$

The Dynamic Heuristic for MCSP. We observe that the static heuristic can sometimes lead us to very bad solutions. For example if $(X, Y) = \{\text{"bceabcd"}, \text{"abcdbec"}\}$ then according to the static heuristic much higher value will be assigned to *edge block* $[0, 0, 1]$ rather than to $[0, 0, 0]$. But if we take $[0, 0, 1]$, we must match it to the block $[1, 1, 2]$ and we further miss the opportunity to take $[0, 3, 6]$ later. The resultant partition will be $\{\text{"bc"}, \text{"e"}, \text{"a"}, \text{"b"}, \text{"c"}, \text{"d"}\}$ but if we would take $[0, 0, 0]$ at first step, then one of the resultant partitions would be $\{\text{"b"}, \text{"c"}, \text{"e"}, \text{"abcd"}\}$. To overcome this shortcoming of the static heuristic we define a dynamic heuristic as follows. The dynamic heuristic (η_d) of an edge block ($B = [id, i, j]$) is inversely proportional to the difference between the length of the block and the minimum span of its correspoding blocks in *matchList*. More formally, $\eta_d(B)$ is defined as follows:

$$\eta_d(B) \propto \frac{1}{|length(B) - minSpan(B)| + 1}, \tag{6}$$

where

$$minSpan(B) = \min\{span(B') \mid B' \in matchList(B)\} \tag{7}$$

In the example, $minSpan([0,0,0])$ is 1 as follows: $matchList([0,0,0]) = \{[1,1,1], [1,4,4]\}$. $span([1,1,1]) = 4$ and $span([1,4,4] = 1)$. On the other hand, $minSpan([0,0,1])$ is 4. So, according to dynamic heuristic much higher numeral will be assigned to block $[0,0,0]$ rather than block $[0,0,1]$.

We define the total heuristic (η) is the linear combination of the static heuristic (η_s) and the dynamic heuristic (η_d). Formally, the total heuristic of an edge block B is, $\eta(B) = a \cdot \eta_s(B) + b \cdot \eta_d(B)$, where a, b are any real valued constant.

3.3 Initialization and Configuration

Given two strings (X, Y), we first construct the common substring graph $G_{cs} = (V, E, id(X))$. We use the following notations. *Local best solution* (L_{LB}) is the best solution found in each iteration. *Global best solution* (L_{GB}) is the best solution found so far among all iterations. The pheromone of the edge block is bounded between τ_{max} and τ_{min}. Like [3], we use the following values for τ_{max} and τ_{min}: $\tau_{max} = \frac{1}{\varepsilon \cdot cost(L_{GB})}$, and $\tau_{min} = \frac{\tau_{max}(1 - \sqrt[n]{P_{best}})}{(avg-1)\sqrt[n]{P_{best}}}$. Here, avg is the average number of choices an ant has in the construction phase. Initially, the pheromone values of all edge blocks (substring) are initialized to *initPheromone* which is a large value to favor the exploration at the first iteration [3].

3.4 Construction of a Solution

Let, *nAnts* denotes the total number of ants in the colony. Each ant is deployed randomly to a vertex v_s of the G_{cs}. A solution for an ant starting at a vertex v_s is constructed by the following steps.

step 1: Let $v_i = v_s$. Choose an *available* edge block starting from v_i by the discrete probability distribution defined below. An edge block is available if its *MatchList* is not empty and inclusion of it to the *partitionList* and *mappedList* obeys Equations 3. The probability for choosing edge block $[0, v_i, v_j]$ is:

$$p([0, v_i, v_j]) = \frac{\tau([0, v_i, v_j])^\alpha \cdot \eta([0, v_i, v_j])^\beta}{\sum_\ell \tau([0, v_i, v_\ell])^\alpha \cdot \eta([0, v_i, v_\ell])^\beta}, \forall \ell \text{ such that } [0, v_i, v_l] \text{ is an available block.} \quad (8)$$

step 2: Suppose, $[0, v_i, v_k]$ is chosen according to Equation 8 above. We choose a match block B_m from the *matchList* of $[0, v_i, v_k]$ and delete B_m from the *matchList*. We also delete every block from every *matchList* of every edge block that overlaps with B_m. Formally we delete a block B if

$$B_m \cap B \neq [] \quad \forall B_i \in E, B \in matchList(B_i)$$

We add $[0, v_i, v_k]$ to the *partitionList* and B_m to the *mappedList*.

step 3: If $(v_k + 1) MOD |X| = v_s$ and the *mappedList* obeys 4, then we have found a common partition of X and Y. The size of the partition is the length of the *partitionList*. Otherwise, we jump to the *step 1*.

3.5 Pheromone Update

When each of the ants in the colony construct a solution (i.e., a common partition), an iteration completes. We set the local best solution as the best partition that is the minimum length partition in an iteration. The global best solution for n iterations is defined as the minimum length common partition over first n iteration.

We define the fitness $F(L)$ of a solution L as the reciprocal of the length of L. The pheromone of each interval of each target string is computed according to:

$$\tau_i \leftarrow (1 - \varepsilon) \cdot \tau_i + \tau_i \cdot \sum_{s \in G_{iter} | c_i \in s} F(s) \cdot \varepsilon, i = 1, 2, ..., n \quad (9)$$

The pheromone are bounded within the range τ_{MIN} and τ_{MAX}. We have updated the pheromone values according to L_{LB} or L_{GB}.

4 Experiments

We have conducted our experiments in a computer with Intel Core 2 Quad CPU 2.33 GHz. The available RAM was 4.00 GB. The operating system was Windows 7. The programming environment was java. The maximum allowed time for each instance was 120 minutes.

4.1 Dataset

We have taken two types of data into consideration: randomly generated DNA sequence and real gene sequence.

Random DNA Sequence: We have generated 30 DNA sequences of length at most 600 randomly using [10]. The fraction of bases A, T, G and C is assumed to be 0.25 each. For each DNA sequence we shuffle it to create a new DNA sequence. The shuffling is done using the online toolbox [11]. The original random DNA sequence and its shuffled pair constitute a single input (X, Y) in our experiment. This dataset is divided into 3 classes. The first 10 have length less than or equal 200 bps (base-pairs), the next 10 have length within $[201, 400]$ and the rest 10 have length within $[401, 600]$ bps.

Real Gene Sequence: We collected the gene sequence data from the NCBI GenBank[1]. For simulation we have taken Bacterial Sequencing (part 14). We have taken the first 15 gene sequences whose lengths are within $[200, 600]$.

4.2 Parameters

The settings of parameters for which we achieved the results is described in Table 1.

Table 1. Parameters

Parameters	Value
α	2.0
β	5.0
Evaporation rate, ε	0.02
$nAnts$	$|X|$
p_{best}	0.09
$initPheromone$	10.0
Maximum Allowed Time	120 min
Coeff. of η_s, a	0.5
Coeff. of η_d, b	0.5

---

[1] http://www.ncbi.nlm.nih.gov

4.3   Results and Analysis

We have compared our approach with the greedy algorithm of [8] because none of the
other algorithms in the literature are for general MCSP: each of the other approximation
algorithms put some restrictions on the parameters.

Random DNA Sequence: Table 2 presents the comparison between our approach and
the greedy approach [8] for the random DNA sequences. For a particular DNA sequence, the experiment was run 4 times and the average result is reported. The first
column under any group reports the partition size computed by the greedy approach,
the second column is the average partition size found by MAX-MIN and the third column represents the difference between the two approaches. A positive (negative) difference indicates that the greedy result is better (worse) than the MAX-MIN result by that
amount. From the table, we can see that out of 30 instances our approach gets better
partition size for 28 cases.

Table 2. Comparison between Greedy approach [8] and MAX-MIN on random DNA sequences

Test No.	Group 1 (200 bps)			Group 2 (400 bps)			Group 3 (600 bps)		
	Greedy	MAX-MIN	Difference	Greedy	MAX-MIN	Difference	Greedy	MAX-MIN	Difference
1.	46	42.75	-3.25	119	114.25	-4.75	182	180.00	-2.00
2.	56	51.50	-4.50	122	119.00	-3.00	175	176.25	1.25
3.	62	56.75	-5.25	114	112.25	-1.75	196	188.00	-8.00
4.	46	43.00	-3.00	116	116.25	0.25	192	184.25	-7.75
5.	44	43.00	-1.00	135	132.25	-2.75	176	171.75	-4.25
6.	48	42.25	-5.75	108	105.5	-2.50	170	163.25	-6.75
7.	65	60.00	-5.00	108	99.00	-9.00	173	168.50	-4.50
8.	51	47.00	-4.00	123	118.00	-5.00	185	176.25	-8.75
9.	46	45.75	-0.25	124	119.50	-4.50	174	172.75	-1.25
10.	63	59.25	-3.75	105	101.75	-3.25	171	167.25	-3.75

Table 3. Comparison between Greedy approach [8] and MAX-MIN on real gene sequence

Test No.	Greedy	MAX-MIN	Difference
1.	95.0000	87.7500	-7.2500
2.	161.0000	158.5000	-2.5000
3.	121.0000	116.5000	-4.5000
4.	172.0000	171.7500	-0.2500
5.	153.0000	146.0000	-7.0000
6.	140.0000	140.7500	0.7500
7.	134.0000	131.0000	-3.0000
8.	149.0000	148.5000	-0.5000
9.	151.0000	149.0000	-2.0000
10.	126.0000	124.5000	-1.5000
11.	143.0000	138.2500	-4.7500
12.	180.0000	181.0000	1.0000
13.	152.0000	147.7500	-4.2500
14.	157.0000	161.2500	4.2500
15.	157.0000	158.7500	1.7500

Real Gene Sequence: Table 3 shows the minimum common partition size found by our approach and the greedy approach for the real gene sequences. Out of the 15 instances we get better results on 11 instances.

5 Conclusion

Minimum Common String Partition problem has important applications in computational biology. In this paper, we have described a metaheuristic approach to solve the problem. We have used static and dynamic heuristic information in this approach. Simulating this algorithm on long DNA sequences would be challenging future improvement.

References

1. Dorigo, M., Di Caro, G., Gambardella, M.L.: Ant algorithms for discrete optimization. J. of Artificial Life 5(2), 137–172 (1999)
2. Dorigo, M., Gambardella, M.L.: Ant Colony System: A Cooperative Learning Approach to the Traveling Salesman Problem. J. of IEEE Transactions on Evolutionary Computation 1 (1996)
3. Stuetzle, T., Hoos, H.H.: MAX-MIN Ant System. J. of Future Gener. Comput. System 16(9), 889–914 (2000)
4. Watterson, G.A., Ewens, W.J., Hall, T.E., Morgan, A.: The chromosome inversion problem. Journal of Theoretical Biology 99, 7 (1982)
5. Goldstein, A., Kolman, P., Zheng, J.: Minimum common string partition problem: Hardness and approximations. In: Fleischer, R., Trippen, G. (eds.) ISAAC 2004. LNCS, vol. 3341, pp. 484–495. Springer, Heidelberg (2004)
6. Chen, X., Zheng, J., Fu, Z., Nan, P., Zhong, Y., Lonardi, S., Jiang, T.: Assignment of orthologous genes via genome rearrangement (2004) (submitted)
7. Damaschke, P.: Minimum Common String Partition Parameterized. In: Crandall, K.A., Lagergren, J. (eds.) WABI 2008. LNCS (LNBI), vol. 5251, pp. 87–98. Springer, Heidelberg (2008)
8. Chrobak, M., Kolman, P., Sgall, J.: The greedy algorithm for the minimum common string partition problem. In: Jansen, K., Khanna, S., Rolim, J.D.P., Ron, D. (eds.) APPROX and RANDOM 2004. LNCS, vol. 3122, pp. 84–95. Springer, Heidelberg (2004)
9. Jiang, H., Zhu, B., Zhu, D., Zhu, H.: Minimum common string partition revisited. In: Lee, D.-T., Chen, D.Z., Ying, S. (eds.) FAW 2010. LNCS, vol. 6213, pp. 45–52. Springer, Heidelberg (2010)
10. Stothard, P.: The Sequence Manipulation Suite: JavaScript programs for analyzing and formatting protein and DNA sequences. Biotechniques 28, 1102–1104 (2000)
11. Villesen, P.: FaBox: an online fasta sequence toolbox (2007), http://www.birc.au.dk/software/fabox

Ant Colony Optimization for Channel Assignment Problem in a Clustered Mobile Ad Hoc Network

Mahboobeh Parsapoor and Urban Bilstrup

School of Information Science, Computer and Electrical Engineering
Halmstad University
{mahboobeh.parsapoor,urban.bilstrup}@hh.se

Abstract. This paper presents an ant colony optimization (ACO) method as a method for channel assignment in a mobile ad hoc network (MANET), where achieving high spectral efficiency necessitates an efficient channel assignment. The suggested algorithm is intended for graph-coloring problems and it is specifically tweaked to the channel assignment problem in MANET with a clustered network topology. A multi-objective function is designed to make a tradeoff between maximizing spectral utilization and minimizing interference. We compare the convergence behavior and performance of ACO-based method with obtained results from a grouping genetic algorithm (GGA).

Keywords: Ant colony optimization, Channel assignment problem, Co-channel Interference, Spectral efficiency.

1 Introduction

Ant colony optimization (ACO) is a type of meta-heuristic algorithm that has been widely used in wireless communication, in particular ad hoc networks. So far, several ACO-based routing algorithms (e.g., ANT_AODV, POSANT) [1], [2] have been proposed to provide efficient routing methods for MANETs. ACO has also been proposed as a basis for clustering algorithms. The ACO-based clustering algorithms have shown the capability to provide a scalable and stable clustered network structure [3], [4]. Using ACO-based channel assignment schemes in cellular networks have also been studied. In wireless communication, finding an optimal channel assignment has been proven as an NP-hard problem [5], [6]. This means that an optimal channel assignment scheme could not be found in the polynomial time by using the traditional exhaustive search methods e.g. branch and bound. In contrast, meta-heuristic methods, e.g. genetic algorithms (GA), swarm intelligence (SI) and in particular ant colony optimization (ACO), can be used to find near optimal solutions in polynomial time [5]-[8]. A channel assignment scheme in a clustered MANET attempts to assign a minimum number of channels to the clusters considering an interference constraint. For this problem, meta-heuristic methods, e.g. grouping genetic algorithm (GGA) and ACO, could provide near optimal solutions. They can converge towards the Pareto front for maximizing spectral efficiency and minimizing the co-channel interference between cluster heads. The rest of this paper is organized as follows: in Section 2,

Y. Tan, Y. Shi, and H. Mo (Eds.): ICSI 2013, Part I, LNCS 7928, pp. 314–322, 2013.

related works in applying ACO on MANETs are briefly reviewed. We also mention related studies in channel assignment problems. Section 3 suggests the use of ACO for channel assignment and illustrates how ACO can be applied as a channel assignment scheme. Section 4 presents the results of the suggested methods for channel assignment in several different scenarios and provides a comparison with another heuristic method. Finally, the paper is concluded with some finale notes in Section 5.

2   Related Studies in Channel Assignment Problem

ACO-based methods have been widely used to provide efficient routing algorithms. They can minimize the delay in routing and control communication overheads [1], [2]. Numerous cluster formation algorithms have also developed combining ACO based algorithms with the traditional clustering algorithms, such as weighted clustering algorithms [5]-[8]. The channel assignment problem was early defined as a frequency assignment problem in cellular communication systems [10]-[13]. However, it is not only a problem of cellular communication systems, it exists in all kinds of wireless networks, e.g., wireless local area network (WLAN), wireless meshes network (WMN), mobile ad hoc network (MANET) and cognitive radio network (CRN). They all require an efficient channel assignment scheme to address scalability, stability, throughput, connectivity, routing, and fault tolerance [10]-[13]. Generally speaking, the channel assignment problem can be defined as finding a desirable scheme to minimize the number of channels needed for maximizing spatial reuse and at the same time satisfying interference constraints (i.e. a co-channel interference constraint). A clustered MANET is partitioned into groups of mobile nodes to provide a well-organized scalable structure for routing algorithms, power control mechanisms and spectrum management methods [10]. A common structure for a clustered network topology is based on defining three types of nodes: cluster head, gateway and ordinary nodes. The cluster head, the master of a cluster, is responsible for allocating resources and coordinates the intra cluster communication. The gateway, which is a common node between two or more clusters, provides the connectivity between clusters. Other nodes are ordinary nodes that determine the boundary of clusters, which depends on the transmission range and the node density [14]. In such a network topology, the channel assignment problem can be defined as finding a desirable scheme for assigning orthogonal channels (time, frequency or code) to the cluster heads [10]. So far, several heuristic methods e.g., greedy algorithms and genetic algorithms have been applied to solve this problem, which has been referred to as a cluster based coloring algorithm [15].

3   Ant Colony Optimization for Cluster- Based MANETs

Ant colony optimization meta-heuristic (ACO_MH) is a collection of algorithms which are inspired by the 'foraging behavior of real ants' [14]. Real ants start from the nest and use both local and global knowledge to construct the shortest path to the source of food. The ACO-based algorithms imitate this behavior to solve optimization problems. For an ACO-based algorithm, the problem is represented by a graph, $G = (V, E)$ where, the nodes, V, represent the components of the problem and the links, E, show the transition paths (i.e., partial parts of the solution) between the nodes. An optimal solution is a sequence of nodes with minimum cost function. The general components of the ACO-based algorithm are summarized as follows [16]-[17]: 1) A graph that represents the optimization problem. 2) A population of ants, $N_{ant} = \{ant_1, ..., ant_{n_{ant}}\}$, where n_{ant} determines the number of ants that traverse the graph; they memorize the traversed paths. 3) A set of feasible nodes, in order to avoid forming a loop during the path construction. This set is represented as N_i^k ; it determines the feasible nodes from the perspective of ant k when it is placed on i th node. 4) Initial states which are assigned to ants and determine the source nodes for the ants (each ant can start from different nodes). 5) A 'probabilistic transition rule' [16] is used by each ant to make a decision to move to the next node. It is defined on the basis of the heuristic information and the pheromone intensity. 6) A Heuristic function, which is a 'problem dependent function' [16], to indicate the desirability of the selected node. 7) Pheromone intensity that represents the desirability of the selected path (i.e., the path between the current node and the next node) from the perspective of other ants. 8) A cost function that is assigned to each ant that constructs a complete path from the source node to destination node. The cost function is defined on the basis of the optimization problem and can be utilized to evaluate the performance of that ant. 9) An updating rule for pheromone intensity that is used to determine the effect of the previous deposited pheromones which is defined on the basis of the cost function.

3.1   An ACO-Based Method for Channel Assignment Problem

To solve the channel assignment problem using ACO, a graph of the problem is represented as $G'' = (V'', E'')$, where V'' represents the cluster heads and E'' represents the adjacency between the clusters, i.e. they have common nodes with each other. Thus, the number of nodes in G'' determines the number of clusters, N_c.

The steps of this algorithm are similar to the steps of the ACO-based graph coloring algorithm, i.e. channels are equivalent with colors. At initialization state, ants are placed on the nodes of G''; they are preferably placed on nodes that have more adjacent nodes, i.e. higher degree. This parameter, $N_{nei,i}$, calculates the number of adjacent nodes of node i. Then, each ant is assigned a set of feasible channels, $N_{ch}^k = \{ch_1, ..., ch_{n_{ch}^k}\}$, that is randomly chosen from the available set of channels, $N_{Avaialble\_channel} = \{ch_1, ..., ch_{n_{ch}}\}$. The parameters n_{ch}^k and n_{ch} determine the number of available channels for k^{th} ant and the number of available channels for the problem. The set of feasible channels is updated when an ant selects a specific channel for the current node. We define two probabilistic transition rules. The first probabilistic transition rule is defined for choosing the next node $p_{ij}^{\prime k}$; it is the probability of choosing node j by the kth ant; while it placed at node i. It is defined on the basis of the heuristic information, $\eta_{ij}^{\prime}$ (t) and pheromone intensity, $\tau_{ij}^{\prime}$ (t). In the equation (1), N_i^k represents the sets of feasible nodes from the perspective of the kth ant at node i. Two parameters, α and β determine the influence of pheromone concentration and heuristic information respectively.

$$p_{ij}^{\prime k}(t) = \frac{\tau_{ij}^{\prime \alpha}(t)\eta_{ij}^{\prime \beta}(t)}{\sum\limits_{j \in N_i^k(t)} \tau_{ij}^{\prime \alpha}(t)\eta_{ij}^{\prime \beta}(t)} \tag{1}$$

The second probabilistic transition rule, $p_{ic}^{\prime\prime k}$ is defined for choosing a channel for the current node. The equation (2) defines the probabilistic function using the heuristic information, $\eta_{ic}^{\prime\prime}$ (t) and pheromone intensity, $\tau_{ic}^{\prime\prime}$ (t). In this equation, $N_{ch,i}^k$ shows the set of feasible colors from the perspective of the kth ant at node i.

$$p_{ic}^{\prime\prime k}(t) = \frac{\tau_{ic}^{\prime\prime \alpha}(t)\eta_{ic}^{\prime\prime \beta}(t)}{\sum\limits_{j \in N_{ch,i}^k(t)} \tau_{ic}^{\prime\prime \alpha}(t)\eta_{ic}^{\prime\prime \beta}(t)} \tag{2}$$

Ants construct the solution by incrementally choosing one node and assign one color to that node. The completed path is a sequence of nodes along their assigned colors while satisfying the constraint that two adjacent nodes should be assigned different colors. The cost function is calculated on the basis of the traversed paths and assigned colors. Using the probabilistic transition rules as equations (1) and (2), ants choose the next node and assign a channel to that node. The ants traverse the graph and assign channels to the nodes of the graph satisfying the co-channel interference requirement.

The two heuristic functions are defined as equations (3) and (4), respectively. The heuristic function $\eta_{ij}^{\prime}$ is defined for choosing nodes and is determined by the equation (3). When an ant is at node i, the heuristic value for choosing the next node j is calculated according to equation (3). The parameter $N_{unallocated,i}^k$ calculates the set of neighbors of node i that have not been allocated channels from the perspective of kth ant.

$$\eta'_{ij} = \frac{1 + \left| N^k_{unallocated,i} \right|}{1 + \left| N_{nei,i} \right|} \qquad (3)$$

The equation (4) defines the heuristic function $\eta''_{ij}(t)$ that is used for choosing the channel for the current node. The equation (4) calculates the desirability of choosing a channel, ch_j, when the ant is placed on node i.

$$\eta''_{ich}(t) = \frac{1 + n_{diff-ch}}{1 + n_{simi-ch}} \qquad (4)$$

The parameter $n_{diff-ch}$ is calculated as the maximum number of different channels for the set $\{ ch_j, N^k_{ch,i} \}$. The parameter $n^k_{simi-ch}$ calculates the maximum number of similar channels for the set $\left\{ ch_j, N^k_{Assigned\_channel,i} \right\}$, where, $N^k_{Assigned\_channel,i}$ is the assigned channels to the neighbors of i from the perspective of kth ant.

As mentioned in the previous section, two rules are utilized to update the pheromone. The first rule is utilized to choose a node and the second rule is used for choosing a channel (i.e., color). The updating rules of nodes are defined according equation (5). The parameter, F^{Best}_{ij}, represents the cost function of the globally best ant, Best, for the path between node i and node j. The parameter N^{Best}_i represents the set of feasible nodes from the perspective of the globally best ant at node i .

$$\Delta \tau'_{i,j} = \frac{F^{Best}_{ij}}{\left| N^{Best}_i \right|} \qquad (5)$$

Another updating rule is defined as equation (6). The parameter $n^{Best}_{simi-ch}$ is the number of similar channels in $\left\{ ch_j, N^{Best}_{Assigned\_channel,i} \right\}$.

$$\Delta \tau''_{ij} = \frac{n^{Best}_{simi-ch}}{\left| N^{Best}_i \right|} \qquad (6)$$

Assuming a channel assignment scheme by ant k is determined as $\mathbf{x}^k$; it is an $N_c \times N_{Avaiable\_ch}$ matrix. If channel ch_q is assigned to the cluster p, $x_{pq} = 1$ otherwise it is 0. The optimization function can be formulated according to equation (7), where, $S_{simi-ch}$ is defined as (8) and, N_{CH} is the number of cluster heads.

$$\arg\max \sum_{p=1}^{N_{Avaiable\_ch}} \sum_{q=1}^{N_c} x_{pq} + S_{simi\_ch} \tag{7}$$

$$S_{simi-ch} = \max_{p=1}^{N_{Avaiable\_ch}} \{ \sum_{q=1}^{N_c} x_{pq} \} \tag{8}$$

4 Simulation

4.1 Simulation Model

For implementing simple scenarios of MANETs, we use a discreet even simulation Implemented in MATLAB. The main goal of this channel assignment algorithm is to assign logical channels (defined as orthogonal frequency hopping sequences) to the clusters for intra-cluster scheduling. The main assumptions of our simulated models are described as follows: 1) The suggested algorithm is implemented for a snapshot of a MANET; thus, there is no change in the network topology during the procedure to select and assign channels to the clusters. 2) The nodes are placed in an area with a 1000 x 1000 meter square and the position of each individual node has two coordinates, x and y, that are drawn from a uniform distribution [0, 1000]. The mobility and traffic of nodes are ignored. 3) All of the nodes are assumed to have omni directional antenna with a similar transmission power. 4) The channel model and interference model are considered as free-space path loss models and disk graph models respectively. The interference range is assumed two times to the transmission range. 5) The nodes are clustered using Lowest ID (LID). 6) There is a centralized controller e.g., base station that senses the available channels; it decides for an on-demand channel assignment scheme on the basis of the unassigned channels; thus, it avoids co-channel interference with previous assigned channels.

4.2 Simulation Results

First, we consider the MANETs that consist of 75 nodes. The MANETs are different in transmission ranges (i.e., TR=100, 200 and 300 meters), transmission powers (i.e., 2, 4 and 6 mill watt) and the number of clusters 31,13 and 3 respectively. The number of allocated channels using ACO and GA is depicted in Figure 1. It becomes clear that the number of required channels is dependent upon the number of clusters and the topology of the network. However, for the same topology, the number of assigned channels is smaller for ACO than GGA. (See Figure 4 the black and grey bars). It can also be observed that the number of assigned channels do not significantly change when the number of clusters increases. Thus, this method would be scalable for a large sized MANET. Figure 2 presents the obtained interference power using an ACO-based scheme and GA-based scheme. In the MANET with 31 clusters (i.e., TR=100), the ACO-based method has smaller value of interference power in comparison with the GA-based method. Using GA-based method in

MANET with a small number of clusters, the interference power becomes very low. In this case, using a GA-based channel assignment scheme, each cluster has been assigned a different channel. As the second test, the convergence behavior of the ACO algorithm is evaluated for the MANET with 31 clusters (the scenario is similar to the previous scenario, here we focus on the number of iterations to achieve a near optimal solution). We compare the results of the ACO with GGA during 200 iterations; while the size of population is considered as 15. Figure 3 depicts the average and minimum values of the objective function. It shows that the ACO-based method converges after approximately 10 iterations, while the GGA converges after 120 iterations. It is also noticeable that the average and minimum values of objective functions using the GGA differ to a large extent and it needs a greater number of iterations to converge to the global minimum.

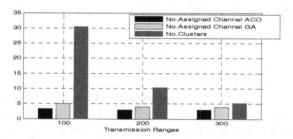

Fig. 1. The average number of assigned channels and clusters for a network with 75 nodes

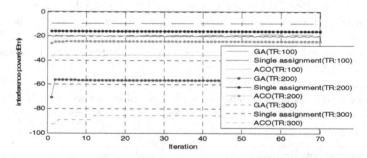

Fig. 2. Demonstrates the interference power between cluster heads for different MANETs

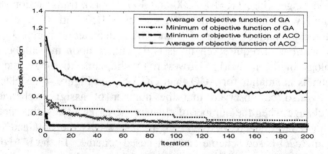

Fig. 3. The minimum and average of objective functions for a network with 30 clusters

5   Conclusion

In this paper, we first develop an ACO-based method for the standard graph coloring problem; then it is extended for the channel assignment problem. We evaluated this method in the MANETs with different numbers of clusters. The results have verified that the proposed ACO algorithm has the capability to find a scheme with a minimum number of assigned channels. The results have also indicated that the ACO-based method provides a stable and scalable solution; the performance of an ACO-based channel allocation scheme does not seem to be dependent on the size of MANETs (e.g., the number of clusters in MANETs). Developing an ACO-based distributed scheme has been considered as future works. We will also replace the lowest ID clustering algorithm with an ACO-based clustering method to effectively address the channel assignment problem.

References

1. Kamali, S., Opatrny, J.: POSANT: A Position Based Ant Colony Routing Algorithm for Mobile Ad-hoc Networks. In: ICWMC 2007 (2007)
2. Jyoti Jain, R.G., Bandhopadhyay, T.K.: Ant Colony Algorithm in MANET-Local link repairing of AODV. In: ICECT (2011)
3. Di Caro, G.A., Ducatelle, F., Gambardella, L.M.: AntHocNet: An ant-based hybrid routing algorithm for mobile ad hoc networks. In: Yao, X., et al. (eds.) PPSN 2004. LNCS, vol. 3242, pp. 461–470. Springer, Heidelberg (2004)
4. Shayeb, I.G., Hussein, A.H., Nasoura, A.B.: A Survey of Clustering Schemes for Mobile Ad-Hoc Network (MANET). American Journal of Scientific Research (2), 135–151 (2011)
5. Kumar, V., Balasubramanie, P.: Ant Colony Optimization Using Hierarchical Clustering in Mobile Ad Hoc Networks. Journal of Scientific Research (4), 549–560 (2011)
6. Yu, J.Y., Chong, P.H.J.: A survey of clustering schemes for mobile ad hoc networks. IEEE Communications Surveys & Tutorials 7, 32–48 (2005)
7. Zhang, J., Wang, B., Zhang, F.: A Distributed Approach of WCA in Ad-Hoc Network. In: Proceeding 6th International Conference on Wireless Communications, Networking and Mobile Computing, pp. 1–5 (2010)
8. Sampath, A., Tripti, C., Thampi, M.: An ACO algorithm for effective cluster head election. International Journal of Advances in Information Technology (JAIT) 2(1), 50–56 (2011)
9. Li, L., Yong, G.: Ant-colony optimization based on cluster routing protocol of Ad Hoc. In: ICCET, pp. 304–308 (2010)
10. Audhya, K., Sinha, S., Ghosh, C., Sinha, B.P.: A survey on the channel assignment problem in wireless networks. Wirel. Commun. Mob. Comput. 11, 583–609 (2010)
11. Wong, S.H.: Channel Allocation for Broadband Fixed Wireless Access Networks. PhD dissertation, University of Cambridge, Department of Engineering, UK (2003)
12. Maniezzo, V., Carbonaro, A.: An ANTS heuristic for the frequency assignment problem. Future Gener. Comp. Sy. 16, 927–935 (2000)
13. Hou, T.C., Tsai, T.J.: On the Cluster Based Dynamic Channel Assignment for Multi hop Ad Hoc Networks. J. Commun. Netw. 4(1), 40–47 (2002)

14. Ephremedis, A., Kutten, S.: Scheduling Broadcasts in Multihop Radio Networks. IEEE Transactions on Communications COM-38, 456–460 (1990)
15. Wu, H., Zhong, Z., Hanzo, L.: A cluster-head selection and update algorithm for ad hoc networks. In: Procedding of the IEEE Globecomm Conference, pp. 1–5 (2010)
16. Engelbrecht, A.P.: Fundamentals of Computational Swarm Intelligence. John Wiley & Sons, UK (2005)
17. Felix, T.S.C., Tiwari, M.K.: Swarm Intelligence: Focus on Ant and Particle Swarm Optimization. I-Tech Education and Publishing (2007)

Constrained Multi-objective Biogeography Optimization Algorithm for Robot Path Planning

Hongwei Mo[1], Zhidan Xu[1,2], and Qirong Tang[3]

[1] Automation College, Harbin Engineering University, Harbin 150001, China
[2] Institute of Basic Science, Harbin University of Commerce, Harbin150028, China
[3] Institute of Engineering and Computational Mechanics, University of Stuttgart.
Pfaffenwaldring 9, 70569 Stuttgart, Germany
{mhonwei,xuzhidanivy}@163.com,
qirong.tang@itm.uni-stuttgart.de

Abstract. Constrained multi-objective optimization involves multiple objectives subjected to some equality or inequality constraints so that it may require search a set of non-dominated feasible solutions. Inspired from this, in this paper, a novel constrained multi-objective biogeography optimization algorithm is proposed and used for solving robot path planning problem since it can be defined as a constrained multi-objective optimization problem. Experimental results compared with Non-dominated Sorting Genetic AlgorithmII show that the proposed algorithm has better performance.

Keywords: constrained multi-objective optimization; differential evolution; biogeography-based optimization; robot path planning.

1 Introduction

Constrained multi-objective optimization problems (CMOPs) are an important area both in research and industry. The CMOPs usually involve optimization of conflicting objectives subject to certain constraints. For example, robot path planning (RPP) needs to search a collision-free optimal or suboptimal path from the start to the target point. Generally, the path is required shorter length, better smoothness, higher security and collision-free. Some multi-objective evolutionary algorithms (MOEAs) [1]-[3] have been developed for RPP.

RPP can be defined as a CMOP because it need optimization multiple objectives such as the length, the smoothness of the path and ease of practical navigation and so on. And collision-free can be seen as a constraint condition. Hu [1] proposed multi-objective mobile robot path planning based on improved genetic algorithm. The algorithm optimizes three objectives: length, smoothness and security. Gong et al [2] proposed multi-objective Particle Swarm Optimization for robot path planning in environment with danger sources.

Biogeography-based optimization (BBO) [4] is a stochastic population-based optimization algorithm that has good exploitation ability of the population information. It has been applied successfully in solving single objective optimization problem with

Y. Tan, Y. Shi, and H. Mo (Eds.): ICSI 2013, Part I, LNCS 7928, pp. 323–329, 2013.

real applications [5]-[7]. In this paper, we conducted a novel biogeography based method, i.e. constrained multi-objective biogeography optimization algorithm (CMBOA) for RPP.

The rest of the paper is arranged as follows. Section II describes the original BBO. The proposed algorithm CMBOA is described in Section III. In Section V, experimental results are discussed. Finally, conclusions are given in Section VI.

2   Constrained Multi-objective Biogeography Optimization

2.1   Individual Fitness

In our algorithm, individuals are classified to ensure the convergence and even distribution of the obtained solutions set, the fitness of feasible individuals is defined as:

$$fit(i) = (1 - \gamma) / c_{ik} + \gamma I_{i.cd} \, , \tag{1}$$

where γ is the proportion of the feasible solutions in the current population, $I_{i,cd}$

and c_{ik} [8]denote crowed-distance and non-dominated rank sort, respectively.

For infeasible individuals, it is considered that the infeasible solutions are able to promote the diversity of solutions on the Pareto front, a novel evaluation method of infeasible solutions is defined as:

$$ifit(i) = \begin{cases} (1 - \gamma)v(i) + \gamma d(i) & \gamma > 0 \\ v(i) & \gamma = 0 \end{cases} \tag{2}$$

where $v(i)$ is constraint violation of the i th individual, $d(i)$ denotes its Euclidean distance from the nearest non-dominated feasible solution. By the new fitness, when the number of non-dominated feasible solutions is small in current population, infeasible solutions with high constraint violation have high fitness and will be selected to evolve towards feasible region. With the number of feasible solutions increasing, infeasible solutions near the feasible region are gained more attention and transformed into feasible ones to enhance the diversity of solutions.

2.2   Disturbance Migration Operator

During species migrating process in BBO, an individual is often affected by other individuals. So we propose a migration operator with disturbance term, and the migration operator is defined as:

$$x_{i,j} = p_{s,j} + Q(t)(p_{s1,j} - p_{s2,j}) \, , \tag{3}$$

where $p_{i,j}$ denotes the j th variable of the i th individual P_i, $Q(t)$ is defined as:

$$Q(t) = \frac{4}{5}(1 - \frac{1}{1 + e^{-0.1(t - g_{max}/2)}}),$$ (4)

where g_{max} denotes the maximum generation, t is the current evolution generation. The amplitude of disturbance factor $Q(t)$ decreases constantly with the increasing of iteration generation t. Various solutions will be created to promote the diversity of population because of difference of $Q(t)$.

2.3   The Procedures of CMBOA

The general process of CMBOA is described in Algorithm 1.

Algorithm 1. The procedures of CMBOA

Step1: Initialization and parameter setting: population size N, the size of feasible elitist archive N_1, the size of infeasible elitist archive N_2, maximum generation g_{max}

Step2: Generate an initial population $A(t)$, set the iterative generation $t = 0$

Step3: Divide the population $A(t)$ into the feasible population $X(t)$ and the infeasible population $\tilde{X}(t)$ based on the constraint violation of individuals where $A(t) = X(t) \cup \tilde{X}(t)$.

Step4: If $X(t) = \Phi$ (empty), then $X'(t) = \Phi$ and perform the mutation operator of DE on infeasible population $\tilde{X}(t)$ to obtain new population $\tilde{X}'(t)$. Otherwise, evaluate individuals of the feasible population $X(t)$ according to the Eq.(1), implement DMI on the population $X(t)$ to obtain new population $X'(t)$. Meanwhile, perform arithmetic crossover operator on infeasible population $\tilde{X}(t)$ to obtain new population $\tilde{X}'(t)$

Step5: Combine the offspring population $X'(t) \cup \tilde{X}'(t)$ and the parent population $A(t)$, and divide them into feasible and infeasible population, and then conserve N_1 feasible populations with small crowded-distance to produce the new feasible population $X(t+1)$, conserve N_2 infeasible population with small constraint violation to gain the population $\tilde{X}(t+1)$

Step6: If $t \geq g_{max}$ is satisfied, export $X(t+1)$ as the output of the algorithm and the algorithm stops; otherwise, $t = t+1$ and go to step4.

In CMBOA, the initial population is produced stochastically in Step2, and then it is classified to feasible and infeasible population. If feasible population is empty, mutation operator of DE is applied to produce them in Step 4. Otherwise, the new operator DMI is performed on feasible population to generate many more non-dominated feasible solutions. Some infeasible solutions nearby feasible regions will recombine non-dominated feasible solutions to approximate the feasibility.

2.4   CMBOA for Robot Path Planning

In robot path planning (RPP), path length and smoothness are considered as optimization objectives and the degree of path blocked by obstacles is a constraint.

In RPP, the environment model of robot is established by using grid method [9]. If a grid is occupied by some obstacles, its attribute value is set one, otherwise, the value is zero. Then the constraint violation of path is defined as:

$$g(P) = \frac{go(P)}{gn(P)},$$
(5)

where $gn(P)$ denotes the number of grids occupied by path P, $go(P)$ is the sum of those grids attribute values.

The length of a path is considered as one optimization objective of the given RPP. It is defined as the sum of all Euclidean distance of path segments as follows:

$$f_1(p) = \sum_{i=0}^{n} |p_i p_{i+1}|,$$
(6)

where p_0 and p_{m+1} are the start S and the target points T of a path, respectively, $|p_i p_{i+1}|$ represents Euclidean distance of the line segment $p_i p_{i+1}$

The smoothness of a path is another optimization objective and is defined as:

$$f_2(p) = \sum_{i=1}^{n-1} (\pi - (\theta_{i+1}(p_i p_{i-1}, p_i p_{i+1})),$$
(7)

where $\theta_{i+1}(p_i p_{i-1}, p_i p_{i+1})$ represents the angle between the vector $p_i p_{i-1}$ and $p_i p_{i+1}$.

So robot path planning can be defined as a constrained multi-objective optimization problem:

$$\min f(P) = [f_1(P), f_2(P)]$$
$$s.t. \quad g(P) = 0$$
(8)

3   Simulation and Results

In order to demonstrate the effectiveness of CMBOA for RPP, CMBOA is compared with NSGA-II [8] for two test problems. In the robot environments, polygons represent the obstacles. The parameters of algorithms are set in Table 1.

Table 1. Parameter settings for test algorithms

Algorithms	Parameter Settings
NSGAII	Population size =100, crossover probability =0.9, mutation probability= $1/n$, $n = 9$ is the length of variable , SBX crossover parameter 20, polynomial mutation parameter 20, the termination generation 200
CMBOA	ize =100, feasible elitist maximum size $N_1 = 100$, infeasible elitist maximum the maximum immigration rate and migration rate E = I = 1, the termination 100, F is a random in the interval $(0.2, 0.8)$ in DE, the dimension of decision

The hypervolume metric(HV)[10] is adopted to evaluate the quality of solutions. In the 30 independent runs, the maximum of HV and the most optimal paths gained by CMBOA and NSGA-II are shown in Table 2.It can be seen that the maximum of HV obtained by CMBOA is larger than that of NSGA-II for the two test problems, which indicates that CMBOA has good convergence and can dominate larger space in objective space.The Pareto fronts with the maximum HV metrics are shown in Fig.1, in which 'o' and '*' denote the Pareto front obtained by NSGA-II and CMBOA, respectively. In Fig.1, we can see that the Pareto fronts of CMBOA dominate those of NSGA-II, which indicates that CMBOA has better convergence than NSGA-II. However, NSGA-II obtains the better spread at the expense of its convergence for test2. In addition, In Table 2, the shortest length and the minimum smoothness of paths gained by CMBOA is smaller than those by NSGA-II, which demonstrates that CMBOA is effective for RPP. The shortest and the smoothest paths are shown in Fig.2-3. From Fig.2-3, it can be seen that the path by CMBOA is shorter than ones by NSGA-II in two environments. For test1, the performance of CMBOA is slightly better than that of NSGA-II. For test2, the path gained by CMBOA is obviously better than NSGA-II, which demonstrates CMBOA is competitive for RPP.

Table 2. The HV and the most excellent objectives gained by CMBOA and NSGA II

		Max HV	Min Length	Min smooth
Test1	CMBOA	1.4202e+004	102.7931	27.9838
	NSGAII	1.4200e+004	102.8171	27.9838
Test2	CMBOA	1.4165e+004	103.3038	27.9834
	NSGAII	1.3598 e+004	111.1323	28.0022

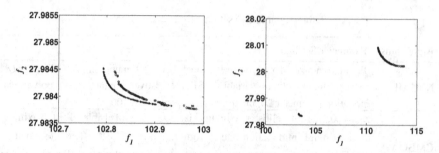

Fig. 1. The Pareto front obtained by CMBOA ('*') and NSGAII('o')

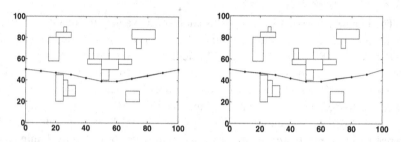

Fig. 2. The shortest length and the most smooth paths gained by CMBOA ('*') and NSGAII('o') for test problem1

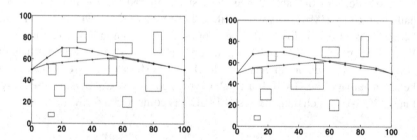

Fig. 3. The shortest length and the most smooth paths gained by CMBOA ('*') and NSGAII('o') for test problem2

4   Conclusions

In this paper, a constrained multi-objective biogeography optimization algorithm (CMBOA) is proposed for solving RPP. In CMBOA, the disturbance migration operator is designed and applied on feasible solutions, while the infeasible solutions nearby feasible regions recombine with the nearest non-dominated feasible solutions to approximate the feasibility. CMBOA makes the set of gained solutions approximate the Pareto front from the inside and outside of feasible region simultaneously.

Compared with NSGA-II for RPP, the simulation results show that the proposed algorithm CMBOA solves RPP more effectively under the given two test situations.

Acknowledgements. This work is supported by the National Natural Science Foundation of China under Grant No.61075113 and the Excellent Youth Foundation of Heilongjiang Province of China under Grant No. JC2012012. The Fundamental Research Funds for the Central Universities, No. HEUCFZ1209.

References

1. Hu, J., Zhu, Q.B.: Multi-objective Mobile Robot Path Planning Based on Improved Genetic Algorithm. In: International Conference on Intelligent Computation Technology and Automation, pp. 752–756. IEEE Computer Society, China (2010)
2. Gong, D.W., Zhang, J.H., Zhang, Y.: Multi-objective Particle Swarm Optimization for Robot Path Planning in Environment with Danger Sources. Journal of Computers 6(8), 1554–1561 (2011)
3. Kala, R., Shukla, A., Tiwari, R.: Robotic Path Planning using Evolutionary Momentum based Exploration. Journal of Experimental and Theoretical Artificial Intelligence 23(4), 469–495 (2011)
4. Simon, D.: Biogeography-Based Optimization. IEEE Transaction Evolutionary Computation 12(6), 702–713 (2008)
5. Cai, Z.H., Gong, W.Y., Charles, X.L.: Research on a Novel Biogeography-based Optimization Algorithm Based on Evolutionary Programming. System Engineering-Theory & Practice 30(6), 1106–1112 (2010)
6. Mo, H., Xu, L.: Biogeography Migration Algorithm for Traveling Salesman Problem. In: Tan, Y., Shi, Y., Tan, K.C. (eds.) ICSI 2010, Part I. LNCS, vol. 6145, pp. 405–414. Springer, Heidelberg (2010)
7. Simon, D., Ergezer, M., Du, D.W.: Population distributions in biogeography-based optimization algorithms with elitism. In: IEEE Conference on Systems, Man, and Cybernetics, San Antonio, TX USA, pp. 1017–1022 (2009)
8. Deb, K.: Multi-Objective Optimization Using Evolutionary Algorithms. Wiley-Interscience Series in Systems and Optimization. John Wiley & Sons, Chichester (2001)
9. Wang, Y., Zhu, Q.B.: Path planning for mobile robot based on binary particle swarm optimization. Journal of Nanjing Normal University (Engineering and Technology Edition) 9(2), 72–78 (2009)
10. Knowles, J.D., Thiele, L.: Zitzler. E.: A Tutorial on the Performance Assessment of Stochastic Multiobjective Optimizers. Technical Report TIK-Report No. 214, ETH Zurich: Computer Engineering and Networks Laboratory (2005)

A Simplified Biogeography-Based Optimization Using a Ring Topology

Yujun Zheng[1], Xiaobei Wu[1], Haifeng Ling[2], and Shengyong Chen[1]

[1] College of Computer Science & Technology, Zhejiang University of Technology,
Hangzhou 310023, China
yujun.zheng@computer.org
[2] College of Field Engineering, PLA University of Science & Technology,
Nanjing 210007, China

Abstract. The paper proposes a new simplified version of biogeography-based optimization (BBO) algorithm. The original BBO is based on a global topology such that migration can occur between any pair of habitats (solutions), and we simplify it by using a local ring topology, where each habitat is only connected to two other habitats and the migration can only occur between neighboring habitats. The new strategy is quite easy to implement, but it contributes significantly to improving the search capability and preventing the habitats from being trapped in local optima. Computational experiment demonstrates the effectiveness of our approach on a set of benchmark problems.

Keywords: Global optimization, biogeography-based optimization (BBO), migration, ring topology.

1 Introduction

The complexity of real-world optimization problems gives rise to various kinds of evolutionary algorithms (EAs), which are stochastic search methods drawing inspiration from biological evolution. Biogeography-Based Optimization (BBO), initially proposed by Simon [1], is a relatively new population-based EA borrowing from biogeography evolution its main principle. In BBO, each individual is considered as a "habitat" or "island" with a habitat suitability index (HSI), based on which an immigration rate and an emigration rate can be calculated. High HSI solutions tend to share their features with low HSI solutions, and low HIS solutions are likely to accept many new features from high HIS solutions. BBO has proven itself a competitive heuristic to other EAs on a wide set of problems [1,2,3]. Moreover, the Markov analysis in [4] proved that BBO outperforms GA on simple unimodal, multimodal and deceptive benchmark functions when used with low mutation rates.

In the original version of BBO, if a given habitat is selected to be immigrated, then any other habitat has a chance to share information with it. That is, BBO uses a global topology of the population of habitats, where each pair of habitats can communicate with each other. We think that such a migration

Y. Tan, Y. Shi, and H. Mo (Eds.): ICSI 2013, Part I, LNCS 7928, pp. 330–337, 2013.

mechanism is computationally intensive. In this paper, we propose a simplified BBO using a local ring topology, namely RBBO, where each habitat can only communicate with two other habitats. The new algorithm does not introduce any more operators and thus is quite easy to implement, but it provides much better performance than the original BBO on a set of benchmark functions.

In the rest of the paper, Section 2 simply introduces BBO, Section 3 discusses related work to BBO, Section 4 describes our RBBO algorithm, Section 5 presents the computational experiment, and Section 6 concludes.

2   Biogeography-Based Optimization

In BBO, each solution is modeled as a habitat, and each habitat feature or solution component is called a suitability index variable (SIV). The habitats are ranked by their habitat suitability index (HSI), a higher value of which signifies a larger number of species, which is analogous to a better fitness to the problem. Each habitat H_i has its own immigration rate λ_i and emigration rate μ_i, which are functions of the HSI. High HSI habitats tend to share their features with low HSI habitats, and low HIS habitats are likely to accept many new features from high HIS habitats. For example, suppose the habitats are ordered according to their fitness, and the immigration and emigration rates of the ith habitat can be calculated as follows:

$$\lambda_i = I(1 - \frac{i}{n}) \tag{1}$$

$$\mu_i = E(\frac{i}{n}) \tag{2}$$

where n is the size of population, and I and E are the maximum possible immigration rate and emigration rate respectively. Such a linear migration model of BBO is illustrated in Fig. 1. However, there are other non-linear mathematical models of biogeography that can be used for calculating the migration rates [1].

Migration is used to modify habitats by mixing features within the population. BBO also has a mutation operator for changing SIV within a habitat itself, which can increase diversity of the population. For each habitat H_i, a species count probability P_i computed from λ_i and μ_i indicates the likelihood that the habitat was expected a priori to exist as a solution for the problem. In this context, very high HSI habitats and very low HSI habitats are both equally improbable, and medium HSI habitats are relatively probable. The mutation rate of habitat H_i is inversely proportional to its probability:

$$\pi_i = \pi_{max}(1 - \frac{P_i}{P_{max}}) \tag{3}$$

where π_{max} is a control parameter and P_{max} is the maximum habitat probability in the population.

Algorithm 1 describes the general framework of the original BBO (where $rand()$ returns a uniformly distributed random number in [0,1]).

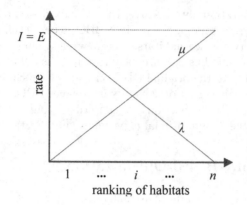

Fig. 1. Illustration of the linear model of emigration rate μ and immigration rate λ

Algorithm 1. The original BBO algorithm.

---

1   Randomly initialize a population of n solutions (habitats) to the problem;
2   **while** stop criterion is not satisfied **do**
3   **for** $i = 1$ **to** n **do**
4   Evaluate the HSI of H_i, according to which calculate λ_i, μ_i, and π_i;
5   **for** $i = 1$ **to** n **do**
6   **for** $d = 1$ **to** D **do**
7   **if** $rand() < \lambda_i$ **then**
8   Select a habitat H_j with probability $\propto \mu_j$;
9   Replace the dth SIV of H_i with the corresponding SIV of H_j;
10   **for** $i=1$ **to** n **do**
11   **for** $d = 1$ **to** D **do**
12   **if** $rand() < \pi_i$ **then**
13   Replace the dth SIV of habitat H_i with a random value;
14   **return** the best solution found so far.

---

3   Related Work

Several work has been devoted to improve the performance of BBO of by adding some features from other heuristics. Du et al. [2] incorporated features from evolutionary strategy to BBO, i.e., at each iteration selecting the best n solutions from the n parents and n children as the population for the next generation. Gong et al. [5] proposed a hybrid DE with BBO, which combines the exploration of DE with the exploitation of BBO effectively. The core idea is to hybridize the DE operator with the migration operator of BBO, such that good solutions would be less destroyed, while poor solutions can accept a lot of new features from good solutions.

The original BBO is for unconstrained optimization. Recently some work has been done to extend BBO for constrained optimization problems. Ma and Simon [6] proposed a blended BBO for dealing with such problems. The algorithm uses a blended migration operator and determines whether a modified solution can enter the population of the next generation by comparing it with its parent. Boussaïd et al. [7] proposed another approach combining BBO and DE, which evolves the population by alternately applying BBO and DE iterations, and at each iteration always selects a fitter one from the updated solution and its parent for the next iteration. In [8] the authors extended the approach for constrained optimization, which replaces the original mutation operator of BBO with the DE mutation operator, and includes the a penalty function to the objective function to handle problem constraints.

4 BBO with a Ring Topology

4.1 Migration on a Local Neighborhood

In the original BBO, if a given habitat is selected to be immigrated, then any other habitat has a chance to be an emigrating habitat for it. In other words, the BBO uses a *global* topology where each pair of habitats can communicate with each other, as shown in Fig. 2(a). However, such a migration mechanism is computationally expensive. In the biogeography theory [9], beside HSI, the immigration and emigration rates of habitats also depend on a number of features (such as distances and capacities) of the migratory routes between the habitats. An obvious example is that, the larger the distance between two habitats, the less the probability of migration between them. However, computing the distances between all pairs of habitats will certainly incur high computational cost.

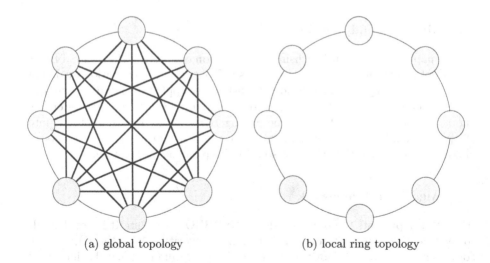

(a) global topology   (b) local ring topology

Fig. 2. Topologies of the population of habitats

Here we use a *local* topology where each habitat can only communicate with its neighbor habitats, in order to not only save computational cost but also avoid premature convergence. Among a number of limited communication topologies, we select one of the simplest forms of local topologies, namely the ring topology, which connects each habitat to only two other habitats in the population, as shown in Fig. 2(b). The ring topology is unrelated to the locality of habitats in the search space as thus is easy to implement. It deserves to note that the local topology has also been used in local best models of PSO, where each particle is attracted toward a fitter local best within its vicinity of the search space, in comparison with the abandoned global best models [10].

Suppose a habitat H_i is selected for immigration and H_j and H_k are two habitats connected to it, for each SIV of H_i we generate a random number between [0,1]: If the number is less than $\mu_j/(\mu_j + \mu_k)$, then the SIV is replaced by the corresponding value of H_j; otherwise it is replaced by the corresponding value of H_k.

One of the most advantages of the local model lies in that, not all the solutions are subject to accept many features from one or several solutions of very high HSI values. In consequence, the diversity of the population is potentially improved, and the whole population is not likely to converge fast to some inferior local optima, especially at the early stage of search. This is similar to the difference between the *gbest* model and *lbest* model used in particle swarm optimization [10,11]. However, the local model is more useful in BBO, since the selection of a solution as an emigration habitat from the whole population is much more computationally expensive than that from two neighbor habitats.

4.2 Mutation and Solution Update

In this paper, we just consider using the mutation operator of the original BBO, but adopt a new solution update mechanism. That is, at each iteration we do not directly modify any existing solution H_i; instead we apply the migration operator to H_i to create a new solution H_i', and then mutate H_i' to H_i''. At the end of the iteration, we select the best one among H_i, H_i', and H_i'' and enter it into the population for the next generation. Note this mechanism also implements an elitism strategy such that the best solution keeps in the population.

4.3 Algorithm Framework

Algorithm 2 presents the pseudo code of the RBBO algorithm. It is not difficult to see that the complexity of each iteration of Algorithm 1 is $O(n^2D)$, but the complexity of that of Algorithm 2 is $O(nD)$ (the difference can be identified by comparing line 8~9 of Algorithm 1 with line 10~13 of Algorithm 2). This is because that, when a habitat is selected for immigration, for each vector component RBBO checks only two other habitats while BBO checks all other $(n-1)$ habitats, and RBBO only generates one random number while BBO needs to generate $(n-1)$ random numbers.

Algorithm 2. The RBBO.

---

1 Randomly initialize a population of solutions (habitats) to the problem;
2 Initialize the ring topology of the population;
3 **while** stop criterion is not satisfied **do**
4 　**for** $i = 1$ **to** n **do**
5 　　Evaluate the HSI of H_i, according to which calculate λ_i, μ_i, and π_i;
6 　**for** $i = 1$ **to** n **do**
7 　　Create an offspring H_i' of H_i;
8 　　**for** $d = 1$ **to** D **do**
9 　　　**if** $rand() < \lambda_i$ **then**
10 　　　　**if** $rand() < \mu_j/(\mu_j + \mu_k)$ **then**
11 　　　　　Migrate the dth SIV of H_i's left neighbor to H_i';
12 　　　　**else**
13 　　　　　Migrate the dth SIV of H_i's right neighbor to H_i';
14 　　Create an offspring H_i'' of H_i';
15 　**for** $i=1$ **to** n **do**
16 　　**for** $d = 1$ **to** D **do**
17 　　　**if** $rand() < \pi_i$ **then**
18 　　　　Replace the dth SIV of habitat H_i'' with a random value;
19 　　Let H_i be the fittest one of H_i, H_i', and H_i'';
20 **return** the best solution found so far.

---

5 Computational Experiment

We made a comparative experiment between the original BBO and our RBBO, which are tested on a set of well-known benchmark functions from [12], denoted as $f_1 \sim f_{13}$, a summary of which is shown in Table 1 (due to length limit, we only consider the problems in 30 dimensions). We set $n = 50$, $I = 1$, $E = 1$ for both the algorithms. π_{max} is set to 0.01 for BBO, but set to 0.02 for RBBO since the ring topology decreases the impact of migration and thus we should increase the chance for mutation.

The experiments are conducted on a computer of 2×2.66GHz AMD Athlon64 X2 processor and 8GB memory. For a fair comparison, on each function we set the same maximum running CPU time for both the algorithms, and compare their mean and standard deviation of the best fitness values and the success rate (with respect to the required accuracies) averaged over 40 runs, the result of which is presented in Table 2.

As we can see from the results, except f_7, on all other benchmark functions, the mean best fitness values of RBBO are always better than the original BBO. In more details, we observe that the mean bests of RBBO are less than 10% of BBO on f_1, f_2, f_3, f_6, f_{10}, f_{12}, and f_{13}, and about 10% $\sim$ 50% of BBO on f_4, f_5, f_8, f_9, and f_{11}. The success rates of RBBO are also better than the BBO on 11 functions. This demonstrates that our method effectively improves the efficiency and robustness of biogeography-based search, since the ring topology limits the range of migration and helps to avoid premature convergence.

Table 1. Detailed information of the benchmark functions used in the paper, where RA denotes "required accuracy" and the running CPU time is in seconds

ID	Fun	Search range	x^*	$f(x^*)$	RA	CPU time
f_1	Sphere	$[-100, 100]^D$	0^D	0	0.01	10
f_2	Schwefel 2.22	$[-10, 10]^D$	0^D	0	0.01	10
f_3	Schwefel 1.2	$[-100, 100]^D$	0^D	0	0.1	20
f_4	Schwefel 2.21	$[-100, 100]^D$	0^D	0	0.1	30
f_5	Rosenbrock	$[-2.048, 2.048]^D$	1^D	0	1	10
f_6	Step	$[-100, 100]^D$	0^D	0	0.001	10
f_7	Quartic	$[-1.28, 1.28]^D$	0^D	0	0.001	10
f_8	Schwefel	$[-500, 500]^D$	420.9687^D	0	0.01	10
f_9	Rastrigin	$[-5.12, 5.12]^D$	0^D	0	0.01	10
f_{10}	Ackley	$[-32.768, 32.768]^D$	0^D	0	0.000001	10
f_{11}	Griewank	$[-600, 600]^D$	0^D	0	0.01	10
f_{12}	Penalized1	$[-50, 50]^D$	1^D	0	0.0001	20
f_{13}	Penalized2	$[-50, 50]^D$	1^D	0	0.001	20

Table 2. The experimental result on test problems

ID	mean best fitness (standard deviation in parentheses)		success rate(%)	
	BBO	RBBO	BBO	RBBO
f_1	1.22E+00 (1.60E-01)	6.26E-02 (6.88E-03)	0	20
f_2	1.12E-01 (1.04E-02)	8.43E-03 (4.77E-04)	0	92.5
f_3	2.45E+00 (2.30E-01)	4.38E-02 (3.31E-03)	0	97.5
f_4	1.29E+00 (7.05E-02)	1.87E-01 (1.02E-02)	0	10
f_5	1.31E+02 (1.43E+01)	6.78E+01 (5.46E+00)	0	0
f_6	2.02E-01 (9.00E-02)	0 (0)	70	100
f_7	1.37E-03 (1.30E-04)	1.81E-03 (2.02E-04)	10	10.5
f_8	1.49E-01 (1.71E-02)	1.33E-02 (1.18E-03)	2.5	40
f_9	1.66E-01 (1.61E-02)	1.12E-02 (1.07E-03)	0	65
f_{10}	3.15E-05 (2.57E-06)	2.01E-06 (1.26E-07)	0	25
f_{11}	1.93E-01 (1.35E-02)	2.38E-02 (2.11E-03)	0	5
f_{12}	2.18E-03 (5.37E-04)	1.03E-06 (2.60E-07)	0	87.5
f_{13}	3.96E-02 (5.38E-03)	3.20E-03 (1.14E-03)	0	20

6 Conclusion

BBO is a new bio-inspired optimization method that has proven its quality and versatility on various problems. Although much research has been devoted to improve the performance of BBO, most of them combine some operators of other heuristic methods or introduce more new operators, and thus increase the difficulty of implementation. The approach proposed in this paper simply replaces the global topology of the original BBO with the local ring topology, which can not only save much computational cost, but also effectively improve the search capability and avoid premature convergence. In ongoing work, we are testing the hybridization of our method with other heuristics. Another direction is the parallelization of RBBO, which is much easier to implement on a local topology than on a global one.

Acknowledgment. This work was supported by grants from National Natural Science Foundation (No. 61105073, 61103140, 61173096, 61020106009) of China.

References

1. Simon, D.: Biogeography-based optimization. IEEE Trans. Evol. Comput. 12(6), 702–713 (2008)
2. Du, D., Simon, D., Ergezer, M.: Biogeography-based optimization combined with evolutionary strategy and immigration refusal. In: Proc. IEEE Conf. Syst. Man Cybern., San Antonio, TX, pp. 997–1002 (2009)
3. Song, Y., Liu, M., Wang, Z.: Biogeography-based optimization for the traveling salesman problems. In: Proc. 3rd Int'l J. Conf. Comput. Sci. Optim., Huangshan, China, pp. 295–299 (2010)
4. Simon, D., Ergezer, M., Du, D., Rarick, R.: Markov models for biogeography-based optimization. IEEE Trans. Syst. Man Cybern. Part B 41(1), 299–306 (2011)
5. Gong, W., Cai, A., Ling, C.X.: DE/BBO: a hybrid differential evolution with biogeography-based optimization for global numerical optimization. Soft Comput. 15(4), 645–665 (2010)
6. Ma, H., Simon, D.: Blended biogeography-based optimization for constrained optimization. Eng. Appl. Artif. Intell. 24(3), 517–525 (2011)
7. Boussaïd, I., Chatterjee, A., Siarry, P., Ahmed-Nacer, M.: Two-stage update biogeography-based optimization using differential evolution algorithm (DBBO). Comput. Oper. Res. 38(8), 1188–1198 (2011)
8. Boussaïd, I., Chatterjee, A., Siarry, P., Ahmed-Nacer, M.: Biogeography-based optimization for constrained optimization problems. Comput. Oper. Res. 39(12), 3293–3304 (2011)
9. MacArthur, R., Wilson, E.: The theory of biogeography. Princeton University Press, Princeton (1967)
10. Li, X.: Niching without niching parameters: Particle swarm optimization using a ring topology. IEEE Trans. Evol. Comput. 14(1), 150–169 (2010)
11. Bratton, D., Kennedy, J.: Defining a standard for particle swarm optimization. In: Proceedings of IEEE Swarm Intelligence Symposium, pp. 120–127 (2007)
12. Yao, X., Liu, Y., Lin, G.: Evolutionary programming made faster. IEEE Trans. Evol. Comput. 3(2), 82–102 (1999)

Optimal Power Flow Solution
Using Self–Evolving Brain–Storming Inclusive
Teaching–Learning–Based Algorithm

K.R. Krishnanand[1], Syed Muhammad Farzan Hasani[1], Bijaya Ketan Panigrahi[2],
and Sanjib Kumar Panda[1]

[1] Electrical and Computer Engineering, National University of Singapore, Singapore
[2] Department of Electrical Engineering, Indian Institute of Technology, Delhi, India
krishkr09@gmail.com, {farzan,eleskp}@nus.edu.sg,
bkpanigrahi@ee.iitd.ac.in

Abstract. In this paper, a new hybrid self-evolving algorithm is presented with its application to a highly nonlinear problem in electrical engineering. The optimal power flow problem described here focuses on the minimization of the fuel costs of the thermal units while maintaining the voltage stability at each of the load buses. There are various restrictions on acceptable voltage levels, capacitance levels of shunt compensation devices and transformer taps making it highly complex and nonlinear. The hybrid algorithm discussed here is a combination of the learning principles from Brain Storming Optimization algorithm and Teaching-Learning-Based Optimization algorithm, along with a self-evolving principle applied to the control parameter. The strategies used in the proposed algorithm makes it self-adaptive in performing the search over the multi-dimensional problem domain. The results on an IEEE 30 Bus system indicate that the proposed algorithm is an excellent candidate in dealing with the optimal power flow problems.

Keywords: Brain-Storming Optimization, Non-dominated sorting, Optimal power flow, Teaching-learning-based optimization.

1 Introduction

Computational intelligence and its derivatives have become very handy tools in the field on engineering, especially for the studies on nonlinear systems. Among the intelligent techniques, evolutionary computation has been of high interest in the field of engineering optimization problems [1]. Various algorithms such as Genetic Algorithm [2], Particle Swarm Optimization [3], Differential Evolution [4], Artificial Bee Colony [5] etc. have been already used in power engineering. The economically optimal power scheduling and stable steady state operation of an electrical grid is one of the major focus areas among them. This problem, which deals with the power flow and voltage states of an electrically connected network during steady state, is often referred to as optimal power flow (OPF) problem. OPF can have multiple objectives for which optimality is to be searched, but has major emphasis on economic schedule

Y. Tan, Y. Shi, and H. Mo (Eds.): ICSI 2013, Part I, LNCS 7928, pp. 338–345, 2013.

of the power generation, along with minimizing the voltage deviations that could occur at the points at which loads to the system are connected. The intricate electrical interconnections, transformer taps for voltage stepping, steady state shunt compensation devices – all of them make the relations between the electrical variables highly complex and nonlinear. This demands requirement of smart methods for deciding the system variables. The conventional gradient-based techniques would end up with suboptimal solutions if applied to such an optimality problem with multiple constraints [1]. Relatively optimal solutions reported so far have been discovered using evolutionary computational techniques [4, 6].

Teaching–Learning–Based Optimization (TLBO) is a recent technique proposed to serve the purpose of nonlinear function optimization [7, 8]. The philosophical essence of the algorithm is based on the interaction of the learners in a class with the teacher of that class and among themselves. For the sake of simplicity, the learner with the highest advantage of knowledge is considered as the teacher for that instance of time and the classroom dynamics is expected to evolve the average level of the learners with respect to the teacher.

Brain Strom Optimization (BSO) is based on the controlled idea generation with the help of some flexible rules. It puts forth the philosophy that improvisation of ideas can be done through brainstorming sessions and piggybacking on existing ideas. Even though the concept of teacher is absent in brainstorming, both the optimization algorithms show mutual compatibility. The TLBO algorithm which has been proposed originally do not stress on controlled cross-fertilization of ideas within the learners. Hybridization of these two algorithms with excellent philosophical bases could give a better algorithm.

Considering the algorithmic sequence of TLBO, it is to be noted that control parameter in the TLBO algorithm is just a singular Teaching Factor. A self–evolving characteristic can be introduced to it to make the algorithm more guided and versatile in its evolution through the iterations. This methodology stresses the idea of self-improvisation in parallel to mutual-improvisations. In this paper the self-evolving hybrid algorithm is put into action by applying it on a multi-objective optimal load flow problem.

The following parts of paper are organized such – section 2 details the objective of optimal power flow and identifies two major objectives. The exact relation of electrical quantities and the constraints imposed are stated. Section 3 explains the algorithmic sequence followed by the proposed algorithm for its execution. Section 4 presents the application of the proposed algorithm on OPF problem along with the simulation results. The conclusions are given in section 5.

2   Objectives of Optimal Power Flow

The primary objective of the OPF problem is to satiate the load demand with minimum possible cost, simultaneously maintaining the voltage levels as seen by the loads around the expected value; all the while satisfying the physical limits of the components involved in generation and transmission. The objective of cost minimization can be expressed as below.

$$Minimize \quad F(P_G) = \sum_{i=1}^{N_G} F_i(P_{G_i}) = \sum_{i=1}^{N_G} \left(c_i.P_{G_i}^2 + b_i.P_{G_i} + a_i \right) \tag{1}$$

The coefficient shown as a accounts for the fixed cost and the coefficients b and c account for the variable cost incurred from power production. P_{G_i} is the power generated by the ith generator and N_G is the total number of generators. The active power supplied should follow the demand constraint given below.

$$\sum_{i=1}^{N_G} P_{G_i} = \sum_{j=1}^{N_D} P_{D_j} + P_{Loss} \tag{2}$$

Where P_{D_j} is the jth load and N_D is the total number of loads. Each generator is restricted physically by a maximum and minimum quantity of power that it is capable of generating. For the ith generator, it can be expressed as given below.

$$P_{G_i}^{Min} \le P_{G_i} \le P_{G_i}^{Max} \tag{3}$$

The total voltage deviation also acts an as an objective to be minimized. The loads require near-per-unit values of voltage for stable operation. The expression of objective meant for minimization is given below.

$$Minimize \quad F(V_L) = \sum_{k=1}^{N_L} |V_{L_k} - 1| \tag{4}$$

where V_{L_k} is the voltage of the kth load bus (expressed in per unit) and N_L is the number of load buses. The voltage states of the network are directly dependent on the generator voltage settings, the network impedances, shunt admittances, transformer tapings which hence are the decision variables. The transformer taps given by T_t and the reactive power compensators given by Q_c are generally discretized values and hence discontinuous in their respective domain. A total of N_T transformer taps and N_C capacitors are considered as decidable parameters. The numerical bounds are expressed as below.

$$V_{G_i}^{Min} \le V_{G_i} \le V_{G_i}^{Max}, \quad T_t^{Min} \le T_t \le T_t^{Max}, \quad Q_c^{Min} \le Q_c \le Q_c^{Max} \tag{5}$$

The voltage states of the network, both magnitude and angle at each node of the network, can be computed using load flow analysis. This information obtained can further be used to calculate the power losses in the network. Below the Root-Mean-Square based phasor matrix calculations which are performed in load flow using the admittance matrix (Y) of an N bus system and the injected power are shown.

$$\left[I^{RMS-Phasor} \right]_{N \times 1} = [Y]_{N \times N} \times \left[V^{RMS-Phasor} \right]_{N \times 1} \tag{6}$$

$$S_i = P_i + jQ_i = V_i^{RMS-Phasor} \times conjugate\left(I_i^{RMS-Phasor}\right) \qquad (7)$$

The powers thus computed are used in the equality constraints to match the injected power with demanded power. Power mismatches beyond tolerance would cause the load flow to fail in obtaining a convergent solution for the local search.

$$\sum_{i=1}^{N}\left|real(S_i^{injected} - S_i^{demanded})\right| = 0,$$

$$\sum_{i=1}^{N}\left|imaginary(S_i^{injected} - S_i^{demanded})\right| = 0 \qquad (8)$$

The objective is a highly nonlinear and highly restricted function of the following decision vector X.

$$X = \begin{bmatrix} P_{G_1} & \cdots & P_{G_{NG}} & V_{G_1} & \cdots & V_{G_{NG}} & T_1 & \cdots & T_{N_T} & Q_1 & \cdots & Q_{N_C} \end{bmatrix} \qquad (9)$$

This gives a total of $(2N_G + N_T + N_C)$ decision variables which belong to four categories. The dimensionality of the problem can be very large as the number of buses increase. This makes the problem all the more challenging to solve.

3   Self–Evolving Brain–Storming Inclusive Teaching–Learning–Based Optimization Algorithm

The Brain Storm Optimization and Teaching–Learning–Based Optimization algorithms have been recently introduced in the research literature [9] and [7] respectively. While TLBO is based on excellence of learners inspired by the teacher, BSO stresses on information interchange through brainstorming. The following guidelines are used with reference to Osborn's Original Rules for Idea Generation in a Brain storming Process [10]: 1) Suspend Judgment 2) Anything Goes 3) Cross–fertilize (Piggyback) 4) Go for Quantity. The rule 3 emphasizes on the key idea of brain storming which is then incorporated into the teacher–learner–based algorithm to cross–fertilize the ideas from teaching phase and learning phase, so as to originate a new set of ideas, which if turns out to be superior, could replace inferior ideas. The Teaching Factor (T_F) used in TLBO can be made self-evolving through its adaptation from the temporal change in consecutive function values. The computational steps are detailed below.

3.1   Initialization

Being a population-based stochastic algorithm, the initialization procedure is done over a matrix of M rows and N columns. M denotes the population size and N is the dimensionality of the problem at hand. The algorithm runs for a total of T iterations after which termination occurs. Considering an element x at m^{th} row and n^{th} column of the said matrix, initialization is done as shown below.

$$x^{(1)}_{(m,n)} = x^{min}_n + \left(x^{max}_n - x^{min}_n\right) \times rand_{(m,n)} \qquad (10)$$

where m and n are indices for row and column respectively. The random value used follows a uniform distribution of randomness within the range (0, 1). The valid existing ideas could be used as a base to start for new idea generation. Similarly, prior known and acknowledged solutions can be used as seeds in the initial matrix. A potential solution X from m^{th} row at time/iteration t is evaluated as shown below.

$$Y^{(t)}_m = f\left(X^{(t)}_m\right) = f\left(\left[x^{(t)}_{(m,1)} \quad x^{(t)}_{(m,2)} \quad \cdots \quad x^{(t)}_{(m,n)} \quad \cdots \quad x^{(t)}_{(m,N-1)} \quad x^{(t)}_{(m,N)}\right]\right) \quad (11)$$

3.2 Teaching Phase

In this phase, the average of the class is considered by taking the mean of each dimension of the population. The mean vector V can be shown as

$$\overset{mean}{V^{(t)}} = \left[\sum_{m=1}^{M}\left(\frac{x^{(t)}_{(m,1)}}{M}\right) \quad \cdots \quad \sum_{m=1}^{M}\left(\frac{x^{(t)}_{(m,n)}}{M}\right) \quad \cdots \quad \sum_{m=1}^{M}\left(\frac{x^{(t)}_{(m,N)}}{M}\right)\right] \qquad (12)$$

The teacher for the current iteration is the best solution of the current population. In a multi-objective problem, the best ranked individual can be obtained by a suitable non-dominated sorting method. The mutation in this phase is as shown below.

$$\overset{new-1}{X^{(t)}_{(m,n)}} = X^{(t)}_{(m,n)} + rand^{(t)}_{(m,n)} \times \left(\overset{teacher}{X^{(t)}_{(n)}} - T_{F_{(m)}}^{(t)} \times \overset{mean}{V^{(t)}_{(n)}}\right) \qquad (13)$$

where T_F shows the vector of teaching factors used for each learner. Since there is only this single type of control parameter present in original TLBO, it is possible to make it adaptive to the situation, making the algorithm self-evolving.

$$T_{F_{(m)}}^{(t)} = T_{F_{(m)}}^{(t-1)} + sign\left(\overset{rank}{Y^{(t)}_m} - \overset{rank}{Y^{(t-1)}_m}\right) \times rand^{(t)}_{(m)} \qquad (14)$$

This is intelligent learning strategy in the proposed algorithm. The improvisation in the rank of the vector with respect to past would lead to reduction in teaching factor, indicating that the learner is performing well and has lesser need to contribute towards driving the class average corresponding to the current teacher. Considering depreciation in the rank, the individual would need to allow higher mutations between teacher and the average vectors and hence higher T_F value is required. When there is no change in the rank, the T_F value for that learner is kept constant.

3.3   Learning Phase

This phase makes the learners undergo self–improvement through differential mutation. The temporal gradients which are present between the learners are used to facilitate the mutation process.

$$
\overset{new-2}{X_{(m,n)}^{(t)}} = X_{(m,n)}^{(t)} + rand_{(m,n)}^{(t)} \times \left(X_{(m,n)}^{(t)} - X_{(\lambda,n)}^{(t)} \right) \times sign\left(\overset{rank}{Y_{\lambda}^{(t)}} - \overset{rank}{Y_{m}^{(t)}} \right) \tag{15}
$$

3.4   Brain–Storming Phase

Here the amalgamation of both new populations occurs. A temporary matrix is created with intermixing from the populations from teaching and learning phases.

$$
\overset{new-3}{X_{(m,n)}^{(t)}} = \left[\alpha_{(n)} \quad \beta_{(n)} \right] \times \left[\overset{new-1}{X_{(m,n)}^{(t)}} \quad \overset{new-2}{X_{(m,n)}^{(t)}} \right]^{T} \tag{16}
$$

where α and β are the multiplication factors which decides the extent of participation of both populations in the brain storming process. The factors are kept within certain limits to assure contribution to brain storming from both the phases. Similar to BSO, the population thus obtained is mutated like in BSO using a smooth and stochastically weighted nonlinear function.

$$
\overset{cross-fertilized}{X_{(m,n)}^{(t)}} = \overset{new-3}{X_{(m,n)}^{(t)}} + \text{logsig}\left(\tfrac{T-2t}{2K} \right) \times rand_{(m,n)}^{(t)} \tag{17}
$$

where K is the slope determining factor of the *logsig* function used. The current population's objective values along with the objective values of new populations are compared together to perform the reinsertion of the superior vectors.

4   Simulation Results

The proposed algorithm is tested so as to obtain optimal power flow solution for an IEEE 30-Bus, 6 generator system. The system data regarding the bus configurations and the line data can be found in [4] and are not presented here due to space constraints. Both objectives are individually optimized first and then used as seeds during consequent initialization. The algorithm is run for 10 trial runs and the best solution is presented in Table 1. All simulations are done in MATLAB R2011b software on a 3.4 GHz Core-i7 device with 8 GB of random access memory. The system is well known and has been studied before even in recent research literature [4, 6]. The real powers and voltages of buses {1, 2, 5, 8, 11, 13} are to be decided.

The transformer taps {11, 12 15, 36} are to be adjusted and the extent of reactive compensations at {10, 12, 15, 17, 20, 21, 23, 24, 29} have to be determined. The fuel cost and voltage deviations are functionally dependent on these 25 parameters. The comparison of results from the proposed algorithm is performed in Table 2. It is very clear from the results that the proposed technique shows comparable performance to recent algorithms and is quite optimal considering the initial states shown in [4, 6].

Table 1. IEEE-30 Bus, 6 generator system – power generataions, voltage magnitudes, transformer taps settings and reactive power values used for compensation

	Minimize $F(P_G)$	Minimize $F(V_L)$		Minimize $F(P_G)$	Minimize $F(V_L)$
P_{G1}	176.6350	171.6105	T_3	0.9755	0.9796
P_{G2}	48.2382	30.1370	T_4	0.9607	0.9659
P_{G3}	21.2394	48.5742	Q_1	4.4348	4.8671
P_{G4}	21.4096	10.0000	Q_2	4.1871	0.3153
P_{G5}	12.0488	16.9136	Q_3	4.2786	4.9994
P_{G6}	12.4177	15.7686	Q_4	4.7454	1.6107
V_{G1}	1.1000	0.9999	Q_5	3.6614	4.9912
V_{G2}	1.0877	1.0211	Q_6	5.0000	4.6549
V_{G3}	1.0622	1.0173	Q_7	4.6468	5.0000
V_{G4}	1.0700	1.0016	Q_8	5.0000	4.9721
V_{G5}	1.0986	1.0341	Q_9	2.1255	2.0984
V_{G6}	1.1000	1.0108	$F(P_G)$	**799.1295**	863.0816
T_1	1.0139	1.0526	$F(V_L)$	1.9557	**0.0884**
T_2	0.9069	0.9034	$Loss$	8.5885	9.6627

Table 2. Comaprison of results based on [4, 6]

Method	Optimal fuel cost (in $/hr)	Optimal total voltage deviations
Gradient-based approach	804.853	-
Improved GA	800.805	-
PSO	800.41	-
DE	799.2891	0.1357
AGAPOP	799.8441	0.1207
Proposed algorithm	799.1295	0.0884

As seen from Table 2, the proposed hybrid algorithm is very effective in finding optimality of the objectives. The time difference in algorithms is negligible owing to high computational capabilities of modern day processors.

5   Conclusion

A hybrid algorithm is proposed based on two algorithms, TLBO and BSO, which have mutually compatible philosophical bases. The proposed algorithm is easily modified to be self-evolving since the original algorithm has only a single type of control parameter. The algorithm is successfully applied to highly complex optimal power flow problem with multiple objectives. The results indicate that the proposed algorithm is an excellent candidate for intelligent decision-making leading to economic-cum-stable operation of the power network. The flexibility and generic nature of the algorithm makes it suitable for optimizations on electrical networks spread over a large area or even those inside a building. The algorithm has higher memory requirement during its intermediate stages, but that disadvantage is trivial in the current world scenario of high end memory availability.

Acknowledgments. This research is funded by the Republic of Singapore's National Research Foundation through a grant to the Berkeley Education Alliance for Research in Singapore (BEARS) for the Singapore-Berkeley Building Efficiency and Sustainability in the Tropics (SinBerBEST) Program. BEARS has been established by the University of California, Berkeley as a center for intellectual excellence in research and education in Singapore.

References

1. Back, T., Fogel, D., Michalewicz, Z.: Handbook of evolutionary computation. Oxford University Press, New York (1997)
2. Bakirtzis, A., Petridis, V., Kazarlis, S.: Genetic algorithm solution to the economic dispatch problem. IEE Gen. Trans. Dist. 141(4), 377–382 (1994)
3. Gaing, Z.L.: Particle swarm optimization to solving the economic dispatch considering the generator constraints. IEEE Trans. on Power Syst. 18(3), 1187–1195 (2003)
4. Abou El Ela, A.A., Abido, M.A., Spea, S.R.: Optimal power flow using differential evolution algorithm. Electric Power Systems Research 80(7), 878–885 (2010)
5. Nayak, S.K., Krishnanand, K.R.: Panigrahi, B.K., Rout, P.K.: Application of Artificial Bee Colony to Economic Load Dispatch Problem with Ramp Rate Limits and Prohibited Operating Zones. In: IEEE Proc. on Nature and Biologically Inspired Computing, pp. 1237–1242 (2009)
6. Attia, A.F., Al-Turki, Y.A., Abusorrah, A.M.: Optimal Power Flow Using Adapted Genetic Algorithm with Adjusting Population Size. Electric Power Components and Systems 40(11), 1285–1299 (2012)
7. Rao, R.V., Savsani, V.J., Vakharia, D.P.: Teaching-learning-based optimization: A novel method for constrained mechanical design optimization problems. Computer-Aided Design 43(3), 303–315 (2011)
8. Krishnanand, K.R., Panigrahi, B.K., Rout, P.K., Mohapatra, A.: Application of Multi-Objective Teaching-Learning-Based Algorithm to an Economic Load Dispatch Problem with Incommensurable Objectives. In: Panigrahi, B.K., Suganthan, P.N., Das, S., Satapathy, S.C. (eds.) SEMCCO 2011, Part I. LNCS, vol. 7076, pp. 697–705. Springer, Heidelberg (2011)
9. Shi, Y.: An Optimization Algorithm Based on Brainstorming Process. Int. J. of Swarm Intel. Res. 2(4), 35–62 (2011)
10. Osborn, A.F.: Applied imagination: Principles and procedures of creative problem solving. Charles Scribner's Son, New York (1963)

Global Optimization Inspired
by Quantum Physics

Xiaofei Huang

eGain Communications, Mountain View, CA 94043, U.S.A.
HuangZeng@yahoo.com

Abstract. Scientists have found that atoms and molecules in nature have an amazing power at finding their global minimal energy states even when their energy landscapes are full of local minima. Recently, the author postulated an optimization algorithm for understanding this fundamental feature of nature. This paper presents a version of this algorithm for attacking continuous optimization problems. On large size benchmark functions, it significantly outperformed the standard particle swarm optimization algorithm.

1 Introduction

Scientists found that atoms and molecules are most often staying at their ground states, i.e., the lowest energy state among the conceivably possible states. Started from any initial state or disturbed by some external actions, the atom or the molecule always falls back to its ground state. Given a molecule consisted of multiple atoms, how could those atoms be capable of finding the global minimal energy state even when the energy function of the molecule is very rugged in the most cases? In other worlds, why a swarm of atoms in a molecule possesses some magic power at finding their social optimal solution? This mysterious property of nature is, as described by Niels Bohr in his 1922 Nobel lecture, peculiar and mechanically unexplainable. It seems to us that nature has implemented a powerful optimization mechanism at the atomic level.

Recently, the author postulated an optimization algorithm [1] to understand this important feature of nature. It is not simply an optimization algorithm inspired by nature. It could be the one deployed by nature at constructing atoms and molecules of the universe for their consistency and stability (which shall be put to the test in the future). Otherwise, the universe will fall into chaos if its building blocks (atoms and molecules) are not stable and consistent. It will be impossible to have life and everything around us.

This algorithm has several distinct features from other natural inspired algorithms. It is directly derived from the laws of quantum physics while others are inspired by other natural phenomena (such as ants, bees, birds, natural evolution, harmony, firefly, bacteria foraging, memetics, monkey, cuckoo, virus, water drops, spiral, galaxy, gravity, immune systems, and bats). Finding the global optimum is critically important in the quantum world, while searching

Y. Tan, Y. Shi, and H. Mo (Eds.): ICSI 2013, Part I, LNCS 7928, pp. 346–352, 2013.

for satisfactory solutions are mostly acceptable in many other cases in nature. It attempts to model the exact optimization process nature deploys at the microscopic level, while many others are loosely modeled after nature. It offers a new mathematical principle for optimization, fundamentally different from existing ones. It is also guaranteed to converge with an exponential rate for any objective function to an optimal solution with the optimality of its solution established in mathematics. However, many others have a hard time to establish the general convergence properties. Their convergence aspects are therefore often demonstrated empirically in the research literature.

Furthermore, none of existing ones deal with an objective function described by a Hamiltonian operator. It is more general than a real-valued, multivariate function. However, dealing with the former case is critical important in physics, chemistry, and biology both in theory for understanding nature and in practice for designing new materials to be environment friendly and new medicines to cure deadly diseases. This algorithm is capable of handling both cases. It is also the only nature-inspired algorithm reported so far capable of solving real-world, hard optimization problems in communications and computer vision. The examples are decoding error correction codes, like Turbo codes and LDPC codes for modern communications (LPDC in China HDTV, WiMax in 4G communications, European satellite communications), and stereo vision and image restoration in computer vision. They are major practical problems from electric engineering department with the number of variables ranging from hundreds of thousands to millions, far more challenge than many sample problems drawn from textbooks. The successful applications of an early version of the algorithm have been reported in [2,3,4].

2   From Quantum Physics to an Optimization Algorithm

To introduce the algorithm, we need to explain some key concepts in quantum mechanics. Scientists found out that the conceivably possible states of an atom or a molecule, called the stationary states in quantum physics, must satisfy the stationary Schrödinger equation:

$$\lambda \Psi(x,t) = H\Psi(x,t) , \tag{1}$$

where $\Psi(x,t)$ is the wavefunction describing the state of an atom or a molecule. It is a function of the space location x and the time t. H is the Hamiltonian operator corresponding to the total energy of the atom or the molecule, and λ is one of the eigenvalues of H which equals to the energy of the atom or the molecule.

The wavefunction $\Psi(x,t)$ is a probability-like function, where the square of the absolute value of $\Psi(x,t)$, $|\Psi(x,t)|^2$, is the probability of detecting the particle at the space location x at the time t.

If the molecule is isolated, does not interact with its environment, its state evolution follows the time-dependent Schrödinger equation:

$$i\hbar\dot{\Psi}(x,t) = H\Psi(x,t) , \tag{2}$$

where $\dot{\Psi}(x,t)$ is the time derivative of $\Psi(x,t)$, i is the imaginary unit ($i = \sqrt{-1}$), and $\hbar$ is the reduced Planck constant.

However, in reality, it is impossible to have an absolutely isolated molecule or atom due to the interactions with its environment and the perturbations caused by quantum fluctuation and background radiation. The molecule or the atom always losses its energy by emitting photons and jumps to a lower stationary state, and eventually to the lowest stationary state, called the ground state. The ground state has the energy level which is the smallest eigenvalue of the Hamiltonian operator H.

Nobody knows the exact process of the quantum jump. There was a famous debate on the subject between two great physicists: Niels Bohr and Erwin Schrödinger (1933 Nobel prize in physics). However, it remains as a mystery of science until recently a mathematical model has been postulated [1] to understand the quantum jump process. The model is described by the following dynamic equation for the evolution of the state of a quantum system:

$$ - \hbar\dot{\Psi}(x,t) = H\Psi(x,t) - \langle H\rangle\Psi(x,t) \ , \tag{3} $$

where $\langle H\rangle$ is the expected value of the Hamiltonian operator H ($\langle H\rangle = (\Psi, H\Psi)$, where (a,b) denotes the inner product of a and b), then the system always converges to an equilibrium satisfying the stationary Schrödinger equation (1). Therefore, it is a stationary state of the system.

Any system following the dynamics described by Eq. (3) always jumps (converges) to a lower energy stationary state started from an arbitrary initial state. Those stationary states are the equilibrium points of the equation. It eventually jumps to the lowest energy stationary state because others are unstable, sensitive to disturbance.

Specifically, the expected value of the Hamiltonian operator H, $\langle H\rangle$, always decreases when time elapses until an equilibrium is reached. That is, if the evolution of the system state $\Psi(x,t)$ is governed by Eq. (3), then the total energy of the system, represented by $\langle H\rangle$, always decreases until an equilibrium is reached. Therefore, Eq. (3) defines an optimization algorithm, called the quantum optimization algorithm here. Also, it defines a new mathematical principle for optimization by minimizing $\langle H\rangle$ instead of the original energy function H.

Experiments have found again and again that the space location of an atomic particle, such as an electron, a nucleus, or an atom, can distribute in space just a cloud does. A particle can appear in multiple locations simultaneously with the probability as $|\Psi(x,t)|^2$ at the space location x as mentioned before. This is a very counter-intuitive feature of nature because our daily life experience tells us that an object should have definite position at any given time. In general, a state variable can only have a single value at any given time instance in classical physics while it can have multiple values in quantum physics. Applying this generalization to classical optimization methods can lead to a paradigm shift.

Traditionally, in an optimization process, such as gradient descent and local search [5], every decision variable has a definite value at any given time, just like the space location of an object in classical physics. By analogy with quantum physics, we can extend that single-value assignment to a multi-value assignment. That is, at any given time, we can assign a decision variable with multiple values, each of them has a probability-like value associated with it, defining the preference of assigning the value to the variable. This can lead to a straightforward generalization of the classical gradient descent algorithm from hard-decision making to soft-decision making. In mathematics, it turns out that the result can be the quantum optimization algorithm defined in (3).

The quantum optimization algorithm (3) has a discrete-time version for combinatorial optimization. That is to minimize a multivariate, real-valued function $E(x_1, x_2, \ldots, x_n)$, where all variables are discrete. Specifically, let the state of variable x_i at time t be defined by a wavefunction $\Psi_i(x_i, t)$. The discrete-time version is defined as repeatedly updating the state of every variable x_i, one at a time, as follows:

$$\Psi_i(x_i, t+1) = \frac{1}{Z_i(t)} e^{-cE_i(x_i,t)} \Psi_i(x_i, t); \qquad (4)$$

where $Z_i(t)$ is a normalization factor such that $\sum_{x_i} \Psi_i^2(x_i, t+1) = 1$, c is a positive constant determining the descending rate, and $E_i(x_i, t)$ is the local energy for variable x_i. The definition of the local energy $E_i(x_i, t)$ is directly copied from quantum physics as follows:

$$E_i(x_i, t) = \sum_{x_1} \Psi_1^2 \cdots \sum_{x_{i-1}} \Psi_{i-1}^2 \sum_{x_{i+1}} \Psi_{i+1}^2 \cdots \sum_{x_n} \Psi_n^2 E(x_1, \ldots, x_n) . \qquad (5)$$

At any time t, the candidate solution for x_i, denoted as $x_i^*(t)$, is defined as the one that has the maximal value of $\Psi_i(x_i, t)$. That is

$$x_i^*(t) = \arg\max_{x_i} \Psi_i(x_i, t) .$$

It is important to note that the discrete-time version (4) of the quantum optimization algorithm (3) always converges to any one of the equilibrium states satisfying the stationary Schrödinger equation (1). In particular, it has the following form

$$\lambda \Psi_i(x_i, t) = E_i(x_i, t) \Psi_i(x_i, t), \quad \text{for } i = 1, 2, \ldots, n .$$

3   A Case Study of Continuous Optimization Problems

The optimization algorithm (4) presented in the previous section works for discrete variables of finite domains. We can adapt the algorithm for continuous variables with three techniques.

The first one is to sample the values for each variable x_i around a center value x_i^c for the variable. One way of sampling in the real domain $\mathcal{R}$ is to uniformly sample around x_i^c as follows:

$$x_i^c - m\Delta x, \ldots, x_i^c - \Delta x, x_i^c, x_i^c + \Delta x, \ldots, x_i^c + m\Delta x.$$

It has $2m + 1$ sampling points with gap Δx, a positive real value. This technique enables the algorithm to explore only a finite number of values for each variable at any movement to control its complexity.

The second technique is to move the sampling center x_i^c to a better location. It enables the algorithm to explore more promising areas in the search space.

The third one is to shrink the sampling gap Δx when there is no better location to move the sampling center x_i^c. It is done by reducing Δx as $\Delta x = \gamma \Delta x$, where the parameter γ is a positive constant less than 1. It enables the algorithm to search for a better solution at a finer scale.

At the very beginning, the algorithm picks up a random solution $x^{(0)}$ in the search space. The i-th component $x_i^{(0)}$ of the initial solution is used as the center value for x_i, i.e., $x_i^c = x_i^0$. At the same time, every wavefunction $\Psi_i(x_i)$ is set to the uniform distribution, i.e., $\Psi_i(x_i) = 1/(2m + 1)$ for all sampling points. Afterward, the optimization equation (4) is applied to update the wavefunction $\Psi_i(x_i)$ for every variable x_i iteratively. At each iteration, the solution for x_i, denoted as x_i^*, is the one which has the highest wavefunction value $\Psi_i(x_i)$. If it is different from the sampling center x_i^c, then the center is moved to x_i^* and $\Psi_i(x_i)$ is reset to the uniform distribution.

Given a maximum iteration number, the iterations are split into two halves. At the first half, there is no shrinking of the sampling gap Δx. At the second half, the gap is shrunk only if the solution of the current iteration is not improved. Putting everything together, we have the following pseudo code.

Generate an initial solution $x^{(0)}$ and set the sampling center $x^c = x^{(0)}$;
Initialize all wavefunctions $\Psi_i(x_i)$ as the uniform distribution;
for $k := 1$ to *max_iteration* **do**
    for each i **do**
        for each sampling point of x_i, update $\Psi_i(x_i)$ as
            $\Psi_i(x_i) = e^{-cE_i(x_i)}\Psi_i(x_i)$; // $E_i(x_i)$ is the quantum local energy (5)
           Normalize $\Psi_i(x_i)$ so that $\sum_i \Psi_i^2(x_i) = 1$;
           Compute the solution for x_i as $x_i^* = \arg\max_{x_i} \Psi_i(x_i)$;
        if x^c (sampling center) $\neq x^*$ (the current solution) **do**
           Move the sampling center to x^*, $x^c = x^*$;
        if the solution x^* is not better than the last one **and**
        $k >$ *max_iteration*/2 **do**
           Shrinking the sampling gap as $\Delta x = \gamma \Delta x$;
        if the sampling gap is shrunk **or** the sampling center is moved **do**
           Reset all wavefunctions $\Psi_i(x_i)$ as the uniform distribution;
Output $(x_1^*, x_2^*, \ldots, x_n^*)$ as the final solution.

Fig. 1. Quantum optimization algorithm

4   Experimental Results

The performance of the quantum optimization for continuous optimization problems is compared with the standard particle swarm optimization algorithm version 2006 (see http://www.swarmintelligence.org for the source code). Particle swarm optimization (PSO) is a population based optimization technique developed by Dr. Eberhart and Dr. Kennedy in 1995 [6]. People have found many ways to twist and modify the standard PSO to improve its performance. However, the standard PSO is chosen here because of its popularity.

In our experiments, the number of evaluations is set to 30, 000. Readers can find other parameters such as the swarm size, topology of the information links of the swarm, the first cognitive/confidence coefficient, and the second cognitive/confidence coefficient in the original source code.

For the quantum optimization algorithm given in Fig. 1, the descending rate $c = 0.7$, the maximum number of iterations is 300, the sampling gap $\Delta x = 1.0$, the number of sampling points for each x_i is 19, and the shrinking rate γ is 0.7.

The four evaluation functions are listed below. The search space is $x \in [-30, 30]^d$.

Eggcrate	$\sum_{i=1}^{d} x_i^2 + 25\sin^2(x_i)$
Rosenbrock	$\sum_{i=1}^{d-1} 100(x_{i+1} - x_i^2)^2 + (x_i - 1)^2$
De Jong	$\sum_{i=1}^{d}(i + 1)x_i^4$
Rastrigin	$\sum_{i=1}^{d} \left(x_i^2 - 10\cos(2\pi x_i) + 10\right)$

The experiments are carried out by comparing the accuracies of the algorithms at finding the optimal value (in our cases, $E^*(x) = 0$). Each algorithm runs 100 times for each test case. The best function values found by each algorithm are averaged. The standard deviation and the average execution time are also calculated. They are done on a Windows Vista 32-bits machine with an AMD Turion 64x2 2.00 GHz CPU and a 3-GB memory. The results are given in the following table.

function(dimension)	PSO	Quantum Optimization
Eggcrate(d=100)	817 ± 153 (0.32s)	0.000 ± 0.000(0.04s)
(d=200)	2,899 ± 358 (0.62s)	0.000 ± 0.000(0.08s)
(d=300)	7,268 ± 565 (0.92s)	0.000 ± 0.000(0.12s)
Rosenbrock(d=100)	2,640 ± 3405 (0.22s)	3.237 ± 8.469 (0.26s)
(d=200)	1,446,311 ± 652,265 (0.42s)	2.529 ± 6.035(0.52s)
(d=300)	$12.27 \times 10^6 \pm 3.8 \times 10^6$ (0.61s)	2.621 ± 8.527(0.87s)
De Jong(d=100)	143 ± 152 (0.21s)	0.000 ± 0.000(0.01s)
(d=200)	1,154,377 ± 534,594 (0.32s)	0.000 ± 0.000(0.03s)
(d=300)	$14.37 \times 10^6 \pm 4.2 \times 10^6$ (0.47s)	0.000 ± 0.000(0.03s)
Rastrigin(d=100)	685 ± 128 (0.31s)	0.000 ± 0.000(0.04s)
(d=200)	3,038 ± 325 (0.64s)	0.000 ± 0.000(0.08s)
(d=300)	6,476 ± 530 (0.93s)	0.000 ± 0.000(0.1s)

From the table we can see that, the quantum optimization algorithm significantly outperformed the standard PSO in terms of the quality of solutions. Both of them have comparable speeds. As the problem size increases from 100, 200, to

300, the quantum optimization algorithm scales much better than the standard PSO in terms of solution quality. For Eggcrate function, De Jong function, and Rastrigin function, the quantum optimization has a near-perfect performance. For Rosenbrock function and De Jong function, the performance of PSO deteriorates very fast with the increase of the variable number. For Rosenbrock function, the quantum optimization algorithm deteriorates just a little bit, insensitive to the problem size.

5   Conclusion

Based on the recent theoretical work on quantum physics at understanding quantum jump, this paper presents a quantum optimization algorithm for attacking hard optimization problems. The underlying optimization principle of this algorithm is the soft-decision based local optimization. It can be viewed as a generalization of the classical gradient descent algorithm from hard decision making to soft decision making at assigning decision variables. It always converges to equilibria described by a key equation in quantum mechanics, called the stationary Schrödinger equation. One version of this algorithm is also offered for solving continuous optimization problems.

References

1. Huang, X.: Dual schrödinger equation as a global optimization algorithm. In: Advances in Quantum Theory, IEEE Computer Society Press, Vaxjo (2010)
2. Huang, X.: Cooperative optimization for energy minimization in computer vision: A case study of stereo matching. In: Rasmussen, C.E., Bülthoff, H.H., Schölkopf, B., Giese, M.A. (eds.) DAGM 2004. LNCS, vol. 3175, pp. 302–309. Springer, Heidelberg (2004)
3. Huang, X.: Cooperative optimization for solving large scale combinatorial problems. In: Theory and Algorithms for Cooperative Systems. Series on Computers and Operations Research, pp. 117–156. World Scientific (2004)
4. Huang, X.: Near perfect decoding of LDPC codes. In: Proceedings of IEEE International Symposium on Information Theory (ISIT), pp. 302–306 (2005)
5. Pardalos, P., Resende, M.: Handbook of Applied Optimization. Oxford University Press, Inc (2002)
6. Eberhart, R.C., Kennedy, J.: A new optimizer using particle swarm theory. In: Proceedings of the Sixth International Symposium on Micro Machine and Human Science, pp. 39–43. IEEE Service Center, Nagoya (1995)

Structural Test Data Generation
Based on Harmony Search

Chengying Mao[1,2]

[1] School of Software and Communication Engineering,
Jiangxi University of Finance and Economics, 330013 Nanchang, China
[2] The State Key Laboratory of Software Engineering,
Wuhan University, 430072 Wuhan, China
maochy@yeah.net

Abstract. Software testing has been validated as an effective way to improve software quality. Among all research topics in software testing, automated test data generation has been viewed as one of the most challenging problems. In recent years, a typical solution is to adopt some meta-heuristic search techniques to automatically tackle this task. In the paper, our main work is to adapt an emerging meta-heuristic search algorithm, i.e. harmony search (HS), to generate test data satisfying branch coverage. Fitness function, also known as the optimization objective, is constructed via the branch distance. In order to verify the effectiveness of our method, eight well-known programs are utilized for experimental evaluation. According to the experimental results, we found that test data produced by HS could achieve higher coverage and shorter search time than two other classic search algorithms (i.e. SA and GA).

Keywords: Test data generation, harmony search, fitness function, branch coverage, experimental evaluation.

1   Introduction

In the past decades, testing is always treated as an effective way to improve software reliability and trustworthiness both in academia and industry. However, how to generate test data which can reveal potential faults in high probability is not an easy task. Apart from the deterministic methods such as symbol execution and constraint solving, a typical solution is to adopt some meta-heuristic algorithms to produce test inputs in software testing activity, that is so-called *search-based software testing* (SBST) [1].

In order to expose the faults in program code with high probability, test data should cover program elements as many as possible. Thus, in software testing, some specific construct elements, such as statements, branches or paths, are usually selected as targets to be covered (a.k.a. coverage criteria). Then, the quality of test data is usually judged by the proportion of covered elements, that is so-called coverage. Therefore, the main objective of search-based testing is to produce test data with much higher coverage.

At present, some meta-heuristic search (MHS) techniques, such as simulated annealing (SA) [2], ant colony optimization (ACO) [3] and genetic algorithm (GA) [4], have been adopted for generating test data. According to the empirical study by Harman and

Y. Tan, Y. Shi, and H. Mo (Eds.): ICSI 2013, Part I, LNCS 7928, pp. 353–360, 2013.

McMinn [5], the above MHS algorithms are suitable for test data generation problem. In recent years, harmony search (HS) [6,7] is a new emerging searching algorithm in the field of evolution computing. In the paper, we attempt to use this algorithm to generate structural test data.

It should be noted that, in the work of Rahman *et al.* [8], HS algorithm is adopted for constructing variable-strength *t*-way strategy, and experimental results demonstrate that HS-based method gives competitive results against most existing AI-based and pure computational counterparts. However, their work belongs to *combinatorial testing*, but not the program code-oriented testing. To the best of our knowledge, this paper is the first attempt to use HS algorithm to solve test data generation problem in software structural testing.

In practice, some well-known programs are utilized for experimental evaluation, and the effectiveness of HS-based method is validated. At the same time, the comparison analysis between HS and other search algorithms is also performed. Experimental results show that HS outperforms other algorithms both in covering effect and searching performance.

2   Background

2.1   Test Data Generation Problem

From a practical point of view, test data generation plays crucial role for the success of software testing activities. In general, the coverage of program constructs is often chosen as an index to determine whether the testing activity is adequate or not. Accordingly, test data generation can be converted into a covering problem for program constructs. In order to realize full coverage of target elements with a test data set as smaller size as possible, it is necessary to introduce some effective search algorithms to solve the difficult problem.

As shown in Figure 1, the basic process of search-based test data generation can be described as below. At the initial stage, the search algorithm generates a basic test suite by the random strategy. Then test engineers or tools seed test inputs from the test suite into the program under test (PUT) and run it. During this process, the instrumented code monitors the PUT and can produce the information about execution results and traces. Based on such trace information, the metric for a specific coverage criterion can be calculated automatically. Subsequently, the metrics are used to update the fitness value of the pre-set coverage criterion. Finally, test data generation algorithm adjusts the searching direction of the next-step iteration according to the fitness information. The whole search process could not be terminated until the generated test suite satisfies the pre-set coverage criterion.

There are three key issues in the above framework: trace collection, fitness function construction and search algorithm selection. In general, execution traces are collected by code instrumentation technology, which can be settled with compiling analysis. While constructing fitness function, testers should determine which kind of coverage will be adopted. Experience tells us that branch coverage is a better cost-effective criterion. Considering easy implementation and high convergence speed, here we adopt HS to produce test data.

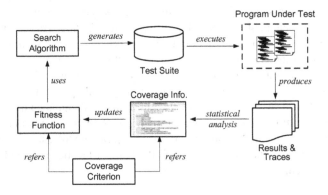

Fig. 1. The basic framework for search-based test data generation

2.2   Brief Review on HS Algorithm

As a new meta-heuristic optimization algorithm, harmony search originally came from the analogy between music improvisation and optimization process. It was inspired by the observation that the aim of music is to search for a perfect state of harmony. The effort to find the harmony in music is analogous to find the optimality in an optimization process [7].

Algorithm . Harmony Search

---

1. define objective function $f(*)$;
2. define harmony memory accepting rate (r_{accept});
3. define pitch adjusting rate (r_{pa}) and other parameters;
4. generate Harmony Memory (HM) with random harmonies;
5. **while** $t <$ max number of iterations **do**
6.   **while** $i <=$ number of variables **do**
7.   **if** $(rand < r_{accept})$ **then**
8.   choose a value from HM for the variable i;
9.   **if** $(rand < r_{pa})$ **then**
10.   adjust the value by adding small random amount;
11.   **end if**
12.   **else**
13.   choose a random value;
14.   **end if**
15.   **end while**
16.   accept the new harmony (solution) if better;
17. **end while**
18. **return** the current best solution;

The pseudo code of harmony search algorithm is listed as follows. It is not hard to find that, the best solution can be produced via the following three operations: (1)

Harmony selection. This operation ensures that the good harmonies in the last iteration can be considered as elements of new solution. (2) *Pitch adjusting.* This action produces a new pitch by adding small random amount to the existing pitch. (3) *Randomization.* This operation is to increase the diversity of the solutions. In general, pitch adjusting can only realize local search, but the use of randomization can drive the system further to explore various diverse solutions so as to attain the global optimality.

In essence, HS is a swarm-based heuristic algorithm, so it can keep the balance between the diversification and convergence through the strategies of swarm intelligence, such as parallel evolution and elitism preserving policy. Meanwhile, simplicity and easy implementation are also the strong points of HS. Thus, it has been applied to various optimization problems.

3   HS-Based Test Data Generation

Given a program under test P, suppose it contains n input variables $x = (x_1, x_2, \cdots, x_n)$. For each input variable x_i, its input domain can be denoted as $R_i (1 \leq i \leq n)$, then the input domain of whole program can be expressed as $R = R_1 \times R_2 \times \cdots \times R_n$. Regarding the testing problem, in fact, it is to select a subset from input domain R according to a specified coverage criterion C, so as to realize full coverage of the corresponding construct elements. In general, the execution status of a program element (i.e., statement, branch or path) is determined by program's input data. Therefore, the satisfaction degree of criterion C can be modeled as a function $f(X)$ of input variables, where X is the set of input vectors, i.e. $X = \{x\}$. Consequently, test data generation is transformed as an optimization problem.

Suppose the test data set satisfying coverage criterion C is $S_{tc} = \{tc_1, tc_2, \cdots, tc_m\}$, here m is the number of test cases. For each test case $tc_i = \{x_1, x_2, \cdots, x_n\}$, $f(tc_i)$ is the fitness of the test case. Meanwhile, the fitness of whole test data set is denoted as $f(S_{tc})$. Then, test data set optimization problem is to find the maximum fitness of $f(S_{tc})$.

Here, a harmony represents a test case, and harmony memory stores the subset of test cases which have much stronger covering ability than the rest in population. For a program with n input variables, S_{hm} test cases with the stronger covering ability can be organized as a harmony memory (HM) as below.

$$HM = \begin{bmatrix} x_1^1, & x_2^1, & \cdots, & x_n^1 \\ x_1^2, & x_2^2, & \cdots, & x_n^2 \\ \vdots & \vdots & \ddots & \vdots \\ x_1^{S_{hm}}, & x_2^{S_{hm}}, & \cdots, & x_n^{S_{hm}} \end{bmatrix} \tag{1}$$

where $(x_1^k, x_2^k, \cdots, x_n^k)(1 \leq k \leq S_{hm})$ represents an elite test case in whole population. In general, S_{hm} varies in the range from 1 to 40.

In order to generate a test suite S_{tc}, we adopt the parallel style to realize population update. A new solution (i.e., test case) $tc_j^{(t+1)}$ (here $j \in [1, m]$) in generation $t + 1$ can be produced by formula (2).

$$tc_j^{(t+1)} = \begin{cases} tc_l^{(t)} = (x_1^l, x_2^l, \cdots, x_n^l), & rnd_j \leq r_{accept} \\ \text{randomly select data from input domain,} & \text{otherwise} \end{cases} \quad (2)$$

where l is a random integer within $[1, S_{hm}]$, rnd_j is a random number from 0 to 1.

If a solution is selected from HM, it will still be conducted local adjustment with the probability r_{pa}. Here, we adopt the linear adjustment shown in formula (3).

$$x_{new} = x_{old} + b_{range} \times \varepsilon \quad (3)$$

where x_{old} is the existing pitch stored in HM, and x_{new} is the new pitch after the pitch adjusting action. b_{range} is the pitch bandwidth, and ε is a random number from uniform distribution with the range of [-1, 1].

For a new generated test case tc_i ($1 \leq i \leq m$), if its fitness is better than the worst harmony (h_{worst}) in HM, i.e. $f(tc_i) > f(h_{worst})$, the worst harmony will be replaced with tc_i. Finally, the stopping rules include: (i) test data set achieve full coverage w.r.t. criterion C, or (ii) the iteration times reach to $maxGen$.

According to the previous testing experiences, branch coverage is the most cost-effective approach in structural testing [9]. In our study, we mainly consider the widely-used branch coverage as searching objective. The fitness function can be constructed in accordance with branch distance. The metrics of branch distance in our experiments refer to the definitions in [10] and [2].

Suppose a program has s branches, the fitness function of the whole program can be defined as below.

$$fitness = 1 \Big/ \left[\theta + \sum_{i=1}^{s} w_i \cdot f(bch_i) \right]^2 \quad (4)$$

where $f(bch_i)$ is the branch distance function for the ith branch in program, θ is a constant with little value and is set to 0.01 in our experiments. w_i is the corresponding weight of this branch. Obviously, $\sum_{i=1}^{s} w_i = 1$. Generally speaking, each branch is assigned to different weight according to its reachable difficulty. The rational way is that, the harder reaching branch (e.g. the equilateral case in triangle classification problem) should be assigned with higher value.

In order to provide reasonable branch weight to construct fitness function, we perform static analysis on program to yield structure information about branch element. Generally speaking, the reachability of a branch body mainly relies on the following two issues: the nesting level of branch and the predicate type in branch. As a consequence, we determine the final branch weight according to the above two factors.

4 Experimental Analysis

4.1 Experimental Setup

In order to validate the effectiveness of our proposed method for generating test data, we use eight programs to perform the comparison analysis, which are the well-known

Table 1. The benchmark programs used for experimental analysis

Program	#Branch	LOC	Description
triangleType	5	31	type classification for a triangle
calDay	11	72	calculate the day of the week
isValidDate	16	59	check whether a date is valid or not
remainder	18	49	calculate the remainder of an integer division
computeTax	11	61	compute the federal personal income tax
bessj	21	245	Bessel J_n function
printCalendar	33	187	print calendar according the input of year and month
line	36	92	check if two rectangles overlap

benchmark programs and have been widely adopted in other researchers' work [11–13]. The details of these programs are listed in Table 1.

Since SA and GA are the two most popular algorithms for settling test data generation problem, we mainly compare our HS-based method with them in this section. The parameter settings of these three algorithms are shown in Table 2. It should be noted that, the parameters of each algorithm are the most appropriate settings for most programs under test in the experiments. The experiment is employed in the environment of MS Windows 7 with 32-bits and runs on Pentium 4 with 2.4 GHz and 2 GB memory, and all three algorithms are implemented in standard C++.

Table 2. The parameter settings of three algorithms

Algorithm	Parameter	Value
SA	initial temperature T_0	1.00
	cooling coefficient c_α	0.95
GA	selection strategy	gambling roulette
	crossover probability p_c	0.90
	mutation probability p_m	0.05
HS	size of harmony memory S_{hm}	2∼20
	harmony choosing rate r_{accept}	0.75
	pitch adjustment rate r_{pa}	0.5

4.2 Comparative Analysis

Here, we define the following two issues as evaluation metrics for experimental analysis. Subsequently, the effectiveness and efficiency of these three algorithms can be compared on them.

(1) *Average coverage* (AC), i.e. the average of achieved branch coverage of all test inputs in 1000 runs.

(2) *Average time* (AT), i.e. the average execution time (millisecond) of realizing all-branch coverage.

The experimental results of three algorithms are shown in Table 3. For the metric AC, the results of HS are always better than those of SA and GA for all eight programs. Apart from the last program `line`, HS can attain 100% branch coverage. Even for program `line`, the AC value of HS algorithm is very close to 100%. On average, HS outperforms SA about 3.85 percentage points w.r.t. metric AC, and is higher than GA about 4.00 percentage points. On the other hand, we find that SA and GA have no significant difference for the metric AC. GA's results are higher than those of SA for program `isValidDate`, `printCalendar` and `line`. For the rest five programs, SA's effect is better than GA w.r.t. metric AC.

Table 3. Comparison analysis on metric AC and AT

Program	Average Coverage (%)			Average Time (ms)		
	SA	GA	HS	SA	GA	HS
triangleType	99.88	95.00	100	3.77	10.83	0.34
calDay	99.97	96.31	100	1.79	35.73	0.44
isValidDate	98.21	99.95	100	2.43	11.68	1.04
remainder	99.85	94.07	100	1.01	6.09	0.52
computeTax	94.44	91.51	100	1.14	18.28	0.53
bessj	99.45	98.61	100	6.10	8.89	2.65
printCalendar	94.31	95.06	100	35.38	35.48	5.11
line	82.86	97.43	99.76	11.00	47.65	5.99

For the metric AT, the mean value of HS-based approach is 2.08 millisecond, and the metrics of SA and GA are 7.83 ms and 21.83 ms respectively. Thus, we can deduce the order of these three algorithms as follows: HS>SA>GA, where symbol '>' means the "faster" relation. On the whole, AT metric of HS is about one-tenth of that of GA, and one-forth of SA's AT. This phenomenon means that the time in each iteration of HS is much shorter than those of SA and GA during the evolution process.

Based on the above analysis, we can conclude that HS is more suitable for test data generation problem than SA and GA, both in test data quality and convergence speed.

5   Conclusions

Test data set usually plays crucial role in the success of software testing activity. Thus, quite a few search techniques are introduced to find the test inputs which can reveal the potential faults in program code with the maximum probability. Harmony search, as a new emerging technique, has shown its strong searching ability for some typical optimization problems. In the paper, we adapt it to generate test data set to cover all branches in program. Meanwhile, eight well-known programs are utilized for experimental evaluation. The results tell us that HS outperforms other two algorithms (i.e., SA and GA) both in test data quality and convergence speed for testing problem. Therefore, we can conclude that HS is suitable to handle test data generation problem in software structural testing.

Acknowledgments. This work was supported in part by the National Natural Science Foundation of China (NSFC) under Grant No. 60803046 and 61063013, the Open Foundation of State Key Laboratory of Software Engineering under Grant No. SKLSE2010-08-23, and the Program for Outstanding Young Academic Talent in Jiangxi University of Finance and Economics. The author is very grateful to Xinxin Yu for providing assistance to the experiments.

References

1. McMinn, P.: Search-Based Software Testing: Past, Present and Future. In: Proc. of the 4th International Workshop on Search-Based Software Testing (SBST 2011), in conjunction with the 4th IEEE International Conference on Software Testing (ICST 2011), pp. 153–163. IEEE Press, New York (2011)
2. Tracey, N., Clark, J., Mander, K., McDermid, J.: An Automated Framework for Structural Test-Data Generation. In: Proc. of the 13th International Conference on Automated Software Engineering (ASE 1998), pp. 285–288. IEEE Press, New York (1998)
3. Ayari, K., Bouktif, S., Antoniol, G.: Automatic Mutation Test Input Data Generation via Ant Colony. In: Proc. of the 9th Annual Conference on Genetic and Evolutionary Computation (GECCO 2007), pp. 1074–1081. ACM Press (2007)
4. Pargas, R.P., Harrold, M.J., Peck, R.: Automated Structural Testing Using Genetic Algorithms. Software Testing, Verification and Reliability 9(4), 263–282 (1999)
5. Harman, M., McMinn, P.: A Theoretical and Empirical Study of Search-Based Testing: Local, Global, and Hybrid Search. IEEE Transactions on Software Engineering 36(2), 226–247 (2010)
6. Geem, Z.M., Kim, J., Loganathan, G.: A New Heuristic Optimization Algorithm: Harmony Search. Simulation 76(2), 60–68 (2001)
7. Geem, Z.M.: Music-Inspired Harmony Search Algorithm: Theory and Applications. Springer, Berlin (2009)
8. Rahman, A., Alsewari, A., Zamli, K.Z.: Design and Implementation of a Harmony-Search-Based Variable-Strength t-Way Testing Strategy with Constraints Support. Information and Software Technology 54(6), 553–568 (2012)
9. Bertolino, A., Mirandola, R., Peciola, E.: A Case Study in Branch Testing Automation. Journal of Systems and Software 38(1), 47–59 (1997)
10. Korel, B.: Automated Software Test Data Generation. IEEE Transactions on Software Engineering 16(8), 870–879 (1990)
11. Bouchachia, A.: An Immune Genetic Algorithm for Software Test Data Generation. In: Proc. of the 7th International Conference on Hybrid Intelligent Systems (HIS 2007), pp. 84–89. IEEE Press, New York (2007)
12. Alba, E., Chicano, F.: Observations in Using Parallel and Sequential Evolutionary Algorithms for Automatic Software Testing. Computers and Operations Research 35, 3161–3183 (2008)
13. Ferrer, J., Chicano, F., Alba, E.: Evolutionary Algorithms for the Multi-Objective Test Data Generation Problem. Software: Practice and Experience 42(11), 1331–1362 (2012)

A Study on an Evaluation Model for Robust Nurse Rostering Based on Heuristics

Ziran Zheng[1] and Xiaoju Gong[2]

[1] School of Management Science and Engineering,
Shandong Normal University, Jinan 250014, P.R. China
[2] Provincial Hospital Affiliated to Shandong University, Jinan 250014, P.R. China

Abstract. Staff scheduling problem has been researched for decades and dozens of approaches have been proposed. Since in the hospital ward, an optimal solution could be changed for the uncertain causes, such as sick leave or other unforeseen events. If these occur, the roster that has been settled as an optimal solution often needs to make changes such as shift moves and others, some of which could have impact on the rosters fitness value. We first investigate the sensitive of an optimal solution under several operations of those types and the result shows that the solutions which are optimal obtained with the searching technique could indeed be affected by those disturbance. Secondly, the evaluation method is used to construct new evaluation function to improve the robustness of a roster. The model could apply to any method such as population-based evolutionary approaches and metaheuristics. Experiments show that it could help generate more robust solutions.

Keywords: Heuristics, nurse rostering, robustness, staff scheduling, metaheuristics.

1 Introduction

Nurse rostering is a class of staff scheduling problem in real life hospital ward, which is usually encountered by the manager and the head of the nurses. This kind of work is often highly constrained and could have large search spaces as a combinatorial optimization problem to find the optimal solution. The major objective is to generate a roster which contains a daily shift to each day during a period, which should be subject to the soft and hard constraints. In the practice, it is common that the head nurses still construct the roster by hand and thus this is of time consuming. Automated nurse rostering present a great challenge both for researchers in universities and personnel managers in hospitals.

The rostering process usually considers issues like coverage demand, workload of nurses, consecutive assignments of shifts, day-off/on requirements, weekend-related requirements, preference or avoidance of certain shift patterns, etc. During the last a few decades, a wide range of effective approaches have been studied and reported. While earlier literatures concentrating on the formulations of the problem and methods with the traditional mathematical programming approaches, the

Y. Tan, Y. Shi, and H. Mo (Eds.): ICSI 2013, Part I, LNCS 7928, pp. 361–368, 2013.

followed researches mainly make use of the heuristics and the population-based search techniques such as evolutionary computation. He et al. [7] developed a hybrid constraint programming based column generation approach. They present a complete to formulate the complex constraints and consists an effective relaxation reasoning linear programming method, and two strategies of column generation searching method are proposed. In [9], Lv et al. reported an adaptive neighborhood searching technique to solve the different scale of period of rostering problem. This method adaptively controls its parameter to solve benchmarks from the NSS dataset. They defined a basic move of neighborhood search and generated moderate competitive results.

Burke et al. [2] proposed a tabu-search based hyperheuristic for both the staff scheduling and timetabling problem. With the developing the hyperheuristic that could automated design the high level heuristics, this approach can produce solutions of at least acceptable quality across a variety of problems and instances. In the literature [1], another hyperheuristic is presented to solve two kinds of scheduling problems usually happen in hospital: nurse rostering and patient addressing scheduling. As for the population-based method, Kawanaca et al. [8] applied the genetic operations to exchange the shifts of the day to search the solution space. Ruibin et al.[11] proposed an hybrid evolutionary approach to the problem. It uses the stochastic ranking method to find the optimal solution and outperforms previously some know results. In [5] [4] more detailed survey are presented.

In recently years, more and more researchers concentrate on the practice aspects of the problem to make the automated method more general for the real use. In [3], Burke et al. introduced the time interval requirements which enables the shift can be split and combined. In [10], Maenhout et al. analyzed the long term nurse scheduling problems and investigate the new shift scheduling approach. In [6], Celia et al. studied structure of the problem to reduce the search space.

In this paper, we study the robustness of a solution under some practical uncertainty events, through analyzing a benchmark problem from the first nurse roster competition. With a heuristic method, we discover that the solution generated based on general searching approach show different performance of the robustness. Some experiments results show that with some tiny changes, the solution could be sensitive and the fitness value could be reduced. Adding the evaluation function to the heuristics, we can conveniently construct more robust solution. This evaluation function can be used in most of the current heuristics and other kinds of techniques.

The remaining part of the paper is organized as follows. Section 2, we propose an evaluation model and study the robustness of the optimal solution that has been found under some changes. In Section 3 we apply this method to the general heuristics and presented several possible results in our approach. The conclusion and future directions will be described in the Section 4.

2   Robustness of a Roster: A Case Study

In this study we use the datasets which were presented in the First International Nurse Rostering Competition (INRC-2010). There are various kinds of datasets which have different time periods, constraints and other parameters. Here we use the Sprint track dataset, where there is two hard constraint and many soft constraints. More detailed description can be found in [9].

The model that concerned is presented here. Given a set of N nurses with each belongs to a working contract and $|N| = 10$; denote D the set of days in a time period with $|D| = 28$; let S be the shift types that contains Night, Early, Late night and Day with the first letter being their abbreviation; we use the C to denote the constraint set and let C_1 be the soft constraints stated above for the hard constraints is always met with certain methods during the initial solution generating process. Let X be an solution and the cost function is:

$$V_{origin}(X) = \sum_{i \in N} \sum_{j \in C_1} w_{i,j} f_{i,j} \tag{1}$$

where w is the weight the f is the soft constraint's fitness function.

We use the direct solution representation for the problem. A candidate solution is represented by an $|N| \times |D|$ matrix X where $x_{i,j}$ corresponds to the shift type assigned for nurse is at day d . If there is no shift assigned to nurse s at day d , then x takes $Null$.

In the practice, some uncertain events could occur. It is sure that when a optimal solution subject to these types of changes, the structure will be changed as well as its evaluation accordingly. To observe the changing degree of this kind of transform, we study these effects under some situations often occurred in the real life scenario. It is apparent that the more the roster is changed the more its value will be influenced. Therefore, we take two relatively simple situations as the examples. Although more situations would appear, most of them are based on some elementary moves in a roster. The two basic and common adjustments are described as follows.

1. Exchanging shift in one day between two nurses. The situation happens when a nurse encounters an unforeseen event and thus would not be available on one shift time of a working day, she need to exchange the shift to another nurse who has work on the same day but has different shift type. In this situation, we assume that the latter nurse is always available to exchange the shift. In a roster, this situation is just applying a swap operation in a column between two nurses.

2. Changing on one day and Changing back on another day between two nurses. This situation is often happened in the condition like this: one nurse could not work on a working day and if other working nurses cannot exchange with her, thus the work shift must be replaced by another nurse who is on off day. In this condition, the two people often exchange back in one of the following days to compensate the latter one. And this leads to a two swapping operation in one column and another column followed.

We use the following algorithm to find the optimal solution. Although this algorithm is currently not the best algorithm among the latest literature, here we just use a simple and fast approach to generate a solution. For the evaluation model could easily be used by any other iterative heuristics, the searching algorithm is not a importance aspects.

The algorithm we use is a iterative-based randomly searching heuristics. At the first, an initial solution which meets the hard constraint is randomly generated. Then the main structure is firstly several days in the roster period is chosen randomly. The number of days should could be any number between 0 and $|D|$. It is usually set by experience and change with the time period. In this case, the Sprint track dataset have four weeks time, we set the number to 7; After the days are chosen, we randomly choose two nurses and swap their shift type which is call a type of metaheuristics. Since any operations in a column could not affect the hard constraint, the operations only search the available spaces. While holding the best solution and the value, every loop the current and the best should be compared and updated. The pseudo code of the algorithm process is as follows:

Algorithm 1. Simple heuristics

Input: problem instance I
Output: roster X_{best}
1 $X = initial\ solution$;
2 $X_{best} = X$;
3 $V_{best} = V_{origin}(X)$;
4 **for** $i = 1; i \leq n_c; i + +$ **do**
5 　　Randomly select n_d days of X to operate the metaheuristics;
6 　　**for** $j = 1; j \leq n_d; j + +$ **do**
7 　　　　Randomly swap two nurses' shifts;
8 　　**if** $V_{origin}(X) < V_{best}$ **then**
9 　　　　$V_{best} = V_{origin}(X)$;
10 　　　　$X_{best} = X$;
11 　　**else**
12 　　　　$X = X_{best}$;

13 return X_{best};

After the optimal solution has been found, we can test the solutions robustness under our assumed uncertain situations. We define the measurements of the optimal solutions robustness in the following way: if a nurse will exchange a shift with another nurse, the accordingly move for the roster is a swap between two shifts in a day. However, for a settled roster, which day will occur the change is uncertain and could not be predicted. Therefore, the probability of every point in a roster can be considered to be equal. In addition, there could be more than one change during an roster in its period. Because of these considerations, we firstly compute the difference of value between exchanged roster and the best one. Then, sum up all these differences and calculate the average penalty of one exchange. At last, multiply the average with the number of exchanges that would happen during a period.

We use $X_{i,j,d}$ to represent roster with exchange happened between nurse i and j in the day d, and $V_{origin}(X_{i,j,d})$ is the fitness value of the current roster. N_c is defined as the number of exchanges, and T represents the number of exchanges. Because of uncertainty, we made by $T = 10$ experience. The penalty value of an roster is defined as:

$$V_{penalty}(X) = (\sum_{d \in D} \sum_{i \in N} \sum_{i \neq j} (V_{origin}(X_{i,j,d}) - V_{best})) \times \frac{1}{N_c} \times T \qquad (2)$$

The overall fitness value with robustness is as follows:

$$V_{total}(X) = V_{origin}(X) + V_{penalty}(X) \qquad (3)$$

The following experiments show the fitness value of difference type with the instance Sprint01 and Sprint02 under the same test parameters. We firstly searched for 20 best solutions for each instance. Then apply our function to compute the penalty value. It is obvious that the results indicate that the original value stays smooth while the new defined robustness shocks, and the total value varies consequently.

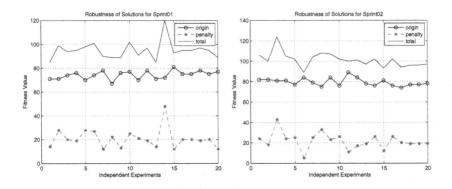

Fig. 1. Uncertainty Impact to the Fitness Value

3   Evaluation Model Experiments

The new defined fitness value can be used with any iterative searching techniques. We apply this to our simple heuristics and made the test of dataset Sprint01 04. Algorithm 2 shows the improved procedure. In this process, we set a flag to record whether the V_{origin} reached a threshold. This means we first take the traditional method to make the roster original value fairly low, then plus the new defined robustness weight. Since the time of compute the $V_{penalty}(X)$ is very slow, if we add it from the very beginning of the iteration, the whole process would stay very inefficient. At the time the threshold is reached, the new evaluation cost function is introduced to choose the roster.

Algorithm 2. heuristics with robustness

    Input: problem instance I
    Output: roster X_{best}
1 $X = initial\ solution$;
2 $X_{best} = X$;
3 $V_{best} = V_{origin}(X)$;
4 $flag = false$;
5 **for** $i = 1; i \leq n_c; i++$ **do**
6   Randomly select n_d days of X to operate the metaheuristics;
7   **for** $j = 1; j \leq n_d; j++$ **do**
8   Randomly swap two nurses' shifts;
9   **if** $V_{best} < V_t$ **then**
10         $flag = true$
11   **if** $flag$ **then**
12         $t1 = V_{penalty}(X)$;
13         $t2 = V_{penalty}(X_{best})$;
14   **else**
15         $t1 = t2 = 0$;
16   **if** $V_{origin}(X) + t1 < V_{best} + t2$ **then**
17         $V_{best} = V_{origin}(X)$;
18         $X_{best} = X$;
19   **else**
20         $X = X_{best}$;
21 **return** X_{best};

Fig. 2 shows the test results which indicate that the new method could reduce the robustness value obviously as well as the total. Note that after applied the penalty value, the new searched rosters' original value could get higher, since it is the tradeoff of the robustness during the searching. However, the total fitness value could also stay lower than the previous method without considered the uncertainty.

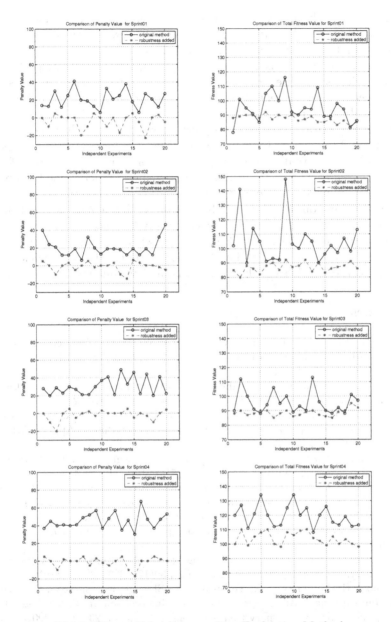

Fig. 2. Fitness Value between Two Evaluation Methods

4   Conclusions and Future Directions

In this research, we study the robustness of rosters under the constraints in the circumstances of uncertainty exchange of shifts frequently appeared in the real life in hospitals. The evaluation model proposed to measure this phenomenon is simple and straightforward, with which the optimal solution could perform different in robustness. After discovered this fact, the new evaluation method can be integrated into any current searching techniques to improve generates more robust nurse roster as well as other staff scheduling and timetabling problem.

There are left some directions to the further research. First one, the situations that occurred in the real life is much more complex. Not all the situations are considered in this paper and some assumption for simplicity also exists. For instance, in some situations that have unforeseen events, the hard constraints may violate and the new roster constraints are hard to formulate. In other situations such as we referred in section 2, when one or more nurse are transferred on leave, the new backup roster is hard to search and in practice the head nurse often completely make a whole new one to use. How to define the robustness of this type should be considered. Secondly, with our evaluation method added, the searching process would take more time since every loop need to compute every change penalty values which take $O(n^2 m)$ steps, where n is the scale of number of nurses and m is the number of day. Thus how to design the more efficient and effective robustness evaluation algorithm is also an open and important problem.

References

1. Bilgin, B., Demeester, P., Misir, M., Vancroonenburg, W., Vanden Berghe, G.: One hyperheuristic approach to two timetabling problems in health care. Journal of Heuristics 18(3), 401–434 (2011)
2. Burke, E.K., Kendall, G., Soubeiga, E.: A tabu-search hyperheuristic for timetabling and rostering. Journal of Heuristics 9(6), 451–470 (2003)
3. Burke, E.K., Causmaecker, P.D., Petrovic, S., Berghe, G.V.: Metaheuristics for handling time interval coverage constraints in nurse scheduling. Applied Artificial Intelligence 20(9), 743–766 (2006)
4. Burke, E., De Causmaecker, P., Berghe, G., Van Landeghem, H.: The state of the art of nurse rostering. Journal of Scheduling 7(6), 441–499 (2004)
5. Cheang, B., Li, H., Lim, A., Rodrigues, B.: Nurse rostering problemsa bibliographic survey. European Journal of Operational Research 151(3), 447–460 (2003)
6. Glass, C.A., Knight, R.A.: The nurse rostering problem: A critical appraisal of the problem structure. European Journal of Operational Research 202(2), 379–389 (2010)
7. He, F., Qu, R.: A constraint programming based column generation approach to nurse rostering problems. Comput. Oper. Res. 39(12), 3331–3343 (2012)
8. Kawanaka, H., Yamamoto, K., Yoshikawa, T., Shinogi, T., Tsuruoka, S.: Genetic algorithm with the constraints for nurse scheduling problem. In: Proceedings of the 2001 Congress on Evolutionary Computation, vol. 2, pp. 1123–1130 (2001)
9. Hao, L.Z., Adaptive, J.K.: neighborhood search for nurse rostering. European Journal of Operational Research 218(3), 865–876 (2012)
10. Maenhout, B., Vanhoucke, M.: An integrated nurse staffing and scheduling analysis for longer-term nursing staff allocation problems. Omega 41(2), 485–499 (2013)
11. Ruibin, B., Burke, E.K., Kendall, G., Jingpeng, L., McCollum, B.: A hybrid evolutionary approach to the nurse rostering problem. IEEE Transactions on Evolutionary Computation 14(4), 580–590 (2010)

A Study of Human Flesh Search
Based on SIR Flooding on Scale-Free Networks

Dawei Meng, Lei Zhang, and Long Cheng

Graduate School at Shenzhen, Tsinghua University, Shenzhen, P.R. China
{mengdawei116,chengchenglonglong}@gmail.com,
zhanglei@sz.tsinghua.edu.cn

Abstract. With the development of social networks, information is shared, amended, and integrated among users. Meanwhile, some questions begin to generate public interests. How does the information propagate in the network? How do human factors affect the spreading patterns of information? How do we construct models to understand the collective group behavior based on probabilistic individual choices? We try to answer the above questions by investigating "Human Flesh Search" (HFS), a phenomenon of digging out full privacy information of a target person with the help of massive collaboration of netizens by integrating information pieces during propagation. SIR model, which is often used to study epidemic diseases, is employed to provide a mathematical explanation of the process of HFS. Experimental results reveal that information entropy has significant influence on the network topology, which in turn affects the probability of affecting network neighbors and finally results in different efficiency of information spreading.

Keywords: Human Flesh Search, information spreading, SIR model, Scale Free Networks.

1   Introduction

Recently in China, there is a very hot phenomenon called "Human Flesh Search" (HFS), which is called "RenRouSouSuo" in Chinese. It is an online activity in which people collaborate to find out the truth of a certain event, a scandal or the full privacy information of a target person who has committed some bad thing.

The most typical processes of a HFS event could be illustrated by the "Guo Meimei" event [1] in Jun. 2011. The event was ignited by a young girl flaunting wealth through her micro-blog. It aroused great public concern and more and more netizens began to participate in the event. In the process of the incident, more information was continuously disclosed with the participation and collaboration of netizens. By putting pieces of information contributed by netizens together, the full jigsaw was accomplished to find out the truth: all her wealth came from a leader from Red Cross China via corruption and even sex bribe. The event showed the great power of crowds of netizens online and caused the donation amount raised by charity organizations to drop significantly by 86% in the following month [8].

Y. Tan, Y. Shi, and H. Mo (Eds.): ICSI 2013, Part I, LNCS 7928, pp. 369–376, 2013.
© Springer-Verlag Berlin Heidelberg 2013

Given the significant efficiency of HFS, hiding personal information on the Web becomes extremely difficult if you unfortunately become the target of a HFS campaign. Therefore, there have been numerous discussions of the legal and privacy issues of the problem. The Times describes it as Chinese vigilantes hunting victims by interviewing some HFS targets who received hate mails and threats after their personal information is exposed on the Web [9]. Xinhua News Agency even views it as a virtual lynching and warns people to behave themselves appropriately to avoid possible HFS punishment [10].CNN goes one step further and considers people participating HFS as lynching mob [11].

In order to acquire better understanding of the structure and evolution procedure of HFS, two mathematical models have been proposed to research on the process. The first mathematical model [6] includes a network modeling the social communities involved in the HFS campaign and a probabilistic flooding algorithm simulating the HFS search process. The second one [7], based on epidemic models, model HFS as a sequence of SIR processes caused by "HFS virus" gene mutation.

There are several flaws in the above two models: First, the network topology of HFS is too ideal in these models. Second, the influence on the process of the person who masters key information of HFS events has not been given sufficient consideration. Therefore, it is necessary to build a new mathematical model to research the HFS events.

After the analysis of the above HFS event, it is obvious that HFS has the following properties:

   a. Adhoc organization
   b. Massive collaboration
   c. Information-people interplay
   d. Online-offline interaction

The aim of HFS is to dig out the hidden truth. The activity is launched by a certain people, but then, a group of people would flock together to set up their organization voluntarily based on their interest. Just because people cooperate with others, the activity could end up with success. Aiming at the truth, people supply new information for the event, which will influence the behavior of people in return. The activity takes place on the web, but people live in the real world, so it is an interaction between online and offline.

HFS is a very interesting phenomenon. It is based on human collaboration to complete jobs, just like computer cluster system. We could regard people as computer nodes in the system, and knowledge owned by people as the data input/output. But the difference is that at the start, it is clear for computer systems to explicitly configure node numbers, network infrastructure and job allocation. While in HFS, all of the above configurations are adhoc.

By considering both the similarities and differences with the two proposed model, our new mathematical model should focus on following aspects:

- First, a scale-free network is to be constructed and the process of HFS should be researched in this network structure.
- Second, several sets of parameters will be determined to control the flow of information from one node to his neighbors in the scale-free networks.

- Third, every time the reveal of key information will involve more people in the process of HFS events. Accordingly the process of HFS event is to be divided into different phases. How to divide the different phases automatically in the mathematical model should be concerned seriously.

The rest of this paper is organized as follows. Section 2 presents an overview of some important related work. Section 3 illustrates the methods. Section 4 shows our experiments, results and analysis of our outcomes. Finally, Section 5 gives out conclusions of our study.

2   Related Work

Due to the interdisciplinary characteristic of HFS, it is attracting research interest from different academic fields. Most of them fall into the following two scopes. One group of scholars study HFS from social field, they aimed at studying the influence, mechanism of HFS. The other one used model and technological methods to study the network structure, information propagation and information interaction with people.

The first group of scholars study on HFS from different social science perspectives. Yaling Li [2] put point on netizens' behavior and tried to use psychology to explain the behavior. HFS is viewed as a "Human-Machine Search" which relies on the interaction between offline and online activities. Psychological harm to HFS target victims is also discussed with some suggestions on how to guide HFS along the right way. Songtao Bu [3] aimed at studying victim's privacy protection in HFS. They regard HFS as a kind of search which starts at punishing one person, but ends up with harming more. Zhuochao Yang [12] studied HFS from jurisprudence point of view and focused on legal issues which are the roots of controversy in HFS. Mengyao Yang [13] studied the development progress of HFS with fruitful case studies. Characteristics of HFS were analyzed by a comparison with traditional media. Yulei Dai et al. [14] discussed psychological factor of HFS. They believe that netizens are driven by their interest and curiosity. It is also pointed out the interesting "Herd Effect" discovered in HFS, where most netizens follow others blindly as sheep following the herd.

Feiyue Wang et al. [4] used many cases to illustrate the significance of studying HFS. They also come outraised some questions to researchers for, asking them to give more reasonable interpretations about the essence of HFS. Qingpeng Zhang et al. [16] focused on the research of the entire HFS communities. By studied on lots of HFS events, they believe that netizens, belonging to different sub-groups that come from different platform such as tianya, mop and son on, cooperate to push forward the process of the same HFS event. A minority of netizens, together with traditional mass media, play a role of "information bridge" in promoting the flow of information between different platform and achieving synchronization of information among all the sub-groups.

Bing Wang et al. [5][15] used mature generating model to simulate a HFS network growth. They used related algorithm to product a grouping network which is very similar with the network observed in real cases. An information aggregation model named "GOSSIP with Feedback" was also proposed. In their model, a new node (a netizen) would supply knowledge to his upper neighbors and the knowledge will be recursively forwarded along the upstream until it reaches the root node who started the HFS.

Probability flooding model [6] divides the HFS campaign into the follow five phases: Ignition, Infection, Fading, Re-ignition, Success / Failure. By modeling the online social community as a network graph, the initiator of a HFS campaign is viewed as the source node and the target becomes a hidden destination node in the graph. Lacking information about the target, the only feasible way of finding it out is to flood the message to all neighbors in the graph. With neighbors forwarding to their neighbors, the solution of HFSE becomes a flooding routing algorithm in the graph.

Long Cheng et al. [7] focused on the interaction between information and people. They believe that information has influences on people's behavior and vice versa. They tried to use variant SIR model to find out relation expression. They did experiments and concluded that the model could help to explain HFS. Nevertheless, they tried to research HFS, but a lot of problems still need to be solved.

All these two groups have not combined network structure, people flow and interaction between people and information. That is what we have done in our paper.

3   Proposed Model

Firstly, we describe the main mind of our mathematical model based on SIR Flooding on Scale-Free network as below: at the start, with information reveal, certain people would share the information in forums based on his interest. As the elapse of time, some other people form a discussion network and information becomes more abundant. Then, new information would attract many new people to attend in the discussion and the discussion network is updated. Meanwhile, some people would quit out. The network would add new certain nodes and lose some old nodes. The change is just like SIR model process. With update of network, more information is developed. But people could just share information with their direct neighbors.

So, we use our new model to describe update of interaction network forming during information propagation and development. It is obvious that certain people existing in the network hold key information about the HFS event. At start, one person would share information with his neighbors with a parameter β which means a probability and could be described by the parameter of SIR model. P_i hold a parameter called $Energy(p_i)$ which means that how much attention the people to share information with his neighbors. So, people pi has his $Energy(p_i)$ as below;

$$Energy(p_i) = Max\{Energy(p_j) * p \mid \langle j,i \rangle\} \tag{1}$$

$<j,i>$ means that people p_j share his information to his neighbor p_i and j maybe equals with i in value; p_i means people who is numbered with "i".

If value of $Energy(p_i)$ is below certain value $Threshold(T)$, the people p_i would stop to share information to his neighbors. There is also parameter called q which means that!! If one people holding key information participate in HFS process, the $Energy(p_i)$ would be multiplied by q as below:

$$Energy(p_j) = \{q * Energy(p_j) \mid p_j hold\_key\_Info.\} \tag{2}$$

In order to investigate people flow, we divide the people in activity into three groups: S which means the number of people who may join in the discussion, I which means the number of people who have already joined in discussion, R which means the number of people who have quit from the discussion. With the information flowing in the network, the numbers of the different groups will change continuously. The update of numbers could be described as below:

$$\begin{cases} \dfrac{dS}{dt} = -\beta IS \\[2mm] \dfrac{dI}{dt} = \beta IS - \alpha IR \\[2mm] \dfrac{dR}{dt} = \alpha IR \end{cases} \tag{3}$$

α:probability one people quit out discussion
β:probability one people join in discussion

4 Experiment and Analysis

We crawl experiment data from websites, including www.tianya.cn, www.mop.com, www.tieba.baidu.com, etc. We take "Guo Meimei" event to study network structure and people flow.

4.1 Network Structure

Firstly, we generate a group of pictures about network in event to display update process of network as Fig. 1.

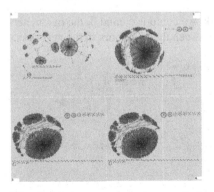

Fig. 1. Update process of network in "Guo Meimei" event

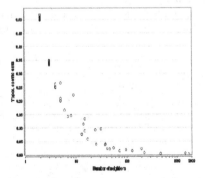

Fig. 2. Topological coefficient distribution of network in "Guo Meimei" event

As shown in Fig. 2, The vast majority of nodes just have a few neighbors while the minority of nodes are connected with a large number of nodes. Fig. 3 shows that the degree distribution of the network of "Guo Meimei" complies with power law. That is why we study the HFS based on scale-free networks. Fig. 4 indicates that the network also shares property of small world on its topology.

Also, it appears local community characteristics as showed in Fig. 1 and this proves our flooding model is reasonable in explaining the network structure to be scale-free.

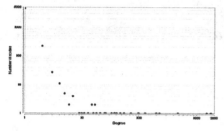
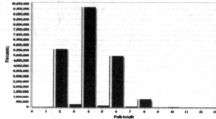

Fig. 3. Node degree distribution of network in "Guo Meimei" event

Fig. 4. Shortest path length distribution of network in "Guo Meimei" event

4.2   People Flow

Based on the data collected from the Web, we use our model to map the behavior of netizens. Fig. 5 shows the comparison results of our model and real data. It indicates that the SIR Flooding on scale-free network model is very useful to study people flow in the activity. With this model, we could find people flow pattern in different activities caused by various information. It could simulate the process of HFS events very well in presentation. Interaction between information and people group would explain more in essence.

Based on SIR Flooding on scale-free network, we could use α and β, the parameters of SIR model, to control the information flow between the neighbors in the networks.

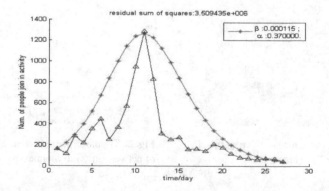

Fig. 5. Fitting trend on number of people joined in "Guo Meimei" event

4.3   Information and Group Interaction

It is clearly obvious that different events result in different parameters, because this phenomenon roots in different information what is implicit in events. In the final analysis, the essence of information decides to last results, and we call the essence as information entropy. Meanwhile, some cases listed in Table 1 just contain one kind of information entropy, but others may contain many information entropies. In the feature, we aim at finding out all information entropy being independent of each other. So, we could find out certain develop pattern of network and people group in HFS for each information entropy.

Table 1. Experiment results of SIR Flooding model on some very famous HFS cases

Event Name	β	α
Guo Meimei	0.000150	0.370000
Chen Guanxi	0.000220	0.007000
Diaoyu Islands Gurd	0.000044	0.003495
Liu Xiang Flopping	0.000015	0.249000
Yao Jiaxin	0.000015	0.000949

Thus, the SIR Flooding on scale-free network model could explain different aspects of HFS in presentation while information is the key factor to understand the process of HFS events better.

5   Conclusion

In this paper, we propose a new combined mathematical model for HFS based on SIR flooding on scale-free network. The contribution of our work is as follows: Through the analyses of our experiments, we could conclude that: First, our model is fit to explain topology of network formed during HFS in presentation and could help to describe people flow pattern in community during process of HFS. Second, the conclusion is drawn that information is the essence thing which decides topology of network and people flow pattern to result in many various appearances with HFS activities.

Although we propose a new mathematical model to simulate the progress of HFS, more work needs to be done to research this massive collaborative phenomenon. It is necessary to get larger datasets to train our model to research the HFS. We could also improve our model to study how to control the different phases of this phenomenon by using the SIR parameters. In the end, how to develop our model and apply it to predict the success/failure of HFS events should be considered in the future work.

References

1. Wikipedia, http://zh.wikipedia.org/wiki/%E9%83%AD%E7%BE%8E%E7%BE%8E%E4%BA%8B%E4%BB%B6
2. Yaling, L.: From Violent "Human Flesh Search Engine" to Friendly "Human computer Search Engine". Press Circles (2008)

3. Songtao, B.: A Study on The Internet Search for Human Flesh Search Engine And The Invasion of Privacy. Journal of Shenyang Normal University (Social Science Edition) (2008)
4. Feiyue, W., Daniel, Z., James, A.H., Qingpeng, Z., Zhuo, F., Yanqing, G., Hui, W., Guanpi, L.: A study of the human flesh search engine: crowd-powered expansion of online knowledge. IEEE Computer Society (2010)
5. Bing, W., Bonan, H., Yiping, Y., Laibin, Y.: Human Flesh Search Model: Incorporating Network Expansion and GOSSIP with Feedback. In: 13th IEEE/ACM International Symposium on Distributed Simulation and Real Time Applications (2009)
6. Zhang, L., Liu, Y., Li, J.: A mathematical model for human flesh search engine. In: Wang, H., Zou, L., Huang, G., He, J., Pang, C., Zhang, H.L., Zhao, D., Yi, Z. (eds.) APWeb 2012 Workshops. LNCS, vol. 7234, pp. 187–198. Springer, Heidelberg (2012)
7. Long, C., Lei, Z., Jinchuan, W.: A Study of Flesh Search with Epidemic Models. In: Proceedings of the 3rd Annual ACM Web Science Conference, WebSci 2012, pp. 67–73 (2012)
8. People's Daily Online, Red Cross Received Fewer Individual Donations This Year, http://finance.people.com.cn/GB/70846/16534289.html
9. Hannah, F.: Human flesh search engines: Chinese vigilantes that hunt victims on the web, The Times (June 25, 2008), http://technology.timesonline.co.uk/tol/news/tech_and_web/article4213681.ece
10. Xu, B., Shaoting, J.: Human Flesh Search Engine": an Internet Lynching, Xinhua English (July 4, 2008), http://english.sina.com/china/1/2008/0704/170016.html
11. CNN, From Flash Mob to Lynch Mob (July 5, 2007)
12. Zhuochao, Y.: The legitimate limit of human flesh search engine. Public Administration & Law (2009)
13. Mengyao, Y.: An exploration of Human Flesh Search Engine in network community. Southwest Communication (2008)
14. Xia, X.: Sociological Reflection about "human-powered search" Phenomenon. China Youth Study (2009)
15. Bing, W., Yiping, Y., Bonan, H., Dongsheng, L., Dan, C.: Knowledge Aggregation in Human Flesh Search. IEEE Computer Society (2010)
16. Qing-peng, Z., Feiyue, W., Daniel, Z., Tao, W.: Understanding Crowd-Powered Search Groups: A Social Network Perspective. Public library of science one (June 27, 2012)

A Modified Artificial Bee Colony Algorithm for Post-enrolment Course Timetabling

Asaju La'aro Bolaji[1,2], Ahamad Tajudin Khader,
Mohammed Azmi Al-Betar[1,3], and Mohammed A. Awadallah

[1] School of Computer Sciences, Universiti Sains Malaysia, Penang Malaysia
[2] Department of Computer Science, University of Ilorin, Ilorin, Nigeria
[3] Department of Computer Science, Jadara University, Irbid, Jordan
{tajudin,mohbetar}@cs.usm.my,
{abl10_sa0739,mama10_com018}@student.usm.my

Abstract. The post-enrolment course timetabling is concern with assigning a set of courses to a set of rooms and timeslots according to the set of constraints. The problem has been tackled using metaheuristic techniques. Artificial Bee Colony (ABC) algorithm has been successfully used for tackling uncapaciated examination and curriculum based course timetabling problems. In this paper ABC is modified for post-enrolment course timetabling problem. The modification is embedded in the onlooker bee where the multiswap algorithm is used to replace its process. The dataset established by Socha including 5 small, 5 medium and one large dataset are used in the evaluation of the proposed technique. Interestingly, the results obtained is highly competitive when compared with those previously published techniques.

Keywords: Artificial Bee Colony, University Course Timetabling, Nature-inspired Computing.

1 Introduction

University Course Timetabling Problems (UCTP) is one of the hardest problems faced by academic institutions throughout the world. The UCTP is known to be difficult combinatorial optimization problems that have been widely studied over the last few decades. The problem involves scheduling a given number of courses to a limited number of periods and rooms subject to satisfying a set of given constraints. The constraints in UCTP are usually classified into hard and soft constraints. The satisfaction of hard constraints is compulsory for the timetabling solutions to be *feasible*, while the satisfaction of the soft constraints is required but not necessary. The quality of the UCTP solution is normally determined by the cost of the violating the soft constraints.

The last four decades have been witnessing introduction of numerous techniques of tackling UCTP by workers in the domain of operational research and Artificial Intelligence [1]. Among these techniques are: early heuristic approaches derived from graph colouring heurstics [2], constraint-based techniques [3], case-based reasoning [4], metaheuristics techniques like local search-based techniques

Y. Tan, Y. Shi, and H. Mo (Eds.): ICSI 2013, Part I, LNCS 7928, pp. 377–386, 2013.

such as tabu search [5], simulated annealing [6], great deluge [7], variable neighbourhood structures (VNS) [8] and population-based techniques such as ant colony optimization [9], genetic algorithm (GA) [10], harmony search algorithm (HSA) [11], particle swarm optimization [12]. In addition, hyper-heuristic approaches and hybrid metaheuristic approaches have also been recently used to tackle course timetabling problems [13,10]. More comprehensive survey of other methodologies used for university course timetabling literatures have been published in [14,15].

Artificial Bee Colony (ABC), a recently proposed algorithm, is a nature-inspired based approach that had been successfully used to tackle several NP-hard problems [16]. In our previous work, ABC was used for both curriculum-based course timetabling and uncapacitated examination timetabling problems with a success stories [17,18,19]. To validate the success of ABC in other scheduling, in this paper, post-enrolment course timetabling problem (PE-CTP) is being tackle with a modified artificial bee colony algorithm. The modification to ABC involves the incorporation of multiswap algorithm in onlooker bee phase of the original ABC. The purpose of incorporating multiswap algorithm is to enhance the power of ABC by searching the PE-CTP solution space rigorously in order to obtain good solution.

2 Post-Enrolment Course Timetabling Problem (PE-CTP)

The PE-CTP involves scheduling lectures of courses to a set of timeslots and a set of rooms subject to satisfying a set of constraints (i.e. hard and soft). The PE-CTP tackled in this paper was originally established by Socha et al in [9] and its descriptions is given as follows:

- A set $C = \{c_1, c_2, ..., c_N\}$ of N courses, each of which consists of specific number of students and requires particular features.
- A set $R = \{r_1, r_2, ..., r_K\}$ of K rooms, each of which has a seat capacity and contains specific features.
- A set $S = \{s_1, s_2, ..., s_L\}$ of L students, each of them assigned to one or more courses.
- A set $F = \{f_1, f_2, ..., f_M\}$ of M features.
- A set $T = \{t1, t_2, ..., t_P\}$ of P timeslots; where $P = 45$ (5 working days, each with 9 timeslots).

The following are the set of the hard constraints that needs to be satisfied:

- H1. Students must not be double booked for courses.
- H2. Room size and features must be suitable for the assigned courses.
- H3. Rooms must not be double booked for courses.

While the set of soft constraints which need to be minimized are :

- S1. A student shall not have a class in the last slot of the day.

- S2. Student shall not have more than two classes in a row.
- S3. Student shall not have a single class in a day.

The objective function to evaluate timetabling solution is given :

$$f(x) = \sum_{s \in \mathbb{S}} (f_1(x, s) + f_2(x, s) + f_3(x, s)) \tag{1}$$

The value of $f(x)$ is referred to as the penalty cost of a feasible timetable. The main objective of the PE-CTP is to produce a feasible solution that minimizes number of soft constraints violations.

3 Artificial Bee Colony Algorithm

One of the recently proposed nature-inspired algorithm is the artificial bee colony (ABC). It was originally developed by Karaboga in 2005 for tackling numerical optimization [20]. It is considered a stochastic optimization technique that is based on the model proposed in [21] for the foraging manners of honey bee in their colonies. The colony of ABC consists of three important operators: employed foraging bees, unemployed foraging bees, and food sources. The first two operators i.e., employed and unemployed forager searches for the third operators (i.e. rich food sources). The model gave insights on the two principal types of behaviour which are necessary for self-organization and collective intelligence. In practice, such behaviour includes the recruitment of foragers to the rich food sources resulting in positive feedback and abandonment of poor food sources by foragers causing negative feedback.

In the colony of ABC there are three groups of bees: employed, onlooker and scout bees. Associated with particular food source is employed bee whose behaviour is studied by the onlookers to select the desired food source while the scout bee searches for new food sources randomly once the old food source is is exhausted. Both onlookers and scouts are considered as unemployed foragers. The position of a food source in ABC corresponds to the possible solution of the problem to be optimized and the nectar amount of a food source represents the fitness (quality) of the associated solution. The number of employed bees is equal to the number of food sources (solutions), since each employed bee is associated with one and only one food source [20].

The key phases of the algorithm as proposed in [22] are as follows:

- Generate the initial population of the food sources randomly.
- REPEAT
    - Send the employed bees onto the food sources and calculate the fitness cost.
    - Evaluate the probability values for the food sources
    - Send the onlooker bees onto the food sources depending on probability and calculate the fitness cost.
    - Abandon the exploitation process, if the sources are exhausted by the bees.

- Send the scouts into the search area for discovering new food sources randomly
- Memorize the best food source found so far.
  - UNTIL (requirements are met).

4 Modified-ABC for Post-enrolment Course Timetabling Problem

The modified-ABC is similar to what was done in our previous works on the basic ABC [19] except for the onlooker phase where in the modified-ABC employs multiswap algorithm to optimize the navigation of the search in its operation. The description of the five stages of modified-ABC with application to the PE-CTP is given in the next subsections. It is worth noting that this research considered the feasible search space region. Therefore, the feasibility of the solutions discuss below is preserved.

4.1 Initialize the Modified-ABC and PE-CTP Parameters

In this step, the three control parameters of modified-ABC that are needed for tackling PE-CTP are initialized. These parameters include solution number (SN) which represent the number of food sources in the population and similar to the population size in GA; maximum cycle number (MCN) refers to the maximum number of iterations; and *limit*, which is normally used to diversify the search, is responsible for the abandonment of solution, if there is no improvement for certain number of iterations. Similarly, the PE-CTP parameters like the set of the courses, set of rooms, set of timeslots and set of constraints are also extracted from the dataset. Note that the courses are the main decision variable that can be scheduled to feasible positions (i.e. rooms and timeslots) in the timetable solution.

4.2 Initialize the Food Source Memory

The food source memory (FSM) is a memory allocation that contains sets of feasible food source (i.e. solutions) vectors which are determined by SN as shown in Eq. 2. Here, the food source vectors are generated with the aid of largest weighted degree (LWD) [23] followed by backtracking algorithm (BA) [24] and they are sorted in ascending order in the FSM in accordance with the objective function values of the, that is $f(x_1) \leq f(x_2) \ldots f(x_{SN})$

$$
\mathbf{FSM} =
\begin{bmatrix}
x_1^1 & x_1^2 & \cdots & x_1^N \\
x_2^1 & x_2^2 & \cdots & x_2^N \\
\vdots & \vdots & \ddots & \vdots \\
x_{SN}^1 & x_{SN}^2 & \cdots & x_{SN}^N
\end{bmatrix}
\begin{bmatrix}
f(x_1) \\
f(x_2) \\
\vdots \\
f(x_{SN})
\end{bmatrix}
\tag{2}
$$

4.3   Send the Employed Bee to the Food Sources

Here, the timetable solutions are sequentially selected from the FSM by the employed bee operator. Each solution is perturbed using three neighbourhood structures to produce a new set of solutions. The three neighbourhood structures used by employed bee are:

1. **Neighbourhood Move (NM)**: moves selected course to a feasible period and room randomly i.e. replace the time period x'_i of course i by another feasible timeslot.
2. **Neighbourhood Swap (NS)**: swap two selected course at random i.e. select course i and event j randomly, swap their time periods (x'_i, x'_j).
3. **Neighbourhood Kempe Chain (NK)**: Firstly, select the timeslot x'_i of course i and randomly select another q' timeslot. Secondly, all courses that have the same timeslot x'_i that are in conflict with one or more courses timetabled in q_i are entered to chain δ . Thirdly, all courses that have the same timeslot q' that are conflicting with one or more courses timetabled in x'_i are entered to a chain δ' and Lastly, simply assign the courses in δ to q' and the courses in δ' to x'_i.

The fitness of each offspring food source is calculated. If it is better than that of parent food source, then it replaces the parent food source in FSM. This process is implemented for all solutions. The detailed algorithm of this process can be found in our previous paper [18].

4.4   Send the Onlooker Bee to the Food Sources

The onlooker bee possesses the same number of food sources (timetabling solutions) as employed bees. The food sources improved by employed bees are evaluated by the onlookers using the selection probability. Then the fittest food sources are selected with roulette wheel selection mechanism. The process of selection in the onlooker phase works as follows:

- Assign to each food source a selection probability as follows:

$$p_j = \frac{f(x^j)}{\sum_{k=1}^{SN} f(x^k)}$$

  Note that the $\sum_{i=1}^{SN} p_i$ is unity.
- The onlooker improve the fittest food sources using the multiswap strategy (MS) proposed in [25,26] for further enhancement. The MS is designed to manage the room operations in the PE-CTP solution. Here, the courses of each timeslot are shuffled randomly to different valid rooms within the same timeslot. It starts with a complete timetable solution and select the timeslot consecutively where P is the set of timeslots. The courses in each timeslot p are shuffled randomly to the appropriate rooms. The output of

this algorithm is another complete timetable solution with different room arrangements for the courses in the same timeslot. It is noteworthy that this strategy is a simple idea of room operations and has a insignificant effect on the computational time. Algorithms 1 give the detailed process of the proposed MS:

Algorithm 1. Multiswap Strategy

---

    for all $p \in P$ **do**
      for all $j \in R$ **do**
        for all $k \in R, j \neq k$, **do**
          begin
            if allocation $a_{j,p}$ contain course c and allocation $a_{k,p}$ is empty **then**
              Note that A is of size PXR and the value of allocation $a_{i,j}$
              contains either the course code or -1 to indicate it is empty
              if room j is suitable room for course c **then**
                move course c to the allocation $a_{k,p}$
            else
              if allocation $a_{j,p}$ contain course c and allocation $a_{k,p}$ contain
              course q **then**
                if room j is suitable for $a_{k,p}$ and room k is suitable for $a_{j,p}$
                then
                  swap the locations $(a_{j,p}a_{k,p})$ of course c and q
              end if
            end if
          end if
        end if
        end
      end for
    end for
  end for

4.5   Send the Scout to Search for Possible New Food Sources

This is known to be the colony explorer. It works once a solution is abandoned, i.e. if a solution in the FSM has not improved for certain number of iterations as determined by the *limit*. The ABC generates a new solution randomly and substitutes the abandoned one in FSM. Furthermore, the best food source x_{best} in FSM is memorized.

4.6   Stopping Criteria

Steps 4.3 to 4.5 are repeated until a stop criterion is met. This is originally determined using maximum cycle number (MCN) value.

5   Experimental Results and Discussions

The modified-ABC is coded in Microsoft Visual C++ 6.0 on Windows 7 platform on Intel 2 GHz Core 2 Quad processor with 2 GB of RAM. The modified-ABC required a maximum of 7 hours to obtain the recorded result, although the computational time is not provided in the literature. The parameter settings for ABC are as follows: MCN=10000; SN=10; limit=100. The performance of the proposed method is tested using the dataset established by Socha et al in [9]. This dataset is available at website[1]. The dataset comprises 100-400 courses that needed to be assigned to a timetable with 45 timeslots corresponding to 5 days of 9 hours per day and at the same time satisfying, room features and room capacity constraints. They are divided into three types: small, medium and large (i.e. 5 small, 5 medium and 1 large instances) as shown in Table 1. Each problem instances of the dataset ran ten times with a different random seed, and the best penalty cost obtained for all dataset is reported in Table 2.

Table 1. Characteristics of Enrolment Course Timetabling Dataset

Class	Small	Medium	Large
number of events	100	400	400
number of rooms	5	10	10
number of features	5	5	10
number of students	80	200	400
number of timeslots	45	45	45
approximate feature per room	3	3	5
percentage of feature use	70	80	90
maximum number of events per student	20	20	20
maximum number of students per event	20	50	100

Similarly, Table 2 shows the modified-ABC results in comparison with previous techniques in the literature. These techniques are modified harmony search algorithm (M1)[25], MAX-MIN Ant System (M2) [9], Hybrid Evolutionary Approach (M3) [27], Evolutionary Non-linear Great Deluge (M4) [28], Extended Guided Search Genetic Algorithm (M5) [10] Guided Search Genetic Algorithm (M6) [29], Hybrid Ant Colony Systems (M7) [30], the electromagnetism-like mechanism approach (M8) [31], and genetic local search algorithm (M9) [32] and incorporating Great deluge with kempe chain (M10) [7]. Note that the best results are presented in bold. It is apparent that the modified-ABC is able to obtain feasible solutions for all problem instances. Similarly, it is able to obtain high quality solutions in comparison with previous methods. For example, the modified-ABC is ranked first in the medium 5 and second in the large datasets respectively which were classified as most difficult datasets. Finally, the proposed method achieved third rank in medium 2 and 3 datasets.

---

[1] http://iridia.ulb.ac.be/~msampels/tt.data/

Table 2. The M-ABC Convergence Cases

Dataset	M-ABC	M1	M2	M3	M4	M5	M6	M7	M8	M9	M10
Small 1	0	0	1	0	0	0	0	0	0	2	0
Small 2	0	0	3	0	1	0	0	0	0	4	0
Small 3	0	0	1	0	0	0	0	0	0	2	0
Small4	0	0	1	0	0	0	0	0	0	0	0
Small5	0	0	0	0	0	0	0	0	0	4	0
Medium 1	129	168	195	221	126	139	240	117	175	226	**98**
Medium 2	119	160	184	147	123	**92**	160	121	197	215	113
Medium 3	137	176	284	246	185	**122**	242	158	216	231	123
Medium 4	146	144	164.5	165	116	**98**	158	124	149	200	100
Medium 5	**63**	71	219.5	130	129	116	124	134	190	195	135
Large	525	**417**	851.5	529	821	615	801	647	912	1012	610

6   Conclusion

In this paper, a modified-ABC to solve the PE-CTP is presented. The modification involves replacing the strategy of onlooker bee with a Mutliswap Algorithm. This modification enables the modified-ABC to explore post-enrolment timetabling search space more efficiently. The dataset introduced by Socha including five small, five medium and one large problem instances were used in the evaluation. In comparison with ten previously used techniques, the modified-ABC achieved exact solutions for all small datasets as achieved by eight others. For medium datasets, it produced overall best results for "medium 5" and ranked among in the first three for others. For the most taxing and large problem instance, the modified-ABC ranked second. Conclusively, the modified-ABC is a suitable optimization technique that is able to provide high quality solution for the NP-hard problem like timetabling. For further extension, a modified-ABC can be hybridized with other techniques, studied for other scheduling problems and simplify to be more general.

Acknowledgement. The first author would like to appreciate Universiti Sains Malaysia for the financial support under IPS-USM fellowship scheme for his PhD study. This research is supported by RU Grant 1001/PKOMP/817047 from the Universiti Sains Malaysia.

References

1. McCollum, B.: A perspective on bridging the gap between theory and practice in university timetabling. In: Burke, E.K., Rudová, H. (eds.) PATAT 2007. LNCS, vol. 3867, pp. 3–23. Springer, Heidelberg (2007)
2. Burke, E., Kendall, G., Mısır, M., Özcan, E., Burke, E., Kendall, G., Özcan, E., Mısır, M.: Applications to timetabling. In: Handbook of Graph Theory, ch.5.6. Citeseer (2004)

3. Daskalaki, S., Birbas, T., Housos, E.: An integer programming formulation for a case study in university timetabling. European Journal of Operational Research 153(1), 117–135 (2004)
4. Burke, E., Petrovic, S., Qu, R.: Case-based heuristic selection for timetabling problems. Journal of Scheduling 9(2), 115–132 (2006)
5. Aladag, C., Hocaoglu, G., Basaran, M.: The effect of neighborhood structures on tabu search algorithm in solving course timetabling problem. Expert Systems with Applications 36(10), 12349–12356 (2009)
6. Kostuch, P.: The university course timetabling problem with a three-phase approach. In: Burke, E.K., Trick, M.A. (eds.) PATAT 2004. LNCS, vol. 3616, pp. 109–125. Springer, Heidelberg (2005)
7. Abdullah, S., Shaker, K., McCollum, B., McMullan, P.: Incorporating great deluge with kempe chain neighbourhood structure for the enrolment-based course timetabling problem. In: Yu, J., Greco, S., Lingras, P., Wang, G., Skowron, A. (eds.) RSKT 2010. LNCS, vol. 6401, pp. 70–77. Springer, Heidelberg (2010)
8. Abdullah, S., Burke, E., Mccollum, B.: An investigation of variable neighbourhood search for university course timetabling. In: The 2nd Multidisciplinary International Conference on Scheduling: Theory and Applications (MISTA 2005), pp. 413–427 (2005)
9. Socha, K., Knowles, J.D., Sampels, M.: A $\mathcal{MAX} - \mathcal{MIN}$ ant system for the university course timetabling problem. In: Dorigo, M., Di Caro, G.A., Sampels, M. (eds.) ANTS 2002. LNCS, vol. 2463, pp. 1–13. Springer, Heidelberg (2002)
10. Yang, S., Jat, S.: Genetic algorithms with guided and local search strategies for university course timetabling. IEEE Transactions on Systems, Man, and Cybernetics, Part C: Applications and Reviews 41(1), 93–106 (2011)
11. Al-Betar, M.A., Khader, A.T., Zaman, M.: University Course Timetabling Using a Hybrid Harmony Search Metaheuristic Algorithm. IEEE Transactions on Systems, Man, and Cybernetics — Part C: Applications and Reviews (2011), doi:10.1109/TSMCC.2011.2174356
12. Irene, S., Deris, S., Zaiton, M.: A study on PSO-based university course timetabling problem. In: International Conference on Advanced Computer Control, pp. 648–651. IEEE (2009)
13. Abdullah, S., Turabieh, H., McCollum, B., McMullan, P.: A hybrid metaheuristic approach to the university course timetabling problem. Journal of Heuristics, 1–23 (2012)
14. Lewis, R.: A survey of metaheuristic-based techniques for university timetabling problems. OR Spectrum 30(1), 167–190 (2008)
15. McCollum, B., Schaerf, A., Paechter, B., McMullan, P., Lewis, R., Parkes, A., Gaspero, L., Qu, R., Burke, E.: Setting the research agenda in automated timetabling: The second international timetabling competition. INFORMS Journal on Computing 22(1), 120–130 (2010)
16. Bolaji, A., Khader, A., Al-Betar, M., Awadallah, M.: Artificial bee colony, its variants and applications: a survey. Journal of Theoretical and Applied Information Technology (JATIT) 47(2), 1–27 (2013)
17. Bolaji, A., Khader, A., Al-Betar, M., Awadallah, M.: An improved artificial bee colony for course timetabling. In: 2011 Sixth International Conference on Bio-Inspired Computing: Theories and Applications (BIC-TA), pp. 9–14. IEEE (2011)
18. Bolaji, A., Khader, A., Al-Betar, M., Awadallah, M.: Artificial bee colony algorithm for solving educational timetabling problems. International Journal of Natural Computing Research 3(2), 1–21 (2012)

19. Bolaji, A., Khader, A., Al-betar, M., Awadallah, M.: The effect of neighborhood structures on examination timetabling with artificial bee colony. In: Practice and Theory of Automated Timetabling IX, pp. 131–144 (2012)
20. Karaboga, D.: An idea based on honey bee swarm for numerical optimization. Techn. Rep. TR06, Erciyes Univ. Press, Erciyes (2005)
21. Teodorović, D., DellOrco, M.: Bee colony optimization–a cooperative learning approach to complex transportation problems. In: Advanced OR and AI Methods in Transportation. Proceedings of the 10th Meeting of the EURO Working Group on Transportation, pp. 51–60. Citeseer, Poznan (2005)
22. Akay, B., Karaboga, D.: Solving integer programming problems by using artificial bee colony algorithm. In: Serra, R., Cucchiara, R. (eds.) AI*IA 2009. LNCS, vol. 5883, pp. 355–364. Springer, Heidelberg (2009)
23. Arani, T., Lotfi, V.: A three phased approach to final exam scheduling. IIE Transactions 21(1), 86–96 (1989)
24. Carter, M., Laporte, G., Lee, S.: Examination timetabling: Algorithmic strategies and applications. Journal of the Operational Research Society 47(3), 373–383 (1996)
25. Al-Betar, M., Khader, A.: A harmony search algorithm for university course timetabling. Annals of Operations Research, 1–29 (2012), doi:10.1007/s10479-010-0769-z
26. Al-Betar, M., Khader, A., Muslih, O.: A multiswap algorithm for the university course timetabling problem. In: 2012 International Conference on Computer & Information Science (ICCIS), vol. 1, pp. 301–306. IEEE (2012)
27. Abdullah, S., Burke, E., McCollum, B.: A hybrid evolutionary approach to the university course timetabling problem. In: IEEE Congress on Evolutionary Computation, CEC 2007, pp. 1764–1768. IEEE (2007)
28. Landa-Silva, D., Obit, J.H.: Evolutionary non-linear great deluge for university course timetabling. In: Corchado, E., Wu, X., Oja, E., Herrero, Á., Baruque, B. (eds.) HAIS 2009. LNCS, vol. 5572, pp. 269–276. Springer, Heidelberg (2009)
29. Jat, S., Yang, S.: A guided search genetic algorithm for the university course timetabling problem. In: The 4th Multidisciplinary International Scheduling Conference: Theory and Applications (MISTA 2009), Dublin, Ireland, August 10-12, pp. 180–191 (2009)
30. Ayob, M., Jaradat, G.: Hybrid ant colony systems for course timetabling problems. In: 2nd Conference on Data Mining and Optimization, DMO 2009, pp. 120–126. IEEE (2009)
31. Turabieh, H., Abdullah, S., McCollum, B.: Electromagnetism-like mechanism with force decay rate great deluge for the course timetabling problem. In: Wen, P., Li, Y., Polkowski, L., Yao, Y., Tsumoto, S., Wang, G. (eds.) RSKT 2009. LNCS, vol. 5589, pp. 497–504. Springer, Heidelberg (2009)
32. Abdullah, S., Turabieh, H.: Generating university course timetable using genetic algorithms and local search. In: Third International Conference onConvergence and Hybrid Information Technology, ICCIT 2008, vol. 1, pp. 254–260. IEEE (2008)

Immune Based Chaotic Artificial Bee Colony Multiobjective Optimization Algorithm

Xia Zhou[1], Jiong Shen[2], and Yiguo Li[2]

[1] School of Mechanical & Electrical Engineering, Jinling Institute of Technology
Nanjing 211169, P.R. China
zenia77@163.com
[2] School of Energy & Environment, Southeast University
Nanjing 210096, P.R. China
{shenj,lyg}@seu.edu.cn

Abstract. This work presents a new multiobjective optimization algorithm based on artificial bee colony, named the ICABCMOA. In order to meet the requirements of Pareto-based approaches, a new fitness assignment function is defined based on the dominated number. In the ICABCMOA, a high-dimension chaotic method based on Tent map is addressed to increase the searching efficiency. Vaccination and gene recombination are adopted to promote the convergence. The experimental results of the ICABCMOA compared with NSGAII and SPEA2 over a set of test functions show that it is an effective method for high-dimension optimization problems.

Keywords: Immune, Multiobjective, Chaotic, Artificial Bee Colony.

1 Introduction

Simulating the forging behavior of honeybee swarm, Artificial Bee Colony(ABC) algorithm is proposed by Karaboga in 2005[1]. The main advantage of the ABC algorithm is that there are few control parameters. Due to its simplicity and ease of implementation, the ABC algorithm is gaining more and more attention and has been used to solve many practical engineering problems, but there is few works focus on multiobjective optimization problems.

In fact, a great number of engineering problems have more than one objective and usually there is no absolutely exclusive best solution for these problems. According to the survey on multiobjectve optimization, the majority of the multiobjective algorithms are concentrated on Pareto-based approaches[2]. In this paper, a new Pareto-based multiobjective ABC algorithm named Immune based Chaotic Artificial Bee Colony Multiobjective Optimization Algorithm (ICABCMOA) is addressed.

The rest of this paper is organized as follows: Section 2 introduces the related background. Section 3 describes the details of the proposed ICABCMOA. Next, the experimental results are discussed in Section 4. Finally, the conclusion of the paper is outlined in section 5.

Y. Tan, Y. Shi, and H. Mo (Eds.): ICSI 2013, Part I, LNCS 7928, pp. 387–395, 2013.

2   Related Background

2.1   Artificial Bee Colony Algorithm

In the ABC algorithm, there are two components: the foraging artificial bees and the food sources. The position of a food source represents a possible solution of the optimization problem and the nectar amount of the food source corresponds to the quality of the solution. The colony of artificial bees are divided into three groups, namely, employed bees, onlookers, and scouts bees. A bee going to the food source exploited by itself is called an employed bee, and the number of employed bees is equal to the number of food sources. A bee waiting on the dance area for watching the waggle dances of the employed bees, and select a food source according to the profitability is named as an onlooker. The number of the onlooker bees is equal to the number of employed bees. For every food source, there is only one employed bee. In every cycle, each employed bee produces a new food source surrounding its food source site and exploits the better one. Each onlooker bee selects a food source according to the quality of each solution and produces a new food source surrounding the selected food source site. An employed bee will become a scout when the corresponded food source has been exhausted, and the scout bee will carry out random search[1].

2.2   Multiobjective Optimization

Generally, a multiobjective optimization problem can be described as:

$$\min \mathbf{f}(X)=(\mathbf{f_1}(X), \mathbf{f_2}(X),...,\mathbf{f_m}(X)) \text{ s.t.} \begin{cases} g_i(X) \geq 0, & i=1,2,...,p \\ h_j(X)=0, & j=1,2,...,q \end{cases} \quad (1)$$

Where $X \in \Omega$ is the decision vector, and Ω is the feasible region. The formal definitions of the "Pareto optimal" are described as follows[3].

Definition 1. $\forall X_i, X_j \in \Omega$, if $\forall k \in \{1,2,...m\}$, $f_k(X_i) \leq f_k(X_j)$; together with $\exists t \in \{1,2,...m\}, f_k(X_i) < f_k(X_j)$, we say that X_i dominates X_j, marked as $X_i \succ X_j$.

Definition 2. $\forall X_i, X_j \in \Omega$, if $\exists k \in \{1,2,...m\}, f_k(X_i) < f_k(X_j)$; together with $\exists t \in \{1,2,...m\}, t \neq k, f_k(X_i) > f_k(X_j)$, we say that X_i has nothing to do with X_j, marked as $X_i \circ X_j$.

Definition 3. $X_i \in \Omega$ is said to be a Pareto optimal solution iff $\neg \exists X_j \in \Omega$, s.t. $X_i \succ X_j$. Denote X^* as the Pareto optimal solutions, $PF = \{ F(X)=(f_1(X), f_2(X),...,f_m(X))|X \in \{X^*\}\}$ is said to be the Pareto-optimal front.

3   Description of ICABCMOA

For convenience of the description, we give some notations in advance.

(1)The size of the employed bees is denoted as n_e. The size of the onlooker bees equals to that of the employed bees, total size of the bees is $2n_e$.

(2)Food sources are denoted as X, in the algorithm it means d-dimension candidate solutions.

(3) X_i is the i-th food source, x_{ij} is the j-th value of $X_i (j = 1, 2..., d)$. The upper and lower bounds of the dimension j are denoted as $x_{j\max}$ and $x_{j\min}$, respectively.

(4)Denote S as the secondary set, and the maximum size of the S is n_s.

(5) 'limit' is used to represent the number of cycles when a food source can't be improved further and it should be abandoned.

3.1   Operator Definition

In the ICABCMOA, the Tent map is introduced to generate the chaotic sequences. Vaccination and gene recombination are adopted to promote the convergence. The new operators are defined as follows.

Definition 4. Tent map

The definition of one-dimension Tent map is as follows:

$$F_a:\ \gamma_i = \begin{cases} \dfrac{\gamma_{i-1}}{a} & 0 \le \gamma_{i-1} \le a \\ \dfrac{1-\gamma_{i-1}}{1-a} & a < \gamma_{i-1} \le 1 \end{cases} ,\quad 0 < a < 1 \tag{2}$$

The chaotic sequences produced by the Tent map have good statistical properties. But because of the limited word-length, after a number of iterations, the result would be equal to 0 or 1, and it would not change in the next iteration. For this reason, we improve the Tent map to avoid the long time 0 or 1. When the result is equal to 0 or 1, a random between 0 and 1 is generated to instead of the 0 or 1.

Definition 5. High-dimension Tent map

On the basis of a two-dimension Tent map proposed in [4], we proposed a high-dimension constructing method as formula (3).

$$F_{a_1, a_2, ..., a_d}:\ (\gamma_{i1}, \gamma_{i2}, ..., \gamma_{id}) = (h(\gamma_{i-1,1}), h(\gamma_{i-1,2}), ...h(\gamma_{i-1,d}))$$

$$h(\gamma_{i-1,j}) = \begin{cases} \dfrac{\gamma_{i-1,j}}{a_j} & 0 < \gamma_{i-1,j} \le a_j \\ \dfrac{1-\gamma_{i-1,j}}{1-a_j} & a_j < \gamma_{i-1,j} \le 1 \\ \text{rand} & \gamma_{i-1,j} = 0 \text{ or } 1 \end{cases} \tag{3}$$

where $0 < a_j < 1$, $j = 1, 2...d$.

Definition 6. Vaccination

Here we extend the meaning of the vaccination to the prior knowledge for the problem. In this work, the vaccination is defined as the mean value of a group of best solutions. Suppose the selected best solutions are composed of : $X_1, X_2, ..., X_c$, and $X_i = (x_{i1}, x_{i2}, ..., x_{id}), i = 1, 2, ...c$. Then the definition of the vaccination can be described as follows:

$$vac(j) = \frac{1}{c} \sum_{i=1}^{c} x_{ij}, \quad j \in \{1, 2, ...d\}. \tag{4}$$

Definition 7. Gene recombination

Gene recombination means generating new genes by recombination of the independent genes. A food source, which will participate in recombination, is treated as an independent gene group. Another food source S_k in the memory set is selected as the other independent gene group.

Denote p_r as the recombination probability, the recombination operator can be described as follows:

$$x_{i+1,j} = \begin{cases} s_{kj} & p_j < p_r \\ x_{ij} & p_j \geq p_r \end{cases}, \quad j \in \{1, 2, ...d\} \tag{5}$$

Definition 8. Fitness assignment method

It is obvious that the fitness assignment method in standard ABC is not fit for multiobjective optimization algorithms. A new fitness function is defined based on the dominated number. Regarding the size of the food source dominate the i-th one as dn(i), and the biggest value of the dn as mdn. The base fitness value of the i-th food source can be evaluated as follows:

$$base\_fit(i) = \frac{mdn - dn(i) + 1}{mdn + 10}, \ dn(i) = \sum_j sig(X_i, X_j) \text{ and } sig(X_i, X_j) = \begin{cases} 1 & X_j \succ X_i \\ 0 & others \end{cases} \tag{6}$$

In order to prevent the numerator or the formula from equaling to zero, two constants are added to the denominator and the numerator in formula (7), respectively.

Generally, the probability of better solutions around good solutions is greater, so we adjust the base fitness value to some extent.

$$fit(i) = \begin{cases} base\_fit(i) & dn(i) \neq 0 \\ 2*base\_fit(i) & dn(i) = 0 \\ 4*base\_fit(i) & dn(i) = 0 \text{ and } (x_{ij} = x_{jmax} \text{ or } x_{ij} = x_{jmin}) \end{cases}, \ j \in \{1, 2..., d\}. \tag{7}$$

3.2   Running Mechanism of ICABCMOA

With the operators described in section 3.1, running mechanism of the ICABCMOA can be described as follows.

```
Initialization
REPEAT
     employed bee optimization;
     onlooker bee optimization;
     scout bee optimization;
     Find the nondominated set;
     Propose of the secondary set;
UNTIL (the stopping criterion is met).
```

Since the nondominated set get method is not the main content of this work, we just adopted the method proposed in [3]. The other procedures are described in detail.

1. Initialization

In the initialization phase, the control parameters and the initial population are initialized. There are only four control parameters in the ICABCMOA, i.e. the size of food sources, the size of the secondary set, the *'limit'* value, and the maximum iteration cycle.

If a food source could not be improved in a number of cycles, its related employed bee will turn out to be a scout. But after doing a random search, the scout will turn back to an employed bee again. So there is no necessary to consider about the scout bees in the initialization. In ICABCMOA, after a high-dimension Tent map sequence is generated, a chaotic based initialization population is generated according to formula(8).

$$x_{ij} = \gamma_{ij} * (x_{j\max} - x_{j\min}) + x_{j\min} \text{ , i=1,2...n}_e \text{ and j=1,2...d} \qquad (8)$$

2. Employed Bee Optimization

The employed bee searching method depends on whether it is nondominated or not. If it is a nondominated one, a food source in the secondary set is selected and gene recombination would be performed. If it is a dominated one, the vaccine would be taken to some randomly selected dimensions.

After performing searching approaches, the new food source will be evaluated and compared with the old one. Here a greedy selection procedure will be performed, and the better food source will be kept in the population.

3. Onlooker Bee Optimization

When all employed bees finished optimizing their food sources, they will go back to the hive to share their information with onlooker bees. Each onlooker bee selects a food source based on the fitness value according to the following equation.

$$P\{T_s(X) = X_i\} = \text{fit}(X_i) / \sum_{i=1}^{n_e} \text{fit}(X_i) \qquad (9)$$

After the food source is selected, each onlooker bee performs the local searching as formula (10), which is the same as the basic ABC presented. And a greedy selection procedure is also performed after the local searching.

$$x_{i+1,j} = x_{ij} + \text{rand}(-1,1) * (x_{ij} - x_{kj}) , \; k \neq i \qquad (10)$$

4. Scouts Bee Optimization

Unlike the definition of the scout bee in standard ABC algorithm, the scout bees refer to two kinds of food sources. One kind is the food sources which have not improved after 'limit' cycles, the other kind is the food sources which have the biggest dominated size. First the selected food source is mapped to [0,1], then the high-dimension constructing method of Tent map will perform k times, after that, the new generated one will be mapped to the real value.

After chaotic searching, different selection method is used to different kinds of scouts. For the scout generated from the food source which has not improved after 'limit' cycles, substitution is performed. But for the other kind of scouts, a greedy selection procedure will be performed.

5. Secondary Set Processing

ICABCMOA uses a fixed size secondary set to hold the best solutions ever found. In each iteration, the new found nondominated solutions are combined with the former secondary set, and the nondominated solutions of the new combination are set are as the new secondary pool candidates. If the size of the new secondary pool exceeds the set limit, the redundant secondary food sources removal operator would be executed. Here the crowded-comparison approach proposed in [5] is adopted. We prefer the solution that is located in a less crowded region.

4   Experimental Results

This section contains the computational results obtained by the ICABCMOA compared to the SPEA2[5] and NSGAII[6] over a set of test problems.

4.1   Test Functions

The four selected test functions are given in Table 1[5]-[6]. There are different decision vector dimensions and different characteristics in the objective space among the four problems. Here we use the measure criterion 'CS' proposed in ref.[7] and 'S' proposed in ref.[8] to provide a quantitative assessment for the performance of the proposed algorithm.

In order to make the NSGAII and SPEA2 work as the presenter described, the parameters of the two algorithms are set originally. The population size and archive size of NSGAII are both set as 100, and the population size and archive size of SPEA2 are set as 200 and 100, respectively. The parameter n_e and n_s of the ICABCMOA are both set as 100, too. In order to make all the three algorithms have the nearly equal searching size, the iteration cycle are set as 100, 50, 50. The 'limit' value of the ICABCMOA is 5.

The other parameters of the NSGAII and SPEA2 are set according to the values suggested by the developers.

Table 1. Mathematical representation of the four test functions

Problem	Variable range	Mathematical representation
Schaffer	$x \in [-10,10]$	$f_1(x) = x^2$ $f2(x) = (x-2)^2$
Deb	$x \in [0,1]$	$f_1(x) = x_1$ $f2(x) = (1+10x_2)[1-(\dfrac{x_1}{1+10x_2})^2 - \dfrac{x_1}{1+10x_2}\sin(8\pi x_1)]$
ZDT1	$m = 30; x_i \in [0,1]$	$f_1(x) = x_1, f2(x) = g(1-\sqrt{f_1/g})$ $g(x) = 1+9\displaystyle\sum_{i=2}^{m}\dfrac{x_i}{m-1}$
ZDT2	$m = 30; x_i \in [0,1]$	$f_1(x) = x_1, f2(x) = g(1-(f_1/g)^2)$ $g(x) = 1+9\displaystyle\sum_{i=2}^{m}\dfrac{x_i}{m-1}$

4.2   Performance Analysis

We performed 10 independent runs on each test problem. Table 2 and 3 show the dominating relations and the distribution of the solutions obtained by the algorithms. The value CS(A,B) reflects the dominating relations between A and B. It is obvious that the less the value, the better the solutions in B. In table 2, 'I' represents ICABCMOA, 'S' represents SPEA2, and 'N' represents NSGAII.

It is shown in table 2, for problem Shaffer, the solutions obtained by the ICABCMOA weakly dominate the NSGAII and weakly dominated by SPEA2. For problem Deb, the solutions obtained by ICABCMOA weakly dominate the others. But for high dimension multiobjective problems ZDT1 and ZDT2, the solutions obtained by the ICABCMOA clearly dominate the solutions obtained by the NSGAII and SPEA2.

Table 3 lists the S value of all the three algorithms. From the definition of S we can know that the less the S value, the better the solution distribution. It is shown in table 3 that the S value got by the ICABCMOA is uniform for all four test functions. The S value got by the ICABCMOA is better than NSGAII and SPEA2 in most tests.

Table 2. Mean, max and min value of CS

		Schaffer	Deb	ZDT1	ZDT2
CS(I,S)	mean	0.02	0.261	0.985	0.918
	max	0.002	0.35	0.99	0.98
	min	0	0.2	0.98	0.8
CS(S,I)	mean	0.06	0.001	0	0
	max	0.02	0.01	0	0
	min	0.04	0	0	0
CS(I,N)	mean	0.01	0.02	0.987	0.921
	max	0.04	0.05	0.99	0.97
	min	0	0.01	0.98	0.82
CS(N,I)	mean	0.006	0.001	0	0
	max	0.02	0.01	0	0
	min	0	0	0	0

Table 3. Mean, max and min value of the metric about the distribution of the solutions--S

		Schaffer	Deb	ZDT1	ZDT2
S_ICABCMOA	mean	0.0603	0.0063	0.0072	0.0062
	max	0.072	0.0069	0.0094	0.008
	min	0.0467	0.0055	0.0052	0.0051
S_SPEA2	mean	0.0311	0.0099	0.0157	0.0063
	max	0.0406	0.0258	0. 0446	0.0099
	min	0.0208	0.0036	0.0071	0.0019
S_NSGAII	mean	0.0906	0.0142	0.0184	0.0139
	max	0.3149	0.0834	0.0455	0.0369
	min	0.0502	0.004	0.0066	0.0041

5   Conclusion

A new multiobjective optimization algorithm named ICABCMOA is addressed in this paper. Fast convergence of ABC, good searching ability of chaotic map, global searching of immune are all integrated into the proposed algorithm and makes the algorithm powerful.

The experimental results of the ICABCMOA compared with the NSGAII and SPEA2 over a set of test functions show that the ICABCMOA is an effective method for high-dimension optimization problems, and the solutions got by the ICABCMOA are uniform for all four test functions.

Acknowledgments. This work is supported by the National Natural Science Foundation of China under No. 51036002, 51076027 and Dr. Start Foundation of Jinling Institute of Technology under No.JIT-B-201218.

References

1. Karaboga, D.: An Idea Based on Honey bee Swarm for Numerical Optimization. Technical Report, Computer Engineering Department. Erciyes University,Turkey (2005)
2. Zhou, A., Qu, B.Y., Li, H., et al.: Multiobjective Evolutionary Algorithms: a Survey of the State-of-the-art. Journal of Swarm and Evolutionary Computation 1(1), 32–49 (2011)
3. Zhou, X., Shen, J., Sheng, J.X.: An Immune Recognition Based Algorithm for Finding Non-dominated Set in Multiobjective Optimization. In: IEEE Pacific-Asia Workshop on Computational Intelligence and Industrial Application, Wuhan, China, pp. 305–310 (2008)
4. Shan, L., Qiang, H., Li, J., et al.: Chaotic Optimization Algorithm Based on Tent Map. Control and Decision 20(2), 179–182 (2005) (in Chinese)
5. Zitzler, E., Laumanns, M., Thiele, L.: SPEA2: Improving the Strength Pareto Evolutionary Algorithm for Multi-objective Optimization. In: Evolutionary Methods for Design, Optimization and Control, Barcelona, Spain, pp. 19–26 (2002)
6. Deb, K., Pratap, A., Agarwal, S., et al.: A Fast and Elitist Multiobjective Genetic Algorithm: NSGA-II. IEEE Transactions on Evolutionary Computation 6(2), 182–197 (2002)
7. Zitzler, E.K., Deb, K., Thiele, L.: Comparison of Multiobjective Evolutionary Algorithms: Empirical Results. IEEE Transactions on Evolutionary Computation 8(2), 173–195 (2000)
8. Schott, J.T.: Fault Tolerant Design Using Single and Multicriteria Genetic Algorithm Optimization. Department of Aeronautics and Astronautics, Massachusetts Institute of Technology (1995)

Using Modular Neural Network
with Artificial Bee Colony Algorithm for Classification

Wei-Xin Ling and Yun-Xia Wang

School of Science, South China University of Technology, Guangzhou, China
lingweixin@21cn.com, yx_wang07@163.com

Abstract. The Artificial bee colony (ABC) algorithm has been used in several optimization problems, including the optimization of synaptic weights from an Artificial Neural Network (ANN). However, it is easy to trap in local minimum and not enough to generate a robust ANN. Modular neural networks (MNNs) are especially efficient for certain classes of regression and classification problems, as compared to the conventional monolithic artificial neural networks. In this paper, we present a model of MNN based on ABC algorithm (ABC-MNN). The experiments show that, compared to the monolithic ABC-NN model, classifier designed in this model has higher training accuracy and generalization performance.

Keywords: Modular Neural Network, Artificial Bee Colony Algorithm, Learning Algorithm.

1   Introduction

Optimization algorithm based on swarm intelligence, known as meta-heuristic algorithms, gained popularity in solving complex and high dimension optimization problems' years ago. The Artificial Bee Colony Algorithm is one of the most popular swarm intelligence algorithm based on the foraging behavior of honey bees for numerical problems, which was proposed by Karaboga [1] in Erciyes University of Turkey in 2005.

Since ABC algorithm is simple in concept, easy to implement, and has fewer control parameters, it also has been widely used in many fields. ABC algorithm has applied successfully to unconstrained numerical optimization problems. Karaboga and Basturk [2] proposed the extended version of the ABC for solving constrained optimization problems in 2007. The experiments show that the extended version of ABC algorithm has better performance than DE and PSO.

ANNs are commonly used in pattern classification, function approximation, optimization, pattern matching, machine learning and associative memories. In [3] the authors apply this algorithm to train a feed-forward Neural Network. In the pattern classification area, other works like [4] ABC algorithm is compared with other evolutionary techniques, while in [5] an ANN is trained with medical pattern classification. It says that ABC algorithm is a good optimization technique. But the

Y. Tan, Y. Shi, and H. Mo (Eds.): ICSI 2013, Part I, LNCS 7928, pp. 396–403, 2013.

monolithic neural network has serious learning problems, easily forget initialization settings and stores the knowledge in a sparsely [6].The retrieval problem with monolithic networks can be solved by proper network design, but such scales very badly with increasing complexity.

In [7], the MNN is applied for the speaker identification. In [8] MNN is used for the biometric recognition. In [9], the topology and parameters of the MNN are optimized with a Hierarchical Genetic Algorithm, and it is used for human recognition. In this paper we want to verify if this algorithm performs in MNNs. As we will see, the MNNs obtained are optimal in the sense that the architecture is simple with high recognition.

The paper is organized as follows: in section 2 we briefly present the basics of ABC. In section 3 we explain the ANN based on the ABC algorithm (ABC-NN) and the MNN based on ABC algorithm (ABC-MNN). In section 4 the experimental results using different classification problems are given. Finally, in section 5 the conclusions of the work are presented.

2   Artificial Bee Colony Algorithm

The ABC algorithm simulates the intelligent foraging behavior of honey bee swarms. In ABC algorithm, the position of a food source represents a possible solution to the optimization problem and the nectar quantity of a food source corresponds to the quality (fitness) of the associated solution. The colony of artificial bees contain three groups of bees: *Employed bees, Onlookers* and *Scout bees*. These bees have got different tasks in the colony, i. e., in the search space.

Employed Bees: Each bee searches for new neighbor food source near of their hive. After that, it compares the food source against the old one using (1). Then, it saves the best food source in their memory.

$$v_{ij} = x_{ij} + rand(0,1)(x_{ij} - x_{kj}) \tag{1}$$

where $k \in \{1,2, \dots, SN\}$ and $j \in \{1,2, \dots, D\}$ are randomly chosen indexes. Although k is determined randomly, it has to be different from i. SN is the number of the *Employed bees* and D is the dimension of the solution.

After that, the bee evaluates the quality of each food source based on the amount of the nectar (the information) i.e. the fitness function is calculated. Providing that its nectar is higher than that of the previous one, the bee memorizes the new position and forgets the old one.Finally, it returns to the dancing area in the hive, where the *Onlooker bees* are.

Onlooker Bees: This kind of bees watch the dancing of the employed bee so as to know where the food source can be found, if the nectar is of high quality, as well as the size of the food source. The *Onlooker bee* chooses a food source depending on the probability value associated with that food source, p_i is calculated by the following expression:

$$p_i = {fit_i} \Big/ {\sum_{k=1}^{SN} fit_k} \tag{2}$$

where fit_i is the fitness value of the solution i which is proportional to the nectar amount of the food source in the position i and SN is the number of food sources which is equal to the number of Employed bees.

Scout Bees: This kind of bees help to abandon the food source which can not be improved further through a predetermined number of cycles and produce a position randomly replacing it with the abandon one. This operation can be defined as in (3).

$$x_{ij} = x_{min}^j + rand(0,1)(x_{max}^j - x_{m\,min}^j) \tag{3}$$

The pseudo-code of the ABC algorithm is next shown:

```
program ABC (globalX)
   const   swarm Size SN; search place [xmin_i,xmax_i];
dimension D; limit; MaxIter;
   var     Iter: 0..MaxIter;
   begin
     Iter := 0;
     Initialize the population of solution x_i by (3) and
evaluate its fitness fit_i, i=1,2,..,SN;
     repeat
       Produce new solution v_i with the employed bees by
(1) and evaluate them, then apply the greedy selection
process;
       Calculate the probability for each solution by (2);
       Produce new solutions for the onlookers from the
solutions selected depending on the probability and
evaluate them ,then apply the greedy selection process;
       Determine the abandoned solution for the scout, if
exist, then replace it with a new randomly produced
solution by (3);
       Memorize the best solution global achieved so far;
     until Iter = MaxIter
end.
```

3   Training Algorithms

3.1   Neural Network Learning Algorithm Based on ABC(ABC-NN)

ANN is widely used in approximation and classification problems. The most common neural network is of a three-layer forward neural network which consists of two fully connected layers of neurons: one hidden layer and one output layer. Thresholds of nodes, the weight values between input and hidden, hidden and output nodes are randomly initialized. For example, the network's structure is n-p-m, where n, p and m are the numbers of the input node, hidden node and out node respectively. The output value of the network can be obtained by the following formula:

$$y_i = f_i(\textstyle\sum_{j=1}^n w_{ij}x_j + b_i) \tag{4}$$

where w_{ij} and b_i are the weight values and thresholds of the network, x_j is the jth input, y_i is the ith output, f_i is the node transfer function. Usually, the node transfer function is a nonlinear function such as: a sigmoid function, a Gaussian functions. Here, we take the sigmoid function as the transfer function. Network error function E will be minimized as

$$E(w(t), b(t)) = \tfrac{1}{N}\textstyle\sum_{i=1}^m \sum_{j=1}^N (y_i - o_i)^2 \tag{5}$$

where $E(w(t), b(t))$ is the error at the tth iteration, $w(t)$ and $b(t)$ are the weights and thresholds in the connections at the tth iteration, y_i and o_i are the actual and desired output of the ith output node, N is the number of patterns.

BP algorithm is a powerful technique applied to train ANN. However, as the problem complicated, the performance of BP falls off rapidly because gradient search techniques tend to get trapped at local minima. When the nearly global minima are well hidden among the local minima, BP can end up bouncing between local minima, especially for those non-linearly separable pattern classification problems or complex function approximation problem [10]. A second shortcoming is that the convergence of the algorithm is very sensitive to the initial value. So, it often converges to an inferior solution and gets trapped in a long training time. Then a powerful swarm intelligence optimization algorithm ABC is introduced to enhance the neural network training.

The weights and thresholds of the networks consist of a bee. The fitness value of each solution corresponds to the value of the error function evaluated at this position. The three-layered structure trained using ABC algorithm by minimizing the error function (5) in [11]. The stop condition is the maxIter or the value of error function.

3.2 Model Neural Network Learning Algorithm Based on ABC(ABC-MNN)

MNN is especially efficient for certain classes of regression and classification problems, as compares to the conventional monolithic artificial neural networks [12]. The ABC algorithm has a strong ability to find global optimistic result. Combining the ABC with the MNN, a new hybrid algorithm (ABC-MNN) is proposed in this paper. The algorithm is made up of Data Division Module (DDM), *ABC-NN$_i$* Module and Integration Module (IM). The structure of the ABC-MNN is showed in figure 1.

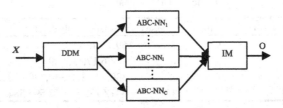

Fig. 1. Architecture of the ABC-MNN

Assume that the input sample is (X, T), where $X = (x_1, x_2, \ldots, x_m)$ is the input vectors, m is the number of the samples, $x \in |R^n$, T is the target output and O is the actual output . The data set has C classes.

Data Division Module (DDM) works to divide the dataset. For each module ABC-NN, the dataset X is divided into two categories, so it produces C datasets according to the number of the categories of the dataset. Each dataset has only two categories, i.e., ith type and non- ith type. The C dataset is noted as $S_1, S_2, \ldots S_C$.

$$S_i = \{(X, T^i)|X \text{ is the input vectors and } T^i \text{ is the new target vectors}\}$$

Each dataset S_i corresponds to a *ABC-NN$_i$* Module. The ABC-NN is an independent ANN which is a three-layer forward neural network like the BP neural network, and its structure is 'n-p-1', p is the number of the hidden layer node. The input of the model *ABC-NN$_i$* is the dataset S_i. At this stage, ABC is used to evolve the synaptic weights of sub-neural network *ABC-NN$_i$* so as to obtain a minimum Mean Square Error (MSE) as well as a minimum classification error (CER) for the ith type of data. Supposing in the well trained network *ABC-NN$_i$*, using y_{ij} to denote the output of the jth sample.

Integration Module (IM) is used to integrate the outputs of all ABC-NN. Using o_j to denote the categories and $o_j = 0$ means the jth sample not belong to any class at first, i.e., if $o_j = i$ then the jth sample belongs to ith class. The following algorithm is used to decide whether this sample belongs to the ith class.

$$o_j = \begin{cases} i, & o_j = 0 \text{ and } y_{ij} \geq 0.5 \\ 0, & o_j > 0 \text{ and } y_{ij} \geq 0.5 \end{cases} \tag{4}$$

It is possible that there are still some samples cannot be classified or be repeated classification, i.e. $o_j = 0$. Here the winner-take-all rule is used. If $o_j = 0$, then $o_j = k$ which $k = \{k|y_{kj} = max(y_{1j}, y_{2j}, \ldots, y_{Cj})\}$.

4   Experiments and Comparison

Several experiments are performed in order to evaluate the accuracy of the ABC-MNN designed by means of the proposal. The accuracy of the ABC-MNN is tested with four pattern classification problems which are taken from UCI machine learning benchmark repository [13]: Wine, Glass, Segment and Optdigits. Their characteristics are given in Table 1.

Table 1. Datasets characteristics

Datasets	Observations	Features	Classes	Respective observations
Wine	178	13	3	59,71,48
Glass	214	9	6	70,76,17,13,9,29
Segment	2310	19	7	330,330,330,330,330,330,330
Optdigits	5620	64	10	541,573,556,552,595,584,549,572,536,562

The parameters of the ABC algorithm and the network are set to the same value for all the dataset problems: Colony size (NP = 40), number of food sources NP/2, limit = 50, the maximum number of cycles is MCN = 500, the minimum of the error function (5) is goal=0.01 and the transfer function is sigmoid function.

Twenty experiments are performed using each dataset, half of them for the case of ABC-NN model and half to the ABC-MNN model. For each experiment, each dataset is randomly divided into two sets: a training set and a testing set, this with the aim to prove the robustness and the performance of the methodology. The same parameters are used through the whole experimentation.

Once generated the NN for each problem, we proceed to test their accuracy and speed. Table 2 shows the best, average and worst percentage of recognition for all the experiments using ABC-NN and ABC-M NN. In this Table, we can observe that the best percentage of recognition for most databases is achieved only during training phase. The accuracy slightly diminish during testing phase, but the Glass problem is more serious. However, the results obtained with the proposed methodology ABC-MNN are highly acceptable and stable.

Table 2. Comparison of training and testing accuracy

Dataset		ABC-NN		ABC-MNN	
		Training(%)	Testing(%)	Training(%)	Testing(%)
Wine	best	100	95.34	100	100
	average	100	94.65	100	97.15
	worst	100	93.02	100	95.34
Glass	best	76.54	75.00	83.95	76.92
	average	75.55	71.15	82.77	74.42
	worst	74.69	63.46	80.86	73.07
Segment	best	90.55	89.72	95.27	94.77
	average	88.69	87.10	94.98	93.36
	worst	86.29	83.97	94.52	91.81
Optdigits	best	81.46	78.46	92.93	89.59
	average	76.67	71.94	92.21	88.19
	worst	68.21	62.66	91.49	87.25

For the best values achieved, there are many hundreds that represent the maximum percentage (100%) of classification that can be achieved by the designed MNNs. This is important because at least, we find one configuration that solves a specific problem without misclassified patterns or with a low percentage of error. For the worst values achieved with the ANN are also represented. Particularly, the dataset that provides the worst results is the Glass problem which is very complicate and imbalance. Nonetheless, the accuracy achieved is highly acceptable.

Table 3 shows the average and standard of the training time. From this table, we can obtain that the speed of the ABC-MNN is more quickly and stable than the ABC-NN methodology for the first three dataset. For the last dataset, although the variance of the ABC-MNN model is large, this is because the training speed is very fast for two times. In general, the training speed of the ABC-MNN model is more quickly and stable.

Table 3. The average and standard of the training time

Dataset	ABC-NN		ABC-MNN	
	Average(s)	Standard	Average(s)	Standard
Wine	0.9592	0.0305	0.4176	0.0034
Glass	5.4566	0.0004	4.4753	0.0001
Segment	73.9704	0.6093	66.9552	0.3948
Optdigits	404.813	122.238	317.9599	850.007

From these experiments, we observe that the ABC algorithm is able to find the best configuration for an NN given a specific set of patterns that define a classification problem for the majority of cases. The experimentation further shows that the design generated by the proposal presents an acceptable percentage of classification for training and testing phase with the MNNs.

5   Conclusions

From the foregoing experimental researches, it is concluded that the MNNs which are evolved by means of ABC algorithm are characterized by satisfying approaching results and high training speed. In this paper we also tested the performance of the ABC algorithm. Although the ABC-MNN has exceeded the traditional algorithm in convergence speed and classification precision, classification problems of the real world may be not fit well with this narrowly-defined model. It still needs to span broad activities and require consideration of multiple aspects.

Acknowledgement. This work is supported by a grant from the Key Projects in the National Science & Technology Pillar Program during the Twelfth Five-year Plan Period(No. 2011BAI08B11), by grant from the Projects in the Science and Technology Program of Guangzhou(No. 2012Y2-00023).

References

1. Karaboga, D.: An idea based on honey bee swarm for numerical optimization. Technical report, Computer Engineering Department, Engineering Faculty, Erciyes University (2005)
2. Karaboga, D., Basturk, B.: Artificial bee colony (ABC) optimization algorithm for solving constrained optimization problems. In: Melin, P., Castillo, O., Aguilar, L.T., Kacprzyk, J., Pedrycz, W. (eds.) IFSA 2007. LNCS (LNAI), vol. 4529, pp. 789–798. Springer, Heidelberg (2007)
3. Karaboga, D., Akay, B., Ozturk, C.: Artificial bee colony (ABC) optimization algorithm for training feed-forward neural networks. In: Torra, V., Narukawa, Y., Yoshida, Y. (eds.) MDAI 2007. LNCS (LNAI), vol. 4617, pp. 318–329. Springer, Heidelberg (2007)
4. Karaboga, D., Ozturk, C.: Neural networks training by artificial bee colony algorithm on pattern classification. Neural Network World 19(10), 279–292 (2009)

5. Karaboga, D., Ozturk, C., Akay, B.: Training neural networks with abc optimization algorithm on medical pattern classification. In: International Conference on Multivariate Statistical Modelling and High Dimensional Data Mining (2008)
6. Auda, G., Kamel, M.: Modualr Neural Network: A Survey. Int. Journal of Neural Systems 9(2), 129–151 (1999)
7. Martinez, G., Melin, P., Castillo, O.: Optimization with Genetic Algorithms of Modular Neural Networks using Hierarichcal Genetic Algorithm Applied to Speech Recognition. In: Proceedings of International Joint Conference on Neural Networks, Canada (2005)
8. Hidalgo, D., Melin, P., Licea, G.: Optimization of modular neural networks with interval type-2 fuzzy logic integration using an evolutionary method with application to multimodal biometry. In: Melin, P., Kacprzyk, J., Pedrycz, W. (eds.) Bio-inspired Hybrid Intelligent Systems for Image Analysis and Pattern Recognition. SCI, vol. 256, pp. 111–121. Springer, Heidelberg (2009)
9. Sánchez, D., Melin, P., Castillo, O.: A New Model of Modular Neural Networks with Fuzzy Granularity for Pattern Recognition and Its Optimization with Hierarchical Genetic Algorithms. In: Batyrshin, I., Sidorov, G. (eds.) MICAI 2011, Part II. LNCS, vol. 7095, pp. 331–342. Springer, Heidelberg (2011)
10. Gori, M., Tesi, A.: On the problem of local minima in back propagation. IEEE Trans. Pattern Anal. Mach. Intell. 14(1), 76–86 (1992)
11. Akay, B., Karaboga, D.: A modified Artificial Bee Colony algorithm for real-parameter optimization. Inform. Sci. (2010)
12. Ling, W.X., Zheng, Q.L., Chen, Q.: GPCMNN: A Parallel Cooperative Modular Neural Network Architecture Based On Gradient. Chinese Journal of Computers 27(9), 1256–1263 (2004)
13. Murphy, P.M., Aha, D.W.: UCI Repository of machine learning databases. Technical report, Irvine, CA, US (1994)

Algorithms and Framework for Comparison of Bee-Intelligence Based Peer-to-Peer Lookup

Vesna Šešum-Čavić and Eva Kühn

Technical University Vienna, Institute of Computer Languages,
Argentinierstrasse 8, 1040 Wien, Austria
{vesna,eva}@complang.tuwien.ac.at

Abstract. Peer-to-peer has proven to be a scalable technology for retrieval of information that is widely spread among distributed sites and that is subject to dynamic changes. However, selection of a right search algorithm depends on many factors related to actual data content and application problem at hand. A comparison of different algorithms is difficult, especially if many different approaches (intelligent or unintelligent ones) shall be evaluated fairly and possibly also in combinations. In this paper, we describe a generic architectural pattern that serves as an overlay network based on autonomous agents and decentralized control. It supports plugging of different algorithms for searching and retrieving data, and thus eases comparison of algorithms in various topology configurations. A further novelty is to use bee intelligence for the lookup problem, spot optimal parameters' settings, and evaluate the bee algorithm by using the architectural pattern to benchmark it with other algorithms.

Keywords: information retrieval, lookup mechanism, bee intelligence, distributed coordination patterns.

1 Introduction

Bio-inspired algorithms play an important role in the design of self-organizing software for distributed systems. Such software is typically characterized by a huge problem size concerning number of computers, clients, requests and size of queries, autonomy and heterogeneity of participating organizations, and dynamic changes of the environment. In such a setting, the common approach of one central coordinator often reaches its technical and conceptual limits. On the technical side, it represents a single point of failure with the risk of becoming a performance and availability bottleneck. On the conceptual site, it is hard to design as it must be aware of the entire business logic, possessing the complete picture of all participants, and being able to cope with all possible dynamics in the environment. Therefore, to cope with the described dynamics and vast number of unpredictable dependencies on participating components, other approaches are demanded like autonomously acting components who are inspired by nature and whose behaviors implement, e.g., bio-inspired algorithms. These components act in a dynamic, ad-hoc way and adapt quickly and self-subsistent to both changing requirements and dynamically evolving system states caused through the interplay and contribution of the many components towards a global goal. On type of

Y. Tan, Y. Shi, and H. Mo (Eds.): ICSI 2013, Part I, LNCS 7928, pp. 404–413, 2013.

bio-inspired algorithms are bee algorithms that have been already applied to several computer science problems, e.g., [13], [16], [18], [19], [21] and [23]. The novelty of this paper is to evaluate the usefulness of bee-intelligence for the problem of information placement and retrieval in distributed systems and to prove its correctness. The use case employed for evaluation is to search information in the Internet. It assumes a highly dynamic setting where new URLs are being added every day. Also, there is a certain amount of invalid links that correspond to old, discarded pages [10], [12], [14], [25].

The *contribution* of this paper is: 1) The adaptation and implementation of an intelligent lookup mechanism based on bee intelligence that is able to cope with complex queries even if the given query is incomplete. 2) The definition and implementation of a generic architectural pattern for searching and retrieving data. This test-bed allows the simple exchange of different algorithms (intelligent and non-intelligent ones) simply through plugging.

The paper is organized as follows: Section 2 gives an overview of already existing approaches of solving the considered problem - location and retrieval of information in the Internet - taking into account P2P computing paradigm with lookup realized by using bee intelligence. Section 3 describes the proposed architectural pattern. Section 4 explains the bee algorithm for searching and retrieving of information. Section 5 presents the best parameters' settings and benchmarks. Section 6 summarizes the results.

2 State-of-the-Art

This section summarizes related work in the context of location and retrieval of information in the Internet that supports "intelligent" lookup mechanisms and that integrates these with the peer-to-peer (P2P) computing paradigm [2]. Table 1 gives an overview of systems that support adaptation and/or provision of a generic software framework pattern. In [4], both adaptation and a framework are offered. Namely, it is proposed a framework that supports an approach for building P2P applications based on the MAS paradigm. It inherits the free search capability of Gnutella, without relying on inefficient broadcasting techniques. However, these systems neither support complex queries (by means of incomplete information[1]), nor flexible plug-ability of different search algorithms, nor the application of different algorithms at the same time. Table 2 depicts P2P systems that support intelligent search, according to their distribution structure. Although unstructured P2P network does not scale well, it supports dynamics very well, and therefore it fits better to our problem.

---

[1] *Example* (the scenario of a crime investigating process): There is some information about the person that the police search for, but unfortunately the complete description and/or the material evidence are not available. According to the police simplified pattern of information would look like (*first_name, last_name, birth_date, ID_number, hair_color, height*). In a real case, the complete information might be missing (e.g., the person under investigation was seen by a witness at a crime scene only shortly) and only 3 entities of data are known (*John, ..., ..., ..., brown, 172*).

Table 1. Features / System

Adaptivity	[4], [6], [22]
Architectural Pattern / Framework	[3], [4], [22]
Supporting complex queries	[22]

Table 2. Classification of P2P systems in related work according to their structure

Structured P2P	[3]
Unstructured P2P	[20], [22]
Loosely structured	[4], [15]
Not defined	[6], [11], [24]

In this paper, we propose an unstructured and completely decentralized P2P based system for searching of possibly incomplete information, that supports plugging of intelligent as well as unintelligent algorithms and that is characterized by a high architectural flexibility. Also, it can serve as a general testbed for comparison of different algorithms in order to fairly compare them under the same conditions. Also, combinations of algorithms can be compared, and the test can be carried out in different topological settings.

3   Architecture and Design

A decentralized and unstructured[2] overlay network is proposed that consists of routing tables realized by means of a tuple-space based middleware [9] that supports subspaces. Each subspace can be published by using one or more names. It is reachable by both its URL and by its published names. The latter, however, must be resolved before access of the container by the corresponding operation (read, write, take) can take place. We use the subspace mechanisms also to implement a so-called *lookup-subspace* where each entry stores a published name and the URL of some container. Different lookup algorithms can be plugged in the proposed overlay network. When some specific subspace is needed to be found, then the searching is done through retrieval of a published name in one of the lookup-subspaces; the result is an URL of a subspace. The relationship between URLs and published names is 1: n.

The architecture is decomposed in different "micro" coordination patterns. A pattern represents a re-usable solution to a recurring distributed coordination problem. In addition, we also provide an implementation for each pattern. A major advantage when implementing the patterns by means of a tuple-space based middleware is that their composition can easily be achieved by just sharing subspaces between software agents that belong to different patterns. The main micro pattern for the lookup framework is termed "local node pattern" and is represented in Fig. 1a). The whole network consists of finite numbers of nodes. The local node pattern is responsible to model the content

---

[2] On the current stage of research, we use the scale-free network approach [1] for an initial construction (with the initial number of subspaces $m_0=2$).

which is the subject of search and provides an environment for its searching by using swarm agents (e.g., ants or bees). The components of a local node pattern are: clients, swarm agents, a swarm subspace, a content subspace and a routing subspace.

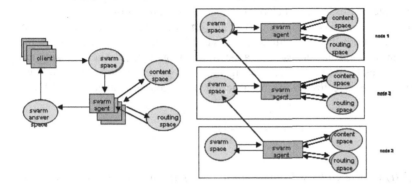

Fig. 1. a) Local Node Pattern (ovals represent subspaces also termed "spaces" for short in the figures, and rectangles represent software agents), b) Pattern composition through the sharing of subspaces. 3 nodes are shown. Each possesses a swarm subspace, a content subspace, and a routing subspace; the violet rectangles represent swarm agents.

Clients supply requests. The swarm subspace maintains these search requests, information about the current status of searching and the current list of visited nodes. A swarm agent picks up the next request, realizes the search and changes the status of the search according to the quality of data contained on a particular node. Swarm agents consult both the content subspace and the routing subspace. The content subspace contains information about public names and real names (URLs). The routing subspace contains the list of neighboring nodes and additional information connected with the type of algorithm (e.g., quantity of pheromones for ant algorithms, duration of waggle dance for bee algorithm on links). When the search is completed, the search result is put in the swarm answer subspace, where it is picked up by clients. Fig. 1b) presents the pattern composition. The elements of the distributed swarm pattern are swarm agents that access also remote swarm subspaces and a data quality policy expressed through the similarity function. The local answer swarm subspaces are not depicted in Fig. 1b).

4   Bee Algorithm for Information Retrieval

4.1   Bee Algorithm

The way of mapping bees' behavior from nature to IT terminology is abstracted in the following way [21]: Software agents represent bees at the particular nodes in the overlay network. A node contains exactly one hive and one flower, where a flower can have many nectar units that can be taken out by a bee. A hive has a finite number of receiver bees and outgoing (i.e., forager plus follower) bees. A task is one nectar unit and represents "the searching through lookup subspaces according to the published name", where the publish name is a string. At the beginning of the process, the

population of bees in a hive is without any information about the environment. Therefore, all outgoing bees are foragers. Foragers navigate and recruit followers[3]. The goal
is to find the most similar or the same information (comparing with the published
name) in the network by taking the best path which is defined to be the shortest one[4].
The two basic phases in bees' behavior are navigation and recruitment.

In the navigation phase, a bee goes (i.e., an agent search) from one lookup subspace to
another until it finds the most similar or the same information in the network or it visits
and examines all lookup subspaces in a network without results. After finishing navigation, a bee goes back to the hive. A navigation strategy determines which node will be
visited next and it is realized by a stochastic state transition rule adopted from [23]:

$$P_{ij}(t) = \frac{[\rho_{ij}(t)]^{\alpha} \cdot [\frac{1}{d_{ij}}]^{\beta}}{\sum\limits_{j \in A_i(t)} [\rho_{ij}(t)]^{\alpha} \cdot [\frac{1}{d_{ij}}]^{\beta}} \tag{1}$$

where $\rho_{ij}(t)$ is the arc fitness from node i to node j at time t and d_{ij} is the heuristic distance between i and j, α is a binary variable that turns on or off the arc fitness influence and β is the parameter that controls the significance of a heuristic distance.

In the recruitment phase, the knowledge about length of the path (distance, which
is expressed as the number of hops) and quality of the solution (measured by the similarity function δ) is communicated and exchanged between bees. A mathematical
description is presented through the fitness function:

$$f_i = \frac{1}{H_i}\delta \tag{2}$$

for a particular bee i, where H_i is the number of hops on the tour, and δ is the similarity function that is new one and specially adapted for this problem. The general form
of the similarity function is: $\delta = \delta$ (current solution, exact solution), that describes
how good (acceptable) solution is found, $\delta \in [0,1]$. The type of the similarity function
δ can be changed, however, its values are normalized into the segment [0,1]. Searching for the specified data can results in the following situations: *no data* found, or
exact data found, or *acceptable data* found with the accuracy/error rate $< \varepsilon$, where ε is
a parameter given in advance, connected to the definition of δ. The comparison of two
strings (that represent two URLs) is done by using a similarity function δ that is
based on the principle of spatial locality. However, a similarity function δ is also
configurable and can be expanded by using temporal locality (they are similar if they
have the same number of access of some processes) or semantic similarity[5].

---

[3] The main actors in the algorithm are foragers and followers, as receivers have no influence on
the algorithm.
[4] The path is measured by means of number of hops. However, the measurement can be
changed (e.g., the quality of links can be used instead).
[5] Generally, the fitness function described by equation (2) as well as the similarity function
δ is use-case specific. However, it can be further extended taking into account some other
parameters connected with another specific use-case (e.g., the quality of links, etc.)

If bee i found a highly similar (i.e., the value of δ is close to 1) or the same string ($\delta=1$) in some lookup container - then its fitness function, f_i will obtain a good value. The colony's fitness function is the average of all fitness functions:

$$f_{colony} = \frac{1}{n}\sum_{i=1}^{n} f_i \qquad (3)$$

where n is the number of outgoing bees. For each outgoing bee, its f_i is compared with f_{colony} which determines how "good it was" and based on that a bee decides its next role (a forager or a follower) [17].

5   Evaluation and Results

Different algorithms for lookup are plugged in order to evaluate the behavior of bee algorithm and compare it with others. The benchmarks are performed on basis of the following criterions: 1) For each intelligent algorithm find out best combination of parameter settings; 2) Compare these optimally tuned swarm based algorithms with Gnutella. Gnutella was chosen for a comparison as it is the most similar to the systems proposed in this paper: unstructured P2P, purely decentralized.

The results are obtained by using swarm intelligence algorithms in different combinations (Table 3). Bee algorithm is combined with random writing of data, as brood sorting mechanisms are characteristic for ants. The test-examples are constructed according to Table 4.

Table 3. Different combinations of algorithms: first column designates ways of writing data, first row designates algorithms for searching data, "*" marks implemented combinations

	MMAS [8]	AntNet [7]	Bee algorithm (from section 4)
Random	*	*	*
Brood Sorting[6] [5]	*	*	

Table 4. Parameters selected according to the best results obtained in [21] and [23]

	bee algorithm parameters	MMAS parameters	AntNet parameters
number of subspaces	40,80,120,160,200	40,80,120,160,200	40,80,120,160,200
α	0,1	0.5, 1	0.2, 0.3, 0.45
β	from 8 to 12 with step 2	from 2 to 5 with step 1	
λ	0.99		
ρ		from 0.2 to 0.9 with step 0.2	
c_2			0.15,0.25,0.3, 0.35

---

[6] It simulates brood sorting mechanism in ant colony from nature; entries are distributed on the basis of their type (similar entries stay closer to each other). As this mechanism comes from ant colony, it is applied in combination with ant-based algorithms only (not applied in combination with bee algorithm).

The rest of parameters for AntNet were based on values recommended in [7] and [8][7]. We do not give an explanation of MMAS and AntNet algorithms and their parameters, as they are implemented using the description from [7] and [8]. The investigation of the best parameters settings for each algorithm led to the following results: 1) for MMAS algorithm: $\alpha = 0.0$, $\beta = 5.0$, $\rho = 0.5$; 2) for AntNet algorithm: $\alpha = 0.2$, $C_2 = 0.35$, if number of subspaces = 40, i.e., $C_2 = 0.25$, if number of subspaces> 40; 3) for Bee algorithm: $\alpha = 1.0$, $\beta = 10.0$, $\lambda = 0.99$.

As the algorithms are non-deterministic, each test is repeated 10 times[8] and the average values are computed. The benchmarks are performed on Amazon Cloud. We used standard instances of 1.7 GB of memory, 1 EC2 Compute Unit (1 virtual core with 1 EC2 Compute Unit), 160 GB of local instance storage, and the 32-bit platform. The comparison of these algorithms' results is shown in Fig. 2, while increasing the number of subspaces.

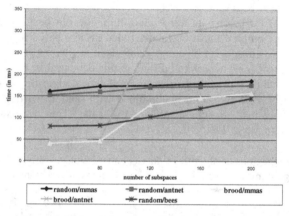

Fig. 2. A comparison of all algorithms (number of queries = 1, number of swarms = 10). The metric used is time (in msec).

Ant algorithms with writing based on brood sorting were successful with small instances. However, increasing the dimensions, brood/Antnet did not obtain good results (possibly an over-clustering affected system's robustness and did not fit to AntNet dynamics), whereas brood/MMAS preserved the obtaining of good results. Bee algorithm introduced in subsection 4, obtained relatively good results on small instances (although not so good as brood based algorithms), but the best results with increasing the dimension (with bigger instances).

The query capability of the system is measured and the different intelligent lookup mechanisms are compared to Gnutella lookup mechanism (implemented by using the

---

[7] The used exponential mean coefficient is 0.005, the time interval between two generations is 0.3 sec, the maximum length of ant's life (in hops) is 15, the maximum length of the observation window is 0.3 and the value of C_1 is 0.7.

[8] The number of sampling of nondeterministic algorithms was chosen according to the fact how quickly the algorithm converges, and whether the obtained results are uniform.

description from [2]). Gnutella can cope with complex queries, but it does not support a case with incomplete information when some search parameters of the query are not specified. For supplied query (with completed information), Gnutella offers the exact result. Our system supports also incomplete information and returns the wanted or a most similar string.

Table 5. and Table 6. present the results obtained by using different intelligent lookup mechanisms, while increasing the number of queries and swarms.

Table 5. A comparison of the performances of different lookup mechanisms (number of subspaces = 80)

Algorithm	# Queries / Swarms				
	1 / 10	2 / 20	3 / 30	4 / 40	5 / 50
random/mmas	172	183	195	205	217
random/antnet	159	167	186	192	201
brood/mmas	46	126	163	194	214
brood/antnet	52	280	309	328	346
random/bees	82	124	160	189	200
Gnutella	474	476	500	502	517

Table 6. A comparison of the performances of different lookup mechanisms (number of subspaces = 120)

Algorithm	# Queries / Swarms				
	1 / 10	2 / 20	3 / 30	4 / 40	5 / 50
random/mmas	174	194	199	210	222
random/antnet	170	182	203	213	234
brood/mmas	130	668	705	726	739
brood/antnet	277	361	379	402	411
random/bees	102	127	144	168	185
Gnutella	539	559	595	607	611

The lookup based on bee intelligence obtained the best results on large instances compared to other algorithms. For example, if number of subspaces is 80 and number of queries is 5, then random/bee algorithm is 60% better than Gnutella, 0.5% better than random/antnet, 42% better than brood/antnet, 7.8% better than random/mmas, 6.5% better than brood/mmas. If number of subspaces is 120 and number of queries is 5, then random/bee algorithm is 69% better than Gnutella, 20% better than random/antnet, 75% better than brood/mmas, 17% better than random/mmas, 55% better than brood/antnet.

The benefit of bee algorithm is visible on large instances. This algorithm differs from ant algorithm especially as it informs the "starting place" of the search directly in a P2P way and, therefore got the better results. Performance of each intelligent algorithm outperforms Gnutella lookup. Namely, the original Gnutella architecture uses a flooding (or broadcast) mechanism that supports an exhaustive optimization.

From the other side, intelligent algorithms focus on important areas of the solutions space. They quickly and efficiently narrow the number of combinations to be calculated by focusing on the areas that are most profitable and most stable. Also, their advantage is that they try to find a global optimum.

Another result of the evaluation is that the generic architectural pattern is robust and allows for plugging of different algorithms without any change in the source code; simply by reconfiguration.

6   Conclusion

Location and retrieval of complex data in the Internet is an important IT problem with an everyday increasing complexity. We propose an intelligent solution: a self-organized architectural pattern that serves as a purely unstructured overlay network and that is based on autonomous agents. Different algorithms for lookup are plugged in: ant algorithms, Gnutella lookup mechanism, and a bee algorithm that is adapted for the first time for this scenario. The proposed bee algorithm simulates the behavior of bees in nature. The benchmarks are performed on Amazon Cloud by using the best parameters' setting. The lookup based on bee intelligence obtained the best results on large instances compared to other algorithms.

Future work will consider plugging of further new swarm-based algorithms, evaluating their behavior also in combination with other algorithms, as well as their application in new use cases.

References

1. Albert R., Barabási A.-L.: Statistical mechanics of complex networks. Reviews of ModernPhysics 74, 47–97 (2002)
2. Androutsellis-Theotokis S., Spinellis D.: A survey of peer-to-peer content distribution technologies. ACM Comput. Surv. 36, 335-371 (2004)
3. Apel S., Buchmann E.: Biology-Inspired Optimizations of Peer-to-Peer Overlay Networks. Quellenangabe Praxis der Informations. und Kommunikation, 28(4) (2005)
4. Babaoglu O., Meling H., Montresor A.: Anthill: A Framework for the Development of Agent-Based Peer-to-Peer Systems. 22th Int. Conf. on Distr.Comp. Systems (2002)
5. Casadei M., Menezes R., Viroli M., et al.: A Self-organizing Approach to Tuple Distribution in Large-Scale Tuple-Space Systems. IWSOS, 146-160 (2007)
6. Dasgupta P: Intelligent agent enabled genetic ant algorithm for P2P resource discovery, 3rd Int. Conf. on Agents and Peer-to-Peer Computing (2004)
7. Di Caro G., Dorigo M.: AntNet: Distributed Stigmergetic Control for Communications Networks. JAIR 9, 317-365 (1998)
8. Dorigo M., Stuetzle T.: Ant Colony Optimization. MIT Press (2005)
9. Gelernter D., Carriero N.: Coordination languages and their significance. ACM Commun 35:97-107 (1992)
10. Gudivada V.N., Raghavan V.V., Grosky W.I., Kasanagottu R., Information Retrieval on the World Wide Web, IEEE Internet Computing 5, 58-68, (1997)

11. Islam M.H., Waheed S., Zubair I: An efficient gossip based overlay network for peer-to-peer networks. 1st Int. Conf. on Ubiquitous and Future Networks. IEEE, 62-67, (2009)
12. Knoblock C.A., Searching the World Wide Web, IEEE Expert: Intelligent Systems and Their Applications 12, 8-14, (1997)
13. Lemmens, N., de Jong, S., Tuyls, K., Nowe, A.: Bee Behaviour in Multi-agent Systems, Adaptive Agents and MAS III, LNAI 4865, 145-156, (2008)
14. Li H., Wu Z., Ji X: Research on the Techniques for Effectively Searching and Retrieving Information from Internet. IEEE Int. Symp. Electronic Commerce and Security, 99-102, (2008)
15. Liang C.Y., Ming L.T.: Small World Bee: Reduce Messages Flooding and Improve Recall Rate for Unstructured P2P System. Int. J. of Computer Science and Network Security, 11(5) (2011)
16. Markovic G., Teodorovic D., Acimovic-Raspopovic V: Routing and wavelength assignment in all-optical networks based on the bee colony optimization. AI Commun. 20(4), 273-285, (2007)
17. Nakrani S., Tovey C.: On honey bees and dynamic server allocation in the Internet hosting centers. Adaptive Behaviour 12:223-240 (2004)
18. Olague G., Puente C: The Honeybee Search Algorithm for Three-Dimensional Reconstruction. 8th European Workshop on Evolutionary Computation in Image Analysis and Signal Processing, LNCS 3907, 427–437, (2006)
19. Pham D.T., Koç E., Lee J.Y. et al: Using the Bees Algorithm to schedule jobs for a machine. 8th Int. Conf. on Laser Metrology, 430–439, (2007)
20. Ren H, Xiao N, Wang Z: An interest-based intelligent link selection algorithm in unstructured P2P environment. 7th Int. Conf. on Algorithms and Architectures for Parallel Processing. Springer Verlag, 326-337, (2007)
21. Šešum-Cavic V., Kühn E., Self-Organized Load Balancing through Swarm Intelligence, Next Generation Data Technologies for Collective Computational Intelligence, Studies in Computational Intelligence, Springer Verlag book chapter, 352, 195-224, (2011)
22. Šešum-Cavic V., Kühn E., A Swarm Intelligence Appliance to the Construction of an Intelligent Peer-to-Peer Overlay Network, Int. Conf. on Complex, Intelligent & Software Intensive Systems (CISIS), IEEE, 1028-1035 (2010)
23. Wong L.P., Low M.Y., Chong C.S: A Bee Colony Optimization for Traveling Salesman Problem. 2nd Asia Int. Conf. on Modelling & Simulation, AMS, IEEE, 818-823, (2008)
24. Yang S.J.H, Zhang J, Lin L, Tsai J. P: Improving peer-to-peer search performance through intelligent social search. Expert Syst. Appl. 36(7), 10312-10324, (2009)
25. Zhao W.: A Novel Approach of Web Search Based on Community Wisdom. 3rd Int. Conf. on Internet and Web Applications and Services, 431 – 436, (2008)

Differential Evolution with Group Crossover for Automatic Synthesis of Analog Circuit

Ting Wu and Jingsong He[*]

Department of Electronic Science and Technology, University of Science and Technology of China, Hefei, China
tingw@mail.ustc.edu.cn, hjss@ustc.edu.cn

Abstract. Analog circuit design is significant and challenging. In this paper, we propose a group-crossover-based variable-length differential evolution (GVDE) for automatic synthesis of analog circuit. We present two experimental results obtained using the proposed GVDE, including a low-pass filter and an inverting amplifier. The results showed that GVDE is able to evolve with variable-length chromosome, which allows both the topology and sizing of analog circuit to be evolved. The proposed GVDE is an efficient algorithmic approach for automatic synthesis of analog circuit.

Keywords: analog circuit design, differential evolution, variable length evolution, group crossover.

1 Introduction

Analog circuit design is known to be challenging and significant. Analog circuit design consists of topology optimization and parameter optimization. The length of chromosome can be variable or fixed during the evolutionary process for topology optimization. Grimblebly[1] proposed a fixed-length chromosome method to evolve topology of analog circuit. In this paper, null type are defined to keep all the chromosome at the fixed length. When translate the chromosome into netlist, null type should be removed. Thus, the size of circuits are changeable. Goh et al. [2] proposed a similar fixed-length chromosome method to evolve both the topology and sizing of analog circuit. Fixed-length chromosome method must predefine the size of the circuit, which neither be too large(search space is too large to find the optimal) nor too small(search space is too small and no optimal can be obtained).

Another approach for topology optimization is evolving with variable-length chromosomes. It can avoid predefining the max size of the circuits to be evolved. Ando et al. [3] proposed a variable-length component-list representation for automatic design of analog circuit. Koza et al. [4] proposed automatic synthesis of a suite of analog circuit by mean of genetic programming. Lohn et al. [5] proposed GA with a linear representation for filter and amplifier design. So in order to optimize topology of circuits, we hope the length of chromosomes can change with the size of circuits.

---

[*] Corresponding author.

Y. Tan, Y. Shi, and H. Mo (Eds.): ICSI 2013, Part I, LNCS 7928, pp. 414–421, 2013.

Parameter optimization operates on circuits of fixed topology. The length of chromosome is fixed during the evolutionary process. Vondras et al. [6] used differential evolution to evolve multi-criterion filter circuit with a given topology. Liu et al. [7] adjusted the amplifier parameter by competitive co-evolutionary differential evolution. Sabat et al. [8] proposed optimizing OTA Miller parameters with differential evolutionary and swarm intelligence techniques.

Differential evolution method is first proposed by Storn and Price in 1995 [9]. It's a powerful and efficient stochastic search technique for global optimization problem. DE gets a good effect on the performance of parameter optimization, such as [6][7][8]. These are also some studies about topology optimization with DE, such as [10][11].

Based on the former analyses, this paper proposes a group-crossover-based variable-length differential evolution method(GVDE). A low-pass filter and an inverting amplifier design task are proposed by the use of GVDE and Winspice simulator. Experimental results show that the proposed GVDE is able to evolve with a variable-length chromosome method, it allows to optimize topology and parameter of analog circuit at the same time. GVDE is an efficient algorithmic approach for automatic synthesis of analog circuit.

2   Classical Differential Evolution for Circuit Design

The basic mutation strategy of DE is called *DE/rand/1* (Eq. (1)), where v is mutation vector, indexes r_1, r_2, r_3 ($r_1 \neq r_2 \neq r_3 \neq i$) are random integers generated in the range [1,NP]. F is the weighted factor. The crossover operation of classical DE is shown in Eq. (2). Where j is the dimension of vector, and $j \in (0, D-1)$. The target vector $x_{i,g}$ and the trial vector $u_{i,g+1}$ are selected by one-to-one greedy strategy.

$$v = F * (x_{r_1,g} - x_{r_2,g}) + x_{r3,g}. \tag{1}$$

$$u_{(i,g+1)j} = \begin{cases} v_{j,g} & rand_j(0,1) < CR \\ x_{(i,g)j} & otherwise \end{cases}. \tag{2}$$

From Eq. (1), it can be seen that vectors participate in differential mutation are at the same dimension. Eq. (2) indicates the crossover method of DE is probability-based stochastic crossover on single dimension. These show that the dimension of the vectors remain the same while using classical DE. Similarly, the chromosome must be at the same length when evolving analog circuit using classical DE. There are some examples for circuit synthesis using DE. Paper [6]~[8] are parameter optimization with a fixed topology structure. [10] and [11] achieve topology optimization and parameter optimization at the same time. The length of chromosome are all the same during the evolutionary process in these papers.

As mentioned in section 1, in order to realize both topology optimization and parameter optimization at the same time. We hope the length of chromosomes can change with the size of circuits. So in this paper, we introduce a group-crossover-based variable-length differential evolution method for analog circuit design.

3   Group-Crossover-Based Variable-Length Differential Evolution

Individuals of new generation come from trial individual. In order to satisfy the variable length characteristic of chromosome, we should guarantee the length of chromosome variable and diverse after crossover operation. We proposed a group crossover strategy to ensure variable-length chromosome evolution in GVDE.

In GVDE, the length of the individuals participate in differential mutation are not exactly the same. This paper proposes a length processing strategy, which can guarantee the implement of differential mutation. Also, it can make the length of mutation individual random and ergodic.

3.1   Group Crossover

The basic idea of group crossover can be briefly stated as follows. Separating both of the two chromosomes (parent individuals participating in crossover operation) into M groups randomly. Then, executing crossover operation between the same groups from the two parent individuals to generate the trial individual. In this paper, we choose the current best individual $xbest,g$ and mutation individual v generated from differential mutation as parent individuals. The trial individual is generated according to Eq. (3).

$$u_{group\_i} = \begin{cases} v_{group\_i} & randj(0,1)<CR \\ x_{(best,g)\ group\_i} & otherwise \end{cases} \quad i = 1,2,\cdots,M. \tag{3}$$

An example of group crossover is showed in Fig. 1. In the proposed group crossover process, M is generated randomly, the length of each groups is random too. It's clearly that the length of trial individual generated from group crossover is random.

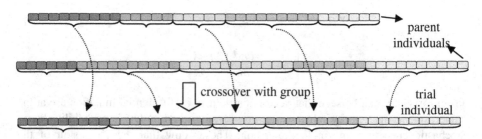

Fig. 1. An example of group crossover

3.2   Differential Mutation with Random Strategy

We deal with individuals participate in mutation operation to make them at the same length, thus easy to execute differential mutation. The proposed operation is called random strategy in this paper. The basic idea is generating a random length for mutation individual, denote as *length_v*. For the individuals participate in mutation, if the length of individuals is larger than *length_v*, then make truncation directly, or conversely complement to *length_v* by random initialization.

```
length_v=ceil(rand*length_max);
IF  length_i>=length_v  THEH
```
$$x_{i,g}= [x_{(i,g)0}, x_{(i,g)1}, \cdots, x_{(i,g)length\_v-1}]^T$$
```
ELSE
```
$$x_{i,g}= [x_{(i,g)0}, x_{(i,g)1}, \cdots, x_{(i,g)length\_i-1}, r_1, r_2, \cdots, r_m]^T$$
```
ENDIF
```

where $m=length\_v-length\_i$, $r_1,r_2,...r_m$ indicate genes generated randomly.

We hope the length after mutation random and ergodic at the same time. Otherwise mutation operation will decrease the diversity of length of chromosomes, further more reducing the diversity of population. Now let's test the ability to generate new length of random strategy for differential mutation. Each chromosome's length is initialized with the same length (e.g. 10), and then we make statistics on the distribution of chromosome's length every 25 generations. As shown in Fig. 2, after differential mutation with random strategy, the length of chromosomes is random and run through all of the available length.

4   GVDE for Analog Circuit Synthesis

4.1   Circuit Representation

Circuit represent method determines the search efficiency and results. Now commonly used circuit representations are linear representation, tree representation and netlist-based representation. In tree representation proposed by Koza et al. [4], circuits are mapped to a program tree composed of function sets and terminal sets. This representation based on genetic programming can achieve topology optimization and parameter optimization at the same time. Lohn et al. [5] proposed a linear representation technique for evolving analog circuits. In paper [5], circuit-constructing robot is used for placing components. Each component in the circuit consists of three genes represent the component type, connection and parameter. Circuit is mapped to the chromosome composed all genes in series. Linear representation in [5] can avoid illegal individuals during encoding process. Besides, Grimbleby [1] and Goh et al. [2] proposed a netlist-based representation scheme. In these two papers, every gene consist of three elements namely nodes, values and type. Chromosomes are made of the genes. In this paper, we choose the linear representation scheme.

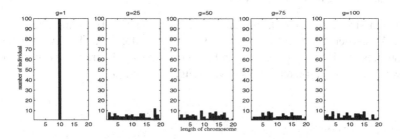

Fig. 2. Statistics on the distribution of the length of chromosomes every 25 generations

4.2   Fitness Measure

The fitness measure guides the evolutionary design directly. A low-pass filter and an inverting amplifier will be designed using GVDE in the following experiment section. Fitness measure method is same as the method presented in Koza [4].

Table 1. Design specification of low-pass filter and weighted factor for fitness measure

frequency(Hz)	attenuation(dB)	weighted factor
<=1000	>-0.26dB	10
1000 to 2000	--	8
>=2000	<-60dB	15

Spice simulation tool is used to perform an AC small signal analysis for low-pass filter. The circuit's behavior is reported for fitness calculation. Fitness is calculated by the sum of errors between the actual value and the ideal value. The design objective of low-pass filter and the weighted penalty are presented in table 1.

A DC sweep analysis is required to describe the amplifier's behavior. An ideal inverting amplifier has a DC bias equal to zero, a linear DC transfer characteristic. The voltage gain equals to the slope of the DC transfer characteristic. When calculate the fitness of an amplifier, we should take voltage gain, DC bias and linearity into account at the same time. In the experiment, the fitness is calculated by summing amplifier penalty, a bias penalty and linearity penalty.

5   Experiments and Results

In this section, we use proposed GVDE to design a low-pass filter and an inverting amplifier. In the evolutionary process, both topology and parameter of circuits are optimized. In the two following experiments, mutation factor F and crossover factor CR are variable during the evolutionary process. CR is a uniform distribution random number between 0 and 1, F is a uniform distribution random number $\in(0,2)$.

5.1   ·Low-Pass Filter Design Task

Population size is set to 200, the maximum generation is 200. Experiment runs 30 times. We chosen three results from the 30 results randomly, and draw the length of the chromosome in every generation in Fig. 3 (a) and (b). The frequency response and fitness track of the best circuit are shown in Fig. 3 (c) and (d).

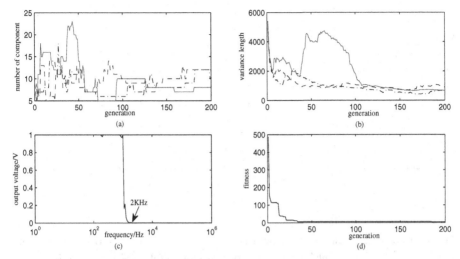

Fig. 3. (a)(b) The length of the chromosome in every generation for 3 results selected random from 30 results. (a) The length of the best chromosome in every generation. (b) The standard deviation of chromosomes in every generation. (c)(d) The best low-pass filter circuit by GVDE. (c) Frequency response of the circuit. (d) Fitness track over optimization process.

As we can see in Fig. 3 (a), the length of the best chromosome is variable during the evolutionary process. It means that chromosome's length is changeable in the evolutionary process. This verifies that GVDE can achieve variable length evolution on the problem of circuit design. In Fig. 3 (b), the standard deviation of chromosomes length in every generation is not monotonically decreasing. But the general trends of it tends to decrease. This denotes that the whole population evolves towards optimal solution as evolution continues. This conforms to the law of convergence of evolution. As shown in Fig. 3 (c), the best low-pass filter circuit meets the design specification (shown in section 4.2) completely. Fig. 3 (b) and (d) together indicate that it is sure to converge when evolving analog circuits with the proposed GVDE.

5.2   Inverting Amplifier Design Task

Ten runs are performed and we present the best performance circuits. It should be emphasized here that the topology and the parameter of the amplifier are evolved at the same time. Population size is set to 1200, the maximum generation is 1000.

The schematic of the best performance amplifier evolved by GVDE is shown in Fig. 5. DC transfer characteristic and AC signal behavior of the result circuit is in Fig.4 (a). Time domain input and output waveform are present in Fig.4 (b). From Fig.4, the DC bias voltage of the inverting amplifier is small to 1.062V, the amplifier has a DC gain of 72.59db(4264). The input signal is a sine waveform with 1uV amplitude and 1KHz frequency. It's obviously that GVDE is efficient for evolving analog circuits.

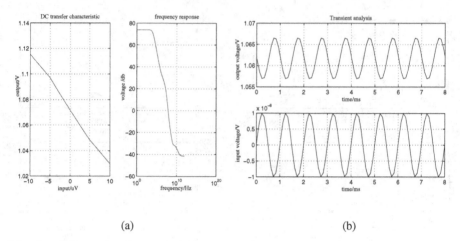

(a)   (b)

Fig. 4. a DC transfer characteristic (left) and AC signal behavior (right) of the inverting amplifier. b Time domain input (down) and output (up) waveform of inverting amplifier.

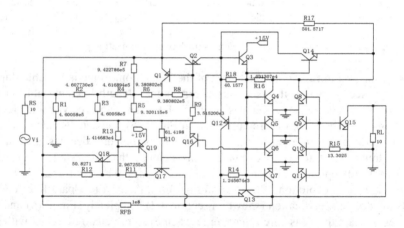

Fig. 5. Circuit schematic of evolved inverting amplifier

6   Conclusion

In this work, we makes an analysis of topology optimization and parameter optimization of evolution design of analog circuit firstly. Group-crossover-based variable-length differential evolution is proposed for analog circuit design. In the experiment, both the topology and the parameter of analog circuits are evolved at the same time. The experiment result show that GVDE is able to evolve analog circuits in the form of variable-length chromosome, GVDE is sure to converge. It's obviously that the proposed GVDE is an efficient algorithmic approach for automatic synthesis of analog circuit.

Acknowledgments. This work is supported by National Nature Science Foundation of China under Grant 60975051 and 61273315.

References

1. Grimbleby, J.B.: Automatic analogue network synthesis using genetic algorithms. In: The First Conference on Genetic Algorithms in Engineering Systems: Innovations and Applications, Sheffield, UK, pp. 53–58 (1995)
2. Goh, C., Li, Y.: GA automated design and synthesis of analog circuits with practical constraints. In: Proc. of the 2001 Congress on Evolutionary Computation, vol. 1, pp. 170–177 (2001)
3. Ando, S., Iba, H.: Analog circuit design with a variable length chromosome. In: Proceedings of the 2000 Congress on Evolutionary Computation, vol. 2, pp. 994–1001 (2000)
4. Koza, J.R., Bennett, F.H., Andre, D., Keane, M.A., Dunlap, F.: Automated synthesis of analog electrical circuits by means of genetic programming. IEEE Transactions on Evolutionary Computation 1, 109–128 (1997)
5. Lohn, J.D., Colonbano, S.P.: A Circuit Representation Techniques for Automated Circuit Design. IEEE Transactions on Automatic Control 3(3), 205–219 (1999)
6. Vondras, J., Martinek, P.: Multi-criterion filter design via differential evolution method for function minimization. In: 1st IEEE International Conference on Circuits and Systems for Communications, pp. 106–109 (2002)
7. Liu, B., et al.: Analog circuit optimization system based on hybrid evolutionary algorithms. Integration, the VLSI Journal 42(2), 137–148 (2008)
8. Sabat, S.L., Kumar, K.S., Udgata, S.K.: Differential evolution and swarm intelligence techniques for analog circuit synthesis. In: Proceedings of 2009 World Congress Nature and Biologically Inspired Computing, NABIC 2009, pp. 469–474 (2009)
9. Storn, R., Price, K.: Differential Evolution – A simple and Efficient Adaptive Scheme for Global Optimization over Continuous Spaces. Technical Report TR-95-012, ICSI (March 1995)
10. Storn, R.: Differential Evolution Design of an IIR-Filter. In: IEEE International Conference on Evolutionary Computation (ICEC 1996), pp. 268–273 (1996)
11. Vondras, J., Martinek, P.: New approach to analog filters and group delay equaliser transfer function design. In: The 8th IEEE International Conference on Electronics, Circuits and System, ICECS 2001, St. Julian's, vol. 1, p. 70 (2001)

MMODE: A Memetic Multiobjective Differential Evolution Algorithm

Zhou Wu, Xiaohua Xia, and Jiangfeng Zhang

Department of Electrical, Electronic and Computer Engineering
University of Pretoria, Pretoria, South Africa
eezhwu@gmail.com, {xxia,jzhang}@up.ac.za

Abstract. For the multiobjective problems, some global search methods may fail to find the Pareto optima with both accuracy and diversity. To pursue the two goals at the same time, a new memetic multiobjective differential evolution algorithm (MMODE) is proposed to hybridize the local search with differential evolution (DE) algorithm. The local search is conducted in an independent population to accelerate the search process, while DE can maintain the diversity. In MMODE, we use a new multiobjective Pareto differential evolution (MOPDE). Experimental results show that the MMODE performs better than other two MODEs in respects of the accuracy and diversity, especially for the multimodal functions.

Keywords: Memetic algorithm, multiobjective optimization, differential evolution, extensive dominance, MMODE.

1 Introduction

Many real world optimization problems, such as networks designing and single processing, have to consider several objectives simultaneously, which are called multiobjective optimization problems (MOOPs)[1]. In MOOPs, there exists no single solution can optimize all objectives, but only some appropriate trade-offs can be found with comprehensive good fitness in each objective. The set of these trade-offs is called the Pareto set[2]. For the multiobjective optimization algorithms, their final solutions are required to be close to the optima and well-distributed in the Pareto set, i.e., accuracy and diversity. Some multiobjective Evolutionary algorithms (MOEAs), such as NSGA-II[3] and SPEA-II[4], have been proposed to solve MOOPs successfully. When applying Differential Evolution (DE) algorithm[5] for MOOPs, several multiobjective Differential Evolution algorithms (MODEs) [6,7,8,9]are proposed with good results.

However, these MODEs tend to require a relatively large number of iterations to find the global optima due to their global explorations. Even worse, they may suffer from the drawback of stagnation, in which case the MODEs stop to converge in the early generations because of the excessive exploration. To find the satisfactory results in the given time, the local search (LS) techniques are usually combined into the MOEAs to accelerate the search process. The hybrid

Y. Tan, Y. Shi, and H. Mo (Eds.): ICSI 2013, Part I, LNCS 7928, pp. 422–430, 2013.

algorithms, which are called the Memetic algorithms (MAs), have balanced the exploration and exploitation throughout the searching process[10]. But there are few MA works applying the basic DE for the MOOPs. The Co-evolutionary Multiobjective Differential Evolution (CMODE) has been proposed to combine co-evolution concept in the MODEs, and a memetic version based on CMODE is mentioned[11]. In the design of the memetic MODEs, three open issues remain. Where should the local search be combined in the basic DE? Which individuals should be fine-tuned by the local search? When should the local refinement be applied? In this paper, a new memetic multiobjective differential evolution algorithm (MMODE) is proposed as a general memetic DE to response these questions.

The contributions of this paper are in three folds. First, MMODE is a new algorithm which can deliver promising results with accuracy and diversity. In MMODE, the DE can maintain the diversity of the population, and the local search can enhance the accuracy while accelerating the convergence rate. Second, a new algorithm multiobjective Pareto differential evolution (MOPDE) is modified for the basic DE for solving MOOPs. MOPDE employs "one-to-many" comparisons, which is less greedy than other MODEs. Third, the MMODE is a general framework based on the DE, which can be replaced by other MODE variants. In this framework, the local search is independent with the global search, so that the local search does not lead the population losing diversity.

The remainder of this paper is organized as follows. In Section 2, we give an overview of some existing MODEs. In Section 3, we introduce the proposed MMODE based on the new MODE variant MOPDE. In Section 4, the experimental results are delivered to compare the propose work with some existing DE algorithms. Section 5will conclude this paper.

2   Related Work

Multiobjective Differential Evolution (MODE) algorithm is an extension of the basic DE algorithm in the MO applications. The basic DE is one of the most recent evolutionary algorithms to solve real-parameter optimization problems[5]. After initializing a population, each individual is evolved by employing three operators - mutation, recombination and selection. At each generation, the mutant vectors are created from their parents. The following are different mutation strategies frequently used in the literature:

DE/rand/1:

$$v_i = x_{r1} + F(x_{r2} - x_{r3}),\qquad(1)$$

DE/best/1:

$$v_i = x_{best} + F(x_{r2} - x_{r3}),\qquad(2)$$

DE/current-to-best/1:

$$v_i = x_i + F \cdot (x_{best} - x_i) + F \cdot (x_{r2} - x_{r3}),\qquad(3)$$

where the indices $r1$, $r2$, $r3$ are distinct integers uniformly chosen from the set $\{1, 2, \ldots, NP\} \setminus \{i\}$. x_{best} is the best solution at the current generation. F is the mutation factor. v_i is the obtained mutant vector of x_i.

After mutation, v_i is then recombined with the target vector x_i with a probability Cr to generate a trial vector $u_i = (u_{1,i}, u_{2,i}, \ldots, u_{D,i})$ as

$$u_{j,i} = \begin{cases} v_{j,i} & \text{if } rand < Cr \text{ or } j = j_r \\ x_{j,i} & \text{otherwise} \end{cases}, \tag{4}$$

where D is the dimension of decision space, and j is the dimension index ($j = 1, 2, \ldots, D$). $rand$ is a uniformly random number in $[0, 1]$. j_r is a randomly chosen from $[1, 2, \ldots, D]$ which ensures that u_r get at least one component from x_r. Then the better solution between x_i and u_i is selected into the next generation. It worth noting that DE's selection is a greedy selection based on the "one-to-one" replacement scheme, which is different with the other EAs.

When applying the basic DE for the MOOPs, two questions need to be answered in designing the MODEs. The first question is how to choose the differential vectors in the mutation? In the MODEs, the whole population is always distributed in one or several Pareto fronts based on the non-dominated sorting[3]. The randomly chosen individuals may lie in the same Pareto front, and then their differential vectors may suffer from the ineptitude of directing toward the superior region. The second question is how to select the offspring between the parent vector and the trail vector. Can MODE greedily select the better one by the "one-to-one" comparison in the basic DE? The parent vector and the trail vector may be non-dominated with each other. It is hard to say which one is better based on the Pareto dominance according to the "one-to-one" comparison. The following MODEs have different ways to answer the above questions.

The Pareto Differential Evolution algorithm (PDE) is an adaptive DE algorithm for multiobjective problems[8]. The mutation factor F is generated from a Gaussian distribution $\mathcal{N}(0, 1)$. First, only the non-dominated solutions are retained for reproduction in PDE. Three random vectors are selected uniformly among the non-dominated set. The size of the set is controlled by pruning the set according to the neighborhood distance function. Second, after the DE/rand/1 mutation and recombination, a trial vector is generated. The selection procedure in PDE is to place the trial vector into the population if it dominates the parent. These two steps are repeated to fill the population until the population size is a predefined size.

The Pareto Differential Evolution Approach (PDEA) is directly inspired by NSGA-II[9]. First, in the mutation three vectors are randomly chosen from the population. Second, the trial vector is directly combined with the existing parent population for the selection. The combined population with the size will be pruned according to the non-dominated sorting and their crowding distances as NSGA-II. The work of PDEA is similar with the basic DE except the selection operation of NSGA-II.

The Generalized Differential Evolution 3 (GDE3) is a MODE algorithm for constrained MO problems[6]. The mutation operations of GDE3 and PDEA are the same, but their selection operations are different. Selection in GDE3 is based

---

1 Definition: PB—the basic population; PE—the elite population; PD—the
non-dominated set of PB.
Input: N—the size of PB and the maximum size of PE, p— the percentage of
     population for local search, G—the maximum number of generations for
     stopping criterion
Output: the non-dominated solutions in PE.
2 Initialization: randomize N individuals in PB uniformly in the search space,
PD is the non-dominated set of PB, PE is initialized as PD;
3 **while** $g = 1$ *to* G **do**
4   **while** $i = 1$ *to* N **do**
5   Randomly choose $x_{r1,g}$, $x_{r2,g}$ from PB and $x_{best,g}$ from PE;
6   Mutate $x_{i,g}$ as $v_{i,g} = x_{i,g} + F(x_{best,g} - x_{i,g}) + F(x_{r1,g} - x_{r2,g})$;
7   Crossover $v_{i,g}$ with $x_{i,g}$ to generation the trial vector $u_{i,g}$;
8   Select the trial vector if it can E-dominate $x_{i,g}$, otherwise select $x_{i,g}$;
9   **end**
10   Do Non-dominated sorting on PB, and update PD;
11   Randomly choose $p * N$ individuals from PE for the local search;
12   **for** *each individual* $x'_{i,g}$ *in the chosen set* **do**
13   Generate a child $u'_{i,g}$ of $x'_{i,g}$ by the local search;
14   Add the child $u'_{i,g}$ in PE;
15   **end**
16   **if** *the size of $PE > N$* **then**
17   Prune PE to the size N according to dominance and density;
18   **end**
19   Update the elite population $PE = PE \cup PD$;
20 **end**
21 return the non-dominated solutions in PE;

Algorithm 1. The pseudo-code of MMODE

on their proposed rules. When both the trial and target vectors are feasible, the
trial is selected if it weakly dominates the target vector. If the target vector
dominates the trial vector, then the target vector is selected. If neither vector
dominated each other, then both vectors are selected for the next generation.
Before continuing the next generation, the population is pruned by the method
of NSGA-II. GDE3 has generalized another similar approach Different Evolution for multiobjective optimization (DEMO)[7], which was proposed without
constraint handling.

3   MEMETIC Multiobjective DE

Our proposed Memetic Multiobjective Differential Evolution (MMODE) is composed of two phases in each iteration, global search and local search. In the two
phases, MMODE deals with two different populations PB and PE. The population PE called the elite population utilizes the local search to generate its offspring, which is independent with the basic population PE. **Algorithm 1** shows
the pseudo-code of MMODE, in which the two phases are performed as follows.

3.1   Phase I: Global Search with MOPDE

In the Phase I, global search is conducted in the basic population. A new multi-objective Pareto DE (MOPDE) algorithm has been used as the search engine to explore the search space. Generally, a D dimensional MOOP for minimization $F(x) = [f_1(x), f_2(x), \ldots, f_M(x)]$ is considered here, where M is number of objectives, $f_i(x)$ is the objective function for the ith objective.

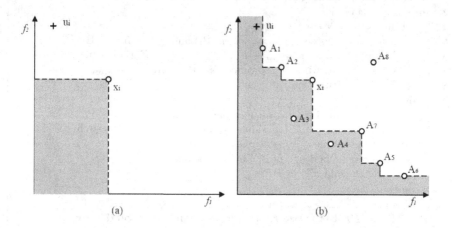

Fig. 1. The comparison of the E-dominance with the normal dominance: (a) the shade region are the set of solutions that can dominate x_i. (b) Considering a population with nine individuals A_1, $A_{2,}$, A_8 and x_i, the shade region is the set of solutions that can E-dominated x_i according to the two conditions.

The procedure of MOPDE is as the following. For each individual in , two solutions x_{r1}, x_{r2} are randomly chosen from $PB(r1 \neq r2 \neq i)$. x_{best} is randomly chosen from the non-dominated solutions. In our memetic application, x_{best} can be randomly chosen from the elite population PE. Then DE/current-to-best/1 strategy is utilized to mutate the target vector as Eq. 3. For crossover operation, the trial vector x_i is generated by recombining the mutant vector with the target vector as Eq. 4. The selection procedure is the main difference from the other MODEs. In the MOPDE, the offspring is not selected as the better solution by "one-to-one" comparison (in GDE3) or the population comparisons (in PDEA), but selected by the "one-to-many" comparisons based on the Pareto extensive dominance (E-dominance) as

$$x_i = \begin{cases} u_i \text{ if } F(u_i) \preceq F(x_i) \\ x_i \text{ otherwise} \end{cases}, \tag{5}$$

$$F(u_i) \preceq F(x_i) \Leftrightarrow \begin{cases} \exists m \in [1, \ldots, M], f_m(u_i) < \min f_m \\ \text{or} \\ \exists x_e \in PB, L(x_e) \leq L(x_i) \text{ and } F(u_i) \prec F(x_e) \end{cases}, \tag{6}$$

where $\preceq$ is the notation of the E-dominance relation that is defined as Eq. 6. In Eq. 6, $\prec$ is the notation of the normal Pareto dominance relation, and $L(x_i)$ means the level of x_i by non-dominate sorting. For example, $L(x_i) = 1$ means that x_i is in the first front. Generally, u_i can E-dominate x_i if u_i has at least one objective function less than the found minimum objective value, or it can dominate any individual in the $L(x_i)$'s and its before fronts. In MOPDE, the trial vector is selected if it can E-dominate the target vector, otherwise the target vector is selected into the next generation.

It is obvious that the E-dominance by the "one-to-many" comparisons can enlarge the search region. Figure 1 shows an example of the E-dominance, in which the shade region represents the search region of x_i. The shade region in the E-dominance is larger than the normal dominance. u_i cannot dominate x_i, but can E-dominate x_i. Therefore, MOPDE turns to be less greedy than other MODEs, and can find more accurate results.

3.2 Phase II: Local Search

The local search (LS) technique is conducted in the elite population PE , which is initializes as the non-dominated solution of the first generated PB. Randomly choose some individuals in for local search. The offspring is inserted into PE regardless of its fitness. As the local search is conducted on a population not on a single best solution in MMODE, the crossover-based LS methods may be more suitable than the gradient-based methods. MMODE utilizes a crossover-based LS method similarly with "DE/current/1". For the chosen parent vector x_i' in the elite population , randomly select two vectors x_{r1}' and $x'r2$ in PE ($r1 \neq r2 \neq i$) and mutate the target vector as

$$v_i' = x_i' + F\left(x_{r1}' - x'r2\right), \tag{7}$$

Like DE, the child u_i' is generated by recombining x_i' with v_i' as Eq. 4. The parameter F and Cr in this local search are set to be the same with those appeared in the Phase I. It is worth noting that p percents of PE for the LS have controlled the LS's ratio in the global search. A proper setting of p is 50%.

Then prune to the size by the method using in GDE3[6]. Delete the identical solutions in PE , and prune the population based on the non-dominated sorting and density estimation. The product distance with the nearest neighbors is used to compute the crowding distance. At the end of the Phase II, the elite population PE is updated by inserting the non-dominated set PD into the original PE.

The running time of MMODE is obviously more than the GDE3 due to the additional local search. For the local search, the Phase II of the MMODE has the computation complexity O(GN). The other parts of MMODE have the same computation complexity with GDE3, which has the computation complexity O$\left(GN\log^{M-1}N\right)$ [6]. Therefore MMODE's computation complexity is the same with GDE3 as O$\left(GN\log^{M-1}N\right)$, where G, N and M are the maximum generations, the population size and the number of objectives.

4 Experimental Results

In order to validate the performance of MOPDE and MMODE, we consider a set of test problems in Table 1, including the frequently used ZDT1-4 and ZDT6 in [12], DTLZ1-7 in [13]. As the proposed algorithms belong to DE algorithm, we compared them with two popular MODEs, i.e. PDEA and GDE3. For PDEA and GDE3, the population size is set to be 100, and the parameters F and Cr are set to be 0.5 and 0.1. For the MMODE, the size of the basic and elite populations is set to be 50. Other parameters are set as $F = 0.5$, $Cr = 0.1$, $p = 0.5$. We also evaluate the MOPDE independently, in which the population size is set to 100 and other parameters are set the same with the MMODE.

We run PDEA, GDE3, MOPDE and MMODE on each function for 50 times. The maximum function evaluation times is set to 300000 for these four algorithms. To evaluate the final results, we use two performance metrics - inverse generational distance (IGD) [12] and hypervolume (HV) difference[13], which can indicate the results' accuracy and diversity. Table 2 and 3 list the mean values of IGD and HV difference of each algorithm. The small IGD and HV difference indicate the good performance. The best results for each function are highlighted with bold in Table 2 and 3.

Table 1. Test Functions

F_i	Name	Dimension (D)	Search space	Objectives
1	ZDT1	30	$[0,1]^D$	2
2	ZDT2	30	$[0,1]^D$	2
3	ZDT3	30	$[0,1]^D$	2
4	ZDT4	10	$[0,1] \times [-5,5]^D$	2
5	ZDT6	10	$[0,1]^D$	2
6	DTLZ1	10	$[0,1]^D$	3
7	DTLZ2	12	$[0,1]^D$	3
8	DTLZ3	12	$[0,1]^D$	3
9	DTLZ4	12	$[0,1]^D$	3
10	DTLZ5	12	$[0,1]^D$	3
11	DTLZ6	12	$[0,1]^D$	3
12	DTLZ7	22	$[0,1]^D$	3

For the IGD, MMODE delivers the best results on six functions F_1, F_2, F_3, F_4, F_8 and F_{12}; GDE3 performs the best on F_5, F_7, F_10 and F_11; PDEA has the best performance on F_9 and F_10. From Table 2, it is obvious that MOPDE cannot surpass GDE3 and PDEA. As MOPDE has not considered the density estimation, MOPDE's results are less diverse than GDE3 and PDEA.

From Table 3, MMODE can deliver the best results of HV difference on the function F_1, F_2, F_3, F_4, F_6 and F_{12}. For the other functions, MMODE shows comparable performance with GDE3 and PDEA. MOPDE has the close performance with MMODE except the function F_9.

Table 2. Mean Values of IGD

	ZDT1	ZDT2	ZDT3	ZDT4	ZDT6	DTLZ1
GDE3	$4.20e^{-3}$	$1.63e^{-2}$	$5.18e^{-3}$	$4.19e^{-3}$	$4.53e^{-2}$	$1.93e^{-1}$
PDEA	$4.19e^{-3}$	$5.10e^{-3}$	$5.20e^{-3}$	$1.24e^{-1}$	$2.90e^{-1}$	$8.80e^{-1}$
MOPDE	$2.56e^{-2}$	$2.79e^{-2}$	$2.58e^{-2}$	$2.04e^{-2}$	$4.47e^{-1}$	$4.38e^{-2}$
MMODE	$\mathbf{4.17e^{-3}}$	$\mathbf{4.31e^{-3}}$	$\mathbf{4.87e^{-3}}$	$\mathbf{4.11e^{-3}}$	$7.85e^{-2}$	$\mathbf{2.27e^{-2}}$
	DTLZ2	DTLZ3	DTLZ4	DTLZ5	DTLZ6	DTLZ7
GDE3	$\mathbf{5.23e^{-2}}$	$1.88e^{-1}$	$5.28e^{-1}$	$4.05e^{-3}$	$\mathbf{2.21e^{-2}}$	$6.02e^{-2}$
PDEA	$\mathbf{5.26e^{-2}}$	$4.96e^{-1}$	$\mathbf{5.27e^{-2}}$	$\mathbf{4.04e^{-3}}$	$1.85e^{-1}$	$\mathbf{6.00e^{-2}}$
MOPDE	$1.27e^{-1}$	$1.49e^{-1}$	$6.12e^{-1}$	$2.50e^{-2}$	$6.28e^{-2}$	$1.18e^{-1}$
MMODE	$5.48e^{-2}$	$\mathbf{5.73e^{-2}}$	$\mathbf{5.27e^{-2}}$	$4.32e^{-3}$	$6.30e^{-1}$	$6.02e^{-2}$

Table 3. Mean Values of HV Difference

	ZDT1	ZDT2	ZDT3	ZDT4	ZDT6	DTLZ1
GDE3	$1.16e^{-2}$	$\mathbf{3.90e^{-3}}$	$1.77e^{-2}$	$1.25e^{-2}$	$4.53e^{-2}$	$1.93e^{-1}$
PDEA	$1.05e^{-2}$	$5.72e^{-3}$	$1.69e^{-2}$	$9.16e^{-2}$	$6.23e^{-3}$	$2.87e^{-2}$
MOPDE	$3.41e^{-2}$	$1.74e^{-2}$	$3.25e^{-2}$	$1.81e^{-2}$	$7.20e^{-2}$	$9.94e^{-3}$
MMODE	$\mathbf{7.66e^{-3}}$	$8.39e^{-3}$	$\mathbf{1.22e^{-2}}$	$\mathbf{3.28e^{-3}}$	$\mathbf{5.26e^{-3}}$	$\mathbf{1.74e^{-3}}$
	DTLZ2	DTLZ3	DTLZ4	DTLZ5	DTLZ6	DTLZ7
GDE3	$5.77e^{-3}$	$\mathbf{1.14e^{-3}}$	$8.96e^{-3}$	$\mathbf{5.77e^{-5}}$	$\mathbf{1.05e^{-3}}$	$2.91e^{-2}$
PDEA	$\mathbf{5.68e^{-3}}$	$1.52e^{-2}$	$\mathbf{5.38e^{-3}}$	$6.45e^{-5}$	$4.89e^{-3}$	$2.83e^{-2}$
MOPDE	$8.35e^{-3}$	$2.27e^{-2}$	$4.79e^{-2}$	$1.39e^{-3}$	$3.63e^{-3}$	$\mathbf{5.85e^{-3}}$
MMODE	$7.10e^{-3}$	$7.45e^{-3}$	$7.48e^{-3}$	$9.48e^{-4}$	$7.28e^{-3}$	$8.86e^{-3}$

In all, MMODE is better than or at least comparable with MOPDE, GDE3 and PDEA for most functions.

5   Conclusion

E-dominance is a general scheme for dominance evaluation, which suits to be applied in the tournament selection of other MOEAs. When the population only has the two solutions, E-dominance can be replaced with the Pareto dominance.

The memetic framework is necessarily used to accelerate the search process of the basic DE algorithm. The proposed MMODE gives reasonable responses of the three open issues mentioned in the introduction. The local search can be performed in a separate population. Some individuals are randomly chosen from the population to do the local search. Following the global search, the local search can fine tune the found optima. The local search can cooperate with the global search to find the accurate and diverse solutions for the MOOPs.

In fact MMODE contains a general framework. We will evaluate other MODEs and other local search strategies using this framework as the future works.

References

1. Wu, Z., Chow, T.W.S.: A local multiobjective optimization algorithm using neighborhood field. Structural and Multidisciplinary Optimization 46(6), 853–871 (2012)
2. Coello, C.A.C., Lamont, G.B. (eds.): Application of Multi-Objective Evolutionary Algorithms (Advances in Natural Computation), vol. 1. World Scientific Publishing Co. Pte. Inc. (2004)
3. Deb, K., Agarwal, S., Pratap, A., Meyarivan, T.: A fast and elitist multiobjective genetic algorithm: NSGA-II. IEEE Trans. on Evolutionary Computation 6(2), 182–197 (2002)
4. Zitzler, E., Laumanns, M., Thiele, L.: SPEA2: Improving the strength pareto evolutionary algorithm. In: Proceedings of Evolutionary Methods for Design, Optimization and Control With Applications to Industrial Problems (EUROGEN), pp. 95–100 (September 2001)
5. Storn, R., Price, K.: Differential Evolution – a simple and efficient heuristic for global optimization over continuous space. J. of Global Optimization 11(4), 341–359 (1997)
6. Kukkonen, S., Lampinen, J.: Gde3: the third evolution step of generalized differential evolution. In: Proceedings of IEEE Congress on Evolutionary Computation (CEC), pp. 443–450 (September 2005)
7. Robič, T., Filipič, B.: DEMO: Differential evolution for multiobjective optimization. In: Coello Coello, C.A., Hernández Aguirre, A., Zitzler, E. (eds.) EMO 2005. LNCS, vol. 3410, pp. 520–533. Springer, Heidelberg (2005)
8. Madavan, N.K.: Multiobjective optimization using a pareto differential evolution approach. In: Proceedings of IEEE Congress on Evolutionary Computation (CEC), pp. 1145–1150 (2002)
9. Abbass, H.: The self-adaptive pareto differential evolution algorithm. In: Proceedings of IEEE Congress on Evolutionary Computation (CEC), pp. 831–836 (2002)
10. Lara, A., Sanchez, G., Coello, C.A.C., Schutze, O.: HCS: A new local search strategy for memetic multiobjective evolutionary algorithms. IEEE Trans. on Evolutionary Computation 14(1), 112–132 (2010)
11. Soliman, O., Bui, L., Abbass, H.: A memetic coevolutionary multi-objective differential evolution algorithm. In: Multi-Objective Memetic Algorithms, pp. 369–388 (2009)
12. Veldhuizen, D.V., Lamont, G.: Multiobjective evolutionary algorithm research: A history and analysis. Technical Report TR-98-03, Air Force Inst. Technol., Dayton, OH (1998)
13. Zhang, Q., Zhou, A., Zhao, S.Z., Suganthan, P.N., Liu, W., Tiwari, S.: Multiobjective optimization test instances for the cec 2009 special session and competition. Technical Report CES-887, University of Essex and Nanyang Technological University (2008)

One Parameter Differential Evolution (OPDE) for Numerical Benchmark Problems

Y. Kang[1], T.O. Ting[1], Xin-She Yang[2], and Shi Cheng[1,3]

[1] Department of Electrical and Electronic Engineering, Xi'an Jiaotong-Liverpool University,
Suzhou, Jiangsu Province, P.R. China
[2] School of Science and Technology, Middlesex University Hendon Campus, London, UK
[3] Department of Electrical Engineering and Electronics, University of Liverpool, Liverpool, UK
yu.kang09@student.xjtlu.edu.cn, toting@xjtlu.edu.cn,
x.yang@mdx.ac.uk, shi.cheng@liverpool.ac.uk

Abstract. Differential Evolution (DE) can be simplified in the sense that the number of existing parameter is decreased from two parameters to only one parameter. We eliminate the scaling factor, F, and replace this by a uniform random number within [0, 1]. As such, it is easy to tune the crossover rate, CR, through parameter sensitivity analysis. In this analysis, the algorithm is run for 50 trials from 0.1 to 1.0 with a step increment of 0.1 on 23 benchmark problems. Results show that using the optimal CR, there is room for improvement in some of the benchmark problems. With the advantage and simplicity of a single parameter, it is significantly easier to tune this parameter and thus take the full advantage of the algorithm. The proposed algorithm here has a significant benefit when applied to real-world problems as it saves time in obtaining the best parameter setting for optimal performance.

Keywords: Benchmark problems, one parameter differential evolution, parameter sensitivity analysis.

1 Introduction

Optimization exists in many scientific, engineering and economic applications that involve optimization of a set of parameters, such as training a neural network to recognize face images or minimizing the losses in a power grid by finding the optimal configuration of the components. As the complexity of the problems that we attempt to solve is ever increasing, there will always be a need for better optimization algorithms.

Evolutionary Algorithms (EAs) have been developed based on the natural selection and survival of the fittest in the biological world to find one or more solutions to minimize or maximize the given objective functions of the optimization problems. Different from the traditional optimization techniques, these algorithm are often agent-based, and thus start the optimization process with a population of potential solutions instead of a single solution point or agent. At each generation, each individual of the population is evaluated. Through different evolution operators, the

Y. Tan, Y. Shi, and H. Mo (Eds.): ICSI 2013, Part I, LNCS 7928, pp. 431–438, 2013.

population moves towards regions of the search space from which good solutions have already been seen or visited. In order to achieve the optimal value for convergence, the optimal setting for control parameters should be used, and this is often achieved by altering these parameters during the evolution process, though we do not know how to tune such parameters so as to produce the best performance [1].

Differential Evolution (DE) which was proposed by Price and Storn [2] in the year 1997 is one of the most competitive EAs. It is a simple yet powerful evolutionary algorithm for global optimization. Recently, DE has become one of the most popular methods; this method has been applied to many real-world problems in different fields, including electromagnetics [3], network system [4], technique system [5] and automatic clustering [6].

DE has some advantages and is a derivative-free algorithm, which only needs the mathematical function and does not require differentiability or continuity on problem landscape. For the DE algorithm, one of the key issue is the so-called parameter setting problem which can be separated into two parts; parameter tuning and parameter control. Parameter tuning means users should find the better setting of parameters before running the algorithm while parameter control is to vary the values of these parameters during running algorithm. Usually, we focused on the former approach for good parameters during running algorithm. According to Eiben et al. [1], three categories can be implemented.

Whenever DE is applied to solve any problem, it is necessary to consider both efficiency and accuracy of the results. In the early stage, most of control parameters, F and CR are fixed, or not changing with rules. Later, in order to increase the efficiency and accuracy of the algorithm, these parameters should be changed during the evolution process.[2]. The function of control parameter is to improve the performance of an optimization algorithm for a given problem by altering different combinations of the control parameters. Researchers have already applied this technique on some evolution algorithms before [7]. Recently, numerous studies have focused on control parameter tuning [8]. Some studies attempted to achieve this by analyzing the mutation vectors by neighborhood mutation in DE algorithm.

There are many studies concerning the control parameter involving both the scaling factor F and crossover rate CR [9], which may significantly influence the performance of conventional DE or modified DE algorithm. However, most of these works on improvement of the algorithm try to introduce a paradigm or framework that requires additional steps and thus increases the complexity to the original algorithm. As opposed to this status quo, we will focus our efforts on the simplification and intend to reduce the number of parameters in DE to only one parameter, which will significantly reduce the efforts for both parameter tuning and parameter control. The eliminated parameter, which is the scaling factor, F, is now replaced by a uniform random number generator within [0, 1]. As such, we have a variant of DE, called one parameter differential evolution (OPDE), and the existence of only one parameter (crossover rate, CR) enables the parameter sensitivity analysis (PSA) to be carried out easily without much hassle as in the case of two parameters.

Hence, in this work, the parameter control is based on a combination of OPDE evolution process. The main purpose of our work is to produce a flexible DE and compare its results with those from literature. The convergence of EAs with OPDE is difficult to prove as the control parameter (F) is now replaced by a random number. It means that F is altered randomly and the value does not influence the evolution process directly [10, 11]. Thus, it is interesting to investigate the performance of the algorithm with random scaling factor, F. The rest of the paper is organized as follows. The canonical DE algorithm is reviewed in Section 2, followed by the proposed strategy in Section 3. Results including the PSA are presented in Section 4. Finally, Section 5 concludes with some remarks and future research directions.

2   Differential Evolution Algorithm

In this work, the most popular strategy, *DE/rand/1/bin* [12] is employed as this is one of the best strategies known. Conceptually, there are D-dimensional parameter vectors. Parameter g is the iteration or generation counter of the vectors. The population size NP represents the number of members in the population with a fixed population size. The ith member in g^{th} generation is usually represented by vector notation of x_i^g, whereby i=1, 2… NP and g=1, 2, 3…$MaxGen$. In DE, there exists three prominent operations, which are mutation, crossover and selection [12]. These manipulating operators are explained in the following sub-sections.

2.1   Mutation

According to DE strategy, the number of indices $r_1 \neq r_2 \neq r_3 \in [1, NP]$. The scaling factor $F \in [1, 2]$, with the suggested choice by Price and Storn [2] to be $F \in [0.5, 1.0]$, is a real number, which is also an important factor affecting the performance of the DE algorithm. This will be investigated further in this work. The vector information for the next generation $g+1$ is updated via:

$$v_{i,j}^{g+1} = x_{r_1,j}^g + F \times (x_{r_2,j}^g - x_{r_3,j}^g), r_1 \neq r_2 \neq r_3 \neq i \qquad (1)$$

2.2   Crossover

The crossover rate is usually set to a value less than unity $CR \in [0,1]$, and the choice also suggested by Price and Storn [2] is $CR \in [0.8, 1]$. The random number generator $rnd \in [0,1]$ generates random numbers uniformly distributed within 0 and 1. This value determines the crossover operation for a given trial vector as follows:

$$T_{ij}^{g+1} = V_{ij}^{g+1} \quad rnd \leq CR$$
$$T_{ij}^{g+1} = V_{ij}^g \quad rnd > CR \qquad (2)$$

2.3   Selection

The selection operation would decide whether trial vector T_{ij}^{g+1} is chosen into the next generation. In the case of minimization, if the fitness $f(T_{ij}^{g+1}) < f(x_{ij}^g)$ then $x_{ij} = T_{ij}$, else the x_{ij} remains intact. In other words, if the trial is found to be better, it will replace the current best solution.

3 The One Parameter Differential Evolution (OPDE)

In order to improve the performance of DE, choosing a suitable value for the control parameter value is crucially important. Better values of these control parameters may lead to individuals, which are more likely to produce offspring and propagate better control parameter value persistently. There are many strategies to adjust the parameter values [2, 13] and they often involve introducing complexity to the framework of the algorithm. In this work, instead of proposing another strategy that complicates these things further, we propose a technique that reduces the complexity of the algorithm so as to simplify the efforts for parameter control. Our approach essentially reduces the two parameters (*CR* and *F*) in DE to only one parameter. The scaling factor *F* is eliminated and thus replaced by a random value as depicted in equation below:

$$v_{i,j}^{g+1} = v_{r_1,j}^{g} + rnd \times (v_{r_2,j}^{g} - v_{r_3,j}^{g})$$

(3)

From the above equation, we observed that the scaling factor *F* is now replaced by a uniform random number generator *rnd* within 0 and 1, or mathematically represented by $rnd \in [0,1]$. The traditional DE has three parameters which need to be adjusted by a user. However, OPDE first reduces the number of parameters to the minimal single parameter, which greatly reduces the computational efforts. This is the main advantage and thus the main contribution of our approach. The rules for OPDE are quite simple and can thus enhance the efficiency in comparison to the classic DE.

4 Experiment Results

In our simulation, a series of benchmark optimization problems [14] have been employed to test the efficiency of OPDE. With a population size of 100, the best value of the *CR* is obtained through PSA. For a fair comparison, 50 trials have been carried out for each benchmark problem, and the mean best value has been calculated. We then ran again the same set of problems with a population size of 30. This enables comparison with other algorithms as many results in literature utilize a population size of 30.

4.1 Parameter Sensitivity Analysis (PSA) for Crossover Rate *CR*

In OPDE, *CR* is increased from 0.1 to 1.0 by a step increment of 0.1. The best way to evaluate the impact of *CR* over the performance is by observing the mean best value during the iterations versus *CR*. Some of these PSA curves are presented in Fig 1, involving f_1, f_5, f_{10} and f_{13} functions. The summary of best *CR* values obtained for all benchmark functions is given in the fifth column of Table 1.

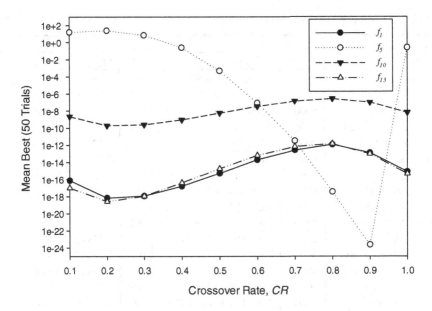

Fig. 1. Parameter Sensitivity Analysis for crossover rate, CR

In Fig 1, f_1, f_{10}, f_{13} depict an 's' like pattern. Interestingly, the optimal CR obtained is 0.2, as recorded in Table 1. For lower dimensional problems such as f_{16}, f_{17} and f_{19}, no significant optimal value is observed. This is due to the simplicity of the problem; they are easily solved problems. For these cases, since the mean best will have the same value, we pick the crossover value with lower standard deviation as the optimal value. For problem featuring a discontinuous landscape such as in the case of Step function f_6, the impact of varying the CR seems insignificant.

Simulation results of both algorithms are tabulated in Table 1. In this paper, according to the literature [2, 14], we have carried out the simulation for conventional DE by using F=0.5 and CR=0.9. Note that OPDE utilizes the best value of CR obtained via PSA. If the performance is the same, the one with smaller standard deviation will be considered as a better solution. Each function is run for 50 trials and the best mean for each of the benchmark problem is calculated. From the results depicted in Table 1, it is observed that OPDE has significant improvement over $f_4, f_5,$ f_8, f_9 and f_{11}. This may imply that there is still room for improvement for many of these benchmark problems. The motivation here is not to prove that OPDE is better than DE, but to prove that we can take the full advantage of OPDE as its PSA contains only one parameter, and thus is more straight forward and much easier compared to DE with two parameters. We believe that careful tuning of CR and F for DE may also result in comparatively similar performance. However, this is a time consuming task, especially for large-scale problem.

Table 1. The mean best values obtained by OPDE and DE out of 50 trials with popsize = 100

$f(x)$	Function name	#Gen	DE (CR=0.9, F=0.5) Mean Best (STD)	OPDE CR	OPDE Mean Best (STD)
f_1	Sphere	1500	$6.33\times10^{-26}\,(5.67\times10^{-26})$	0.2	$6.73\times10^{-19}\,(2.51\times10^{-19})$
f_2	Schwefel 2.22	2000	$1.64\times10^{-18}\,(8.31\times10^{-19})$	0.2	$5.59\times10^{-17}\,(1.19\times10^{-17})$
f_3	Schwefel 1.2	5000	$5.07\times10^{-16}\,(6.91\times10^{-16})$	1.0	$3.73\times10^{-11}\,(4.94\times10^{-11})$
f_4	Schwefel 2.21	5000	2.03(1.59)	0.8	$1.13\times10^{-06}\,(5.01\times10^{-10})$
f_5	Rosenbrock	20000	0.24 (0.94)	0.9	$2.22\times10^{-24}\,(3.67\times10^{-24})$
f_6	Step	1500	0 (0)	0.7	0 (0)
f_7	Quartic	3000	$2.44\times10^{-03}\,(7.74\times10^{-04})$	1.0	$3.92\times10^{-03}\,(1.35\times10^{-03})$
f_8	Schwefel	9000	-12333 (183.25)	0.1	-12569.5 (5.71×10^{-12})
f_9	Rastrigin	5000	13.96(4.47)	0.2	0 (0)
f_{10}	Ackley	1500	$6.49\times10^{-14}\,(3.05\times10^{-14})$	0.2	$1.95\times10^{-10}\,(4.01\times10^{-11})$
f_{11}	Girewank	2000	$2.96\times10^{-04}\,(1.45\times10^{-03})$	0.5	$3.25\times10^{-20}\,(2.66\times10^{-20})$
f_{12}	Penalized P8	1500	$7.45\times10^{-27}\,(3.72\times10^{-26})$	0.2	$4.29\times10^{-20}\,(1.44\times10^{-20})$
f_{13}	Penalized P16	1500	$3.38\times10^{-26}\,(6.25\times10^{-26})$	0.2	$2.60\times10^{-19}\,(9.07\times10^{-20})$
f_{14}	Foxholes	100	$1.00\,(1.96\times10^{-16})$	0.8	$1.00\,(1.11\times10^{-16})$
f_{15}	Kowalik	4000	$3.07\times10^{-04}\,(2.82\times10^{-19})$	0.6	$3.07\times10^{-04}\,(9.64\times10^{-20})$
f_{16}	6H Camel-Back	100	-1.03 (4.04×10^{-14})	1.0	-1.03 (2.51×10^{-15})
f_{17}	Brain	100	0.40 (3.82×10^{-12})	0.8	0.40 (1.98×10^{-10})
f_{18}	Goldstein-Price	100	3.00 (1.53×10^{-15})	0.8	3.00 (1.08×10^{-15})
f_{19}	Hartman-3	100	-3.86 (2.66×10^{-15})	1.0	-3.86 (2.66×10^{-15})
F_{20}	Hartman-6	200	-3.32 (5.82×10^{-02})	0.4	-3.32 (4.66×10^{-05})
f_{21}	Shekel-5	100	-10.15 (5.23×10^{-07})	0.7	-9.64 (1.29)
f_{22}	Shekel-7	100	-10.40 (5.70×10^{-08})	0.7	-10.40 (0.03)
f_{23}	Shekel-10	100	-10.54 (4.91×10^{-09})	0.9	-10.54 (7.62×10^{-05})

4.2 F and CR Properties

The value of scaling factor F may influence the population diversity. When F has a larger value, the track of best solutions during the search/iterations seems to extend over larger scales since a greater F will result in larger step sizes. As a result, the algorithm will be incapable of doing local search efficiently, and therefore lacks the accuracy, which also slow down the search process for a given accuracy or tolerance. As this parameter has been replaced by a random value, the tradeoff between global and local search capabilities may have been taken care of automatically. The impact of this randomness is further compensated by tuning the CR to an optimal value through PSA. Therefore, this strategy is efficient both on unimodal and multimodal functions as we have demonstrated using the chosen set of test functions.

4.3 Effect of the Population Size

In this study, we have carried out numerous experiments involving different population size. With the similar CR values, the same problem is now solved using a population size of 30. It is interesting to note that results are slightly better for some of these problems although the population size used is lower. This is true for f_1, f_2, f_{10}, f_{12} and f_{13}. When comparing to the original results with population size of 100, generally, it is well accepted that a larger population size will result in better performance. Our results agree with this fact. However, one important finding from our numerous experiments is that the results in the right column of Table 2 can be further improved by carrying out independent PSA involving for a relevant population size.

Table 2. Comparing OPDE with different population sizes

Popsize	100	30
$f(x)$	Mean Best (Std)	Mean Best (Std)
f_1	$6.73\times10^{-19}(2.51\times10^{-19})$	$1.29\times10^{-23}(1.52\times10^{-23})$
f_2	$5.59\times10^{-17}(1.19\times10^{-17})$	$1.92\times10^{-20}(2.05\times10^{-20})$
f_3	$3.73\times10^{-11}(4.94\times10^{-11})$	$1.32\times10^{-06}(2.14\times10^{-06})$
f_4	$1.13\times10^{-06}(5.01\times10^{-10})$	$1.29\times10^{-06}(5.83\times10^{-07})$
f_5	$2.22\times10^{-24}(3.67\times10^{-24})$	$2.87 (1.90)$
f_6	$0 (0)$	$0 (0)$
f_7	$3.92\times10^{-03}(1.35\times10^{-03})$	$5.30\times10^{-03}(1.45\times10^{-03})$
f_8	$-12569.5 (5.71\times10^{-12})$	$-12548.17 (51.30)$
f_9	$0 (0)$	$0.16 (0.42)$
f_{10}	$1.95\times10^{-10}(4.01\times10^{-11})$	$7.73\times10^{-13}(4.22\times10^{-13})$
f_{11}	$3.25\times10^{-20}(2.66\times10^{-20})$	$4.12\times10^{-20}(2.32\times10^{-20})$
f_{12}	$4.29\times10^{-20}(1.44\times10^{-20})$	$9.81\times10^{-23}(1.27\times10^{-22})$
f_{13}	$2.60\times10^{-19}(9.07\times10^{-20})$	$3.59\times10^{-22}(4.88\times10^{-22})$
f_{14}	$1.00 (1.11\times10^{-16})$	$1.00 (9.06\times10^{-16})$
f_{15}	$3.07\times10^{-04}(9.64\times10^{-20})$	$3.26\times10^{-04}(1.28\times10^{-04})$
f_{16}	$-1.03 (2.51\times10^{-15})$	$-1.03 (4.26\times10^{-06})$
f_{17}	$0.40 (1.98\times10^{-10})$	$0.40 (6.99\times10^{-08})$
f_{18}	$3.00 (1.08\times10^{-15})$	$3.00 (1.61\times10^{-15})$
f_{19}	$-3.86 (2.66\times10^{-15})$	$-3.86278(1.18\times10^{-13})$
f_{20}	$-3.32 (4.66\times10^{-05})$	$-3.32184(3.38\times10^{-04})$
f_{21}	$-9.64 (1.29)$	$-8.25 (2.35)$
f_{22}	$-10.40 (0.03)$	$-9.25 (2.01)$
f_{23}	$-10.54 (7.62\times10^{-05})$	$-9.48 (2.32)$

5   Conclusion

By replacing the scaling factor with a uniformly distributed random number, we have reduced the number of key parameters in DE to a single parameter, apart from the population size. This essentially leads to the one-parameter differential evolution. We have shown that DE can work effectively without the concern of setting a suitable value for scaling factor, F. With the presence of one parameter, users can take the full advantage of the OPDE algorithm by tuning the CR value through parameter sensitivity analysis. This usually saves a great amount of time, compared to PSA involving two parameters. In future studies, different distributions such as Cauchy or Gaussian may be incorporated for further analysis. It may be useful to use chaotic maps. It can be expected that the proposed OPDE may perform efficiently for large-scale optimization problems as the PSA can be easily done, compared with two-parameter DE variants. Further studies using large-scale test functions will be highly recommended.

Acknowledgement. The authors would like to acknowledge the SURF sponsorship granted by EEE department to support this research works.

References

1. Eiben, A.E., Hinterding, R., Michalewicz, Z.: Parameter control in evolutionary algorithms. IEEE Trans. Evol. Comput. 3, 124–141 (1999)
2. Price, K.V., Storn, R.: Differential evolution – A simple and efficient heruistic for global optimization over continuous spaces. J. Global Optim. 11, 341–359 (1997)
3. Anyong, Q.: Electromagnetic inverse scattering of multiple two-dimensional perfectly conducting objects by the differential evolution strategy. IEEE Trans. Antennas Propag. 51, 1251–1262 (2003)
4. Ching-Tzong, S., Chu-Sheng, L.: Network reconfiguration of distribution systems using improved mixed-integer hybrid differential evolution. IEEE Trans. Power Delivery 18, 1022–1027 (2003)
5. Storn, R.: System design by constraint adaptation and differential evolution. IEEE Trans. Evol. Comput. 3, 22–34 (1999)
6. Das, S., Abraham, A., Konar, A.: Automatic Clustering Using an Improved Differential Evolution Algorithm. IEEE Trans. Syst. Man Cybern. Part A Syst. Humans 38, 218–237 (2008)
7. Grefenstette, J.J.: Optimization of Control Parameters for Genetic Algorithms. IEEE Transactions on Systems, Man and Cybernetics 16, 122–128 (1986)
8. Smit, S.K., Eiben, A.E.: Comparing parameter tuning methods for evolutionary algorithms. In: IEEE Congress on Evolutionary Computation, CEC 2009, pp. 399–406 (2009)
9. Price, K.V.: Differential evolution: a fast and simple numerical optimizer. In: 1996 Biennial Conference of the North American Fuzzy Information Processing Society, NAFIPS, pp. 524–527 (1996)
10. Terkel, D.A., Semenov, M.A.: Analysis of convergence of an evolutionary algorthm with self-adaptation using a stochastic Lyapunov function. Evol. Comput. 11(4), 363–379 (2003)
11. Yao, X., He, J.: Toward an analytic framework for analysing the computation time of evolutionary algorithms. Artificial Intell. 145(1-2) (2003)
12. Brest, J., Greiner, S., Boskovic, B., Mernik, M., Zumer, V.: Self-Adapting Control Parameters in Differential Evolution: A Comparative Study on Numerical Benchmark Problems. IEEE Transactions on Evolutionary Computation 10, 646–657 (2006)
13. Das, S., Abraham, A., Chakraborty, U.K., Konar, A.: Differential Evolution Using a Neighborhood-Based Mutation Operator. IEEE Transactions on Evolutionary Computation 13, 526–553 (2009)
14. Vesterstrom, J., Thomsen, R.: A comparative study of differential evolution, particle swarm optimization, and evolutionary algorithms on numerical benchmark problems. In: Congress on Evolutionary Computation, CEC 2004, vol. 1982, pp. 1980–1987 (2004)

Parameter Optimization of Local-Concentration Model for Spam Detection by Using Fireworks Algorithm

Wenrui He, Guyue Mi, and Ying Tan

Key Laboratory of Machine Perception (Ministry of Education),
Department of Machine Intelligence, School of Electronics Engineering and Computer Science,
Peking University, Beijing, 100871 P.R. China
{wenrui.he,gymi,ytan}@pku.edu.cn

Abstract. This paper proposes a new framework that optimizes anti-spam model with heuristic swarm intelligence optimization algorithms, and this framework could integrate various classifiers and feature extraction methods. In this framework, a swarm intelligence algorithm is utilized to optimize a parameter vector, which is composed of parameters of a feature extraction method and parameters of a classifier, considering the spam detection problem as an optimization process which aims to achieve the lowest error rate. Also, 2 experimental strategies were designed to objectively reflect the performance of the framework. Then, experiments were conducted, using the Fireworks Algorithm (FWA) as the swarm intelligence algorithm, the Local Concentration (LC) approach as the feature extraction method, and SVM as the classifier. Experimental results demonstrate that the framework improves the performance on the corpora PU1, PU2, PU3 and PUA, while the computational efficiency is applicable in real world.

Keywords: Spam Detection, Fireworks Algorithm, Parameter Optimization, Local Concentration Approach.

1 Introduction

Spam, defined as Unsolicited Commercial E-mails (UCE) or Unsolicited Bulk E-mails (UBE), has become a significant problem for both recipients and Internet Service Providers (ISPs). For recipients, coping with spam is time-consuming; furthermore, spam frequently contains images that recipients find offensive, or attached malicious programs that attack recipients' computers. For ISPs, large scale of spam is a considerable burden on their systems. Commtouch reported that in Q4 2012, the average daily spam level was 90 billion messages per day, which is a slight increase over Q3 2012. [1] Ferris Research revealed that spam cost $130 billion worldwide in 2009, which was a 30% raise over the 2007 estimates. [2] Therefore, it is necessary to find an effective method for the spam detection.

Many approaches were proposed to handle the problem. In fact, Spam detection involves mainly three research fields, namely term selection, feature extraction, and classifier design. In the classifier design field, many machine learning (ML) methods

Y. Tan, Y. Shi, and H. Mo (Eds.): ICSI 2013, Part I, LNCS 7928, pp. 439–450, 2013.

were adopted to classify emails, such as Support Vector Machine (SVM) [3]–[6], k-Nearest Neighbor (k-NN), Naive Bayes (NB), Artificial Neural Network (ANN), Boosting, and Artificial Immune System (AIS). As the performance of an ML method depends on the extraction of discriminative feature vectors, feature extraction methods are crucial to the process of spam filtering. Commonly used feature extraction methods are, for example, Concentration-based Feature Construction (CFC) [4], Local concentration (LC) [7] and Bag-of-Words (BoW). The researches of term selection have also attracted much attention from researchers all over the world, widely utilized methods including Information Gain (IG) [8], Term Frequency Variance (TFV) [9] and Document Frequency (DF).

In previous research, parameters in the anti-spam process are set simply and manually. However, the manual setting might cause several problems. For instance, lack of prior knowledge may lead to improper parameter setting, repeated attempts of users cost overmuch human effort, and the inflexibility of the dataset-relevant parameters should also be taken into counted.

To solve the problems, this paper proposes a new framework that automatically optimizes parameters in anti-spam model with heuristic swarm intelligence optimization algorithms, and this framework could integrate various classifiers and feature extraction methods. 2 experimental strategies were designed to objectively reflect framework performance. Then, experiments are conducted, using the Fireworks Algorithm (FWA) as the Swarm Intelligence algorithm, the Local Concentration approach as the feature extraction method, and SVM as the classifier. Experimental results demonstrate that the framework improved the performance on the corpora PU1, PU2, PU3 and PUA, and the computational efficiency is applicable in real world.

The remainder of the paper proceeds as follows. To begin with, we will provide a brief background on the LC approach and the FWA in Section II. The proposed framework for anti-spam is presented in Section III. In Section IV, the corpora, the criteria and the experimental setup are described, and experiments results are analyzed in detail. Section V concludes the paper.

2   Related Works

2.1   Local Concentration (LC) Based Feature Extraction Approach for Anti-spam

In an anti-spam model, feature extraction is an essential step. The feature extraction method decides spatial distribution characteristics of email sample points, influencing construction of a specific email classification model and the final classification performance. An effective feature extraction method is able to extract extinguishing features of emails, endowing different kinds of emails possessing obvious spatial distribution difference. Moreover, it should be capable of reducing the complexity and difficulty of classification, so as to improve overall performance of the anti-spam model. The Local-concentration (LC) approach is proved to meet both of the requirements mentioned above. It not only greatly reduces feature dimensionality by remaining the position-correlated information of emails, but also performs better in terms of both accuracy and measure compared to the BoW approach and the GC approach.

Inspired from the biological immune system, the LC feature extraction approach is able to extract position-correlated information from messages by transforming each area of a message to a corresponding LC feature effectively. Two implementation strategies of the LC approach were designed by using a fixed-length sliding window and a variable-length sliding window. To incorporate the LC approach into the whole process of spam filtering, a generic LC model is designed. In the LC model, two types of detector sets are at first generated by using term selection methods and a well-defined tendency threshold. Then a sliding window is adopted to divide the message into individual areas. After segmentation of the message, the concentration of detectors is calculated and taken as the feature for each local area. Finally, all the features of local areas are combined as a feature vector of the message.

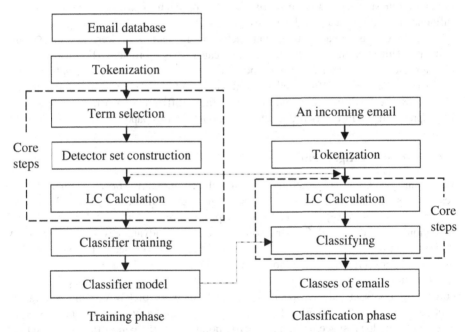

Fig. 1. Training and classification phases of the LC model

The generic structure of the LC model is shown in Fig. 1. The tokenization is a simple step, where messages are tokenized into words (terms), while term selection, detector set construction and LC calculation are quite essential to the model.

In the term selection step, terms are sorted in the order of importance and the top m% of the terms are selected to form the gene library. The term selection rate parameter, m%, decides the size of the gene library, influencing the computational complexity of the detector construction algorithm and distinguishability of detectors in the next step. An optimal value of m% is supposed to effectively screen out noise terms, while guarantee the existence of the informative terms.

In the detector construction step, the tendency of each detector, namely the difference between a term's posterior probability of presence in normal emails and that in spams, is calculated. If the tendency of a term exceeds θ, the term will be added into the detector set. This parameter, θ, as the standard of detector set construction, is capable of controlling significance of detector matching, yet can't be set too high, so as not to cause loss of information.

In the LC calculation step, the number of sliding windows, N, no matter in fix-length LC approach or in variable-length LC approach, is an important parameter, since it decides the size of a single sliding window and defines the local region,. As a result, it has a great impact on the dimensionality of LC feature vectors, and also performance of the algorithm.

The above three parameters, as well as the parameters of classifiers in the classification step, are fairly essential in LC approach. They, as a whole, heavily influence the performance of the anti-spam model.

In the previous research, these parameters in LC approach were set simply and manually. However, the manual setting might cause several problems. For instance, lack of prior knowledge may lead to improper parameter setting, repeated attempts of users cost overmuch human effort, and the inflexibility of the dataset-relevant parameters should also be taken into counted. To solve these difficulties, a parameter-optimized LC approach using Fireworks Algorithm is proposed in this paper.

2.2   Fireworks Algorithm

In recent years, swarm intelligence (SI) algorithms have been popular among researchers who are working on optimization problems. SI algorithms, e.g. Fireworks Algorithm (FWA) [10], Particle Swarm Optimization (PSO), Ant System, Clonal Selection Algorithm, and Swarm Robots, etc., have advantages in solving many optimization problems. Among all the SI algorithms, FWA is one of the most popular algorithms for searching optimal locations in a D-dimensional space.

Like most swarm intelligence algorithms, FWA is inspired by some intelligent colony behaviors in nature. Specifically, the framework of FWA is mimicking the process of setting off fireworks. The explosion process of a firework can be viewed as a search in the local space around a specific point where the parent firework is set off through the offspring sparks generated in the explosion.

Assume the population size of fireworks is N and the population size of generated spark is M. Each fireworks $i(i = 1, 2, \cdots, N)$ in a population has the following properties: a current position x_i, a current explosion amplitude A_i and the amount of the generated sparks s_i. Each firework generates a number of sparks within a fixed explosion amplitude. In each generation, N fireworks set off within a feasible bounds within explosion amplitude A_i and spark size s_i, then the spark are generated. In addition, the fireworks algorithm also takes Gaussian mutation operators to enhance local search capability.

The best firework is kept for the next generation, and the other $N - 1$ fireworks for the next generation are selected based on their distance to other fireworks or randomly as to keep the diversity in the set, which includes the N fireworks, the generated sparks and Gaussian mutation fireworks. The fireworks algorithm continues conducting these operations till the termination criteria is satisfied.

As to the optimization problem f, a point with better fitness is considered as a potential solution, which the optima locate nearby with high chance, vice versa. Suppose FWA is utilized to solve a general optimization problem:

$$Minimize\ f(x) \in R, \qquad x \in R^n \tag{1}$$

where $x = x_1, x_2, \cdots, x_d$ denotes a location in the potential space, $f(x)$ is an objective function, and R^n denotes the potential space. Then the FWA is implemented to find a point $x \in R^n$, which has the minimal fitness value. This is also how the optimization of the anti-spam process is implemented.

3   Parameter Optimization of Local-Concentration Model for Spam Detection by Using Fireworks Algorithm

The classification problem that whether an email is spam or a normal email, is here considered as an optimization problem, that is, to achieve the lowest error rate by finding the optimal parameter vector in the potential search space.

The optimal vector $P^* =< F_1^*, F_2^*,\ \cdots, F_n^*, C_1^*, C_2^*,\ \cdots, C_m^* >$, composes of 2 parts: the first part is the feature calculation relevant parameters $F_1^*, F_2^*,\ \cdots, F_n^*$, and the second part is the classifier relevant parameters $C_1^*, C_2^*,\ \cdots, C_m^*$. The optimal vector P^* is the vector whose cost function $CF(P)$ associated with classification achieves the lowest value, with

$$CF(P) = Err(P) \tag{2}$$

where $Err(P)$ is the classification error measured by 10-fold cross validation on the training set. Input vector P consists of two parts – parameters $F_1^*, F_2^*,\ \cdots, F_n^*$ associated with a certain feature extraction method and $C_1^*, C_2^*,\ \cdots, C_m^*$ associated with a certain classifier. $F_1^*, F_2^*,\ \cdots, F_n^*$ uniquely determine the performance of feature construction, while $C_1^*, C_2^*,\ \cdots, C_m^*$ influence the performance of a certain classifier. Different feature extraction methods hold different parameters and lead to different performance. For LC approach, specifically, m, the Term Selection Rate, helps select the top m % terms with descending importance in term set, which determines the term pool size. θ, the Proclivity Threshold, the minimal difference of a term's frequency in non-spam e-mails minus that in spam e-mails, has an assistant function in   screening out terms with greater discrimination. N, the number of sliding windows, determines the dimensionality of the feature vector of emails. Different classifiers hold different parameters and also lead to different performance. Parameters associated with neural network, which determine the structure of the network, include number of layers, number of nodes within a layer and each connection weight between two nodes. SVM-related parameters that determine the position of optimal hyper-plane in feature space, include cost parameter C and kernel parameters, just to name a few.

The vector P is the optimization objective whose performance is measured by $CF(P)$. Therefore, the optimization of concentrations can be formulated as follows.

Finding $P^* =< F_1^*, F_2^*, \cdots, F_n^*, C_1^*, C_2^*, \cdots, C_m^* >$, so that

$$CF(P^*) = \min_{\{F_1, F_2, \cdots, F_m, C_1, C_2, \cdots, C_m\}} CF(P) \tag{3}$$

Several optimization approaches not demanding an analytical expression of the objective function such as particle swarm optimization (PSO), genetic algorithms (GA) and so forth can be employed for the optimization process. Fireworks Algorithm was used to design concentrations.

Figure 2 shows the optimization process of Parameter Optimization of Local-Concentration Model for Spam Detection Using Fireworks Algorithm.

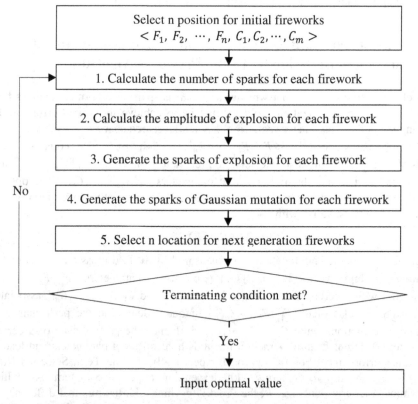

Fig. 2. Process of the Parameter Optimization of Local-Concentration Model for Spam Detection by Using Fireworks Algorithm

This framework utilizes the Fireworks Algorithm to optimize parameters in the Local Concentration approach. Not only the essential parameters in the LC approach, but also the classifier-relevant parameters are optimized in this framework, so that the whole anti-spam process gets optimized.

This framework optimizes anti-spam model with heuristic Swarm Intelligence optimization algorithms, which could integrate various classifiers and feature extraction methods.

4   Experiments

4.1   Experimental Corpora

The experiments were conducted on four benchmark corpora PU1, PU2, PU3, and PUA, using 10-fold cross validation. The corpora have been preprocessed with removal of attachments, HTML tags, and header fields except for the subject. The duplicates were removed from the corpora in that they may lead to over-optimistic conclusions in experiments. In PU1 and PU2, only the duplicate spam, which arrived on the same day, are deleted. While in PU3 and PUA, all duplicates (both spam and legitimate e-mail) are removed, even if they arrived on different days. Different from the former PU1 corpus (the one released in 2000), the corpora are not processed with removal of stop words, and no lemmatization method is adopted. The details of the corpora are given as follows.

1) **PU1:** The corpus includes 1099 messages, 481 messages of which are spam. The ratio of legitimate e-mail to spam is 1.28. The preprocessed legitimate messages and spam are all English messages, received over 36 months and 22 months, respectively.

2) **PU2:** The corpus includes 721 messages, 142 messages of which are spam. The ratio of legitimate e-mail to spam is 4.01. Similar to PU1, the preprocessed legitimate messages and spam are all English messages, received for over 22 months.

3) **PU3:** The corpus includes 4139 messages, 1826 messages of which are spam. The ratio of legitimate e-mail to spam is1.27. Unlike PU1 and PU2, the legitimate messages contain both English and non-English ones. While spam are derived from PU1, Spam Assassin corpus and other sources.

4) **PUA:** The corpus includes 1142 messages, 572 messages of which are spam. The ratio of legitimate e-mail to spam is 1. Similar to PU3, the legitimate e-mail contain both English and non-English messages, and spam is also derived from the same sources.

4.2   Evaluation Criteria

In spam filtering, many evaluation methods or criteria have been designed for comparing performance of different algorithms [12], [13]. We adopted four evaluation criteria, which were spam recall, spam precision, accuracy, and F_β measure, in all our experiments to do a before-and-after comparison. Among the criteria, accuracy and F_β measure are more important, for accuracy measures the total number of messages correctly classified, and F_β is a combination of spam recall and spam precision.

1) **Spam recall:** It measures the percentage of spam that can be filtered by an algorithm or model. High spam recall ensures that the filter can protect the users from spam effectively. It is defined as follows:

$$R_S = \frac{n_{s \to s}}{n_{s \to s} + n_{s \to l}} \tag{4}$$

where $n_{s \to s}$ is the number of spam correctly classified, and $n_{s \to l}$ is the number of spam mistakenly classified as legitimate e-mail.

2) **Spam precision:** It measures how many messages, classified as spam, are truly spam. This also reflects the amount of legitimate e-mail mistakenly classified as spam. The higher the spam precision is, the fewer legitimate e-mail have been mistakenly filtered. It is defined as follows:

$$P_S = \frac{n_{s \to s}}{n_{s \to s} + n_{l \to s}} \tag{5}$$

where $n_{l \to s}$ is the number of legitimate e-mail mistakenly classified as spam, and $n_{s \to s}$ has the same definition as in (4).

3) **Accuracy:** To some extent, it can reflect the overall erformance of filters. It measures the percentage of messages (including both spam and legitimate e-mail) correctly classified. It is defined as follows:

$$A = \frac{n_{l \to l} + n_{s \to s}}{n_l + n_s} \tag{6}$$

where $n_{l \to l}$ is the number of legitimate e-mail correctly classified, $n_{s \to s}$ has the same definition as in (4), and n_l and n_s are, respectively, the number of legitimate e-mail and the number of spam in the corpus.

4) **F_β measure:** It is a combination of R_s and P_s, assigning a weight β to P_s. It reflects the overall performance in another aspect. F_β measure is defined as follows:

$$F_\beta = (1 + \beta^3) \frac{R_s + P_s}{\beta^2 P_s + R_s} \tag{7}$$

In our experiments, we adopted $\beta = 1$ as done in most approaches [12]. In this case, it is referred to as F_1 measure. In the experiments, the values of the four measures were all calculated. However, only accuracy and F_1 measure are used for parameter selection and comparison of different approaches. Because they can reflect overall performance of different approaches, and F_1 combines both R_s and P_s. In addition, R_s and P_s, respectively, reflect different aspects of the performance, and they cannot reflect the overall performances of approaches, separately. That is also the reason why the F_β is proposed. We calculated them just to show the components of F_1 in detail.

4.3   Experimental Setup

All the experiments were conducted on a PC with Intel Core i5-2300 CPU and 4G RAM. The LC-based model with variable-length sliding window was optimized and the term selection method utilized was information gain. SVM was employed as classifier and LIBSVM was applied for the implementation of the SVM.10-fold cross validation was utilized on each corpora. Since FWA is a stochastic algorithm, the experimental results we present are average results under ten independent runs. Accuracy, recall, precision and F1 measure were selected as evaluation criteria, in which accuracy and F1 measure are main ones since they can reflect the overall performance of spam filtering.

4.4   Experimental Results and Analysis

Two strategies for experiments were designed to investigate the effectiveness of the proposed optimization process of LC model. In both strategies, optimization of the LC model is conducted on the training set and finally examined on the testing set in each fold. In this case, the original training set is further divided into a new training set and a testing set for computing the fitness to evaluate the LC model that the current spark is corresponding to.

For the consideration of efficiency, the first strategy (strategy-1) is designed by defining a validation set on the original training set and making it independent from the original training set, e.g. the original training set is divided into a new training set and a validation set. The fitness of each spark is independently computed on the validation set after a corresponding classifier is trained on the new training set. The optimal model that corresponding to the optimal spark achieved and trained on the new training set is finally examined on the testing set in each fold. In this strategy, fitness of each spark is evaluated on an independent validation set in each fold, thus the computational complexity is relatively low and the optimization process of the LC model could be finished quickly.

Table 1. Performance comparison of LC before and after optimization with strategy-1

Corpus	Approach	Precision (%)	Recall (%)	Accuracy (%)	F1 (%)
PU1	LC	94.85	95.63	95.87	95.21
	Strategy-1	96.55	95.21	**96.33**	**95.81**
PU2	LC	95.74	77.86	94.79	85.16
	Strategy-1	95.15	80.71	**95.35**	**86.65**
PU3	LC	96.68	94.34	96.03	95.45
	Strategy-1	95.81	95.71	**96.18**	**95.69**
PUA	LC	95.60	94.56	94.91	94.94
	Strategy-1	96.63	94.56	**95.53**	**95.49**

Experiments were conducted on the original PU1, PU2, PU3 and PUA corpus to verify the effectiveness of strategy-1. Table 1 shows the optimization results with strategy-1 as well as the performance of the original LC model. It is clear that the performance of the LC model is improved with the optimization process defined by strategy-1, indicating that strategy-1, e.g. the FWA-based optimization process, is effective to improve the performance of the original LC model. On the other hand, as shown in Table 1, the performance improvement of the LC model with strategy-1 is limited due to that the validation set cannot well reflect the data distribution of the testing set all the time.

For the consideration of robustness, the second strategy (strategy-2) is designed based on strategy-1. Different from strategy-1, the fitness of each spark in this strategy is not simply computed on an independent validation set. Instead, 10-fold cross validation mechanism is employed in the process of computing fitness of each spark, where the original training set is divided into ten parts and one of them is defined as the validation set and others are defined as the new training set in each fold. The current spark is evaluated by training a corresponding model on the new training set and computing fitness on the validation set ten times. In this case, each spark is comprehensively evaluated by the performance on 10 folds. The optimal model that is corresponding to the optimal spark achieved and trained on the original training set is finally examined on the testing set. In this strategy, fitness of each spark is evaluated on the training set   by 10-fold cross validation, overcoming the shortage of strategy-1 that the performance improvement of LC model is totally dependent on the consistency of data distribution in validation set and testing test. Strategy-2 enhances the robustness of the optimization process and is considered to achieve the improvements, with great performance, of the LC model.

Table 2. Performance comparison of LC before and after optimization with strategy-2

Corpus	Approach	Precision (%)	Recall (%)	Accuracy (%)	F1 (%)
PU1s	LC	100	92.36	96.67	95.88
	Strategy-2	100	96.64	**98.57**	**98.22**
PU2s	LC	100	64.00	90.71	74.62
	Strategy-2	100	94.17	**98.57**	**96.57**
PU3s	LC	97.84	91.30	95.37	94.34
	Strategy-2	98.25	95.91	**97.56**	**97.02**
PUAs	LC	95.78	90.72	93.64	92.68
	Strategy-2	98.75	96.44	**97.73**	**97.42**

Considering the efficiency of experiments, we randomly selected part of each corpora instead of the original corpus to investigate the effectiveness of strategy-2, e.g. 20% samples of PU1, PU2 and PUA were selected to form PU1s, PU2s and PUAs, and 10% samples of PU3 were selected to form PU3s. Table 2 presents the comparison of LC model before and after the optimization with strategy-2. It is notable that strategy-2 indeed brings a great improvement to the performance of the LC model, validating the effectiveness (taken the precision, recall, accuracy and F1 into account) of this strategy as well as the FWA-based optimization process. But the drawback of strategy-2 is that employing 10-fold cross validating in computing the fitness of sparks is time consuming. However, in fact, the usual offline training of the spam filters in the real world endows this strategy with usability.

5   Conclusion

This paper proposes a new framework that optimizes anti-spam model with heuristic swarm intelligence optimization algorithms, and this framework could integrate various classifiers and feature extraction methods. 2 experimental strategies were designed to objectively reflect the performance of the framework. Then, experiments are conducted, using the Fireworks Algorithm (FWA) as the Swarm Intelligence algorithm, the Local Concentration approach as the feature extraction method, and SVM as the classifier. During the experiments, 3 core parameters of the LC approach and 2 core parameters of SVM were optimized by using FWA. Experimental results demonstrated that the framework improved the performance on the corpora PU1, PU2, PU3 and PUA, and the computational efficiency is applicable in real world.

In future work, we intend to incorporate other swarm intelligence algorithms, feature extraction methods and classifiers into the framework, and investigate their performance under these configurations.

Acknowledgements. This work is supported by the National Natural Science Foundation of China (NSFC), under grant number 61170057 and 60875080.

References

1. Commtouch,: Internet threats trend report-February 2013. Tech. rep. (2013)
2. Cost of spam,
   http://www.ferris.com/2009/01/28/
   cost-of-spam-is-flattening-our-2009-predictions/
3. Drucker, H., Wu, D., Vapnik, V.N.: Support vector machines for spam categorization. IEEE Transactions on Neural Network 10, 1048–1054 (1999)
4. Ruan, G., Tan, Y.: Intelligent detection approaches for spam. In: Proceedings of International Conference on Natural Computation, pp. 1–7 (2007)
5. Bickel, S., Scheffer, T.: Dirichlet-enhanced spam filtering based on biased samples. Adv. Neural Inf. Process. Syst. 19, 161–168 (2007)
6. Kanaris, I., Kanaris, K., Houvardas, I., Stamatatos, E.: Words versus character N-grams for anti-spam filtering. Int. J. Artif. Intell. T. 16(6), 1047–1067 (2007)
7. Zhu, Y.C., Tan, Y.: A local-concentration-based feature extraction approach for spam filtering. IEEE Transactions on Information Forensics and Security 6(2), 1–12 (2011)
8. Information gain, http://en.wikipedia.org/wiki/Information_gain
9. Koprinska, I., Poon, J., Clark, J., Chan, J.: Learning to classify e-mail. Inform. Sci. 177, 2167–2187 (2007)
10. Tan, Y., Zhu, Y.: Fireworks algorithm for optimization. In: Tan, Y., Shi, Y., Tan, K.C. (eds.) ICSI 2010, Part I. LNCS, vol. 6145, pp. 355–364. Springer, Heidelberg (2010)
11. Dasgupta, D.: Advances in artificial immune systems. IEEE Computational Intelligence Magazine, 40–49 (2006)
12. Guzella, T.S., Caminhas, M.: A review of machine learning approaches to spam filtering. Expert Syst. Appl. 36, 10206–10222 (2009)
13. Blanzieri, E., Bryl, A.: A Survey of Learning-Based Techniques of e-mail Spam Filtering. Tech. Rep. 1 DIT-06-065 (2008)

14. Timmis, J.: Artificial immune systems—today and tomorrow. Nat. Comput., 1–18 (2007)
15. Oda, T., White, T.: Developing an immunity to spam. In: Cantú-Paz, E., et al. (eds.) GECCO 2003. LNCS, vol. 2723, pp. 231–242. Springer, Heidelberg (2003)
16. Tan, Y., Deng, C., Ruan, G.: Concentration based feature construction approach for spam detection. In: Proceedings of International Joint Conference on Neural Networks, pp. 3088–3093 (2009)
17. Ruan, G., Tan, Y.: Intelligent detection approaches for spam. In: Proceedings of International Conference on Natural Computation, pp. 1–7 (2007)
18. Tan, Y.: Multiple-point bit mutation method of detector generation for SNSD model. In: Wang, J., Yi, Z., Żurada, J.M., Lu, B.-L., Yin, H. (eds.) ISNN 2006. LNCS, vol. 3973, pp. 340–345. Springer, Heidelberg (2006)
19. Tan, Y., Xiao, Z.: Clonal particle swarm optimization and its applications. In: Proceedings of IEEE Congress on Evolutionary Computation, pp. 2303–2309 (2007)
20. Tan, Y., Wang, J.: A support vector network with hybrid kernel and minimal vapnik-chervonenkis dimension. IEEE Trans. Knowl. Data Eng. 26, 385–395 (2004)
21. Stuart, I., Cha, S.-H., Tappert, C.C.: A neural network classifier for junk E-mail. In: Marinai, S., Dengel, A.R. (eds.) DAS 2004. LNCS, vol. 3163, pp. 442–450. Springer, Heidelberg (2004)
22. Zhu, Y., Tan, Y.: A danger theory inspired learning model and its application to spam detection. In: Tan, Y., Shi, Y., Chai, Y., Wang, G. (eds.) ICSI 2011, Part I. LNCS, vol. 6728, pp. 382–389. Springer, Heidelberg (2011)
23. Ruan, G., Tan, Y.: A three-layer back-propagation neural network for spam detection using artificial immune concentration. Soft Comput. 14, 139–150 (2010)
24. Zhu, Y., Tan, Y.: Extracting discriminative information from E-mail for spam detection inspired by immune system. In: Proceedings of IEEE Congress on Evolutionary Computation, pp. 2491–2497 (2010)
25. Wu, C.-H.: Behavior-based spam detection using a hybrid method of rule-based techniques and neural networks. Expert Syst. Appl. 36, 4321–4330 (2009)
26. Siefkes, C., Assis, F., Chhabra, S., Yerazunis, W.S.: Combining winnow and orthogonal sparse bigrams for incremental spam filtering. In: Boulicaut, J.-F., Esposito, F., Giannotti, F., Pedreschi, D. (eds.) PKDD 2004. LNCS (LNAI), vol. 3202, pp. 410–421. Springer, Heidelberg (2004)

Parameter Optimization for Bezier Curve Fitting Based on Genetic Algorithm[*]

Linghui Zhao[1], Jingqing Jiang[1,2,**], Chuyi Song[1], Lanying Bao[1], and Jingying Gao[1]

[1] College of Mathematics, Inner Mongolia University for Nationalities,
Tongliao Inner Mongolia 028043, China
[2] College of Computer Science and Technology, Inner Mongolia University for Nationalities,
Tongliao Inner Mongolia 028043, China
jiangjingqing@yahoo.com.cn

Abstract. Fitting is one of the most important methods for free curve and surface modeling. This paper constructs the least squares fitting mathematical model for Bezier curve to fit the given data points on two-dimensional space. The genetic algorithm is applied to optimize the parameters of Bernstein basis function. The chromosomes are coded using real numbers. The fitness function is the reverse of the sum of the squared error. The simulation results show the feasibility and efficiency of the proposed method.

Keywords: Genetic Algorithm, Bezier Curve, Curve Fitting.

1 Introduction

With the rapid development of computer technology and measure technology, it has become convenient to obtain lots of measured data points. One of the important researching works for computer aided geometry design (CAGD) is to obtain the best curve representation to fit the measured data points. This is called curve fitting. In the process of measure, there are some errors caused by the influence of varies factors. So, the fitting curve or the fitting surface does not need to cross each data point. Recently, researchers have paid more and more attention to curve fitting. Curve fitting has been widely used in the fields of medicine, chemistry, biology and industry [1, 2]. There are many methods to deal with the curve fitting such as Gauss-Newton iteration method, L-M method, quasi-Newton method, B-spline curve fitting method and Bezier curve fitting method etc. In recent years, with the development and wide application of intelligence computational algorithms, some researchers have proposed curve fitting methods based on different intelligence computational algorithms. A curve fitting method based on BP artificial neural networks was presented by Bao [3]. Zhu proposed a curve fitting method of B-spline based on particle swarm optimization [4]. Genetic algorithm is a stochastic optimization method which is

---

[*] The authors are grateful to the support of the National Natural Science Foundation of China (Grant 61163034).
[**] Corresponding author.

developed based on natural selection principles and biological evolution mechanisms. It is suitable to process with complexity non-linear optimization problems which are difficult to do with traditional searching algorithm. Some researchers have applied it to solve curve fitting problems [5-13]. Bai combined genetic algorithm and steepest descent algorithm to fit Bezier curve [7]. Zhang performed B-spline curve fitting based on genetic algorithm and simulated annealing [8]. Renner described least square fitting B-spline curve based on genetic algorithm [9]. Zhou presented least squares fitting B-spline and Bezier curve based on genetic algorithm [10]. Sun optimized the parameters of B-spline curve by genetic algorithm[11]. Inspired by Sun [11], this paper constructs a least squares fitting mathematical model for Bezier curve, and optimizes the parameters of Bernstein basis function using genetic algorithm. There are some methods for the choice of parameters, such as uniform parameterization, cumulative chord length parameterization [14], centripetal model parameterization [15] and gradient parameterization [16-17].

2   Least Squares Fitting Mathematical Model for Bezier Curve

Bezier curve is a polynomial curve driven by a set of control points. It is named after their inventor Dr. Pierre Bezier. A Bezier curve is a parametric curve frequently used in computer graphics and related fields. Bezier curve has the ability to represent and design the free curve and free surface. We use the Bezier curve to fit the ordered data points so that the least squares error is minimized.

The least squares fitting mathematical model for Bezier curve is described as follows:

Given r ordered data points $Q = \{q_j(x_j, y_j), j = 1,...r\}$ on two-dimensional space and the corresponding ordered parameter vector $T = \{t_j, j = 1,..r\}$. The n order Bezier curve is defined as follows:

$$B(t) = \sum_{i=0}^{n} B_i^n(t) P_i \tag{1}$$

where $P = \{P_i, i = 0,...,n\}$ represent the control point and $B_i^n(t)$ is the Bernstein basis function. The definition of $B_i^n(t)$ is:

$$B_i^n(t) = \binom{n}{i} (1-t)^{n-i} t^i \tag{2}$$

While utilizing Bezier curve to fit the ordered data points on two-dimensional space, the least squares fitting error should be minimized. If a data point lies on Bezier curve, it should satisfy:

$$q_j = B(t_j) = \sum_{i=0}^{n} B_i^n(t_j) P_i; j = 1,...,r \tag{3}$$

Eq.(3) can be written in matrix form:

$$Q = BP \tag{4}$$

where B is Bernstein basis function and Q is the given data points on two-dimension space. If the parameter vector T is given, the approximate solution of the control points can be obtained:

$$P = \left(B^T B\right)^{-1} B^T Q \tag{5}$$

Eq.(5) is fitted into in Eq.(4) . Then the fitting curve equation can be written as:

$$Q^c = B\left(B^T B\right)^{-1} B^T Q \tag{6}$$

q_j^c denotes the point on the fitting curve that is related to the parameter t_j . The sum of the squared error is:

$$SSE = \sum_{j=0}^{r} \left| q_j - q_j^c \right|^2 \tag{7}$$

It can be seen from Eq.(7) that the choice of the parameter vector $T = \{t_j, j=1,...,r\}$ for Bernstein basis function will affect the error of the fitting. This paper will apply genetic algorithm to obtain the optimal parameter vector.

3   Bezier Curve Fitting Process Based on Genetic Algorithm

Genetic algorithm is a computer simulation of the biological evolution process. The population consists of some individuals (chromosomes). Each chromosome is composed of some genes. A possible solution of a problem is considered a chromosome by the genetic algorithm. The algorithm calculates the fitness function value of each individual according to the given fitness function and then performs the searching process in terms of the fitness function value. The searching process consists of a selection operation, a crossover operation and a mutation operation. Genetic algorithm does well in global and parallel searching. This paper applies genetic algorithm to optimize the parameter vector in Bernstein basis function.

3.1   Coding and Genetic Operation

Coding and Initialization. The parameter vector $T = \{t_j, j=1,...,r\}$ of Bezier curve to be optimized is considered as a chromosome. The chromosome is coded using real number. The selection, crossover and mutation operations are applied on the population iteratively. Finally, the parameter vector which maximizes the fitness function is obtained.

In the initialization step, each chromosome in the population is initialized randomly. Random real numbers between 0 and 1 are produced for r-1 times. These numbers should be sorted to keep the parameters orderly. The first parameter is set as 0 so that the left point of the Bezier curve coincides with the first given data point. There are r parameters in Bernstein basis function. So each chromosome corresponds to a parameter vector and there are r genes in each chromosome. The samples of two initial chromosomes are given below.

$$G_i : 0 \quad 0.0680 \quad 0.1537 \quad 0.2407 \quad 0.2665 \quad 0.2810 \quad \ldots\ldots \quad 0.9577$$

$$G_j : 0 \quad 0.0566 \quad 0.1247 \quad 0.2555 \quad 0.2766 \quad 0.2820 \quad \ldots\ldots \quad 0.9566$$

Fitness Function. In order to obtain the best fitting curve, the sum of the squared error should be minimized. So the fitness function can be chosen as

$$\text{Fitness} = 1 / SSE$$

When the value of SSE is smaller the value of the fitness function is bigger. The points on Bezier curve will better fit the given data points on two-dimensional space.

Selection Operation. Genetic algorithm selects the better chromosome through the selection operation. The roulette method is used to select the better chromosome with a higher fitness function value. If f_i denotes the fitness function value of chromosome G_i, the probability that G_i is selected is

$$P_i = \frac{f_i}{\sum\limits_{i=1}^{N} f_i}$$

Crossover Operation. According to the mathematic model of Bezier curve, the elements of the parameter vector are ordered increasingly. That is the genes in each chromosome are ordered ascendingly. The crossover operation should satisfy the ordered feature. The chromosomes that generated by single-point crossover and multi-point crossover do not satisfy the ordered feature. In order to solve this problem, the chromosomes are coded using real numbers and the linear combination crossover operation is applied. The process of crossover is shown as follows:

G_i and G_j denote parent chromosomes. A random real number that lies between 0 and 1 is h. The generated chromosomes after the crossover operation are:

$$G_i' = hG_i + (1-h)G_j$$
$$G_j' = (1-h)G_i + hG_j$$

The offspring chromosomes generated from linear combination crossover operation continues the increasing feature.

Mutation Operation. In order to obtain the global optimization solution, genetic algorithm utilizes the mutation operation to expand the search region. Any gene on a chromosome may mutate. The mutation could destroy the increasing order of genes on a chromosome and the chromosome will no longer be a possible solution for the parameter vector. To keep the increasing order of the genes, the mutation operation is adopted as follows in this paper.

Set $G = (g_1, g_2, \ldots, g_r)$ is a chromosome that will mutate. A random integer number which lies between 1 and r is k. The kth gene on the chromosome is denoted by g_k'. The mutation operation is show as follows:

$$\begin{cases} g_k' = hg_2 & k = 1 \\ g_k' = g_{k-1} + h(g_k - g_{k-1}) & k \neq 1 \end{cases}$$

where h is a random real number that lies between 0 and 1.

3.2 Flow of Algorithm

The steps of the optimizing parameters of Bezier curve based on genetic algorithm are as follows:

Step1: Set the order of Bezier curve, the data points on two-dimensional space, the size of the population and the maximum iteration generation.
Step2: Initialize the chromosomes.
Step3: Calculate the fitness function value for each chromosome. Record the best fitness function value and the corresponding chromosome.
Step4: Apply the selection operation. Select the chromosomes with higher fitness function value using roulette selection.
Step5: Match the chromosomes randomly. Apply the linear combination crossover operation to each pair of chromosomes.
Step6: Mutate the chromosome according to mutation probability.
Step7: Calculate the fitness function value for each chromosome. Update the best fitness function value and the corresponding chromosome.
Step8: Repeat step 4 through step 7 until the maximum iteration generation is reached. Record the best fitness function value and the best chromosome

4   Simulation Experiment

There are two examples to test the proposed algorithm in this paper. The first example is to fit some points from sin function. The second example is to fit some given data points on two-dimensional space. The program is written in Matlab. The parameters used in the algorithm are as follows: the order of Bezier curve is 5, the size of the population is 40, the maximum iteration generation is 200, the crossover probability is 0.6, the mutation probability is 0.01.

Example 1: Given 20 ordered data points. The x-coordinates are 20 real numbers from 0 to 2π. The interval is 0.1π. The y-coordinates are the function value of *sin* corresponding to x-coordinates. Fitting these 20 data points the least SSE is 0.0446. Fig. 1 shows the best value and average value of SSE in each generation. Fig. 2 shows the fitting curve and the given data points.

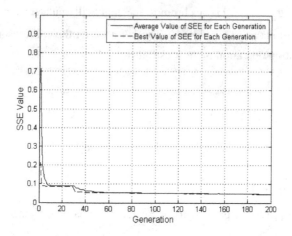

Fig. 1. SSE value of example 1 in each generation

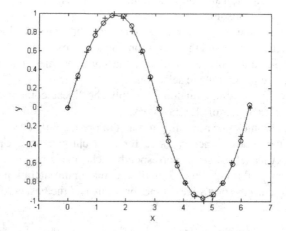

Fig. 2. The fitting curve for example 1. 'o' represents the fitting data points and '+' represents the given data points.

Example 2: Given 20 ordered data points on two-dimensional space. Fitting these 20 data points the least SSE is 0.0446. Fig. 3 shows the best value and average value of SSE in each generation. Fig. 4 shows the fitting curve and the given data points.

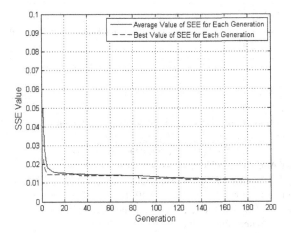

Fig. 3. SSE value of example 2 in each generation

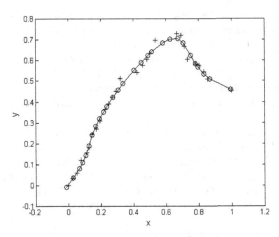

Fig. 4. The fitting curve for example 2. 'o' represents the fitting data points and '+' represents the given data points.

It can be seen from the two examples that the genetic algorithm is efficient to optimize the parameters of Bezier curve to fit the given data points. The fitting precision is comparable.

5   Conclusion

This paper constructs the least squares fitting mathematical model of Bezier curve based on given data points on two-dimensional space. The genetic algorithm is applied to optimize the parameters of this model. The chromosomes are coded using real numbers. The fitness function is the reverse of the sum of squared error. The selection operation is the roulette selection. The crossover operation is the linear combination crossover. The mutation operation is selected in accordance with the location of the gene. The simulation results show that the proposed method is feasible and efficient for curve fitting.

References

1. Huang, W.: Blurred Medical Image Segmentation Based on Curve Fitting. Computer Engineering and Applications 45(18), 194–197 (2009)
2. Shao, M., Li, L.: Oil Well Output Forecast Based on Least Squares Curve Fit. Microcomputer Applications 25(12), 13–14 (2009)
3. Bao, J., Zhao, J., Zhou, H.: Study on Method of Curve Simulation Based on BP Network. Computer Engineering and Design 26(7), 1840–1841 (2005)
4. Zhu, Q., Zeng, L., Qu, H., Liu, J.: Curve Fitting of B-spline Based on Particle Swarm Optimization. Computer Science 36(10), 289–291 (2009)
5. Renner, G., Ekart, A.: Genetic Algorithms in Computer Aided Design. Computer-Aided Design 35, 709–726 (2003)
6. Yoshimoto, F., Harada, T., Yoshimoto, Y.: Data Fitting With a Spline Using a Real-Coded Genetic Algorithm. Computer-Aided Design 35, 751–760 (2003)
7. Bai, X., Peng, G., Chen, X.: Bezier Curve Fitting Based on Genetic Algorithm and Steepest Descent Algorithm. Computer Engineering and Design 30(1), 194–196 (2009)
8. Zhang, J., Wang, H.: B-Spline Curve Fitting Based on Genetic Algorithms and The Simulated Annealing Algorithm. Computer Engineering & Science 33(3), 191–193 (2011)
9. Renner, G., Ekart, A.: Genetic Algorithms in CAD. Computer-Aided Design 35, 707–708 (2003)
10. Zhou, M., Wang, G.: Genetic Algorithm-Based Least Square Fitting of B-Spline and Bezier Curves. Journal of Computer Research and Development 42(1), 134–143 (2005)
11. Sun, Y., Wei, J., Xia, D.: Parameter Optimization for B-Spline Curve Fitting Based on Adaptive Genetic Algorithm. Journal of Computer Applications 30(7), 1878–1882 (2010)
12. Balu, R., Selvakumar, U.: Optimum Hierarchical Bezier Parameterization of Arbitrary Curves and Surfaces. In: 11th Annual CFD Symposium, Bangalore, August 11-12, pp. 46–48 (2009)
13. Prasanth, V., Lal, S.A.: Bezier Parameterization of an airfoil using genetic algorithm, http://hdl.handle.net/123456789/1026
14. Grossman, M.: Parametric curve fitting. The Computer Journal 17(2), 169–172 (1971)
15. Lee, E.T.Y.: Choosing Nodes in Parametric Curve Interpolation. Computer-Aided Design 21(6), 363–370 (1989)
16. Plass, M., Stone, M.: Curve-Fitting with Piecewise Parametric Cubics. Computer Graphics 17(3), 229–239 (1983)
17. Borges, C.F., Pastva, T.: Total Least Squares Fitting of Bezier and B-Spline Curves to Ordered Data. Computer-Aided Geometric Design 19(4), 275–289 (2002)

OFDM System with Reduce Peak-to-Average Power Ratio Using Optimum Combination of Partial Transmit Sequences

Yung-Cheng Yao[1], Ho-Lung Hung[2], and Jyh-Horng Wen[3]

[1] Department of Electrical Engineering, National Chung Cheng University, Taiwan
yaoyc@thu.edu.tw
[2] Department of Electrical Engineering, Chienkuo Technology University, Taiwan
hlh@ctu.edu.tw
[3] Department of Electrical Engineering, Tunghai University, Taiwan
jhwen@thu.edu.tw

Abstract. In this paper, we propose a new peak-to-average power ratio (PAPR) reduction scheme of orthogonal frequency division multiplexing (OFDM) system, called invasive weed optimization (IWO) scheme, which considerably reduces the computational complexity with keeping the similar PAPR reduction performance compared with the conventional partial transmit sequences (PTS) scheme. PTS is a distortionless PAPR reduction technique, but its high search complexity for finding optimal phase factors must be reduced for usable applications. The proposed scheme is analytically and numerically evaluated for the OFDM system specified in the IEEE 802.16 standard. IWO based PTS is compared to different PTS schemes for PAPR reduction and search complexity performances. The simulation results show that the proposed IWO-based PTS method provides good PAPR reduction and bit error rate (BER) performances.

Keywords: Orthogonal frequency division multiplexing, Peak-to-average power ratio reduction, Partial transmit sequence and Invasive weed optimization.

1   Introduction

As the bandwidth demand in communication systems is increased, the new transmission formats like orthogonal frequency division multiplexing (OFDM) is applied in modern communication systems like worldwide interoperability for microwave access (WiMAX). Despite the advantages of OFDM signals like high spectral efficiency and robustness against inter-symbol interference (ISI), the OFDM signals suffer from disadvantages in which the main one is high peak-to-average power ratio (PAPR) [1-2]. The reason for high PAPR is that in time domain the OFDM signal is actually sum of many narrowband signals. A major drawback of orthogonal frequency division multiplexing (OFDM) is the high peak-to-average power ratio (PAPR) of the transmitted signal. However, a major problem associated with multicarrier modulation is its large peak-to-average power ratio (PAPR) [3-4], which makes system performance very sensitive to distortion introduced by nonlinear

Y. Tan, Y. Shi, and H. Mo (Eds.): ICSI 2013, Part I, LNCS 7928, pp. 459–466, 2013.

devices such as power amplifiers. Soft computing techniques are useful in communication field [5-7]. Partial transmit sequence (PTS) technique use iterative routine similar to the trial-and-error method for finding the optimum phase factors that leads to lower PAPR. It is distortionless but is time-consuming, and needs large number of computations [8-11].

In this paper, we take a fresh look at PTS for PAPR reduction and propose solutions for both the above-mentioned problems. To tackle the complexity issue of PTS, we formulate the sequence search of PTS as a particular combinatorial optimization (CO) problem. To reduce complexity for phase weight searches, some stochastic search techniques [10-12] have recently been proposed because they could obtain the desirable PAPR reduction with a low computational complexity.

In recent past several intelligences nature-inspired metaheuristics like the Artificial bee colony algorithm [9], Electromagnetism-like Method (EM) method [10], genetic algorithm (GA) [11] and particle swarm optimization (PSO) [12], etc. have been applied to solve the PAPR reduction problems. However, to our knowledge, invasive weed optimization (IWO) [13-14] has not yet been used for the same purpose till date. In this paper, we propose a novel solution to reduce the complexity while keeping the optimal combination of the phase factors to reduce the PAPR largely. Specifically, we apply the IWO to search the optimal combination of phase factors with largely reduced complexity. In simulations the fixed WiMAX signal based on IEEE 802.16-2004 standard is applied for demonstrating the effectiveness of the proposed technique. The rest of this paper is organized as follows. In Section 2, typical OFDM system is given and the PAPR problem is formulated and then PTS is explained. Then, IWO is proposed to search the optimal combination of phase factors for PTS in section 3. Section 4 and 5 discuss the simulation results and conclusions respectively.

2   System Model and Problem Definition

In OFDM systems, a fixed number of successive input data samples are modulated first, and then jointly correlated together using IFFT at the transmitter side. IFFT is used to produce orthogonal data subcarriers. Mathematically, IFFT combines all the input signals to produce each element (signal) of the output OFDM symbol. The time domain complex baseband OFDM signal can be represented as [2]:

$$s_n = \frac{1}{\sqrt{N}} \sum_{i=0}^{N-1} S_i e^{j2\pi t i / N}, \quad n = 0, 1, 2, \ldots, N-1, \tag{1}$$

where S_i is i-th data modulated symbol in OFDM frequency domain, N is the number of subcarriers and s_n is the n-th signal component in OFDM output symbol. However, OFDM output symbols typically have large dynamic envelope range due to the superposition process performed at the IFFT stage in the transmitter. PAPR is widely used to evaluate the variation of the output envelope. PAPR is an important factor in the design of both high power amplifier (PA) and DAC and for generating error-free (or with minimum errors) transmitted OFDM symbols and also preventing the PA to work in nonlinearity region. The PAPR of the transmitted signal in Eq.(1) could be defined as

$$PAPR(s) = 10\log_{10} \frac{\max\limits_{0 \le n \le N-1} |s_n|^2}{E\left[|s|^2\right]}, \qquad (2)$$

where $\max|s_n|^2$ is the maximum values of the OFDM signal power, $s=[s_0,s_1,...,s_{N-1}]$ and $E[\cdot]$ denotes the expected value operation. In principle, PAPR reduction techniques are concerned for reducing $\max|s_n|^2$.

In the PTS technique, an input data block of N symbols is partitioned into disjoint sub-blocks. The subcarriers in each sub-block are weighted by a phase weighting factor for the sub-block. The phase weighting factors are selected such that the PAPR of the combined signal is minimized. The block diagram of IWO-based PTS scheme is shown in Fig. 1 [3]. In the scheme, S is partitioned into M disjoint sub-blocks such that

$$S = \sum_{i=1}^{M} S_i \qquad (3)$$

Here, it is assumed that the clusters S_i consist of a set of sub-blocks with equal size. The objective is to find sets of phase weighting factors $\mathbf{b}$. Then, the weighted sum combination of the M sub-blocks which could be written as

$$Z(b) = \sum_{i=1}^{M} b_i S_i, \quad b_i = e^{j\varphi_i} \cdot \qquad (4)$$

where b_i, $i=1, 2,...., M$ is the phase weighting factor. In general, the selection of the phase weighting factors is limited to a set with finite number of elements to reduce the search complexity. After transforming to the time domain, the new time domain vector becomes

$$\mathbf{s} = IFFT\left\{\sum_{i=1}^{M} b_i \mathbf{S}_i\right\} = \sum_{i=1}^{M} b_i IFFT\{\mathbf{S}_i\}, \qquad (5)$$

The optimization process is to find phase weighting factor that minimize the PAPR. The optimal phase weighting factor b_i that minimizes the PAPR can be obtained from a comprehensive simulation of all possible combination. The objective of the proposed method is to choose a phase weighting vector $\mathbf{b}=\{b_1,b_2,...,b_i\}$ to reduce the PAPR of $Z(b)$, and the cost function is defined as

$$\hat{\mathbf{b}} = \arg\min_{b}\left\{\max\left|\sum_{i=1}^{M} b_i \mathbf{S}_i\right|\right\} \cdot \qquad (6)$$

The sub-block partition for PTS scheme is one of the effective PAPR reduction techniques of division on sub-carriers into multiple disjoint sub-blocks [15]. However, the computation complexity C shows a corresponding exponential increase, which is related as $C=2^{M-1}$. With the viewpoint from PAPR reduction, pseudo-random sub-block partitioning has a better performance than that with the other methods.

In this paper, a sub-blocks partition method is implemented initially. In the following, the sub-blocks partition method is used into the PTS scheme with applying the IWO optimization algorithm to reduce the PAPR in the OFDM systems.

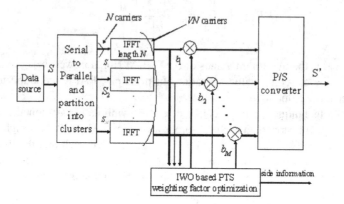

Fig. 1. The block diagram of the IWO-based PTS technique

3   The Invasive Weed Optimization-Based PTS Scheme for Reduction

In a D-dimensional search space, a weed which represents a potential solution of the objective function is represented by $\mathbf{b}=(b_1,b_2,...,b_m)$. Firstly, P weeds, called a population of plants, are initialized with random growth position, and then each weed produces seeds depending on its fitness and the colony's lowest fitness and highest fitness to simulate the natural survival of the fittest process. The number of seeds each plant produce increases linearly from minimum possible seed production to its maximum. The generated seeds are being distribution randomly in the search area by normal distribution with mean equal to zero and a variance parameter decreasing over the number of iteration. By setting the mean parameter equal to zero, the seeds are distributed randomly such that they locate near to the parent plant and by decreasing the variance over time, the fitter plants are grouped together and inappropriate plants are eliminated over times. The general scheme for the IWO algorithm is shown in Algorithm 1, which consists of four main procedures: ***Initialization, reproduction, Spatial dispersal and Competitive exclusion*** operator, respectively.

Algorithm 1. Differential Evolution algorithm

---

1.   Initialize population space( ); **select an initial population of N candidate solution,** Evaluate fitness
2.   **while** termination criteria are not satisfied **do**
3.   **Reproduction ( )**
4.   **Spatial dispersal ( )**
5.   **Competitive exclusion ()**
6. **end while**

To model and simulate the colonizing behavior of weeds in order to introduce a novel optimization algorithm, some basic properties of the colonization process are considered:

Step1. *Initializing a population*: A population of initial solutions is being spread out over the D-dimensional problem space with random positions.

Step 2. *Reproduction*: Each member of the population is allowed to produce seeds depending on its own, as well as the colony's lowest and highest fitness, such that, the number of seeds produced by a weed increases linearly from lowest possible seed for a weed with worst fitness to the maximum number of seeds for a plant with best fitness.

Step 3. *Spatial dispersal*: The produced seeds in this step are being dispread over the search space by normally distributed random numbers with mean equal to the location of the producing plants and varying standard deviations. Thus, seeds will be randomly distributed such that they abide near the parent plant.

Step 4. *Competitive exclusion*: This process continues until the maximum number of plants is attained by fast reproduction. At this stage, only the plants with higher fitness can survive and produce seeds, whereas others are eliminated (competitive exclusion).

4   Results and Discussions

In this section, the system performance with proposed PTS scheme is evaluated based on the PAPR complementary cumulative distribution function (CCDF) and the bit error rate (BER) by computer simulation. The modulation is chosen as QPSK scheme and the number of sub-carriers is assumed to be $N=1024$. In the simulations, the random sub-blocks partitioning is used both in the conventional PTS and the proposed PTS schemes. The complexity of those two techniques with several of number of sub-blocks M is also considered in this paper. The parameters, the number of clusters and the number of allowed phase weighing factors W for transmit sequences, are also considered in the simulations. The cumulative distribution function (CDF) of the amplitude of a sampling signal is computed by $CDF=1-\exp(PAPR_0)$, and the CCDF [2] could defined as

$$CCDF = P_r(PAPR > PAPR_0)$$
$$= 1 - P_r(PAPR \leq PAPR_0) = 1 - (1 - \exp(-PAPR_0))^N. \tag{7}$$

It is assumed that the N time domain signal samples are mutually independent and uncorrelated.

Fig. 2 depicts the CCDF of the PAPR with the PTS sequence search by IWO technique with $M=2$, 4 and $b=4$. Just as expected, the PAPR performance of our proposed IWO-based PTS scheme with $(M,b)=(2,2)$, is not only almost the same as that of the IWO -based PTS scheme with$(M,b)=(4,4)$, but also having much lower computational complexity.

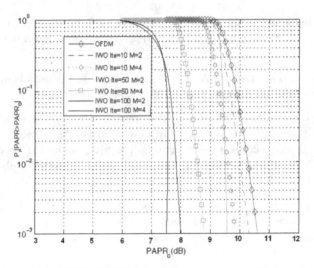

Fig. 2. Comparison of the PAPR CCDF of the different numbers of the maximum number of iterations of the IWO method for $N=1024$, $W=4$, $M=2$ and 4

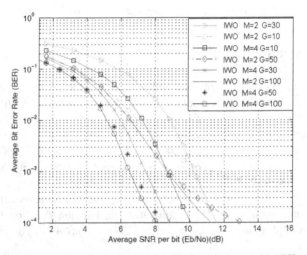

Fig. 3. Comparison of BER performance of the IWO technique with different numbers of iterations in AWGN channels

Fig. 3 shows a comparison of Bit Error Rate (BER) performance of the IWO method in Additive White Gaussian Noise (AWGN) channels. From this figure, we can see that the BER is slightly increased when IWO-PTS method is applied as compared to conventional PTS, but PAPR is much improved according to the result of Fig. 2. The performance of system shows improvement at the cost of BER. In the IWO method, the population size is assumed to be $P = 20$; and the corresponding maximum number of iterations are $G = 10, 30, 50$ and 100, respectively.

5   Conclusions

This paper proposes an IWO technique to obtain the optimal phase weighting factor for the PTS scheme to reduce computational complexity and improve PAPR performance. The searches on phase weighting factors of the PTS technique is formulated as a global optimization problem with bound constraints. The computer simulation results show that the proposed IWO technique obtained the desirable PAPR reduction with low computational complexity when compared with the various stochastic search techniques. The performance of system shows improvement at the cost of BER. The application of the proposed method will be useful in WiMAX applications.

Acknowledgment. This work is partially supported by National Science Council, Taiwan, under Grant NSC 101-2221-E-029-020-MY3 and NSC 101-3113-P-033-003.

References

1. Muller, S.H., Huber, J.B.: OFDM with reduced peak-to-average power ratio by optimum combination of partial transmit sequences. IEE Electron. Lett. 33(5), 368–369 (1997)
2. Jiang, T., Wu, Y.: An overview:peak-to-average power ratio reduction techniques for OFDMsignals. IEEE Trans. on Broadcasting 54(2), 257–268 (2008)
3. Tellambura, C.: Improved phase factor computation for the PAR reduction of an OFDM signal using PTS. IEEE Commun. Lett. 5(4), 135–137 (2001)
4. Kwon, O.J., Ha, Y.H.: Multi-carrier PAPR reduction method using sub-optimal PTS with threshold. IEEE Trans. Broadcast. 49(2), 232–236 (2003)
5. Soontornpipit, P., Furse, C.M., Chung, Y.C., Lin, B.M.: Optimization of a buried microstrip antenna for simultaneous communication and sensing of soil moisture. IEEE Trans. Antennas and Propagation 54, 797–800 (2006)
6. Wang, L.P. (ed.): Soft Computing in Communications. Springer, Berlin (2003)
7. Wang, L.P., Shi, H.X.: A gradual noisy chaotic neural network for solving the broadcast scheduling problem in packet radio networks. IEEE Trans. Neural Networks 17, 989–1000 (2006)
8. Ho, W.S., Madhukumar, A., Chin, F.: Peak-to-average power reduction using partial transmit sequences: a suboptimal approach based on dual layered phase sequencing. IEEE Trans. Broadcast. 49, 225–231 (2003)
9. Wang, Y., Chen, W., Tellambura, C.: A PAPR Reduction Method Based on Artificial Bee Colony Algorithm for OFDM Signals. IEEE Trans. on Wireless Commun. 9(10), 2994–2999 (2010)
10. Chen, J.C.: Partial transmit sequences for PAPR reduction of OFDM signals with stochastic optimization techniques. IEEE Transactions on Consumer Electronics 56(3), 1229–1234 (2010)
11. Zhang, Y., Ni, Q., Chen, H.H.: A new partial transmit sequence scheme using genetic algorithm for peak-to-average power ratio reduction in a multi-carrier code division multiple access wireless system. International J. Autonomous Adaptive Commun. Systems 2(1), 40–57 (2009)

12. Wen, J.H., Lee, S.H., Huang, Y.F., Hung, H.L.: A sub-optimal PTS algorithm based on particle swarm optimization technique for PAPR reduction in OFDM systems. EURASIP Journal on Wireless Communications and Networking 2008 (2008)
13. Monavar, F.M., Komjani, N., Mousavi, P.: Application of Invasive Weed Optimization to Design a Broadband Patch Antenna With Symmetric Radiation Pattern. IEEE Antennas and Wireless Propagation Letters 10, 1369–1372 (2011)
14. Roy, S., Islam, S.M., Das, S., Ghosh, S.: Multimodal optimization by artificial weed colonies enhanced with localized group search optimizers. Applied Soft Computing 13(1), 27–46 (2013)
15. Kang, S.G., Kim, J.G., Joo, E.K.: A novel subblock partition scheme for partial transmit sequence OFDM. IEEE Trans. Broadcasting 45(3), 333–338 (1999)

Evolutionary Three-Stage Approach for Designing of Neural Networks Ensembles for Classification Problems

Vladimir Bukhtoyarov[1] and Eugene Semenkin[2]

[1] Siberian State Aerospace University, Department of Information Technologies Security,
Krasnoyarsky Rabochy Av. 31, 660014 Krasnoyarsk, Russia
vladber@list.ru
[2] Siberian State Aerospace University,
Department of System Analysis and Operational Research,
Krasnoyarsky Rabochy Av. 31, 660014 Krasnoyarsk, Russia
eugenesemenkin@yandex.ru

Abstract. The use of the neural network ensemble approach for solving classification problems is discussed. Methods for forming ensembles of neural networks and methods for combining solutions in ensembles of classifiers are reviewed briefly. The main ideas of comprehensive evolutionary approach for automatic design of neural network ensembles are described. A new variant of a three-stage evolutionary approach to decision making in ensembles of neural networks is proposed for classification problems. The technique and results of a comparative statistical investigation of various methods for producing of ensembles decisions on several well-known test problems are given.

Keywords: classifiers, neural networks, ensembles, combining strategies.

1 Introduction

Artificial neural networks and, in particular, multilayer perceptrons, are one of the most popular machine learning techniques applied to classification problems. The popularity of the neural network approach is mainly due to its high efficiency, proven for a wide range of practical problems. However, the constant desire to improve the quality of classification, the increasing scale of problems, the increasing demands for performance and the need for the development and use of parallel computing systems lead to the need for developing approaches that are significantly different from those that use a single neural network. Therefore, one of the most promising and popular approaches to solve classification problems in recent years is the neural network ensemble approach. The development of this approach was initiated in the article of Hansen and Salomon [1], although a collective approach using other techniques for classification was known before [2]. Examples of successful solutions of various problems with the use of neural network ensemble methods can be found in [3-5].

In this article we will focus on the classification problems and the development and analysis of appropriate neural network ensemble methods. We suppose that the improvement of the technique used to combine the single expert (participants of the

Y. Tan, Y. Shi, and H. Mo (Eds.): ICSI 2013, Part I, LNCS 7928, pp. 467–477, 2013.
© Springer-Verlag Berlin Heidelberg 2013

ensemble) decisions may be one of the ways for further development of the ensemble approach, particularly for solving problems of classification. Generally, there are many variants of strategies for combining the individual classifier's decisions. In this paper we propose a new three-stage approach, which extends the idea of a stacked generalized method developed by Wolpert [6]. In the proposed approach trained classifiers with continuous outputs placed on the second stage are used for the aggregation of individual experts' decisions. They are formed using genetic programming (in general, any other appropriate technology, such as fuzzy logic or ANN classifiers, can be used as second stage classifiers). The final decision about the class for each input pattern is made on the basis of an analysis of the decisions of all classifiers placed on the second stage. The proposed approach is described in detail in the last part of Section 1.

Section 2 is devoted to a statistical investigation of the performance of the proposed approach. Also, in Section 2 a method for evaluating the effectiveness of the investigated approaches is described and test problems from the UCI Machine Learning Repository [7], used to evaluate the effectiveness of approaches, are listed and characterized. The results of the study are also presented at the end of Section 2.

Finally, we discuss the results of our statistical investigation.

2   Neural Network Ensembles

Generally, the ensemble of neural networks is characterized by a pair (N, D), where $N=(N_1, N_2, ..., N_n)$ is a set of n neural networks (experts), whose solutions are taken into account in the ensemble decision evaluation procedure, and D is a method of obtaining the ensemble decision from the individual neural network decisions (the combining strategy). Thus, to use a neural network ensemble as a problem solving technology in each particular situation, two problems must be solved: a set of neural networks N must be formed and a method of deriving the common ensemble decision $D=f(N)$ from the individual expert decisions must be chosen. Each of these stages constitutes a separate problem, whose solution can be obtained in different ways.

The most resource-intensive stage is the first stage, during which the structure and the parameters of the neural networks used as classifiers are determined. Often, a relatively simple neural network structures are used because each of them theoretically solves a simpler problem obtained by decomposing the original problem (in an explicit or implicit form) and the division of the original set into subsets for each of the classifiers.

The second stage, which includes the choice of method for combining the single classifier decisions, usually requires less computational resources. Its complexity depends on the chosen combining strategy and the complexity of the problem being solved. Choosing the simplest combining strategy takes considerably less time compared to the first stage. However, using a complex combining strategy is one of the main ways of increasing the efficiency of problems solving with ensemble approach. In this regard improving the existing and developing new strategies for effective combination, which require intensive use of computing resources, becomes a more urgent issue to increase the efficiency of classification problems solving.

2.1   Forming a Set of Neural Networks

The first stage is equivalent to multiple solutions of the problem of designing intelligent systems based on the use of only one neural network. The complexity and resource consumption of this problem depends on the "deepness" of the network configuration during the formation of an artificial neural network. In the simplest case (or rather – in most cases) developers are limited to setting only the weights for some a priori given neural network structure. The more general case of neural networks design also involves the formation of a network structure i.e. determining the number of network layers, the number of hidden units (neurons) in each layer and the type of activation function in each hidden unit. Due to the complexity of designing the neural network structure, the development and application of effective methods for solving this problem in an automated mode are very important. Recently, approaches which utilize a genetic algorithm (or various modifications of this) have become widely used to design neural network structures [8]. Despite the fact that in many cases this approach is extremely effective, it includes a time-consuming and not formally defined stage, which is the setting of the genetic algorithm parameters. Therefore a new method with relatively small number of adjustable parameters which is based on a probability estimation evolutionary algorithm (named PGNS: Probability Generator for Networks Structures) was developed to effectively implement this stage of the design of neural network ensembles. The proposed approach automatically designs the structures of neural networks. This method is described in details in [9] where we proposed a neural network design paradigm based on estimated probability of the presence of different types neurons on the neural network layers. It is also useful to collect and process information on the presence or absence of neuron at each available place in the neural network structure. By including these estimated probabilities into our approach we can process the information about an optimal structure of the neural network in terms of mathematical statistics. This helps us to generate good neural networks. Apparently, the proposed method should provide neural networks which are simple and have good generalization ability. To find more effective ensemble solutions by forming more complex (compared with simple or weighted average or voting) mixtures of the component network decisions we used the genetic programming approach for automated generation of symbolic regression formulae.

2.2   Combining Strategies

Choosing an effective way to calculate the ensemble decision is a very important stage when using ensembles of neural networks. An inefficient combining strategy or a weak adaptation of it to a particular situation may have a significant negative impact on the effectiveness of the ensemble solution. Therefore, one of the most important areas for research relates to approaches for choosing combining strategies.

In most studies one of two main combining strategies is used: either selection or fusion [10]. The selection combining strategies are based on the assumption that each of the base experts is specialized in a particular local area of the problem space [11]. Expert fusion paradigm assumes that all experts are trained over the whole problem space, and are therefore considered as competitive rather than complementary [12].

Combining strategies can be divided into two groups concerning the use of classifiers' input when estimating an ensemble decision. The first group includes approaches that use "static" structures to evaluate ensemble prediction. The schemes for decision-making in these approaches are static and do not depend on the values of input variables. The second group includes methods that also operate with input variables to estimate effectiveness of ensemble decisions. Such methods are called "dynamic". There are adaptive and non-adaptive static combining strategies. Non-adaptive schemes include such traditional approaches as averaging, maximum, median, product rules, as well as voting and Borda rule [13]. Adaptive approaches use different methods for the adaptation and configuration of schemes for ensemble member interaction. Such approaches are of particular interest when considering possible options for their implementation and the possibility of improving the quality of the classification with "small losses", i.e. without re-executing the most costly stage which is designing the individual experts (neural networks or other technology). Well-known approaches are decision templates method by Kuncheva[14], weighted averaging and stacked generalization method (SGM) proposed by Wolpert [6].

Proposed Approach
A three-stage evolutionary approach with the explicit decomposition of the problem on the second stage was proposed to improve the efficiency of combining decisions in ensembles of neural networks. The method is based on the well-known stacked generalization method [6] and partly on the decision templates method [14].

The basic scheme for combining solutions using SGM is shown below:

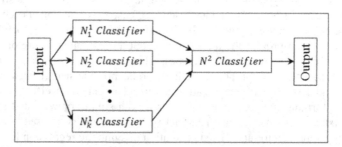

Fig. 1. Obtaining solutions using the stacked generalization method. Here k is the number of neural networks (classifiers) in the ensemble.

This scheme is modified in the proposed approach. Obviously at the first stage of the SGM the decomposition of the problem is done in implicit form. The kind of the decomposition that is used at this stage is the sample decomposition. In sample decomposition (also known as partitioning), the goal is to partition the training set into several sample sets, such that each sub-learning task considers the entire space [15].

At the second stage, there is an aggregation of the individual decisions of the classifiers in order to effectively map the solutions into a single target area. The idea is to supplement the implicit decomposition of the first stage of the SGM (the first stage of decision-making), with the decomposition of the problem in explicit form on the second stage (the second stage of decision-making).

The kind of the decomposition that is used at this stage is the space decomposition. In space decomposition the original instance space is divided into several sub-spaces. Each sub-space is considered independently and the total model is a (possibly soft) union of such simpler models [16]. We use the approach named "one-vs-all". This approach learns a classifier for each class, where the class is distinguished from all other classes, so the base classifier giving a positive answer indicates the output class [17-19].

We suggest to add the third stage to the basic scheme. In the third stage of decision-making the aggregation of decisions is carried out. The general scheme of the proposed approach is presented below:

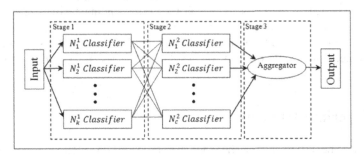

Fig. 2. Decision-making process with the proposed approach. Here k is a number of neural networks (classifiers) in the ensemble, c – is a number of classes.

The following describes the stages of the proposed evolutionary method of designing ensemble classifiers.

Stage 1
At this stage a set of classifiers is formed (a pool of neural networks), whose decisions would be involved in the evaluation of a generalized solution. This stage is common to most ensemble approaches. We use the PGNS method [9] for this stage using neural networks at the first-stage. In general, any available effective method of obtaining individual classifiers of the selected type can be used. The amount of computational resources available for use at this stage is determined on the basis of the requirements for decision-making time, the required accuracy and available computational capabilities. Note that, in general, the classifiers at the first stage can be not only neural networks, but also any other classifiers.

Stage 2
At the second stage a set of c classifiers of stage 2 are formed independently of each other. Here c is equal to the number of classes in the problem. The inputs to the classifiers at this stage are the values produced by the first-stage classifiers. And for each j-th classifier ($j = \overline{1,c}$) on the second stage, training is performed according to the rule:

— Target output value of the classifier is equal to 1 for all examples of the class j;
— Target output value of the classifier is equal to 0 for all other examples.

Thus, at the second phase the decomposition of the problem is performed. Each classifier on the second stage forms a surface in the space and this surface cuts off objects of the appropriate class from objects belonging to any of the other classes. We propose a method based on a genetic programming ensembling [9, 21] for combining the individual decisions of neural networks to perform this stage. Also any other kind of binary classifiers can be used to perform this stage.

Stage 3
At the third stage an aggregation of classifier outputs takes place to provide an evaluation of the generalized decisions. The output of the aggregator is the target class identifier. In the proposed approach we used maximum confidence strategy - the object is classified to the class for which the corresponding second-stage classifier produces the maximum value of the output [19].

Also any other kind of ensemble methods for binary classifiers in multi-class problems (i.e. dynamically ordered "one-vs-all" [20, 22]) can be used to perform this stage.

3   Experimental Study

To evaluate the efficiency of the approach proposed in Section 1 a number of numerical experiments were carried out. The investigation also involves some other approaches based on ensemble classifiers or on other classifying techniques. Results for ensemble approaches have been obtained by performing statistical tests using a software system IT-Pegas developed by the authors. We carried out experiments with the following methods for combining classifier decisions: simple averaging, majority voting, Borda rule and stacked generalization method. For other approaches results have been taken from [23]. The full list of methods is in the first column of Table 2.

In our experiments to investigate efficiency, we perform 5-fold cross validation on each data set, where 5 neural network ensembles are trained using the proposed approach in each fold. As a measurement of the effectiveness, we used the average value of the reliability of the classification, which was calculated as the ratio of the correctly classified patterns to the total number of patterns in the test sample. We use ANOVA tests [24] to evaluate the statistical robustness of the results obtained using the proposed approach.

To design the first stage classifiers we used the following parameters for the proposed method: the ensemble size is 10, the used neural network topology is the multilayer perceptron, the maximum number of layers in the neural networks is 3 and the maximum number of units in hidden layers is equal to 5.

We use the following parameters to generate the second stage classifiers for the stacked generalization method: the neural network topology used is the multilayer perceptron, the maximum number of layers in the neural networks is 3 and the maximum number of units in hidden layers is 5. The number of generations for forming the structure of the second layer neural network classifier is 200 and the number of individuals that encode a network is 50. For each individual its weights are tuned using a genetic algorithm with 100 individuals during 100 generations. To generate the second stage classifiers using the proposed approach in the second stage we obtained symbolic expression classifiers using genetic programming method. The number of classifiers is equal to the number of classes for the particular problem. The number of generations to form each classifier is 200 and the number of individuals is 100.

Test Problems

In our experimental study we used three well-known classification problems from the Machine Learning Repository [7]. The information on the data sets used in our experiments is presented in Table 1.

Table 1. Data Sets Used for Classification

Data Set	Number of Attributes	Size (number of instances)
Credit Australia-1	14	690
Credit Germany	20	1000
Liver Disorder	6	345

Results

The experimental results (average classification reliability) found for the proposed approach, as well as the results of competing approaches, are presented in Table 2.

As for variation analysis, our ANOVA tests show that there is a significant difference between the approach proposed here and other neural network ensemble approaches on Liver Disorder and Credit Australia-1 problems and there is no significant difference between the stacked generalization method and our proposed approach on Credit Germany problem. ANOVA tests also show that both our proposed approach and the SGM outperform all other competitive neural network ensemble approaches on the Credit Germany problem. Note that we have no opportunity to perform ANOVA tests for approaches which are not implemented in the IT-Pegas program system and for which results were taken from other papers.

Table 2. Comparative Results for Classification Problems

Classifiers	Credit Australia-1	Credit Germany	Liver Disorder
Ensemble of fuzzy classifiers approach	0.921	0.821	0.757
Fuzzy classifier	0.891	0.794	0.725
Bayes approach	0.847	0.679	0.629
Single multilayer perceptron	0.833	0.716	0.693
Boosting	0.760	0.700	0.656
Bagging	0.847	0.684	0.630
Random subspaces approach	0.852	0.677	0.632
ANN ensemble with simple averaging	0.892	0.805	0.740
ANN ensemble with majority voting	0.918	0.815	0.783
ANN ensemble with Borda rule	0.905	0.831	0.772
ANN ensemble with SGM	0.925	0.852	0.785
Proposed Approach	*0.947*	*0.857*	*0.804*

In general, the results of the numerical experiments show that the effectiveness of the proposed approach is not lower than the effectiveness of most of other methods. It shows promising results on Liver Disorder and Credit Australia-1 problems but we need more information about the conditions and the amount of computational resources used for obtaining the solutions by other non-ensemble competitive approaches.

In the second stage of our study, we decided to find (if possible) the best solution results for 5 known classification problems and compare them with the results obtained using our approach.

The list of tasks for which the comparison was done is shown in the Table 3. There are brief descriptions of the selected test data sets and references to papers from which the results were taken.

Table 3. The second Test Data

Data Set	Number of Attributes	Number of classes	Size	Reference
Cancer	9	2	699	[25]
Glass	9	6	214	[26]
Heart	13	2	303	[27]
Pima	8	2	768	[28]
Sonar	60	2	208	[27]

In our experimental study we divided the original data set in accordance with the scheme of statistical tests into two sub-samples: training and validation.

The partitions were made in accordance with those used in the works of which were the results obtained. A brief description of the test problems can be found in [27].

To obtain statistically robust results and to determine the statistical parameters of solution quality 30 test runs were accomplished for each data set. The results are shown in the Table 4.

Table 4. Best Found and Obtained Results on 5 Well-known Test Data Set

Data Set	Best Found Result	Obtained Result
Cancer	0.120	0.108
Glass	0.226	0.211
Heart	0.119	0.120
Pima	0.196	0.172
Sonar	0.144	0.159

In addition to evaluating classification error, the proposed approach is also achieving statistical evaluation the following parameters:

1. The average number of neural network classifier used in the second stage.
2. The average complexity of the network in the team at the first stage.

The results are shown in Table 5.

Table 5. Average Parameters of Neural Network Ensemble Classifiers

Data Set	Average number of neural networks in the ensemble	Average number of hidden nodes in neural networks
Cancer	4	5
Glass	6	6
Heart	5	6
Pima	4	5
Sonar	4	7

The number in the second column indicates how many solutions of the first-stage classifiers on average accounted for design of a class decision by the symbolic regression model formed with genetic programming. Apparently, not all classifiers solutions are usually used (the initial total number of classifiers on the first stage is 10). In this case, each second-stage classifier automatically generates a pool of input variables (outputs of the first-stage classifiers) based on the criterion of minimizing the classification error. Respectively the second-stage classifiers may involve different first-stage classifiers. It is possible that several second-stage classifiers can use to generate solutions all the classifiers of the first stage, and it is the subject of a more detailed study.

4 Conclusion

In the paper the basic ideas for the neural network ensemble approach for classification problems are reviewed and some of the well-known methods of combining classifier decisions in ensembles are described. We proposed a three-stage approach for solving classification problems using ensembles of neural networks. The methods of decision-making process at every stage of the proposed approach are explained.

The results of numerical studies of the effectiveness of the proposed approach on some well-known classification problems are given. The results show that on a number of the problems the proposed approach performs classification with a higher reliability compared with approaches that use other combination strategies and with some well-known non-ensemble approaches.

In future we intend to use some other large-scale data sets to test our approach and tune its performance. Then we hope to apply this method to a broad variety of practical problems.

References

[1] Hansen, L.K., Salamon, P.: Neural network ensembles. IEEE Transactions on Pattern Analysis and Machine Intelligence 12, 993–1001 (1990)
[2] Rastrigin, L.A., Erenstein, R.H.: Method of collective recognition. Energoizdat, Moscow (1981)

[3] Javadi, M., Ebrahimpour, R., Sajedin, A., Faridi, S., Zakernejad, S.: Improving ECG Classification Accuracy Using an Ensemble of Neural Network Modules. PLoS One 6 (2011)

[4] Perrone, M.P., Cooper, L.N.: When networks disagree: ensemble method for neural networks. In: Mammone, R.J. (ed.) Artificial Neural Networks for Speech and Vision, pp. 126–142. Chapman & Hall, New York (1993)

[5] Shimshoni, Y., Intrator, N.: Classification of seismic signals by integrating ensembles of neural networks. IEEE Transactions on Signal Processing 46(5), 1194–1201 (1998)

[6] Wolpert, D.H.: Stacked generalization. Neural Networks 5, 241–259 (1992)

[7] Frank, A., Asuncion, A.: UCI Machine Learning Repository. University of California, School of Information and Computer Science, Irvine (2010), http://archive.ics.uci.edu/ml

[8] Belew, R.K.: Evolving networks: Using genetic algorithm with connectionist learning. Technical report CS90-174, Computer Science and Engineering Department. University of California, San Diego (1991)

[9] Bukhtoyarov, V., Semenkina, O.: Comprehensive evolutionary approach for neural network ensemble automatic design. In: Proceedings of 2010 IEEE World Congress on Computational Intelligence, Barcelona, pp. 1640–1645 (2010)

[10] Woods, K., Kegelmeyer, W.P., Bowyer, K.: Combination of multiple classifiers using local accuracy estimates. IEEE Transactions on Pattern Analysis and Machine Intelligence 19, 405–410 (1997)

[11] Jacobs, R.A., Jordan, M.I., Nowlan, S.J., Hinton, G.E.: Adaptive mixtures of local experts. Neural Computation 3, 79–87 (1991)

[12] Xu, L., Krzyzak, A., Suen, C.Y.: Methods of combining multiple classifiers and their applications to handwriting recognition. Transactions on Systems, Man, and Cybernetics 22, 418–435 (1992)

[13] Polikar, R.: Ensemble based systems in decision making. IEEE Circuits and Systems Magazine 6, 21–45 (2006)

[14] Kuncheva, L.I., Bezdek, J.C., Duin, R.P.W.: Decision templates for multiple classifier fusion: an experimental comparison. Pattern Recognition 34, 299–314 (2001)

[15] Rokach, L., Maimon, O., Arad, O.: Improving supervised learning by sample decomposition. Int. J. Comput. Intell. Appl. 5(1), 37–54 (2005)

[16] Rokach, L., Maimon, O., Lavi, I.: Space decomposition in data mining: A clustering approach. In: Zhong, N., Raś, Z.W., Tsumoto, S., Suzuki, E. (eds.) ISMIS 2003. LNCS (LNAI), vol. 2871, pp. 24–31. Springer, Heidelberg (2003)

[17] Lorena, A.C., Carvalho, A.C., Gama, J.M.: A review on the combination of binary classifiers in multiclass problems. Artificial Intelligence Review 30(1–4), 19–37 (2008)

[18] Rifkin, R., Klautau, A.: In defense of one-vs-all classification. Journal of Machine Learning Research 5, 101–141 (2004)

[19] Sun, Y., Wong, A.C., Kamel, M.S.: Classification of imbalanced data: a review. International Journal of Pattern Recognition and Artificial Intelligence 23(4), 687–719 (2009)

[20] Hong, J.H., Min, J.K., Cho, U.K., Cho, S.B.: Fingerprint classification using one-vs-all support vector machines dynamically ordered with Naive Bayes classifiers. Pattern Recognition 41(2), 662–671 (2008)

[21] Koza, J.R.: The Genetic Programming Paradigm: Genetically Breeding Populations of Computer Programs to Solve Problems. MIT Press, Cambridge (1992)

[22] Galar, M., Fernandez, A., Barrenechea, E., Bustince, H., Herrera, F.: An overview of ensemble methods for binary classifiers in multi-class problems: Experimental study on one-vs-one and one-vs-all schemes. Pattern Recognition 44(8), 1761–1776 (2011)

[23] Sergienko, R.B., Semenkin, E.S., Bukhtoyarov, V.V.: Michigan and Pittsburgh Methods Combining for Fuzzy Classifier Generating with Coevolutionary Algorithm for Strategy Adaptation. In: Proceedings of 2011 IEEE Congress on Evolutionary Computation, New Orleans, LA, USA (2011)

[24] Bailey, R.A.: Design of Comparative Experiments. Cambridge University Press (2008)

[25] Islam, M.M., Yao, X., Murase, K.: A constructive algorithm for training cooperative neural network ensembles. IEEE Trans. Neural Netw. 14, 820–834 (2003)

[26] Breiman, L.: Randomizing outputs to increase prediction accuracy. Mach. Learn. 40, 229–242 (2000)

[27] Garcia, N., Hervas, C.: Ortiz. D.: Cooperative coevolution of artificial neural network ensembles for pattern classification. IEEE Transactions on Evolutionary Computation 9(3), 271–302 (2005)

Evolved Neural Network
Based Intelligent Trading System for Stock Market

Lifeng Zhang[*] and Yifan Sun

School of Information, Renmin University of China
59, Zhongguancun Street, Haidian, Beijing, P.R. China, 100872
lifeng.zhang@hotmail.co.uk

Abstract. In the present study, evolved neural network is applied to construct a new intelligent stock trading system. First, heterogeneous double populations based hybrid genetic algorithm is adopted to optimize the connection weights of feedforward neural networks. Second, a new intelligent stock trading system is proposed to generates buy and sell signals automatically through predicting a new technical indicator called medium term trend. Compared to traditional NN, the new model provides an enhanced generalization capability that both the average return and variance of performance are significantly improved.

Keywords: genetic algorithm, neural network, network training, stock trading system.

1   Introduction

Intelligent modelling for stock market, especially stock price prediction, has drawn considerable attention in scientific studies and real world applications due to its potential profits. Neural networks (NNs), as a powerful nonlinear model, have been extensively studied in dealing with finical issues [1, 2, 3, 4]. It has been proved that even a one hidden layer feedforward NN is capable of approximating uniformly any continuous multivariate function, to any desired degree of accuracy. However, stock market is a complex and dynamic system with noisy, non-stationary and chaotic data series [1] so that it is always difficult to learn the regularity of stock movement from the use of such corrupted data. The random factors in stock data can be viewed as noise which may massively reduce the effectiveness of prediction models. Moreover, NNs have a widely accepted drawback that the training of NN is more sensitive than the parameter estimation of mathematical models. Traditionally, gradient based search techniques are used to find out the optimal set of weight values for NN in order to match the known target values at the output layer. These approaches, however, sometimes may result in inconsistent and unpredictable performance of the networks due to the danger of local convergence. In this case, the generalization of NN will be reduced, and it makes lower the credibility of its prediction results.

As a type of well-known global searching methodologies, evolution computation (EC) techniques, such as genetic algorithm (GA) and evolutionary programming (EP),

---

[*] Corresponding author.

Y. Tan, Y. Shi, and H. Mo (Eds.): ICSI 2013, Part I, LNCS 7928, pp. 478–488, 2013.

have been successfully applied to aid in training and designing feedforward networks [5, 6, 7]. It is widely accepted that the performance of the direct use of conventional GA in adapting NN weights is not satisfactory due to the complexity of network learning, and therefore previous literatures have proposed many NN specific evolution based training methods. Some of these approaches adopt GA to search initial weights of NN, then, followed by deterministic methods to finalize the training process [8]. Some of them integrate different hill climbing techniques, including both conventional and heuristic approaches, into EC framework to enhance the effectiveness of evolution based training [9]. In addition, the previous studies have proposed a number of modified GA algorithms together with specific genotype representations and genetic operators. The genetic representation schemes of these approaches, generally, can be classified into two broad categories, i.e., weight based and neuron based codifications. In some of these approaches, binary or real valued weights are codified as the alleles of gene [10], and in others neurons with associated weight are codified as the representational components of gene [11]. Different genetic variation operators, such as combinatorial crossover, removing or adding neurons, and stochastic or scheduled mutations, have been developed based on these codification schemes to meet the nature of NN training [11].

In our recent study of [12], a novel evolved NN had been developed. This method adopts two separate optimization processes, GA and least squares, to determine the connection weights of hidden layer and output layer respectively. In addition, a binary representation and a neuron based real number representation are constructed for separately codifying hidden layers and hidden neurons. Then, two heterogeneous populations are initialized and evolve together in a single evolution procedure. In this study, a new intelligent stock trading system is developed based on the evolved NN. The NN model is applied to predict the movement trend of stock price for a medium term, subsequently, a series of trading strategies are established to automatically generate buy and sell signals. Experimental studies suggest that the new intelligent trading system performs much better than traditional NN models.

This study is organised as follows. Sections 2 and 3 respectively present the evolved NN model and the new automatic trading system. In section 4, experimental studies are carried out to test the performance of the new method. Finally, in Section 5 conclusions are drawn to summarise the study.

2   Heterogeneous Double Populations Based GA for NN Training

Consider an underlying system expressed as follows.

$$\mathbf{Y} = f(\mathbf{X}) + \mathbf{E} \qquad (1)$$

where

$$\begin{cases} \mathbf{Y} = [\mathbf{y}_1,...,\mathbf{y}_m] \\ \mathbf{E} = [\mathbf{e}_1,...,\mathbf{e}_m] \\ \mathbf{X} = [\mathbf{x}_1,...,\mathbf{x}_n] \end{cases} \qquad (2)$$

denote measured output, noise, and input signals respectively. Feedforward NN with single hidden layer can be mathematically derived as follows.

$$\begin{cases} \hat{\mathbf{y}}_j^H = F^H\left(\sum_{i=1}^{n} w_{ij}\mathbf{x}_i + b_j\right) \\ \hat{\mathbf{y}}_k = \sum_{j=1}^{q} v_{jk}\hat{\mathbf{y}}_j^H + a_k \end{cases} \tag{3}$$

where $\hat{\mathbf{y}}_j^H$ and $\hat{\mathbf{y}}_k$ respectively denote the outputs of hidden neurons and predicted outputs of NN. F^H is the nonlinear activation of hidden layer, and w_{ij}, v_{jk}, b_j, and a_k denote the connection weights and biases of hidden and output neurons respectively. The activation function of hidden layer is selected as tan-sigmoid and the activation function of output layer is fixed to be a linear combination.

2.1   NN Based Stock Price Prediction

Hidden layer essentially determines the nonlinear mapping relationship that a NN could have, GA, then, is adopted to optimize the parameters of hidden neuron by making use of its advantage of global searching. Then, once a hidden layer is determined the unique optimal output layer will be also fixed, and the optimization procedure becomes a linear parameter estimation problem. LS estimator is applied in the new method to yield the best parameters for output neurons. Consider an output neuron k, it is derived as.

$$\mathbf{v}_k = \left(\mathbf{X}^{O^T}\mathbf{X}^O\right)^{-1}\mathbf{X}^{O^T}\mathbf{y}_k \tag{4}$$

where

$$\begin{cases} \mathbf{X}^O = [\mathbf{y}_1^H,...,\mathbf{y}_q^H,-1] \\ \mathbf{v}_k = [v_{1k},...,v_{qk},a_k] \end{cases} \tag{5}$$

LS will be embedded in GA's iterative procedure. On the one hand, the prediction results obtained from the use of LS reveals the best performance which a hidden layer could have, i.e., a fair evaluation of the hidden layer. On the other hand, LS is a one-step parameter estimator so that it would not considerable increase the computational cost of GA.

Subsequently, two different codified populations, including a neuron population and a network population, are constructed to separately present and learn the connection weights and combinations of hidden neurons. That is to say, the evolution of hidden layer is broke up into two separate parts. The first one is a real number optimization problem for optimizing the parameter values of each neuron in the neuron population, and the second one is a combinatorial problem, which can be viewed as a special type of knapsack problem, for finding the best hidden layer which is codified as a subset of neuron population.

2.2   Algorithm Design

The computational procedure of the new GA design is presented as follows.

Evolved NN training

Begin
$t \leftarrow 0$

  Initialize populations $\mathbf{S}^{(0)} = \{\mathbf{w}_1^{(0)}, \cdots, \mathbf{w}_{Kq}^{(0)}\}$ and $\mathbf{P}^{(0)} = \{\mathbf{c}_1^{(0)}, \cdots, \mathbf{c}_K^{(0)}\}$ // $\mathbf{w}$ and
$\mathbf{c}$ respectively represent neuron and hidden layer

  $J^* \leftarrow \infty$, $\mathbf{S}^* \leftarrow \{\}$, and $\mathbf{c}^* \leftarrow \{\}$ // Define best solution

  While Termination condition is not met

    Variation on $\mathbf{P}^{(t)}$ // crossover and mutation for reproducing networks

    Variation on $\mathbf{S}^{(t)}$ // mutation for reproducing neurons

  Implement $(\mu + \lambda)$-selection by using $\mathbf{r}$ to yield $\mathbf{S}^{(t+1)}$

    LS parameter estimation   // to yield $\mathbf{V}$ and $\mathbf{a}$ in output layer

    Calculate MSE $\mathbf{J}^T$, $\mathbf{J}^V$ and $\mathbf{r}$ // to yield fitness values for both $\mathbf{S}^{(t)}$ and
$\mathbf{P}^{(t)}$

    If $\min(\mathbf{J}^V) < J^*$

      $J^* \leftarrow \min(\mathbf{J}^V)$

      $\mathbf{S}^* \leftarrow \mathbf{S}^{(t)}$ and $\mathbf{c}^* \leftarrow \mathbf{c}_i^{(t)}$, where $J_i = \min(\mathbf{J}^V)$

    End if

    Implement rank selection by using $\mathbf{J}^T$, and elitist preservation to yield $\mathbf{P}^{(t+1)}$

    $t \leftarrow t + 1$

  End While

  Return the best solution $\mathbf{S}^*$ and $\mathbf{c}^*$

End Begin

Genetic Representations: In this method, two populations are involved in the evolution procedure, and defined as a neuron population and a network population.

$$\begin{cases} \mathbf{S}^{(t)} = \{\mathbf{w}_1^{(t)}, \cdots, \mathbf{w}_{qK}^{(t)}\} \\ \mathbf{P}^{(t)} = \{\mathbf{c}_1^{(t)}, \cdots, \mathbf{c}_K^{(t)}\} \end{cases} \tag{6}$$

where

$$\begin{cases} \mathbf{w}_i^{(t)} = [w_{i,1}, ..., w_{i,n}, b_i] \\ \mathbf{c}_j^{(t)} = [c_{j,1}, ..., c_{j,qK}] \end{cases} \tag{7}$$

Each $\mathbf{w}_i^{(t)}$ presents a hidden neuron consists of a set of real coded connection weights, and each $\mathbf{p}_j^{(t)}$ presents a binary coded hidden layer corresponds to the neurons defined in $\mathbf{S}^{(t)}$. In $\mathbf{p}_j^{(t)}$, if a component $c_{j,i}$ is 1, the neuron $\mathbf{w}_i^{(t)}$ is selected to be included in the network $\mathbf{c}_j^{(t)}$, and vice versa.

The size of hidden layer, q, is pre-specified and will not change during the evolution process. For initializing $\mathbf{S}^{(0)}$, all the weights and biases are randomly generated with uniform distribution, zero mean and limited amplitude. During the evolution, there is no limit restricts the value of these parameters.

The first advantage of the new genetic codification scheme is that by using two populations both combinations and parameters of hidden neuron can be optimized individually, so that various operations, such as exchanging neurons and modifying connection weights, can be easily implemented. The second advantage is that since networks are codified in a binary form and neuron parameters are stored in a separate population, it completely avoids the permutation problem of NN's genetic representation that each $\mathbf{p}_j^{(t)}$ presents a unique network without worrying about permutation of neurons, and each $\mathbf{w}_i^{(t)}$ presents a unique neuron and can easily prevent the problem of redundancy.

Genetic Variations: Different genetic variations are applied for separately evolving hidden layers and neurons. For network population, uniform crossover and random mutation are employed in this study. These operators produce offspring by exchanging or changing binary coded neurons of the parents, with the aim of finding the best subset of neuron population to collect meaningful nonlinear regression information as much as possible. For the neuron population, Gaussian mutation is used to make random changes on the connection weights of selected neuron.

Remarkably, the proposed GA only uses traditional genetic operators, which are simple and easy to be achieved, and have been widely used in combinatorial and real number optimization. It guarantees a much lower computational time and better algorithm efficiency.

Evaluation and Selection: For network population, NNs are evaluated by the mean squared error (MSE) of the prediction outputs obtained from using LS estimator, which is derived as follows.

$$\hat{\mathbf{y}}_k = \mathbf{X}^O \mathbf{v}_k \tag{8}$$

$$J_k = \frac{1}{N} \sum_{t=1}^{N} (y_k(t) - \hat{y}_k(t))^2 \tag{9}$$

Fitness value of a network is then calculated as the mean of $\{J_k\}$, and rank selection is employed here to samples the population to produce copies for new generation.

For neuron population, individuals cannot be evaluated directly. In this study, $\mathbf{w}_i^{(t)}$ are evaluated based on $J_{j,k}$ of the networks which contain or used to contain such neuron. For the neurons in initial population or newly generated in mutation process, neuron fitness r_i can be computed as follows

$$r_i = \frac{1}{m}\sum_{k=1}^{m} J_{j,k} \tag{10}$$

For other neurons,

$$r_i = 0.5r_i^{(t-1)} + \frac{0.5}{gm}\sum_{j=1}^{g}\sum_{k=1}^{m} J_{j,k} \tag{11}$$

where g is the number of the networks in $\mathbf{P}^{(t)}$ which include $\mathbf{w}_i^{(t)}$. $r_i^{(t-1)}$ denotes the fitness assigned in the last generation. It is noted that the average performance of the newest generated networks determines half of the fitness, and all the previous fitnesses jointly decide the other half. For neuron population selection, a deterministic mechanism called ($\mu+\lambda$)-selection is applied to make newly generated neurons to replace the inferior free neurons in $\mathbf{S}^{(t)}$, which are not occupied by any network in $\mathbf{P}^{(t)}$, and with the lowest r_i.

Termination Condition: Additionally, the new method prevents the problem of overfitting by means of early stopping method that for each learning trial, the modelling data set is randomly split into a training set and a validation set. MSE for training set, J_i^T, is used to calculate the fitnesses for implementing the two selection and elitist preservation. MSE for validation set, J_i^V, is only used for externally choosing the best solution $\mathbf{S}^*$ and $\mathbf{c}^*$, and would not be involved in the evolution process at all. Finally, the evolution procedure will be terminated if maximum generation is met or J_i^V stops improving for a certain generation number.

3 Evolved NN Based Intelligent Trading System

In this study, a new intelligent stock trading system is developed to automatically generate buy and sell signals based on the prediction of the medium term trend of stock price.

3.1 NN Based Stock Price Prediction

The evolved NN introduced in the last section is adopted here to produce the mapping from past stock price to future directions to make good use of its advantage in generalization. First, a new indicator is proposed to measure the future trend, and will be used as the output signal of NN model. Consider at day i, future trend indicator (FT) is formulated as follows.

$$FT = \frac{1}{k} \sum_{\tau=1}^{k} P_{i+\tau} - P_i \qquad (12)$$

where P_i is the closing price, and k denotes the medium term which can be viewed as the length of time window of interest. FT shows the difference between the stock price at a given time and the average price in the following period. In other words, a FT value larger than zero indicates an upward tendency of the given stock, otherwise, the price is likely fall down in near future. Compared to short term prediction such as one step ahead forecasting, medium term prediction eliminates the randomness of daily stock price. Moreover, it is also easy to induce proper trading operations as it looks ahead for longer term.

For the input signals of NN, there are many well-established technical indicators that contain refined and distinguishable information about the stock, rather than using raw price data which is always stochastic, collinear, and corrupted. According to the previous studies, 10 indicators were selected, and they are given in table 1.

Table 1. Technical indicators used as NN inputs

CCI	Commodity Channel Index
Di10	10 Days' Disparity
MA5	5 Days' Moving Average
MA10	10 Days' Moving Average
MA20	20 Days' Moving Average
Mo	Momentum
PO	Price Oscillator
RSI	Relative Strength Index
SD	Stochastic D Index
SK	Stochastic K Index

3.2   Automatic Trading System

In this study, A series of trading strategies are developed for deriving daily trading operations, i.e. at most one of the three actions, buy, sell or hold could be taken automatically within each trading day.

As aforementioned, the output of NN is a signal that predicts the market trend which could be either positive or negative. To avoid imprudent and too frequent trading operation, two thresholds are used to filter out the FT signal. When the signal is higher than up threshold $L1$, the system regards it as a significant uprising signal and take buy action. On the contrary, when the signal is lower than down threshold $L2$, it is likely to have a market downtrend and sell action will be carried out. The output signals falls between $(L2, L1)$ will be considered as small fluctuations, then takes hold action. The computation procedure of the new method is presented as follows.

Step 1: Set *Hold* as 0.
Step 2: Input stock price data day by day.
Step 3: Calculate technical indicators.
Step 4: Predict *FT* by using evolved NN.
Step 5: If *FT*>*L1* and *Hold*=0, generate a buy signal and set *Hold* as 1; if *FT*<*L2* and *Hold*=1, generate a sell signal and set *Hold* as 0.
Step 6: Take action as the signal suggested.
Step 7: Go to step 2.

It should be noted that the thresholds, *L1* and *L2*, can be determined according to the risk appetite of decision makers. Risk seeking people may prefer thresholds close to zeros as the trading could be carried out more frequently, while risk aversion people prefer larger thresholds in order to avoid incorrect trading caused by the randomness of stock price. Actually too frequently trading will increase the variance of the return even though the mean may go up at the same time, i.e. the higher the risk it takes, the higher the expected return it gets.

4   Case Study

S&P 500 index from January 2nd 2001 to February 24th 2012, with length of totally 2800 data points, was used to demonstrate the performance of the new method.

First, the entire data set was divided into two parts that the first part with length of 2600 was applied as modeling data for NN training and validation, and the last 200 daily prices were used to test the effectiveness of the automatic trading system. Fig. 1 shows closing price data. In should be noted that there are both rise and fall trends emerge during the test period so that the index maintains almost the same value at the end of the 200 days.

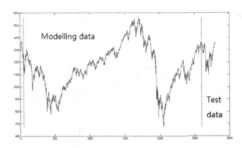

Fig. 1. S&P 500 index from January 2nd 2001 to February 24th 2012 with data length of 2800

Second, Three performance measurements, which are more intuitive and easy to understand, are adopted in this study in place of the common used statistics such as mean absolute percentage error (MAPE) and root mean square error (RMSE).

It is because that the new method is an automatic trading system rather than a common prediction model. The three indexes, cumulated return (Creturn), correct rate (CRate), and correct value (CValue), are derived as follows

$$CReturn = \frac{Money_{end} - Money_{Initial}}{Money_{Initial}} \times 100\% \tag{13}$$

$$CRate = \frac{1}{n}\sum_{i=1}^{n} d_i \times 100\% \tag{14}$$

$$CValue = \frac{\sum_{i=1}^{n} d_i \times |P_{i+1} - P_i|}{\sum_{i=1}^{n} |P_{i+1} - P_i|} \times 100\% \tag{15}$$

$$d_i = \begin{cases} 1 & (P_{i+1} - P_i)S_i > 0 \\ 0 & \text{otherwise} \end{cases} \tag{16}$$

where S_i is the trading signal at time i obtained from using the new method. $S_i = 1$ denotes a buy signal, and $S_i = -1$ denotes a sell signal.

Third, traditional neural network was also implemented by using Matlab toolbox for comparing the performance of prediction model especially generalization capability. In addition, each experiment was independently carried out for 100 times for the sake of eliminating the random nature of both data collection and training algorithms.

Finally, the statistical results for the 100 runs are presented in table 2, and the corresponding hypothesis tests are also adopted. Table 3 gives the hypothesis for each statistics.

Table 2. Statistics and hypothesis test of evaluation indexes

Measure		Model		Hypothesis Test	
		GANN	ANN	H0	P
Return	Min	93.04	77.28	-	
	Max	118.35	135.81	-	
	Mean	104.77	99.98	1	0.00
	Std	5.68	10.66	1	0.00
CRate	Min	44.00	4.00	-	
	Max	57.50	50.50	-	
	Mean	50.33	33.65	1	0.00
	Std	2.41	8.19	1	0.00
CValue	Min	45.77	0.78	-	
	Max	57.96	44.61	-	
	Mean	50.50	24.31	1	0.00
	Std	2.06	8.90	1	0.00

Table 3. Null and alternative hypothesizes for evaluation indexes

	H0	H1
Mean	MENN<=MNN	MENN>MNN
Std	SENN>=SNN	SENN<SNN

MENN and MNN respectively denote the mean value of the experimental results obtained from using evolved NN and traditional NN. SENN and SNN denote the standard deviations for the two methods respectively.

The experimental results clearly suggest that the average return obtained by the use of evolved NN over 100 runs is significantly larger than that of traditional method, more importantly, it also displays a significantly smaller standard deviation which means the system is more reliable and riskless. The hypothesis tests also show the CRate and CValue of the new method are much better than those of the traditional method, which proves the enhancement in the prediction accuracy.

5   Conclusions

I the present study, a new intelligent trading system was proposed based on stock trend prediction for medium term by the use of heterogeneous double population based evolved NN. First the evolved NN provides a more stable and accurate nonlinear prediction compared to the traditional NNs. Second the new trading system generate buy and sell signals automatically without any human interactive operation in which to provide the stock traders a technical and independent decision support. Experiment results demonstrate the effectiveness of the system that it displays a much higher average return and a significantly reduced variance.

References

1. Lin, X., Yang, Z., Song, Y.: Intelligent Stock Trading System Based on Improved Technical Analysis and Echo State Network. Expert Systems with Applications 38, 11347–11354 (2011)
2. Nunez-Letamendia, L.: Fitting the Control Parameters of A Genetic Algorithm: an Application on Technical Trading Systems Design. European Journal of Operational Research 179, 847–868 (2007)
3. Teixeira, L.A., Oliveira, A.L.I.D.: A Method for Automatic Stock Trading Combining Technical Analysis and Nearest Neighbor Classification. Expert Systems with Applications 37, 6885–6890 (2010)
4. Izumi, K., Toriumi, F., Matsui, H.: Evaluation of Automated-Trading Strategies Using an Artificial Market. NeuroComputing 72, 3469–3476 (2009)
5. Van Rooij, A.J.F., Jain, L.C., Johnson, R.P.: Neural networks training using genetic algorithms. In: Machine Perception and Artificial Intelligence, vol. 26. World Scientific, Singapore (1996)
6. Yao, X., Liu, Y.: A new evolutionary system for evolving artificial neural networks. IEEE Transactions on Neural Networks 8(3), 694–713 (1997)

7. Maniezzo, V.: Genetic evolution of the topology and weight distribution of neural networks. IEEE Transactions on Neural Networks 5(1), 39–53 (1994)
8. Su, C.-L., Yang, S.M., Huang, W.L.: A two-stage algorithm integrating genetic algorithm and modified Newton method for neural network training in engineering systems. Expert Systems with Applications 38, 12189–12194 (2011)
9. Teixeira, L.A., Oliveira, A.L.I.D.: A Method for Automatic Stock Trading Combining Technical Analysis and Nearest Neighbor Classification. Expert Systems with Applications 37, 6885–6890 (2010)
10. Zhao, S., Xu, G., Tao, T., Liang, L.: Real-coded chaotic quantum-inspired genetic algorithm for training of fuzzy neural networks. Computers and Mathematics with Applications 57, 2009–2015 (2009)
11. Balakrishnan, K., Honavar, V.: Evolutionary design of neural architectures—a pleliminary taxonomy and guide to literature, technical report CS TR#95-01. Artificial Intelligence Research Group, Department of Computer Science, Iowa State University, Ames, Iowa 50011-1040, USA (1995)
12. Zhang, L.F., He, R., Yan, M.L.: Heterogeneous double populations based hybrid genetic algorithm design for training feedforward neural networks. In: Proceedings of IEEE Congress on Evolutionary Computation (IEEE CEC 2012), Australia, Brisbane, June 10-15, p. 492 (2012)

The Growing Radial Basis Function (RBF) Neural Network and Its Applications

Yan Li[1], Hui Wang[2,*], Jiwei Jia[3], and Lei Li[3]

[1] School of Insurance and Economics,
University of International Business and Economics, Beijing, China
[2] School of Banking and Finance,
University of International Business and Economics, Beijing, China
oceanwanghui@163.com
[3] BGP INC., China National Petroleum Corporation

Abstract. This paper proposes a framework based on the cross-validation methods for constructing and training radial basis function (RBF) neural networks. The proposed growing RBF (GRBF) neural network begins with initial number of hidden units. In the process of training, the GRBF network adjusts the hidden neurons by eliminating some "small" hidden units and splitting one "large" hidden unit at the same cycle. If the prediction error in the system is not less than the pre-given threshold, the proposed method increases hidden units to re-estimate the parameters in the next process of training, until the stop criterion is satisfied. In practice, the proposed GRBF network are evaluated and tested on two real 3D seismic data sets with very favorable self-adaptive ability and satisfactory results.

Keywords: Radial Basis Function (RBF) neural network, Parameter learning, Cross-validation method, Geological characteristics.

1 Introduction

Along with the multilayer perceptron (MLP), radial basis function (RBF) networks hold much interest in the fields of modern finance, signal processing and seismic exploration [1], [2], [3]. The radial basis function (RBF) networks can be regarded as one kind of the feed-forward neural networks with a single layer of hidden units. Compared with a conventional neural network, Radial basis function (RBF) networks have simpler topological structure and faster convergent speed (using a linearly weighted combination of single hidden-layer neurons). Moreover, RBF networks use multivariate radial basis functions (particularly the Gaussian functions) as hidden units, which improve the precision and accuracy of prediction of complex nonlinear mappings problems.

Generally, the performance of an RBF network depends on the number and positions parameters of hidden units. The value of the hidden units in the RBF network can be decided by unsupervised/supervised clustering procedures. However,

---

* Corresponding author.

Y. Tan, Y. Shi, and H. Mo (Eds.): ICSI 2013, Part I, LNCS 7928, pp. 489–496, 2013.
© Springer-Verlag Berlin Heidelberg 2013

like many other ANN algorithms, the number of radial basis function in the RBF networks must be fixed or known before the parameter learning process. To overcome this problem, an easy method is to set a different number of hidden units for the RBF networks, and then make several attempts to find the "best" results. However, this kind of "interactive" operation may be very complex and time-consuming.

In this paper, we attempt to use a growing radial basis function (GRBF) network to predict the distribution of reservoir characteristic by combining the well-log data with seismic attributes. Firstly, the method first trains all the parameters of the RBF network with initial given values. Secondly, the cross-validation method is used to calculate the prediction accuracy. If the prediction accuracy is less than a predetermined threshold, eliminating or splitting operations are implement to decrease or increase the number of hidden units. Finally, the parameter self-learning process is applied to determine the relationship between the reservoir characteristic and seismic attributes.

The remaining of this paper is organized as follows. Section 2 reviews the basic principles of RBF neural network. Section 3 focuses on the parameter learning theory of growing RBF network. Section 4 tackles the description of datasets used to perform our empirical study. Section 5 provides two examples to interpret the theory the RBF neural network. Section 6 makes a summary about our growing RBF neural network.

2   RBF Neural Networks

As mentioned above, the RBF network is embedded in a three-layer neural network, which can be considered as a mapping: $R^n \rightarrow R^m$. The input layer consists of source nodes. hidden unit layer implements a radial basis function and the output units layer implement a weighted sum of hidden unit outputs. Generally, we consider just the case of the RBF network with one single output unit. In fact, the problem of a multi-output RBF network can be transformed into many of one-output RBF network learning. The output function of the RBF network, $f(x)$, with the input vector, $x \in R^n$, can be expressed as follows:

$$f(x) = \sum_{i=1}^{K} w_i \phi_i (\|x - c_i\|).$$ (1)

Where w_i is the connection weight and c_i is known as the RBF center, which can be obtained by using the linear least squares method and clustering methods, respectively. In addition, the value K is the number of hidden units, determined by realistic problems. Function $\phi_i(\bullet)$ is called the radial basis function. Although the choice of the basis function is crucial to the performance of the RBF network, the most commonly used radial basis functions are Gaussian functions, expressed as:

$$\phi_i(x) = \exp(-\frac{\|x - c_i\|^2}{\sigma_i^2}).$$ (2)

Where $c_i \in R^m$ and σ_j are the center and the width of ith Gaussian hidden units. The distance between the input point, x, and the center of the jth node, c_j, as measured by Euclidean norm, is given as $\|\bullet\|$. From the above equations, we can find that the outputs of RBF neural network strongly depend on the link weights, the number and initial positions of hidden units. Since the training process is based on adjusting the parameters of the network to reproduce a set of input-output patterns. There are three types of parameters: the weight w between the hidden nodes and the output nodes, the center $\{c_j\}_{j=1}^{n}$ of unit of the hidden layer and unit width $\{\sigma_j\}_{j=1}^{n}$.

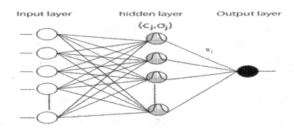

Fig. 1. Radial Basis Function Network architecture

3   Parameter Learning

Usually, the RBF network training process is composed of two stages. The first stage involves constructing the RBF network structure including the number of RBF centers, the initial value of the hidden units and the unit widths The second stage involves optimizing the weights between hidden units and the output layer. In the first stage, the RBF centers can be obtained by many clustering algorithms. These algorithms consist of unsupervised clustering algorithms (such as k-means [4], fuzzy c-means [5], enhanced LBG [6]), and supervised clustering algorithms (such as Fuzzy Clustering [2] and the Alternating Cluster Estimation [7]). In the second stage, the least mean square error (LMS) method is usually used to train the link weights between the hidden units and the output. Given the structure of RBF network, the least mean square error on the sample set, $S = \{(x_i, y_i)\}_{i=1}^{N}$, can be expressed as follows:

$$E = \frac{1}{2}\sum_{t=1}^{N}[y_t - f(x_t)]^2 = \frac{1}{2}\sum_{t=1}^{N}[y_t - \sum_{i=1}^{n}\lambda_i\phi_i(x_t)]^2 \cdot \tag{3}$$

In this function, y_t and $f(x_t)$ are real values and the prediction for the sample set S. By computing partial derivatives for function (3) and setting the equation equal to zero, we obtain the gradient learning rules for each parameter in the RBF network. The process of parameter training is very quick, because it does not have to back-propagated an error through multiple layers.

In addition, the number size of the hidden units is very importance. If the chosen number size of the hidden units is too large, the computation loading is heavy so that it is not suitable for practical applications. In contract, if the chosen number size of hidden units is too small, the performance of the RBF network may not be good enough to the applicable problems.

4   Proposed Approach

To solve the problem, we propose a growing RBF network based on the cross-validation method. Firstly, training samples are divided into several groups, each of which contains the same samples. In the process of training, each group of samples will be regarded as the test samples and other training samples are used to train the RBF network. The "optimal" RBF network performs well for each group of test samples.

Secondly, according to the pre-defined number, n, of hidden units, we perform a k-means[4] algorithm to determine the centers, $\{c_i\}_{i=1}^n$, and the initial unit widths, $\{\sigma_i\}_{i=1}^n$, of the hidden units. Then, LMS is implemented to update all the parameters in the RBF network until the learning process converges. Note that the current structure of the RBF network may not be be the "best" for the input samples.

Thirdly, a cross-validation method is used to verify whether the appropriate hidden units, receptive fields and link weights have been achieved or not. The total error rate will be calculated with current the RBF network for each group of test samples. If the total error rate is larger than the pre-defined threshold δ_r, the number of hidden units will be adjusted by the merging or splitting of some hidden units. The splitting hidden unit ϕ_j is defined as the hidden units with the maximum link weight $\|w_j\|$. The parameters associated with the new hidden units (ϕ_p and ϕ_q) are defined as:

$$w_p = \gamma w_j , w_q = (1-\gamma) \tag{4}$$

$$c_p = c_j - \frac{\gamma}{(1-\gamma)}\sigma_j \quad c_p = c_j + \frac{(1-\gamma)}{\gamma}\sigma_j \tag{5}$$

$$\sigma_p = \frac{\gamma}{(1-\gamma)}\sigma_j \quad \sigma_q = \frac{(1-\gamma)}{\gamma}\sigma_j \tag{6}$$

On the other hand, the hidden unit ϕ_j will be discarded from the current RBF network, once the link weight $\|w_j\|$ is less than a pre-set threshold δ_M. By the way, the proposed method can realize determining the correct number of hidden units in RBF network quickly.

Finally, we return to the second step in the algorithm to re-calculate the centers of the hidden units, receptive fields and link weights again. Otherwise, when the total error rate is less than some pre-defined threshold δ_T, the method will be stopped and the "optimal" structure of the RBF network has been established at this time.

5   Examples

The reservoir characteristic is an important piece of quantitative information for seismic exploration and reservoir analysis. In the past several years, artificial neural networks (ANNs) have been used to unravel and forecast complex nonlinear relationships between seismic data, well-log data and the reservoir characteristic.

Generally, geologists and geophysicists can interpret the target zone with well-log data. However, well-log data will only be able to predict the geological information within a limited range. Limited well-log data may be inefficient at predicting the latent variations in the reservoir characteristic. On the other hand, seismic attributes such as amplitude, coherency and frequency, provide much more horizontal and lateral variations in geological characteristic than well-log data. Many researchers have proposed that to predict effective porosity, well-log data and seismic attributes should be combined in the fields of reservoir analysis and geological explanation [9] [10], [11],[12]. To test the efficiency of the proposed method, we apply it to estimate the reservoir porosity distribution and the effective thickness of the sand case, respectively. In both cases, the results proved that the prediction results are similar to the actual seismic log data.

In this example, we implement the GRBF network to predict effective porosity volume using well-log data sets as training samples. All of well-log data sets are extracted from three actual logs in the survey. The target reservoir ranges from 600 ms to 1300 ms in the seismic data. With a sampling interval of 4 ms, we can obtain 1124 training samples $S = \{(x_i, y_i)\}_{i=1}^{N}$ from these three wells. Each training sample point consists of seven uncorrelated variables, of which the first six seismic attributes variables are used as characteristic vectors $x_i = (x_{i1}, x_{i2}, ..., x_{i6})$ and the last one is the effective porosity value y_i.

In the data processing phase, all of the seismic attributes are normalized between 0 and 1 to eliminate the differences in each seismic attribute. At the beginning of the proposed algorithm, ten samples are divided from the input sample set as the self-testing samples at each time and other samples are used to train the GRBF network as before. We set the size of the initial hidden units to be 5. Then, the k-means algorithm [4] is used to cluster the training samples. After the clustering process, the centers of each cluster are regarded as the hidden units and the unit widths of the RBF basis function are calculated. After that, in the parameter learning stage, the link weights between hidden units and the output units are adjusted automatically. Based on the trend of the total error rate, we adjust the number of hidden units by the increasing and eliminating operation. Finally, the GRBF algorithm stops when the total error rate is less than 30 and the number of hidden units equals 40.

The comparison results of the prediction (red) using the proposed algorithm and the three real well-log data sets (black) are shown in Fig 2(a), (b) and (c), respectively. In each figure, the horizontal axis gives the porosity values and the vertical axis is in time (in ms). A close correlation exists between the predicted effective porosity and current well-log data in these figures. In addition, a single well profile extracted from the original amplitude profile is shown in Figure 3(a). Correspondingly, the profile of effective porosity using the GRBF network is shown in Figure 3(b). From the above discussion, we determine that the GRBF algorithm results in a good porosity prediction.

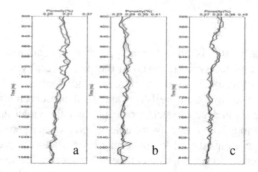

Fig. 2. Application of the GRBF method to predicting porosity. Three different results (a, b, c) are compared using the GRBF method (red) and the actual porosity curve (black). We find that the GRBF method successfully predicts the trends in the actual porosity curve.

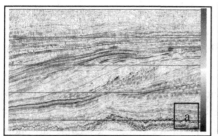 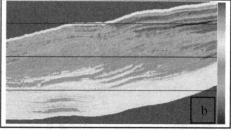

Fig. 3. (a) A 2d line extracted from the original seismic amplitude volume. (b). The same 2D line extracted from the porosity volume that was predicted using the GRBF algorithm. (c). The same 2D line extracted from the porosity volume that was predicted using the BP network algorithm.

In the second example, we use the proposed algorithm to predict the effective reservoir thickness of the sand case study from western China. In this study, the target zone ranges from 800 to 1540 in the inline direction and 650 to 1550 in the cross-line direction. In Figure 4, the 32 wells marked with red points were used in the case study. There are nine sampling points around each well, leading to 32*9 = 288 training sample points as the training samples. Each training sample was composed of 12 seismic attributes, such as amplitude, phase, and frequency, and one output effective thickness of the sand case). The Principal Components Analysis method [8] was used to reduce the relativities among the seismic attributes.

In the process of parameter learning, the initial number of hidden units was 15 and the other parameters were obtained by cluster analysis. After training the structure of the GRBF network, we then used the multiple seismic attributes to predict the reservoir effective thickness of the sand case. At this time, a self-testing example was used to test the validity of the GRBF network. In the example, we removed the reservoir thickness from well-log data and used the remaining seismic attributes as the input sample to calculate the "predicted" value. The comparison between the original

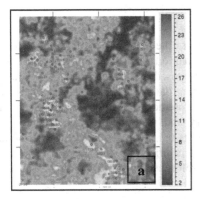 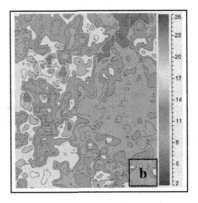

Fig. 4. The prediction results for the efficient thickness of the sand case. (a. The predicted results of GRBF algorithm and b. the predicted results of Back-Propagation (BP) network.

Table 1. Comparison of the prediction value between BP network and GRBF

Number (well)	Well data	Predicted value	BP algorithm	Number (well)	Well data	Predicted value	BP algorithm
1	16.7	16.69	16.70	17	4.7	4.69	4.65
2	15.2	15.23	15.1	18	27	26.85	26.85
3	48.5	48.48	48.4	19	10	10.10	10.09
4	46.1	46.10	46.3	20	8.7	8.72	8.80
5	50.4	50.38	50.37	21	6.5	6.52	6.51
6	18.6	18.59	18.6	22	31	31.00	31.10
7	26.8	26.78	26.82	23	26	25.90	25.90
8	29.9	29.80	29.6	24	6.2	6.21	6.05
9	31	30.90	30.85	25	36.6	36.40	36.80
10	19	19.20	19.30	26	27.4	27.40	27.40
11	41.8	41.60	41.60	27	4.1	4.10	4.50
12	24.4	24.70	24.80	28	21	20.90	21.30
13	31.5	31.40	31.30	29	19.5	19.29	19.70
14	29.1	29.30	29.10	30	9.4	9.43	9.48
15	23.6	23.60	23.65	31	19.9	19.90	20.05
16	13.9	14.10	14.12	32	3.4	3.45	3.46

well-log reservoir thickness, the predicted results and the results of Back-Propagation (BP) [13] neural network are listed in Table 1. From table 1, we find that the results of the prediction show minor relative errors compared with the available well-log data (less than 2%). And, Fig.4 shows the application of the proposed method and BP neural network to the target reservoir. Through the analysis of seismic interpreters, the results of GRBF are superior to those of BP network.

6   Conclusion

In this paper, we proposed the growing RBF neural network and applied the method to predict the reservoir characteristics, which has advantages in dealing with the complex nonlinear problem. In addition, the proposed algorithm adaptively adjusts the structure of the RBF network with the training data, which not only reduces the workload of the interpreter, but also enhance the predictive accuracy of the reservoir characteristic. After a discussion of the theory and methodology, we applied the method to two real seismic data sets to illustrate its feasibility. However, we also find that the prediction result of the reservoir characteristic is very sensitive to the selection of the multiple seismic attributes. Therefore, some additional processes and analysis should be made in future studies.

References

1. Broomhead, D.S., Lowe, D.: Multivariable Functional Interpolation and Adaptive Networks. Complex System 2, 321–355 (1988)
2. Moody, J., Darken, C.: Fast Learning in Networks of Locally Tuned Processing Units. Neural Computation 1, 281–294 (1989)
3. Poggio, T., Girosi, F.: Regularization Algorithms for learning that are equivalent to multiplayer networks. Science 247, 978–982 (1990)
4. Davis, J.: Statistics and data analysis in geology, 2nd edn. Wiley (1986)
5. Bezdek, C.: Pattern Recognition with Fuzzy Objective Function Algorithms. Plenum, New York (1981)
6. Russo, M., Patanè, G.: Improving the LBG Algorithm. In: Mira, J. (ed.) IWANN 1999. LNCS, vol. 1606, pp. 621–630. Springer, Heidelberg (1999)
7. Runkler, A., Bezdek, C.: Alternating Cluster Estimation: A New Tool for Clustering and Function Approximation. IEEE Transactions on Fuzzy Systems 7(3), 377–393 (1999)
8. Scheevel, J.R., Payrazyan, K.: Principal Component Analysis Applied to 3D Seismic Data for Reservoir Property Estimation. Paper 56734, SPE Reservoir Evaluation & Engineering, 64–72 (2001)
9. Schultz, P.S., Ronen, S., Hattori, M., Corbett, C.: Seismic Guided Estimation of Log Properties, Part 1: A Data-driven Interpretation Technology. The Leading Edge 13, 305–315 (1994)
10. Ronen, S., Schultz, P.S., Hattori, M., Corbett, C.: Seismic Guided Rstimation of Log Properties, Part 2: Using Artificial Neural Networks for Nonlinear Attribute Calibration. The Leading Edge 13, 674–678 (1994)
11. Russell, B., Hampson, D., Schuelke, J., Quirein, J.: Multiattribute Seismic Analysis. The Leading Edge 16, 1439–1443 (1997)
12. Schuelke, J.S., Quirein, J.A.: Validation: A Technique for Selecting Seismic Attributes and Verifying Results. In: 68th Ann. Internat. Mtg., Soc. Expl. Geophys., Expanded Abstracts, pp. 936–939. (1998)
13. Horikawa, S.I., Furuhashi, T., Uchikawa, Y.: On Fuzzy Modeling using Fuzzy Neural Networks with the Back-propagation Algorithm. IEEE Transactions on Neural Networks 3(5), 801–806 (1992)

Network-Based Neural Adaptive Sliding Mode Controller for the Ship Steering Problem

Guoqing Xia and Huiyong Wu

College of Automation, Harbin Engineering University,
Harbin, China 150001
wuhyong1688@163.com

Abstract. In this paper, the concept of networked control system (NCS) is introduced into the course autopilot of the ship. A network-based neural adaptive sliding mode controller is designed for the ship steering in waves. The unknown term, including the wave disturbances and the unmodeled dynamics, is approximated by the RBF neural network. The sliding mode controller is designed to compensate the neural network approximation error besides of the network-induced delay. The stability of the closed-loop system is proven and the neural network weight is updated using the Lyapunov theory. It indicates that the designed controller can guarantee the system state tracks the desired state asymptotically. Finally, a simulation on a Mariner class vessel in waves is carried out to demonstrate the effectiveness of the proposed control scheme.

Keywords: Networked Control System (NCS), Autopilot, RBF Neural Network, Sliding Mode Controller.

1 Introduction

Navigation of ships has been a major concern for sailors since humans took to the waters. In recent years, sophisticated ship autopilots have been proposed, based on advanced control engineering concepts [1]. A predictive controller is designed to steer the ship sailing forward with the constant velocity along the predefined reference path in [2]. In [3, 4], Ming-Chung Fang and Jhih-Hong Luo adopted sliding mode control technique and line-of-sight guidance technique to navigate the ship. Backstepping is also frequently used in autopilot system of ships [5, 6]. Meanwhile, due to the better approximate ability of nonlinear function and faster learning ability, the ship course control by using neural network architecture for complex and unknown nonlinear dynamic systems has appeared in many literatures [7, 8]. Such control systems are able to alter the course of the vessel in the desired manner by regulating the deflection of the rudder. However, with the advanced electronic devices and smart units widely used on the ship, the connection between the units has become increasingly complex, which therefore increases the maintenance costs and the possibility of faults.

In order to address the challenges, the network control technique is introduced into the control system of the ship. Feedback control systems wherein the control loops are closed through a real-time network are called networked control systems (NCS) [9].

Y. Tan, Y. Shi, and H. Mo (Eds.): ICSI 2013, Part I, LNCS 7928, pp. 497–505, 2013.

The defining feature of a NCS is that information is exchanged using a network among control system components. Fig.1 illustrates a typical setup of the NCS. The primary advantages of the NCS are reduced system wiring, ease of system diagnosis and maintenance, high reliability and increased system flexibility. System with this configuration (depicted in Fig.1) can be found in a variety of settings, including spacecraft, automobiles, and groups of autonomous vehicles, to name a few [10].

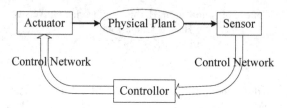

Fig. 1. A typical setup of the NCS

However, the insertion of the communication network in the feed-back control loop makes the analysis and design of NCS complex, and some special issues need to be addressed. The research on the delay is one of the hot topics of the NCS. A robust control approach is proposed to solve the stabilization problem for NCS with short time-varying delays in [11]. An adaptive fuzzy sliding mode control method is designed to compensate the network-induced delay and the uncertain of the system in [12]. In [13], a networked predictive control scheme is employed to compensate for communication delay and data loss actively rather than passively.

In this note, a network-based adaptive sliding mode controller is designed for the ship tracking in waves. For the unknown term in controller, an adaptive method based on the RBF neural network is proposed to approximate the unknown term. The controller is therefore able to compensate the unknown term besides of the network-induced delay, and the chattering of the controller also can be reduced.

The structure of the paper is as follows. In section 2, the control objective is formulated. Section 3 contains the design and stability analysis of the network-based neural adaptive sliding mode controller. Section 4 provides simulation results for the proposed control scheme. Finally, conclusions are made in the last section.

2   Problem Formulation

For the horizontal motion of a vessel moving in waves, the kinematics and dynamics models are described by the following equations [4]:

$$\dot{\boldsymbol{\eta}} = \mathbf{R}(\boldsymbol{\eta})\mathbf{v} \tag{1}$$

$$\mathbf{M}\dot{\mathbf{v}} + \mathbf{C}(\mathbf{v})\mathbf{v} + \mathbf{D}(\mathbf{v})\mathbf{v} = \boldsymbol{\tau} + \mathbf{R}^{\mathrm{T}}(\boldsymbol{\eta})\boldsymbol{\varepsilon} \tag{2}$$

Where $\boldsymbol{\eta} = [x, y, \psi]^{\mathrm{T}}$, $\mathbf{v} = [u, v, r]^{\mathrm{T}}$, (x, y, ψ) denote the coordinates and heading of the ship in the earth-fixed frame; u, v and r denote the speeds in surge, sway, and yaw respectively in the body-fixed reference frame; $\boldsymbol{\tau} = \mathbf{B}\mathbf{u}$ denotes the control inputs, $\mathbf{R}(\boldsymbol{\eta})$ is a state dependent transformation matrix, $\mathbf{M}$ is the mass and inertia matrix, $\mathbf{C}(\mathbf{v})$ is the Coriolis-centripetal matrix, $\mathbf{D}(\mathbf{v}) = \mathbf{D} + \mathbf{D}_n(\mathbf{v})$ is the hydrodynamic damping matrix including the linear damping matrix $\mathbf{D}$ and the nonlinear damping matrix $\mathbf{D}_n(\mathbf{v})$, and $\boldsymbol{\varepsilon}$ is a bias term representing slowly varying environmental forces and moments.

Assume that the ship is controlled by a single rudder; the rolling mode is negligible. Moreover, since the surge speed is not the controlled variable, the speed in surge can be treated as a constant value consequently. Then, equation (2) can be reduced as:

$$\mathbf{M}_1 \dot{\mathbf{v}}_1 = -\mathbf{N}(u_0)\mathbf{v}_1 + \mathbf{b}_1 \delta + \mathbf{f}_1(\mathbf{v}_1) \tag{3}$$

Where $\mathbf{v}_1 = \begin{bmatrix} v & r \end{bmatrix}^T$, $\mathbf{b}_1 = \begin{bmatrix} -Y_\delta & -N_\delta \end{bmatrix}^T$, δ is the rudder angle, $\mathbf{M}_1$ and $\mathbf{N}(u_0)$ are derived from $\mathbf{M}$, $\mathbf{C}(\mathbf{v})$, $\mathbf{R}(\psi)$ and $\mathbf{D}$, and $\mathbf{f}_1(\mathbf{v}_1) = \begin{bmatrix} f_v & f_r \end{bmatrix}^T$ denotes the environmental disturbances and unmodeled dynamics including the nonlinear damping term in sway and yaw.

Combining $\dot{\psi} = r$ and (3), yields

$$\dot{\mathbf{x}} = \mathbf{A}\mathbf{x} + \mathbf{b}u + \mathbf{f}(\mathbf{x}) \tag{4}$$

Where $\mathbf{x} = \begin{bmatrix} v & r & \psi \end{bmatrix}^T$, $u = \delta$ is the control input, $\mathbf{f}(\mathbf{x}) = \begin{bmatrix} f_v & f_r & 0 \end{bmatrix}^T$ is the unknown term, and $\mathbf{A}$, $\mathbf{b}$ can be calculated by $\mathbf{M}_1$, $\mathbf{N}(u_0)$, and $\mathbf{b}_1$.

As mentioned in last section, the concept of NCS is introduced to the ship course control system. Any controller computational delay can be absorbed into controller-to-actuator delay τ_{ca} without loss of generality [9].

Meanwhile, for convenience of analysis, we assume that the sensor is time driven with a sampling period h; the controller and actuator are event driven; the networks and communications are error-free; the sensor-to-controller delay τ_{sc} can be ignored;

Under the above assumptions, the ship model with delay can be expressed as:

$$\dot{\mathbf{x}} = \mathbf{A}\mathbf{x} + \mathbf{b}u(t-\tau) + \mathbf{f}(\mathbf{x}) \tag{5}$$

Where $\tau = \tau_{ca}$ is the network-induced delay.

Control objective: Using the LOS guidance, design the control input u to guide the ship (5) to pass through the commanded waypoints only with the position measurements, and ensure the tracking error converges to zero.

3   Control Systems for Ship Course Tracking

In this section, the heading autopilot is incorporated with equation (5) to simulate the ship course tracking which is composed of several waypoints. According to the line-of-sight (LOS) guidance, the desired heading angle is calculated as follows:

$$\psi_d = \text{atan2}(y_{wp} - y_p, x_{wp} - x_p) \tag{6}$$

Where (x_p, y_p) is the coordinate of the ship position, and (x_{wp}, y_{wp}) is the coordinate of the waypoint position.

3.1   Network-Based Sliding Mode Control

In order to compensate the effect of the controller-to-actuator delay τ_{ca} and the unknown term, a sliding mode controller is applied because of its good performance and robustness. This control theory has a switching action, which provides a robustness to match uncertainties.

Considering the network-induced delay, the sliding surface is given as:

$$S = \mathbf{c}^T \tilde{\mathbf{x}} + \mathbf{c}^T \mathbf{b} \int_{-\tau}^{t} u(\alpha) d\alpha \tag{7}$$

Where $\tilde{\mathbf{x}} = \mathbf{x} - \mathbf{x}_d$ is the state tracking error, τ is the network-induced delay, $\mathbf{c} \in \Re^3$ is a design vector to be chosen such that $S \to 0$, implying convergence of the state tracking error $\tilde{\mathbf{x}} \to 0$, and $\mathbf{c}^T \mathbf{b}$ is nonsingular.

The time derivative of S is

$$\dot{S} = \mathbf{c}^T (\dot{\mathbf{x}} - \dot{\mathbf{x}}_d) + \mathbf{c}^T \mathbf{b}[u(t) - u(t - \tau)] \tag{8}$$

Substituting (5) into (8), yields the dynamic of S:

$$\begin{aligned} \dot{S} &= \mathbf{c}^T (\mathbf{A}\mathbf{x} + \mathbf{b}u(t - \tau) + \mathbf{f}(\mathbf{x}) - \dot{\mathbf{x}}_d) + \mathbf{c}^T \mathbf{b}[u(t) - u(t - \tau)] \\ &= \mathbf{c}^T \mathbf{A}\mathbf{x} + \mathbf{c}^T \mathbf{b}u(t) + \mathbf{c}^T \mathbf{f}(\mathbf{x}) - \mathbf{c}^T \dot{\mathbf{x}}_d . \end{aligned} \tag{9}$$

Where $\dot{\mathbf{x}}_d = \begin{bmatrix} v_d & r_d & \psi_d \end{bmatrix}^T = \begin{bmatrix} 0 & \dot{\psi}_d & \psi_d \end{bmatrix}^T$, and ψ_d can be calculated via Eq.(6).

Let $\dot{S} = 0$, then the best approximation of the control law can be obtained

$$\hat{u}(t) = -(\mathbf{c}^T \mathbf{b})^{-1}[\mathbf{c}^T \mathbf{A}\mathbf{x} + \mathbf{c}^T \hat{\mathbf{f}}(\mathbf{x}) - \mathbf{c}^T \dot{\mathbf{x}}_d] \tag{10}$$

Where $\hat{\mathbf{f}}(\mathbf{x})$ is an estimate of $\mathbf{f}(\mathbf{x})$.

Considering the uncertainty of $\mathbf{f}(\mathbf{x})$, an additional non-linear controller term is added to $\hat{u}(t)$, then we have

$$u(t) = -(\mathbf{c}^T\mathbf{b})^{-1}[\mathbf{c}^T\mathbf{A}\mathbf{x} + \mathbf{c}^T\hat{\mathbf{f}}(\mathbf{x}) - \mathbf{c}^T\dot{\mathbf{x}}_d + k\,\mathrm{sgn}(S)] \qquad (11)$$

Where $k > 0$ is the switching gain, and sgn is the sign function.

In practice, it is difficult to get the unknown term $\mathbf{f}(\mathbf{x})$ exactly, therefore the control law (11) is hard to realize. Considering the nonlinear function approximate ability of neural network, we can use a multilayered neural network to model the unknown function $\mathbf{f}(\mathbf{x})$.

3.2   RBF Neural Adaptive Controller Design

In this section, the RBF neural network is used to approximate the unknown term $\mathbf{f}(\mathbf{x})$, and the network weight is updated according to the Lyapunov method.

Based on the neural network approximation theory, there exists weight $\mathbf{W}$ such that $\hat{\mathbf{f}}(\mathbf{x},\mathbf{w})$ approximates the continuous function $\mathbf{f}(\mathbf{x})$, with accuracy ε over a compact subset $Q \subset \Re^3$, that is, $\exists \mathbf{w}$ such that [7]

$$\max \left\| \hat{\mathbf{f}}(\mathbf{x},\mathbf{w}) - \mathbf{f}(\mathbf{x}) \right\| \leq \varepsilon \quad \forall \mathbf{x} \in Q \qquad (12)$$

Let $\mathbf{w}_t$ denotes the estimate of $\mathbf{W}$ at time t, then the control law become as

$$u(t) = -(\mathbf{c}^T\mathbf{b})^{-1}[\mathbf{c}^T\mathbf{A}\mathbf{x} + \mathbf{c}^T\hat{\mathbf{f}}(\mathbf{x},\mathbf{w}_t) - \mathbf{c}^T\dot{\mathbf{x}}_d + k\,\mathrm{sgn}(S)] \qquad (13)$$

Where $\hat{\mathbf{f}}(\mathbf{x},\mathbf{w}_t)$ is the estimate of $\mathbf{f}(\mathbf{x})$ at time t, and

$$\hat{\mathbf{f}}(\mathbf{x},\mathbf{w}_t) = \mathbf{w}_t^T\phi(\mathbf{x}) \qquad (14)$$

Where the system state $\mathbf{x}$ is chosen as the input of RBF network, and $\phi(\mathbf{x})$ is a Gaussian function denoting the output of the hidden layer.

The block diagram of the presented controller is shown in Fig.2. Substitute the new control law (13) into (9), then the dynamic of the sliding surface becomes:

$$\begin{aligned}
\dot{S} &= \mathbf{c}^T\mathbf{A}\mathbf{x} + \mathbf{c}^T\mathbf{b}u(t) + \mathbf{c}^T\mathbf{f}(\mathbf{x}) - \mathbf{c}^T\dot{\mathbf{x}}_d \\
&= \mathbf{c}^T[(\hat{\mathbf{f}}(\mathbf{x},\mathbf{w}) - \hat{\mathbf{f}}(\mathbf{x},\mathbf{w}_t)) + (\mathbf{f}(\mathbf{x}) - \hat{\mathbf{f}}(\mathbf{x},\mathbf{w}))] - k\,\mathrm{sgn}(S) \\
&= \mathbf{c}^T\left[-\frac{\partial\hat{\mathbf{f}}(\mathbf{x},\mathbf{w})}{\partial\mathbf{w}}\bigg|_{\mathbf{w}_t} \mathbf{w}_t + \frac{\partial\hat{\mathbf{f}}(\mathbf{x},\mathbf{w})}{\partial\mathbf{w}}\bigg|_{\mathbf{w}} \mathbf{w} \right] - k\,\mathrm{sgn}(S)
\end{aligned} \qquad (15)$$

Define the weight error as $\tilde{\mathbf{w}}(t) = \mathbf{w}_t - \mathbf{w}$, then (15) can be represented as:

$$\dot{S} = \mathbf{c}^{\mathrm{T}}(-\tilde{\mathbf{w}}^{\mathrm{T}}\phi(\mathbf{x}) + \eta(t)) - k\,\mathrm{sgn}(S) \tag{16}$$

Where $\eta(t) = \mathrm{O}\left(\|\tilde{\mathbf{w}}\|^2\right) + \mathrm{O}(\varepsilon)$.

The updating law of the neural network weight is

$$\dot{\mathbf{w}}_t = \dot{\tilde{\mathbf{w}}}(t) = \phi(\mathbf{x})\mathbf{c}^{\mathrm{T}}S \tag{17}$$

Theorem 1. For system (5), under the designed control law (13) and adaptive updating law (17), the state vector $\mathbf{x}$ of the closed-loop system asymptotically tracks the desired state $\mathbf{x}_d$ with the weight of neural network converging to its best approximation, when choosing the switching gain to satisfy

$$k > \left\|\mathbf{c}^{\mathrm{T}}\right\| \cdot \left\|\eta(t)\right\| \tag{18}$$

Proof. Define the Lyapunov function candidate as:

$$V = \frac{1}{2}S^2 + \frac{1}{2}(\mathbf{c}^{\mathrm{T}}\mathbf{c})^{-1}\Delta^{\mathrm{T}}(t)\Delta(t) \tag{19}$$

Where $\Delta(t) = \tilde{\mathbf{w}}(t)\mathbf{c}$ is a new defined vector denoting the variety of $\tilde{\mathbf{w}}(t)$.

Compute the derivative of V to obtain

$$
\begin{aligned}
\dot{V} &= S\dot{S} + (\mathbf{c}^{\mathrm{T}}\mathbf{c})^{-1}\Delta^{\mathrm{T}}(t)\dot{\Delta}(t) \\
&= S[\mathbf{c}^{\mathrm{T}}(-\tilde{\mathbf{w}}^{\mathrm{T}}\phi(\mathbf{x}) + \eta(t)) - k\,\mathrm{sgn}(S)] + (\mathbf{c}^{\mathrm{T}}\mathbf{c})^{-1}\mathbf{c}^{\mathrm{T}}\tilde{\mathbf{w}}^{\mathrm{T}}(t)\phi(\mathbf{x})\mathbf{c}^{\mathrm{T}}S\mathbf{c} \\
&= S\mathbf{c}^{\mathrm{T}}\eta(t) - k|S| \le |S|(\mathbf{c}^{\mathrm{T}}\eta(t) - k) \le 0 .
\end{aligned}
\tag{20}
$$

Obviously, V is a positive definite and decrescent function, and by application of Barbalat's lemma, S and $\Delta(t)$ will be uniformly stable at the equilibrium point $S = 0, \mathbf{w}_t = \mathbf{w}$, implying that the state tracking error will converge to zero in finite time while the switching gain is chosen to be large enough to satisfy the condition (18). This completes the proof.

Note that the magnitude of k will be a trade-off between robustness and performance. Meanwhile, chattering can be reduced by replacing $\mathrm{sgn}(S)$ with $\tanh(S/\varphi)$, where φ is the sliding surface boundary layer thickness.

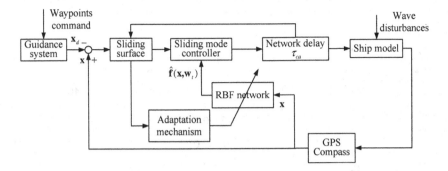

Fig. 2. Architecture of the closed-loop control scheme

4   Simulation Results

In this section, a numerical simulation on a Mariner class vessel in waves is carried out to demonstrate the proposed control scheme. To provide the network environment, an ideal simulation platform of NCS, Truetime toolbox, is introduced [14].

Here the commanded course is composed of three waypoints, and each acceptance radius is 200m. The initial velocity of the ship is 7.72m/s, the maximum rudder deflection is 30° and the rate limit is set to be 3°/s. The principal parameters of the ship are given in [15] in detail. The network-induced delay can be treated as a constant by setting a buffer between the actuator and the controller.

The simulation results are shown in Fig.3-4. In Fig.3, we can see that the proposed controller is able to force the ship to pass through the acceptance regions of all waypoints despite of waves acting on the ship.

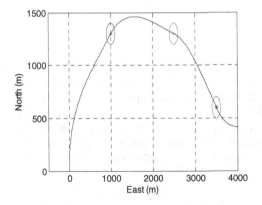

Fig. 3. The time domain simulation of ship trajectories

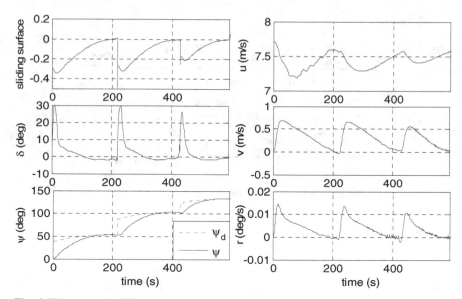

Fig. 4. The curves of sliding surface, rudder angle and yaw angle (the left three plots); the time responses of the velocity of the ship (the right three plots)

In Fig.4, the left three plots show the time responses of sliding surface, the rudder angle and the yaw angle. As seen from the figure, the dynamic of sliding surface asymptotically converges to zero while the actual yaw angle also tracks the desired yaw angle, and the rudder angle response is rather smooth but with some sharp lines due to the maximum rudder and rate limit. The right three plots indicate that all the velocities in surge, sway and yaw are also smooth and satisfied, and the sway and yaw speed can be controlled to the desired values, while the surge speed is smaller than the initial value but reasonable without the control effect.

5   Conclusion

The concept of networked control is introduced to the ship course control system, and the considered ship model includes the unmodeled dynamics and the nonlinear wave disturbances. Due to the good nonlinear approximation ability of RBF neural network and good robustness of sliding mode control, a neural adaptive sliding mode controller is designed. The performance of the aforementioned control strategy was tested through simulation. The network environment was provided by the Truetime toolbox. Simulation results showed the effectiveness of the proposed approach.

References

1. Rigatos, G., Tzafestas, S.: Adaptive fuzzy control for the ship steering problem. Mechatronics 16, 479–489 (2006)
2. Wu, J., Peng, H., Ohtsu, K., Kitagawa, G., Itoh, T.: Ship's tracking control based on nonlinear time series model. Applied Ocean Research 36, 1–11 (2012)

3. Fang, M.C., Luo, J.H.: On the track keeping and roll reduction of the ship in random waves using different sliding mode controllers. Ocean Engineering 34, 479–488 (2007)
4. Fang, M.C., Luo, J.H.: The nonlinear hydrodynamic model for simulating a ship steering in waves with autopilot system. Ocean Engineering 32, 1486–1502 (2005)
5. Yang, Y., Du, J.L., Guo, C., Li, G.Q.: Trajectory Tracking Control of Nonlinear Full Actuated Ship with Disturbances. In: Proceedings of the 2011 International Conference of Soft Computing and Pattern Recognition, pp. 318–323 (2011)
6. Witkowska, A., Śmierzchalski, R.: Nonlinear Backstepping Ship Course Controller. International Journal of Automation and Computing 6, 277–284 (2009)
7. Horng, J.H.: Neural adaptive tracking control of DC motor. Information Sciences 118, 1–13 (1999)
8. Zhang, L.J., Jia, H.M., Qi, X.: NNFFC-adaptive output feedback trajectory tracking control for a surface ship at high speed. Ocean Engineering 38, 1430–1438 (2011)
9. Zhang, W., Branicky, M.S., Phillips, S.M.: Stability of networked control systems. IEEE Control Systems Magazine 21, 84–99 (2001)
10. Zhang, L., Hristu-Varsakelis, D.: Communication and control co-design for networked control systems. Automatica 42, 953–958 (2006)
11. Zhang, W.A., Yu, L.: A Robust Control Approach to Stabilization of Networked Control Systems with Short Time-varying Delays. Acta Automatica Sinica 36, 87–91 (2010)
12. Xu, L.J., Dong, C.Y., Chen, Y.: An Adaptive Fuzzy Sliding Mode Control for Networked Control Systems. In: Proceedings of the 2007 IEEE International Conference on Mechatronics and Automation, pp. 1190–1195 (2007)
13. Liu, G.P.: Predictive Controller Design of Networked Systems with Communication Delays and Data Loss. IEEE Transactions on Circuits and Systems 57, 481–485 (2010)
14. Department of Automatic Control, Lund University, http://www.control.lth.se/truetime
15. Chislett, M.S., Strom-tejsen, J.: Planar motion mechanism tests and full-scale steering and manoeuvring predictions for MARINER class vessel. International Shipbuilding Progress 12, 201–224 (1965)

A Behavior Modeling Method of Virtual Characters Approximated by Artificial Neural Network

Ya Zhao[1,*], Xianmei Liu[1], and Qiong Wu[2]

[1] Institute of Computer & Information Technology, North-East Petroleum University,
Daqing Heilongjiang 163318, China
[2] Exploration and Development Research Institute, Daqing Oilfield Company,
Daqing Heilongjiang 163712, China
zhaoyaya1980@163.com

Abstract. In this paper, behavior model is established by adopting artificial neural network for virtual characters to resolve the reality of virtual characters behavior modeling in intelligent virtual environments, including acquiring training samples, data standardization, neural network training and application. This method improves on running performance, modeling efficiency and complexity of traditional cognitive model, which makes virtual characters adapting to changeability of virtual environments better, and plans behaviors according to diversity of virtual environments intelligently and autonomously.

Keywords: intelligent virtual environments, virtual characters, behavioral model, artificial neural network.

1 Introduction

In order to simulate the real world realistically, make the participators get more immersion, we construct the IVE (Intelligent Virtual Environment) [1] through adding one or many entity objects with life features into the VE (Virtual Environment). These virtual entity objects with life features are also known as virtual characters, which play important roles in the IVE.

How to activate the animations of these characters (also known as behavior animations) is the problem that we should consider first. The certainty model of human behaviors is still simple now [2], this makes the behaviors of virtual characters seem lack reality, even influences the simulation application effects [3].In 2003, Toni Conde realized behavior animations using reinforcement learning firstly [4].

In this paper, we use ANN (Artificial Neural Network) model replace certainty model, establish a complex behavior system applying AI (Artificial Intelligence) technology. In this system, virtual characters generate similar behaviors through autonomous learning, and do intelligent decision through the received actions of observing. The core of it is allowing the characters decide their behaviors themselves

---

* Corresponding author.

Y. Tan, Y. Shi, and H. Mo (Eds.): ICSI 2013, Part I, LNCS 7928, pp. 506–513, 2013.

without the users' intervention. On the base of learning and researching the reference [4-5], we realize a behavior model modeling method of virtual characters approximated by ANN in this paper. The method presented in this paper belongs to the field of realizing behavior animations using learning technologies.

2   Behavior Model and Modeling Methods

2.1   Behavior Model

Behavior model is a model of researching how to describe the virtual characters' behaviors really as far as possible, which makes the programmer who constructing a virtual character with real behaviors conveniently according to the behavior model [6].

The virtual character layer model structure is divided to 5 layers bottom-up [3]. They respectively are: geometry layer, the forward and reverse motion layer, physical layer, behavior layer and cognitive layer.

Behavior layer and cognitive layer both belong to behavior models. The behavior planning of behavior layer has no the process of "thought", virtual characters confirm next behavior planning and execute immediately according to current environment state and internal stimulus. Cognitive layer controls the whole process of virtual characters' perception, behaviors and actions, is a kind of model that in order to accomplish long-term goal which makes long-term planning to virtual characters' behaviors and actions, it takes the way of "thought" [7].

2.2   Behavior Model Modeling Method

Now the main behavior model modeling methods include: the method based on computer animation, the method based on FSM (finite state machine), the method based on artificial life and the method based on artificial intelligence [8]. All of these traditional methods exists certain shortcomings and insufficiency. We try to realize a behavior model modeling method which simulates the cognitive process of virtual characters in VEs in this paper. The process includes behavior planning and behavior choosing of virtual characters. This method can realize the automatic generation of the behavior animations, reduce the animators' work burden and generate computer animations with vivid lifelike figures.

2.3   The Application of BP Algorithm in Behavior Modeling

We approximate virtual characters' behavior model in cognitive layer by ANN, realize it by applying BP (Back Propagation) algorithm. BP algorithm is a kind of study altorithms used in multilayer feedforward neural network, which includes input layer, output layer and hidden layer. The hidden layer can be multistory structure. The learning process of BP neural network includes two stages: calculating feedforward output and adjusting connection weight matrix from the direction. In the stage of feedforward propagation, input information is processed layer-by-layer from input layer via hidden layer and translated to output layer, neuron's output in each layer as the neuron's input upper layer. If the actual output value has error with the expected value in the output layer, we should modify connection weight coefficient layer-by-layer using error signal from the reverse and iterate repeatedly, make the mean-square deviation least between actual and expect output. Gradient descent algorithm usually is taken when correcting connection weight coefficient.

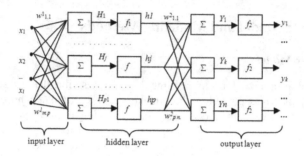

Fig. 1. The topological structure of BP neural network algorithm

The topological structure of BP neural network algorithm is shown in figure 1.

In the practical application of BP algorithm, study result is received by presenting established training instances to neural network repeatedly. To a given training set, BP algorithm has two basic training modes: serial training mode and centralized training mode. The realization of serial training mode is relatively easy, easy understanding, small memory space and fast calculating speed. So, we take this mode in prototype realization to train neural network in this paper.

In the practical application standard BP algorithm can't do, so many improved algorithms emerging. The improvement of BP algorithm has two ways mainly, one is adapting elicitation method of studying method, another is adapting more effective numerical optimization algorithm. In this paper we use the BP algorithm with the momentum vector method and adaptive adjustment learning rate mainly, so to improve the learning rate and increase the reliability of the algorithm.

3   Constructing Virtual Characters' Behavior Model Instance

We adopt the instance of social and group of drive to check behavior model modeling algorithm based on ANN, simulate the model using Delta3D SDK and C++. Our experiment test scene as follows: in a virtual 3D environment, has two types virtual characters, the first one is group virtual characters, that is, the be pursuers, represented by crowd, the other one is individuality virtual character, that is, the driver, represented by a solider with a gun. The state and behaviors of group virtual characters can be confirmed, and the number are finite, so we can do behavior model by adopting the method of FSM. Driver virtual character chooses actions according to the state change of VE, is asked for intellectuality and independency, so it do decision by adopting cognize layer behavior model, do behavior model by adopting ANN algorithm.

Figure 2 is the sketch map of the driver driving group virtual characters. Destination indicates the drive destination, Barrier indicates the barriers, C_1, C_2, C3 and C4 indicate group virtual characters. The arrow in the figure indicates the forward direction of Expler(driver virtual character), α indicates the rotation angle of the driver in VE during the motor process, β indicates the direction between the driver and the goal(presented by the intersection angle between the driver and the goal), ExplertoBar indicates the distance between the driver and the barriers, BestNeartoExpler and BetterNeartoExpler separately indicate the relevant parameters of the two group virtual characters nearest the driver.

From the example analysis we know that: the driver virtual character's states and behaviors can be described by septet: (DirtoDes, DistoBar, BestNeartoExpler_x, BestNeartoExpler_y, BetterNeartoExpler_x, BetterNeartoExpler_y, ExplerAction), the dimensions of state space have 6, the dimension of behavior space has 1.

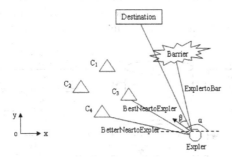

Fig. 2. The behavior model sketch map of the driver driving group virtual character

DirtoDes indicates the angular separation of the driver and the goal, is presented by β in figure 2; DistoBar indicates the distance between the driver and the barriers; BestNeartoExpler_x indicates the x coordinate difference between the driver and the nearest group virtual characters; BestNeartoExpler_y indicates the y coordinate difference between the driver and the nearest group virtual characters; BetterNeartoExpler_x indicates the x coordinate difference between the driver and the nearer group virtual characters; BetterNeartoExpler_y indicates the y coordinate difference between the driver and the nearer group virtual characters; ExplerAction indicates the adopting behavior the driver takes in current state, the behavior's output result is the current rotation angle of the virtual character in VE, is indicated by α value in figure 2.

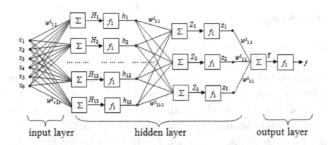

Fig. 3. The structural design of neural network

In the cognitive layer behavior model, the behaviors of the driver is a kind of mapping of the states, the mapping relation is get by ANN calculation. According to the description of the driver's behaviors states above, the number of the ANN's input parameters are 6, output parameter is 1 in this paper. The structural design is shown as figure 3.

4   The Establishment of the Virtual Characters' Behavior Model Approximated by Neural Network

In this paper we use a ANN study the decision planned by cognitive model, accomplish a goal, then recall these decision rapidly by carrying out trained neural network. Training is off-line, use is online. Using fewer CPU loops get intelligent virtual characters' target behaviors in real time.

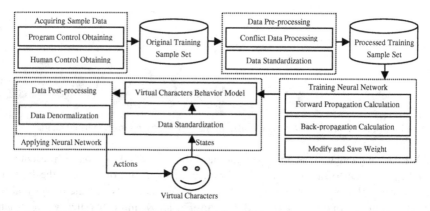

Fig. 4. The whole flow chart of behavior model modeling and application based on neural network

The process of virtual characters' behavior model approximated by ANN algorithm includes: acquiring sample data, sample data pre-processing, sample data post-processing, training neural network and applying neural network, etc,. The whole flow of its establishment and application is shown in Figure 4.

4.1   Acquiring Sample Data

The establishment of behavior model approximated by ANN need the support of sample data, accomplish by the function module of "acquiring sample data". This process can be proceeded through two ways, the first is acquiring training sample data automatically by program control, this way is on the base of establishment of virtual characters' specific behavior model, acquiring some parameters in the process of behavior motion through defining specific rules for virtual characters, saving these parameters and using as training samples; another way is acquiring training sample data by manual control, virtual characters as the controller's agent exist in VE under this way, their behaviors and actions reflect the controllers' behaviors and actions, the controllers acquire the virtual characters' states (actions group) by guiding virtual character act, learn state (actions mapping) from demonstration, the sample frequency of this way is the frequency of the controllers guiding virtual characters. We realize it through the virtual characters' specific behavior model and acquiring training sample data by manual control.

4.2   Sample Data Pre-processing and Post-processing

Noisy data may appear in the training sample data during the process of acquiring training sample data, which will influence the training effect, will cover main information content or make pick up useful information more difficult. So we should process training sample data before training neural network by the methods of conflict data processing method and data standardization method.

①   The definition of conflict:

$$\text{if } \|si - sj\| < v \text{ and } \|ai - aj\| > v, \text{then conflicting [9]} \tag{1}$$

In which s presents the current state of virtual characters, a presents the picked actions in this state, v presents a threshold (states and actions' thresholds may be different), that's to say: if the data of two examples has close states but distinctively different behaviors and actions, we think they are conflicting. But eliminating the conflict easily will lead to the behaviors' high frequency. The algorithm's target is eliminating the high frequency in data. Every example in the state space should promise be not relevant to the examples of adjacent area. To the current example we define L examples around it as its adjacent area. Uncorrelated indicates the behaviors of the example be clearly different from the median data in its adjacent area.

②   Because the curve is quite gentle, the variation speed is quite slow when the Sigmoid function value in the algorithm is close to 0 or 1. In order to reduce the learning time on network, we control the input and output data change between [0.1~0.9] or [0.2~0.8], thus the variation gradient of the Sigmoid function is greater and the network convergence time shorten greatly, change the performance of the network, define standardization function as [10]:

$$x^* = 0.1 + \frac{0.8(x - x_{min})}{x_{max} - x_{min}}, x^* \in [0.1, 0.9] \tag{2}$$

$$x^* = 0.2 + \frac{0.6(x - x_{min})}{x_{max} - x_{min}}, x^* \in [0.2, 0.8] \tag{3}$$

Worth attention is, if the training sample set used by training neural network is the data processed after standardization treatment, then the neural network after learning applying new test samples still should adopt normalized sample set, of course all the data used must adopt the relevant data in learning sample set, such as: maximum value, minimum value, mean value and standard deviation. The new sample data applied to the trained neural network also need do the same pre-processing to the input data. The output data of the neural network must do post-processing. The post-processing of the data is the inverse operation of the pre-processing.

4.3   Training Neural Network

The training neural network module's functions are divided into forward propagation calculation, back-propagation calculation, modifying and saving weight. Forward propagation calculation and back-propagation calculation accomplish the neural network's calculation process. Modifying and saving weight accomplish modifying the weight during the calculation process, save the final modifying result into weight data files for the calling of neural network application.

4.4   Applying Neural Network

The applying neural network module includes data normalization, virtual character behavior model and data post-processing. Data normalization does normalizing process of the state's information which acquired by the virtual characters, prepares the calculation of neural network calculation. Virtual character behavior model calculates normalized output result through neural network. Data post-processing denormalizes the calculation result of neural network as the output behaviors of the virtual characters.

5   Result Analysis of the Virtual Characters' Behavior Model Approximated by ANN

The paper establishes behavior model for the driver virtual character based on the method of neural network. According to the example of social and group of drive, we analyze and compare the behavior model approximated by ANN and specific behavior model.

During the process of simulating virtual characters' behaviors by specific behavior model, the virtual characters' behaviors are get through the judging of current state by program. The example chosen in this paper, the states and behaviors of virtual characters can be analyzed easily, be defined unequivocally. But to some virtual characters with more complex behaviors such as: the states and behaviors are far harder to determine, the scene virtual characters in changing constantly, specific behavior model be inapplicable obviously.

Behavior model approximated by ANN is the simulation of virtual character reasoning, analysis and decision process. We take ANN algorithm do the driver virtual characters' cognitive layer behavior model modeling, its process is more complex relative to the specific behavior model, includes the process bellow: the confirmation of the states and behaviors of the virtual characters, the acquisition of the sample(In this paper it is realized through the virtual characters' obvious behavior model and the training sample data acquired by manual control), the dispose of sample data training and the training of the ANN. Though the process is complex, but this method approximates the process of virtual characters' behavior decision.

6   Conclusions

We establish virtual characters' behavior model by taking the method of artificial neural network in this paper. The experiment result states clearly the virtual characters can adapt the VEs' change best, has better autonomy and intellectuality, can generate a good deal of behavior animations steadily, achieving the expectant goal. Artificial neural network method is a method of off-line learning and on line using, need the support of the training sample data in applying. But not all virtual characters are fit to this method, for example: the virtual characters' states and behaviors are uncertain, the virtual characters can't model specifically, etc. All of these problems above can be solved by applying on line learning way.

Acknowledgments. This work is supported by the Education Science Research Project #HGJXH B1110131, the Education Science Research Project # GBC1211028, the Scientific and Technological Project # 12511011 & # 12521050.

References

1. Ruth, A., Marc, C.: Intelligent Virtual Environments: a State-of-the-Art Report. Technical report, Eurographics (2001)
2. Funge, J., Tu, X., Terzopoulos, D.: Cognitive Modeling: Knowledge, Reasoning and Planning for Intelligent Characters. In: 26th Annual Conference on Computer Graphics and Interactive Techniques, pp. 29–38. ACM Press, New York (1999)
3. Yoh, E., Yuk, I., Toru, I.: Modeling human behavior for virtual training systems. In: 20th National Conference on Artificial Intelligence, pp. 127–132. AAAI Press, Pittsburgh (2005)
4. Janathan, D., Parris, K.E., Hugo, D.G., Nelson, D.: Fast and learnable behavioral and cognitive modeling for virtual character animation. Computer Animation and Virtual Worlds 15, 95–108 (2004)
5. Toni, C., Daniel, T.: An Artificial Life Environment for Autonomous Virtual Agents with Multi-sensorial and Multi-perceptive Features. Computer Animation and Virtual Worlds 15, 311–318 (2004)
6. Lucio, I., Luca, C.: A Virtual Human Architecture that Integrates Kinematic, Physical and Behavioral Aspects to Control H-Anim Characters. In: 10th international conference on 3D Web Technology, pp. 69–77. ACM Press, New York (2005)
7. Libo, S.: Behavior Modeling and Simulation Technology for Virtual Crowds, pp. 19–22. Tianjin University, Tianjin (2011) (in Chinese)
8. Xudong, J., Fangzheng, X.: Artificial Life Model Based on Bionic Fish. Chongqing University of Technology (Natural Science) 26, 54–60 (2012) (in Chinese)
9. Jonathan, D., Trent, C., Parris, K.E.: Intelligence Capture. Technical Report. Brigham Young University (2005)
10. Zhen, L., Weigang, X.: Method of Autonomous Path Planning for Virtual Character. Journal of System Simulation 24, 104–107 (2012) (in Chinese)

A New Hybrid Fuzzy-Rough Dendritic Cell Immune Classifier

Zeineb Chelly and Zied Elouedi

LARODEC, Institut Supérieur de Gestion de Tunis, Tunis, Tunisia
zeinebchelly@yahoo.fr, zied.elouedi@gmx.fr

Abstract. The Dendritic Cell Algorithm (DCA) is an immune-inspired classification algorithm based on the behavior of natural dendritic cells (DC). This paper proposes a novel version of the DCA based on a two-level hybrid fuzzy-rough model. In the top-level, the proposed algorithm, named RST-MFDCM, applies rough set theory to build a solid data pre-processing phase. In the second level, RST-MFDCM applies fuzzy set theory to smooth the crisp separation between the DC's semi-mature and mature contexts. The experimental results show that RST-MFDCM succeeds in obtaining significantly improved classification accuracy.

Keywords: Dendritic cell algorithm, Rough sets, Fuzzy sets, Hybrid model.

1 Introduction

The Dendritic Cell Algorithm (DCA) [1] is derived from behavioral models of dendritic cells (DCs). DCA has been successfully applied to various applications. However, it was noticed that DCA is sensitive to the input class data order [2]. Such a drawback is the result of an environment characterized by a crisp separation between the DC semi-mature context and the DC mature context. Hence, in [3], a first work named the Modified Fuzzy Dendritic Cell Method (MFDCM) was developed to solve this issue. MFDCM is based on the fact of smoothing the mentioned crisp separation between the DCs' contexts. This was handled by the use of Fuzzy Set Theory (FST). However, MFDCM suffers from some limitations as its data pre-processing phase, which is divided into feature selection and signal categorization, is based on the use of the Principal Component Analysis (PCA). More precisely, MFDCM uses PCA to automatically select features and to categorize them to their specific signal types; as danger signals (DS), as safe signals (SS) or as pathogen-associated molecular patterns (PAMP). Using PCA for the MFDCM feature reduction step presents a drawback as it is not necessarily true that the first selected components will be the adequate features to retain [4]. Thus, the choice of these components for the MFDCM can influence its classification task by producing unreliable results. As for feature categorization, MFDCM uses the generated PCA ordered list of standard deviation values to assign for each selected attribute its signal type. However, this categorization process does not make "sense" as a coherent process which can influence

Y. Tan, Y. Shi, and H. Mo (Eds.): ICSI 2013, Part I, LNCS 7928, pp. 514–521, 2013.
© Springer-Verlag Berlin Heidelberg 2013

negatively the MFDCM functioning. Thus, it is clearly seen that there is a need to develop a new MFDCM model with a more robust data pre-processing phase. On the other hand, in [5], a new DCA model based on a rough data pre-processing technique has been developed. The work, named RST-DCA, aims at applying Rough Set Theory (RST) [6] for feature selection and signal categorization. It was shown that using RST, instead of PCA, for the DCA data pre-processing phase yields better performance in terms of classification accuracy. However, it is important to note that RST was only applied to the standard DCA which is sensitive to the class data order. Thus, in this paper, we propose to hybridize the works of [3] and [5] in order to obtain a robust dendritic cell stable classifier. Our hybrid model, named RST-MFDCM, is built as a two-level hybrid immune model. In the top-level, RST-MFDCM uses RST to ensure a more rigorous data pre-processing phase. In the second level, RST-MFDCM uses FST to ensure a non-sensitivity to the input class data order.

2   The Dendritic Cell Algorithm

The initial step of the DCA is data pre-processing where PCA is applied. After features are selected and mapped to their signal categories DCA adheres these signals and antigen to fix the context of each object (DC) which is the step of Signal Processing [7]. The algorithm processes its input signals in order to get three output signals: costimulation signal (Csm), semi-mature signal ($Semi$) and mature signal (Mat). A migration threshold is incorporated into the DCA in order to determine the lifespan of a DC. As soon as the Csm exceeds the migration threshold; the DC ceases to sample signals and antigens. The migration state of a DC to the semi-mature state or to the mature state is determined by the comparison between cumulative $Semi$ and cumulative Mat. If the cumulative $Semi$ is greater than the cumulative Mat, then the DC goes to the semi-mature context, which implies that the antigen data was collected under normal conditions. Otherwise, the DC goes to the mature context, signifying a potentially anomalous data item. This step is known to be the Context Assessment phase. The nature of the response is determined by measuring the number of DCs that are fully mature and is represented by the Mature Context Antigen Value (MCAV). $MCAV$ is applied in the DCA final step which is the Classification step and used to assess the degree of anomaly of a given antigen. The closer the $MCAV$ is to 1, the greater the probability that the antigen is anomalous. Those antigens whose $MCAV$ are greater than the anomalous threshold are classified as anomalous while the others are classified as normal.

3   RST-MFDCM: The Hybrid Approach

In this Section, we present our two-level hybrid model which combines the theory of rough sets for a robust data pre-processing phase, which is the top-level, and fuzzy set theory in order to get a stable DCA classifier, which is the RST-MFDCM second level.

3.1   RST-MFDCM Data Pre-processing Phase

The data pre-processing phase of our RST-MFDCM includes two sub-steps which are feature selection and signal categorization; both based RST.

1) Feature Selection Process: Our learning problem is to select high discriminating features for antigen classification from the original input data set which corresponds to the antigen information database. We may formalize this problem as an information table, where universe $U = \{x_1, x_2, \ldots, x_N\}$ is a set of antigen identifiers, the conditional attribute set $C = \{c_1, c_2, \ldots, c_A\}$ contains each feature to select and the decision attribute D corresponds to the class label of each sample. The decision attribute D has binary values d: either the antigen is classified as normal or as anomalous. RST-MFDCM computes, first of all, the positive region for the whole attribute set C for both label classes of D: $POS_C(\{d\})$. Secondly, RST-MFDCM computes the positive region of each feature c and the positive region of all the composed features $C - \{c\}$ (when discarding each time one feature c from C) defined respectively as $POS_c(\{d\})$ and $POS_{C-\{c\}}(\{d\})$, until finding the minimal subset of attributes R from C that preserves the positive region as the whole attribute set C does. In fact, RST-MFDCM removes in each computation level the unnecessary features that may affect negatively the accuracy of the RST-MFDCM. The result of these computations is either one reduct $R = RED_D(C)$ or a family of reducts $RED_D^F(C)$. Any reduct of $RED_D^F(C)$ can be used to replace the original antigen information table. Consequently, if the RST-MFDCM generates only one reduct $R = RED_D(C)$ then for the feature selection process, RST-MFDCM chooses this specific R. If the RST-MFDCM generates a family of reducts then RST-MFDCM chooses randomly one reduct R among $RED_D^F(C)$ to represent the original input antigen information table. By using the REDUCT, our method can guarantee that the selected attributes will be the most relevant for its classification task.

2) Signal Categorization Process: RST-MFDCM has to assign, now, for each selected attribute, its specific signal type. In biology, both PAMP and SS have a certain final context (either an anomalous or a normal behavior) while the DS cannot specify exactly the final context to assign to the collected antigen as the DS may or may not indicate an anomalous situation. This problem can be formulated as follows: Both PAMP and SS are more informative than DS which means that both of these signals can be seen as indispensable attributes. To define this level of importance, our method uses the CORE RST concept. As for DS, it is less informative than PAMP and SS. Therefore, RST-MFDCM uses the rest of the REDUCT attributes to represent the DS. As stated in the previous step, our method may either produce only one reduct or a family of reducts. In case where our RST-MFDCM generates only one reduct, RST-MFDCM selects randomly one attribute c from $CORE_D(C)$ and assigns it to both PAMP and SS as they are the most informative signals. Using one attribute for these two signals requires a threshold level to be set: values greater than this can be classed as SS, otherwise as a PAMP signal. The rest of the attributes $CORE_D(C) - \{c\}$ are combined and the resulting value is assigned to the DS as it is less than certain to be anomalous. In case where our RST-MFDCM produces a family of reducts, the RST-MFDCM selects, randomly, one attribute c among the features in $CORE_D(C)$ and assigns it to both PAMP and SS. As for the DS signal assignment, RST-MFDCM chooses, randomly, a reduct $RED_D(C)$ among $RED_D^F(C)$. Then, RST-MFDCM combines all the $RED_D(C)$ features except that c attribute already chosen and assigns the resulting value to the DS.

3.2 RST-MFDCM Fuzzy Classification Process

This second level of our RST-MFDCM hybrid model is composed of five main sub-steps and is based on FST.

1) Fuzzy System Inputs-Output Variables: Once the signal database is ready, our RST-MFDCM processes these signals to get the semi-mature and the mature signals values. To do so and in order to describe each of these two object contexts, we use linguistic variables [8]. Two inputs (one for each context) and one output are defined. The semi-mature context and the mature context denoted respectively C_s and C_m are considered as the input variables to the fuzzy system. The final state "maturity" of a DC, S_{mat}, is chosen as the output variable. They are defined as:

$C_s = \{\mu_{C_s}(c_{s_j})/c_{s_j} \in X_{C_s}\}$, $C_m = \{\mu_{C_m}(c_{m_j})/c_{m_j} \in X_{C_m}\}$ and $S_{mat} = \{S_{mat}(s_{mat_j})/s_{mat_j} \in X_{S_{mat}}\}$; where c_{s_j}, c_{m_j} and s_{mat_j} are, respectively, the elements of the discrete universe of discourse X_{C_s}, X_{C_m} and $X_{S_{mat}}$. μ_{C_s}, μ_{C_m} and $\mu_{S_{mat}}$ are, respectively, the corresponding membership functions.

2) Defining Linguistic Variables: The term set $T(S_{mat})$ interpreting S_{mat} is defined as: $T(S_{mat}) = \{Semi - mature, Mature\}$. Each term in $T(S_{mat})$ is characterized by a fuzzy subset in a universe of discourse $X_{S_{mat}}$. Semi-mature might be interpreted as a data instance collected under safe circumstances, reflecting a normal behavior and Mature reflecting an anomalous behavior. Similarly, the input variables C_s and C_m are interpreted as linguistic variables with: $T(Q) = \{Low, Medium, High\}$, where $Q = C_s$ and C_m respectively.

3) Fuzzy and Membership Functions Construction: In order to specify the range of each linguistic variable, we have run the RST-MFDCM and we have recorded both semi-mature and mature values which reflect the (Semi) and (Mat) outputs generated by the algorithm. Then, we picked up the minimum and maximum values of each of the two generated values to fix the borders of the range which are:

$$min(range(S_{mat})) = min(min(range[C_m]), min(range[C_s]))$$

$$max(range(S_{mat})) = max(max(range[C_m]), max(range[C_s]))$$

The parameters of our RST-MFDCM fuzzy process, the extents and midpoints of each membership function, are generated automatically from data by applying the fuzzy Gustafson-Kessel clustering algorithm. Each cluster reflects a membership function. The number of clusters is relative to the number of the membership functions of each variable (inputs and output).

4) The Fuzzy Rules Set Description: A knowledge base, comprising rules, is built to support the fuzzy inference. The different rules of the fuzzy system are extracted from the information reflecting the effect of each input signal on the state of a dendritic cell [3].

1. If (C_m is Low) and (C_s is Low) then (S_{mat} is Mature)
2. If (C_m is Low) and (C_s is Medium) then (S_{mat} is Semi-mature)
3. If (C_m is Low) and (C_s is High) then (S_{mat} is Semi-mature)
4. If (C_m is Medium) and (C_s is Low) then (S_{mat} is Mature)
5. If (C_m is Medium) and (C_s is Medium) then (S_{mat} is Semi-mature)
6. If (C_m is Medium) and (C_s is High) then (S_{mat} is Semi-mature)
7. If (C_m is High) and (C_s is Low) then (S_{mat} is Mature)
8. If (C_m is High) and (C_s is Medium) then (S_{mat} is Mature)
9. If (C_m is High) and (C_s is High) then (S_{mat} is Mature)

5) The Fuzzy Context Assessment: Our RST-MFDCM is based on the "Mamdani" composition method and the "centroid" defuzzification mechanism. Once the inputs are fuzzified and the output (centroid value) is generated, the cell context has to be fixed by comparing the output value to the middle of the S_{mat} range. In fact, if the centroid value generated is greater than the middle of the output range then the final context of the object is "Mature" indicating that the collected antigen may be anomalous; else the antigen collected is classified as normal.

4   Experimental Setup and Results

1) Experimental Setup: To test the validity of our RST-MFDCM hybrid model, our experiments are performed on two-class databases from [9] described in Table 1.

Table 1. Description of Databases

Database	Ref	♯ Instances	♯ Attributes
Spambase	SP	4601	58
SPECTF Heart	SPECTF	267	45
Cylinder Bands	CylB	540	40
Chess	Ch	3196	37
Ionosphere	IONO	351	35
Mushroom	Mash	8124	23
Horse Colic	HC	368	23
Hepatitis	HE	155	20

For data pre-processing, the standard MFDCM uses PCA to automatically select and categorize signals. As for our method, it uses RST as explained in Section 3. For both MFDCM methods, each data item is mapped as an antigen, with the value of the antigen equal to the data ID of the item. The migration threshold of an individual DC is set to 10. To perform anomaly detection, a threshold which is automatically generated from the data is applied to the MCAVs. Items below the threshold are classified as class one and above as class two. For each experiment, the results presented are based on mean MCAV values generated across 10 runs. We evaluate the performance of our RST-MFDCM method in terms of number of extracted features, sensitivity, specificity and accuracy which are defined as: $Sensitivity = TP/(TP + FN)$; $Specificity = TN/(TN + FP)$; $Accuracy = (TP + TN)/(TP + TN + FN + FP)$; where TP, FP, TN, and FN refer respectively to: true positive, false positive, true negative and false negative.

2) Results and Analysis: In this Section, we show that using RST instead of PCA is more convenient for the MFDCM data pre-processing phase as it improves its classification performance. This is confirmed by the results given in Table 2. Let us remind that for data pre-processing, MFDCM applies PCA where it reduces data dimension and categorizes the obtained features to their signal types. As previously shown, in [5], using PCA for the DCA data pre-processing is not convenient. Thus, in this Section, we show that the use of RST instead of PCA is more adequate for our new RST-MFDCM algorithm to process.

Table 2. PCA-MFDCM and RST-MFDCM Comparison Results

Database	Sensitivity (%) MFDCM		Specificity (%) MFDCM		Accuracy (%) MFDCM		♯ Attributes MFDCM	
	PCA	RST	PCA	RST	PCA	RST	PCA	RST
SP	88.25	97.07	93.25	96.12	91.28	96.50	14	8
SPECTF	81.60	87.26	70.90	81.81	79.40	86.14	11	4
CylB	92.50	97.00	94.55	97.43	93.75	97.26	16	7
Ch	94.66	98.20	94.95	99.21	94.80	98.68	14	11
IONO	94.44	96.03	95.55	98.22	95.15	97.43	24	19
Mash	99.46	99.87	99.33	99.80	99.39	99.84	7	6
HC	92.59	96.75	85.52	92.10	89.67	94.83	19	14
HE	90.62	93.75	95.93	98.37	94.83	97.41	10	4

From Table 2, we can notice that the number of features selected by our RST-MFDCM is less than the one generated by the standard MFDCM when applying PCA (PCA-MFDCM). This can be explained by the appropriate use of RST for feature selection. In fact, RST-MFDCM, by using the REDUCT concept, keeps only the most informative features from the whole set of features. For instance, by applying our RST-MFDCM method to the CylB data set, the number of selected features is only 7 attributes. However, when applying the PCA-MFDCM to the same database (CylB), the number of the retained features is 16. We can notice that PCA preserves additional features than necessary which leads to affect the PCA-MFDCM classification task by producing less accuracy in comparison to the RST-MFDCM results. On the other hand, RST-MFDCM based on the REDUCT concept, selects the minimal set of features from the original database and can guarantee that the reduct attributes will be the most relevant for its classification task. As for the classification accuracy, from Table 2, we notice that the classification accuracy of our RST-MFDCM is notably better than the one given by the PCA-MFDCM. For instance, when applying the RST-MFDCM to the CylB database, the classification accuracy is set to 97.26%. However, when applying the PCA-MFDCM to the same database, the accuracy is set to 93.75%. Same remark is noticed for both the sensitivity and the specificity criteria. These encouraging RST-MFDCM results are explained by the appropriate set of features selected and their categorization to their right and specific signal types. RST-MFDCM uses the REDUCT concept to select only the essential part of the original database. This pertinent set of minimal features can guarantee a solid base for the signal categorization step. The RST-MFDCM good classification results are also explained by the appropriate categorization of each selected signal to its right signal type by using both the REDUCT and the CORE concepts. We have also compared the performance of our RST-MFDCM to other classifiers including the standard DCA when applying PCA (PCA-DCA), the standard DCA when applying RST (RST-DCA), MFDCM when applying PCA (PCA-MFDCM), SVM, ANN and DT. The comparison made is in terms of the average of accuracies on the databases presented in Table 1.

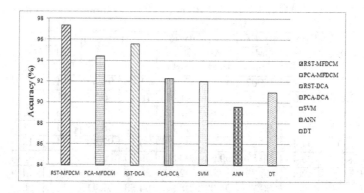

Fig. 1. Comparison of Classifiers' Average Accuracies

Figure 1 shows that the standard DCA, PCA-DCA, when applied to the ordered datasets, has nearly the same classification accuracy as SVM and a better performance in comparison to ANN and DT. Figure 1, also, shows that when applying RST, instead of PCA, to the standard DCA, the classification accuracy of RST-DCA is notably better than PCA-DCA and also better than the mentioned classifiers including SVM, ANN and DT. This confirms that applying RST is more convenient to the DCA. Furthermore, from Figure 1, we can notice that applying the fuzzy process to the DCA leads to better classification results. We can see that the classification accuracy of PCA-MFDCM is better than the one generated by PCA-DCA. This confirms that applying a fuzzy process to the DCA is more convenient for the algorithm leading to a more stable classifier. From these remarks, we can conclude that if we hybridize both RST and the fuzzy process to the DCA, we will obtain a better classifier. We can clearly notice that our hybrid RST-MFDCM developed model outperforms both PCA-MFDCM, which only applies FST, and RST-DCA, which only applies RST. In addition, our RST-MFDCM immune hybrid model outperforms the rest of the classifiers including PCA-DCA, SVM, DT and ANN.

5 Conclusion and Further Works

In this paper, we have developed a new fuzzy rough hybrid immune model. Our method aims at combining the rough set theory to select the right set of features and their categorization to their right signal types and fuzzy set theory to smooth the crisp separation between the two DC contexts leading to better results in terms of classification accuracy. Our RST-MFDCM hybrid model is characterized by its robust data pre-processing phase as it is based on RST. It is also characterized by its stability as a binary classifier as it is based on the DCA fuzzy version, MFDCM. As future work, we intend to further explore this new instantiation of our RST-MFDCM by introducing new methods in the algorithm data pre-processing phase.

References

1. Greensmith, J., Aickelin, U., Cayzer, S.: Introducing dendritic cells as a novel immune-inspired algorithm for anomaly detection. In: Jacob, C., Pilat, M.L., Bentley, P.J., Timmis, J.I. (eds.) ICARIS 2005. LNCS, vol. 3627, pp. 153–167. Springer, Heidelberg (2005)
2. Greensmith, J., Aickelin, U.: The deterministic dendritic cell algorithm. In: Bentley, P.J., Lee, D., Jung, S. (eds.) ICARIS 2008. LNCS, vol. 5132, pp. 291–302. Springer, Heidelberg (2008)
3. Chelly, Z., Elouedi, Z.: Further exploration of the fuzzy dendritic cell method. In: Liò, P., Nicosia, G., Stibor, T. (eds.) ICARIS 2011. LNCS, vol. 6825, pp. 419–432. Springer, Heidelberg (2011)
4. Kaiser, H.: A note on guttman's lower bound for the number of common factors. British Journal of Mathematical and Statistical Psychology 14, 1–2 (1961)
5. Chelly, Z., Elouedi, Z.: RST-DCA: A dendritic cell algorithm based on rough set theory. In: Huang, T., Zeng, Z., Li, C., Leung, C.S. (eds.) ICONIP 2012, Part III. LNCS, vol. 7665, pp. 480–487. Springer, Heidelberg (2012)
6. Pawlak, Z.: Rough sets. International Journal of Computer and Information Science 11, 341–356 (1982)
7. Gu, F., Greensmith, J., Oates, R., Aickelin, U.: Pca 4 dca: The application of principal component analysis to the dendritic cell algorithm. In: Proceedings of the 9th Annual Workshop on Computational Intelligence (2009)
8. Zimmermann, J.: Fuzzy set theory and its applications. European Journal of Operational Research 1, 227–228 (1996)
9. Asuncion, A., Newman, D.J.: UCI machine learning repository, (2007), http://mlearn.ics.uci.edu/mlrepository.html

Multi Objective Swarm Optimization Design Fuzzy Controller to Adjust Speed of AC Motor Drive

Nasseer K. Bachache and Jinyu Wen

College of Electrical and Electronic Engineering,
Huazhong University of Science and Technology (HUST), Wuhan 430074, China
tech_n2008@yahoo.com, jinyu.wen@mail.hust.edu.cn

Abstract. In this paper a Multi-Objective Particle Swarm Optimization (MOPSO) is utilized to design sets of linguistic Fuzzy Logic Controller (FLC) type Mamdani to govern the speed of Induction Motor (IM). The first objective function is the error between the actual speed and desired speed, and the second function is the energy dissipated during (10 Sec). PSO are implemented in M-file/MATLAB, but when the algorithm reaches the step of assessing the "fitness functions", this program linked with SIMULINK-MATLAB to evaluate these values. This simulation includes the complete mathematical model of IM and the inverter. The simulation results show the proposed controller offers an optimized speed behavior as possible with a low-slung of energy along the points of Pareto front.

Keywords: Multiobjective Particle Swarm Optimization MOPSO, Fuzzy Logic Control FLC, Induction Motor IM.

1 Introduction

Electric vehicles are desperately needed for the applications of variable speed under a high quality controller, this controller must be designed for a specified performance, with a limited power supply such as battery, the controller should be maintained the energy storage; Nowadays the ac drives have been one of the most important strategies in speed control due to a high reliability of IM and the development took place in fields of power electronic; Ac drives has a wide spread applications to adjust motor speed, the typical drivers include a poly-phase inverter works as a Voltage Source Inverter VSI, VSI produce Pulse Width Modulation (PWM) to get a sinusoidal signal at a lower gauge of harmonics which it's power losses. The mathematical model of system bases on a stator flux orientation of Ac motor [1]. However, of its highly nonlinear system is difficult to obtain an optimal controller. Fuzzy logic control FLC is a powerful controller tool; even if the system is non-linear and an accurate mathematical model is unavailable, the FLC is a robust controller. But, FLC suffers from drawbacks of how can tuning its parameters (number of membership functions and its type, rule number, and formulating rules). The most tuning of these parameters is done by either interactively method (trial and error) or by a human

Y. Tan, Y. Shi, and H. Mo (Eds.): ICSI 2013, Part I, LNCS 7928, pp. 522–529, 2013.

expert [2]. Therefore, the tuning FLC parameters are necessary look for effective methods. Nowadays, several new intelligent optimization techniques have been emerged, such as Genetic Algorithms (GA), Simulated Annealing (SA), Ant Colony Optimization (ACO) and Bacteria Foraging Optimization (BFO) among these nature-inspired strategies the Particle Swarm Optimization (PSO) algorithm is relatively novel [3], PSO has received a great attention in the control system. In [4] the fuzzy controller parameters generated by PSO for an AC-Drive speed controller, in [5] designed an AC-Drive system with an adaptive fuzzy controller using Reference Mode (RF), the reference [6] described a FLC based on a "floating membership function". All this research didn't care to calculate the energy dissipated or the power losses. In this paper, we propose a multi-objective function optimization for AC Drive works under a powerful controller FLC, the main question is: how can get an optimized dynamic speed behavior with a limited power? Pareto diagram has two functions; the first is reducing the error between the actual speed and desired speed while the other must preserve the energy dissipated from the electrical supply.

2   Modeling and Simulation of Three Phases I.M.

The most popular induction motor's models are based on applying the axes transformation named (Krause's model) according to d-q axes (stationary frame), assuming d-q axes are rotated synchronously with the rotor speed. The complete mathematical model and equivalent circuit diagrams of an I.M are illustrated in [4].

3   Multi-Objective Particle Swarm Optimization

The most nature problems in the real world are boosted on a multi - objective investigation which can confront effectively, Multi-objective Particle Swarm Optimization MOPSO is one of the popular algorithms. Primarily Particle Swarm Optimization PSO is a computation technique had been proposed in 1995 by Kennedy and Eberhart [7, 8], PSO simulates a dynamic population behavior of fish swarm or bird flocks. Due to PSO simplicity, effectiveness, high convergence rate and efficiency this algorithm can solve a wide variety of single-objective optimization also non-linearity or non-differentiability problems [4], this algorithm explains briefly in as follows:

1. Evaluate the fitness of each particle
2. Update individual and global best fitness's and positions
3. Update velocity and position of each particle according the equations (1, 2):
4. Repeat step 1

$$V_{i,m}^{(It.+1)} = W * V_{i,m}^{(It.)} + c1.r.\left(P_{best_{i,m}} - x_{i,m}^{(It.)}\right) + c2.r.\left(g_{best_m} - x_{i,m}^{(It.)}\right) \qquad (1)$$

$$x_{i,m}^{(It.+1)} = x_{i,m}^{(It.)} + v_{i,m}^{(It.)} \qquad (2)$$

Where:

i=1, 2… Number of particles.
m=1, 2… Dimension.
$It.$: Iterations pointer.
$V_{i,m}^{(It.)}$: Velocity of particle no. i at iteration It.
W: Inertia weight factor.
$c1$, $c2$: Acceleration constant.
r: Random number between(0-1).
$x_{i,m}^{(It.)}$: Current position of particle i at iteration It.
P_{best_i} : Best previous position of ith particle.
g_{best_m}: Global best particle among all the particles in the population.

However, Multi-objective optimization is quite different from single-objective optimization, the latter has one objective function so it is easy to calculate its (p_{best}, g_{best}) values. But, the former has many objective functions so it is difficult to calculate (p_{best}, g_{best}); the basic version PSO algorithm is not effective to solve multi objective problem directly, because there are many objectives then the global best particle is incommensurable, and impossible to create all objective functions reach their minimum value at the same time [9]. Weighted Sum Approach (WSA) is the effective method to solve the desired solution by converting Multiobjective problem into a Single-objective problem, WSA represented in an equation (3)

$$\min F(x) = \sum_{j=1}^{m} w_j \cdot f_j(k) \tag{3}$$

Where: $w_m \in [0,1]$, $\sum_{j=1}^{m} w_m = 1$

$f_j(k)$ representing the best optimal finesses value of the j-th objective function in the k-th generation.

Hopefully by expression (3) we can get the Pareto solutions along 23 points against 23 Global best particles presented by the $w_1 = [0.04\ 0.08\ 0.12\ ………\ 0.96\ ]$ and $w_2 = [0.96\ 0.92\ 0.88………\ 0.04]$ making the current solutions to move toward the direction of the minimum distance from current position to each objective's best optimal value.

4　MOPSO Implemented with FLC

Owning for one doesn't have the experience of the system behavior, it is very difficult to forming the center and width of the triangle Membership Fuzzy Functions MFFs for (inputs and outputs), and the challenge is how to get the optimized MFFs. There is no formal framework for choosing the parameters of FLC; hence the means of tuning them and learning models in general have become an important subject of fuzzy control. The function of the fuzzy controller is to observe the pattern of the speed loop error signal then the control signal can updated, A simple fuzzy logic controller of two inputs and one output can be designed in our work, the two inputs are: error between actual speed and desired speed (E) and change of error (CE), CE means derivative of the error (dE/dt); these MFFs have seven triangle membership for each input and nine memberships for output. MOPSO can design positions of triangle memberships for input/output implementing by MATLAB/m-file program summarized in the flowchart shown in figure (3); this figure illustrates the linked with MATLAB/SIMULINK the system simulation program by AC-drive model presented by (IM, inverter and FLC block sets) shown in Figure (1).The optimization criteria (Integrated of Time Weight Square Error ITSE) equation (4) is used to evaluate accuracy performance of the speed (FF1) to minimize the first function which evaluated in the variable (e2) exposed in fig (1).

$$FF1 = ITSE = \int_0^t t * e^2(t)dt \qquad (4)$$

But, the optimization of the other function FF2 is evaluated by the variable (e) expressed the energy delivered to the motor; e is assessed by the integration of the power input (multiplying the voltage and the current). A set of good controller parameters can yield a good step response.

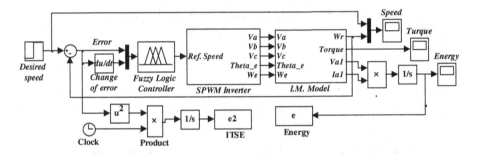

Fig. 1. A simulation of an AC-derive with FLC

5   Simulation Result

Figure 2 shows the space of operating points (dominated, non-dominated) and determined the Pareto curve of the multi objective optimization of the two functions the ITSE of the difference between the desired speed and the actual speed on the x axis and the other function on y-axis is the energy dissipated and shows the Excellent Dynamic Speed Behavior (EDSB) is spend the largest scale of energy (49.93 Pu) and (ITSE (speed) =1. 322pu). Figure 4 shows a two step response of speed governed by (EDSB) controller which needs large energy and the Lower Dissipated of Energy (LDOE) which has the worst dynamic speed behavior. Figure 5 shows the energy dissipated of (EDSB) and (LDOE). Figures 6, figure7 FLC memberships in two inputs and one output designed by MOPSO for excellent speed dynamic behavior and the Lower Dissipated of Energy (LDOE) respectively.

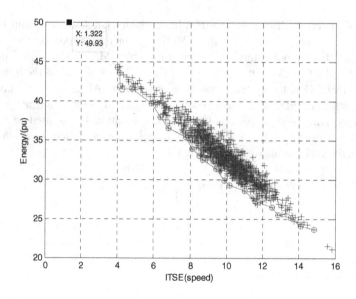

Fig. 2. Pareto of MOPSO and the all dominated points

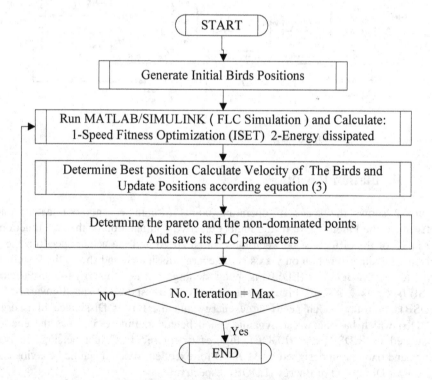

Fig. 3. A flowchart of PSO algorithm

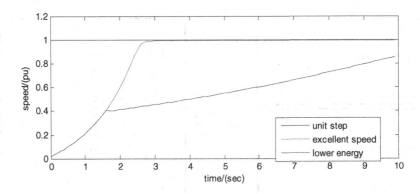

Fig. 4. Step response of speed for two FLC

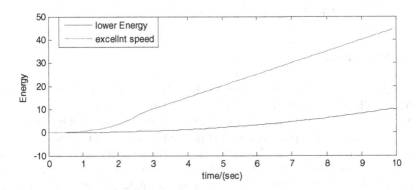

Fig. 5. Energy disputed Step response for two FLC

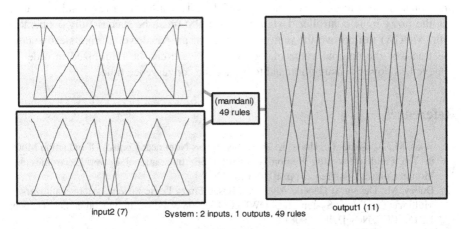

Fig. 6. FLC designed by MOPSO for excellent speed dynamic behavior

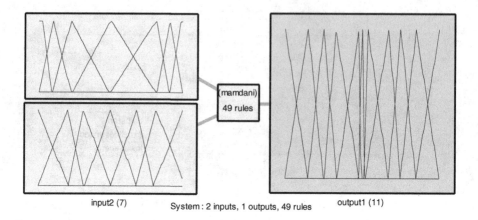

input2 (7)   System: 2 inputs, 1 outputs, 49 rules   output1 (11)

Fig. 7. FLC designed by MOGA for lower energy dissipated

6   Conclusions

In this paper, we have proposed MOPSO, which is an evolutionary multiobjective algorithm developed to generate a set of FLCs with different function's accuracy of dynamic speed and energy, which is measure a numerous of membership (MF) antecedent conditions. The proposed method is able to learn the MF parameters. This allows us to take the better controller during the life of the battery or just before it will be discharged. A multi-objective PSO algorithm MOPSO can solve the engineering constrained optimization problem by converting into a single objective problem using Weighted Sum Approach (WSA) so it is easier to implement, and give the result sometime better than Genetic Algorithm (MOGA) and MOPSO algorithm is able to find competitive solutions and converge quickly. The simulation test results validate that the fuzzy logic controller designed by (MOPSO) can be successfully cooperated with (MOGA) to improve the motor efficiency of an AC drive. During steady states and transient the optimal value of the favorable case can be minimized in order to achieve less power consumption and to reduce the convergence time.

References

1. Wlas, M., Abu-Rub, H., Holtz, J.: Speed Sensorless Nonlinear Control Of Induction Motor in The Field Weakening Region. In: IEEE 13th International Power Electronics and Motion Control Conference, pp. 1084–1089 (2008)
2. Dubey, M.: Design of Genetic Algorithm Based Fuzzy Logic Power System Stabilizers in Multimachine Power System. In: POWERCON 2008 & Power India Conference, October 12-15, IEEE, New Delhi (2008)
3. Rapaic, M.R., Kanovic, Z., Jelicic, Z.D.: A Theoretical and Empirical Analysis of Convergence Related Particle Swarm Optimization. Wseas Transactions on Systems and Control 11(4) (November 2009)

4. Bachache, N.K., Wen, J.: Particle swarm optimize fuzzy logic memberships of AC-drive. In: Tan, Y., Shi, Y., Ji, Z. (eds.) ICSI 2012, Part I. LNCS, vol. 7331, pp. 460–469. Springer, Heidelberg (2012)
5. Kung, Y.-S., Wang, M.-S., Huang, C.-C.: Digital Hardware Implementation of Adaptive Fuzzy Controller for AC Motor Drive. In: The 33rd Annual Conference of the IEEE Industrial Electronics Society (IECON), Taipei, Taiwan (2007)
6. Keskar, A.G., Asanare, K.L.: Floating Membership Fuzzy Logic Controller For Adaptive Control Of AC Drive. In: IEEE Catalog Number: 97TH8280ISIE 1997 - Guhariies, Portugal (1997)
7. Martinez, F., Gastiblanco, M.: Increase the Boost Converter Performance Using Genetic Algorithms. Online Journal on Electronics and Electrical Engineering 2(1) (2008)
8. Allaoua, B., Laaoufi, A.: Intelligent Controller Design for DC Motor Speed Control Based on Fuzzy Logic-Genetic Algorithms Optimization. Leonardo Journal of Sience (13), 90–102 (2008)
9. Benachaiba, C., Abdel Khalek, O., Dib, S.: Optimization of Parameters of the Unified Power Quality Conditioner Using Genetic Algorithm Method. Information Technology and Control 36(2) (2007) ISSN 1392
10. Davis, L.: A Handbook of Genetic Algorithm. Van Nostrand Reinhold, New York (1990)

Rough Soft Sets in Fuzzy Setting

Xueyou Chen

School of Mathematics, Shandong University of Technology,
Zibo, Shandong 255049, P.R. China
math-chen@qq.com

Abstract. Fuzzy set theory, soft set theory and rough set theory are mathematical tools for dealing with uncertainties and are closely related. In the paper, we define the notion of a soft set in **L**-set theory, introduce several operators for **L**-soft set theory, and investigate the rough operators on the set of all **L**-soft sets induced by the rough operators on L^X.

Keywords: Rough set, L-set, soft set, approximations.

1 Introduction

In 1982, Z.Pawlak initiated the rough set theory [15], D.Dubois, H.Prade combined fuzzy sets and rough sets all together [4]. In 1999, D.Molodtsov introduced the concept of soft sets to solve complicated problems and various types of uncertainties [11]. P. K. Maji et al. studied the (Zadeh's) fuzzification of the soft set theory [8].

In the paper, we focus on **L**-fuzzification of the soft set theory, construct a framework to combine **L**-sets, rough sets, and soft sets all together, which leads to many interesting new concepts and results.

The paper is arranged into three parts, Section 3: **L**-soft sets, and Section 4: Rough **L**-soft sets. In Section 2, we give an overview of **L**-sets, soft sets and rough sets, which surveys Preliminaries.

2 Preliminaries

2.1 L-Sets

Suppose $\mathbf{L} = \langle L, \vee, \wedge, \otimes, \rightarrow, 0, 1 \rangle$ is a complete residuated lattice, X a universe set, an **L**-set in X is a mapping $\tilde{A} : X \rightarrow L$. $\tilde{A}(x)$ indicates the truth degree of "x belongs to $\tilde{A}$". We use the symbol L^X to denote the set of all **L**-sets in X. For instance: $\tilde{1}_X : X \rightarrow L$, $\tilde{0}_X : X \rightarrow L$ are defined as: for all $x \in X$, $\tilde{1}_X(x) = 1, \tilde{0}_X(x) = 0$, respectively. The negation operator is defined: for $\tilde{A} \in L^X$, $\tilde{A}^*(x) = \tilde{A}(x) \rightarrow 0$ for every $x \in X$. For more details, see [2].

Y. Tan, Y. Shi, and H. Mo (Eds.): ICSI 2013, Part I, LNCS 7928, pp. 530–539, 2013.

2.2 Soft Sets and Fuzzy Soft Sets

A soft set is an approximate description of an object precisely consisting of two parts, namely predicate and approximate value set.

Let X be an initial universe set and E_X (simple E) be a collection of all possible parameters with respect to X. Usually, parameters are attributes, characteristics, or properties of objects in X.

In [11], D.Molodtsov introduced the notion of a soft set as follows.

Definition 1. *A pair (F, A) is called a soft set over X if $A \subseteq E$, and $F : A \to 2^X$, where 2^X is the power set of X.*

Some researchers have studied soft sets, rough sets and fuzzy sets, see [1, 5, 6, 8, 9, 10, 11, 12, 13, 14]. In [8], P.K.Maji et al. defined the notion of a fuzzy soft set.

Definition 2. *A pair (F, A) is called a fuzzy soft set over X if $A \subseteq E$, and $F : A \to [0, 1]^X$, where $[0, 1]^X$ is the collection of all fuzzy sets on X.*

2.3 Rough Sets

In [7], suppose $(P, 0, 1, \vee, \wedge, ')$ is an atomic Boolean lattice, Q is the set of all atoms. For an arbitrary mapping $\varphi : Q \to P$, J. Järinven defined two rough approximation operators, for every $a \in P, x \in Q$,

$$N(a) = \vee\{x \mid \varphi(x) \leq a\}, H(a) = \vee\{x \mid a \wedge \varphi(x) \neq 0\}. \tag{1}$$

We generalized the method in [3], which includes [15] as a special case.

Definition 3. *Suppose X is a universe set, L^X is the set of all **L**-sets on X, $M = \{\{a/x\} \mid a \in L, a > 0, x \in X\}$ is the set of all singletons, $\varphi : M \to L^X$ is an arbitrary mapping, then we obtain two **L**-rough operators N_φ and H_φ: for every $\tilde{A} \in L^X, x \in X$,*

$$N_\varphi(\tilde{A})(x) = \bigvee_{\{a/x\} \in M} a \otimes S(\varphi(\{a/x\}), \tilde{A}),$$

$$H_\varphi(\tilde{A})(x) = \bigvee_{\{a/x\} \in M} a \otimes \rho(\varphi(\{a/x\}), \tilde{A}). \tag{2}$$

Where $\rho(\tilde{A}, \tilde{B}) = \bigvee_{x \in X} \tilde{A}(x) \otimes \tilde{B}(x)$.

Example 1. Suppose $X = \{x_1, x_2, x_3\}$, and $L = [0, 1]$ with $a \otimes b = min(a, b)$, $a \to b = 1$, if $a \leq b$; $a \to b = b$, if $a > b$. (Gödel Structure). Let $\varphi(\{a/x_i\}) = \{a \otimes 0.5/x_i\}$ for $i = 1, 2, 3$, then for $i = 1, 2, 3$, we have

$$S(\varphi(\{a/x_i\}), \tilde{A}) = \bigwedge_{y \in X} \varphi(\{a/x_i\})(y) \to \tilde{A}(y) = a \otimes 0.5 \to \tilde{A}(x_i),$$

$$\rho(\varphi(\{a/x_i\}), \tilde{A}) = \bigvee_{y \in X} \varphi(\{a/x_i\})(y) \otimes \tilde{A}(y) = [a \otimes 0.5] \otimes \tilde{A}(x_i). \tag{3}$$

For $\tilde{A} = \{0.6/x_1, 0.2/x_2, 0.7/x_3\}$,

$$
\begin{aligned}
N_\varphi(\tilde{A})(x_1) &= \bigvee_{\{a/x_1\}\in M} a \otimes S(\varphi(\{a/x_1\}), \tilde{A}) \\
&= \bigvee_{\{a/x_1\}\in M} a \otimes [a \otimes 0.5 \to \tilde{A}(x_1)] \\
&= \bigvee_{\{a/x_1\}\in M} a \otimes [a \otimes 0.5 \to 0.6] \\
&= \bigvee_{a\in L} a \otimes [a \otimes 0.5 \to 0.6] \\
&= 1.
\end{aligned}
\tag{4}
$$

Similarly, we obtain

$$
N_\varphi(\tilde{A}) = \{1/x_1, 0.2/x_2, 1/x_3\}, and\ H_\varphi(\tilde{A}) = \{0.5/x_1, 0.2/x_2, 0.5/x_3\}. \tag{5}
$$

3   L-Soft Sets

In the section, we generalize the notion of a soft set in fuzzy setting, define **L**-order, **L**-equivalence relation, and several operators on the set of all **L**-soft sets over X. The definitions are accompanied by examples.

Suppose X is a universe set, L^X is the set of all **L**-sets in X. Let E be a collection of all possible parameters with respect to X.

First, we define the notion of a soft set in fuzzy setting.

Definition 4. *A pair (F, A) is called an **L**-soft set over X if $A \subseteq E$, and $F : A \to L^X$, denoted by $\theta = (F, A)$.*

Clearly, when **L=2**, the above definition coincides with Definition 1; when **L=[0, 1]**, the above definition coincides with Definition 2.

Example 2. Follows Example 1. Let $E = \{t_1, t_2, t_3, t_4\}$, $A_1 = \{t_1, t_2, t_3\}$. $F_1 : A_1 \to L^X$, where $F_1(t_1) = \{0.7/x_1\}$, $F_1(t_2) = \{1/x_1, 0.5/x_2\}$, $F_1(t_3) = \{0.6/x_1, 0.2/x_2, 0.7/x_3\}$, clearly (F_1, A_1) is an **L**-soft set.

Let LS(X) be the set of all **L**-soft sets over X. On which, there exist two kinds of special elements: one is called a absolute soft set $(1_A, A)$, $\forall t \in A$, $1_A(t) = \tilde{1}_X$, denoted by $\Gamma_A = (1_A, A)$; the other is called a null soft set $(0_A, A)$, $\forall t \in A$, $0_A(t) = \tilde{0}_X$, denoted by $\Phi_A = (0_A, A)$.

Second, we introduce the relation **L**-order $\preceq$, and **L**-equivalence relation $\approx$ which correspond the relations $\tilde{\subseteq}, =$ in classical case [1, 8, 9, 11]. For two **L**-soft sets $\theta_1 = (F, A)$, $\theta_2 = (G, B) \in$LS(X),

$$
(\theta_1 \preceq \theta_2) = S(\theta_1, \theta_2) = \bigwedge_{t\in A} S(F(t), G(t)), (\theta_1 \approx \theta_2) = S(\theta_1, \theta_2) \wedge S(\theta_2, \theta_1). \tag{6}
$$

Example 3. Follows Example 2, (F_1, A_1) is an **L**-soft set. Let $A_2 = \{t_1, t_2, t_3, t_4\}$, $F_2 : A_2 \to L^X$, where $F_2(t_1) = \{0.4/x_1\}$, $F_2(t_2) = \{0.9/x_1, 0.5/x_2, 0.3/x_3\}$,

$F_2(t_3) = \{0.4/x_1, 0.2/x_2, 0.5/x_3\}$, $F_2(t_4) = \{1/x_1, 0.7/x_2, 0.6/x_3\}$, thus (F_2, A_2) is also an **L**-soft set. We obtain

$$S((F_1, A_1), (F_2, A_2)) = \bigwedge_{t \in A_1} S(F_1(t), F_2(t)) = 0.4,$$

$$S((F_2, A_2), (F_1, A_1)) = \bigwedge_{t \in A_2} S((F_2(t), F_1(t)) = 0. \tag{7}$$

Clearly, we have

$$\theta_1 \tilde{\subseteq} \theta_2 \Leftrightarrow S(\theta_1, \theta_2) = 1 \Leftrightarrow A \subseteq B, \text{ and } \forall t \in A, F(t) \subseteq G(t);$$

$$\theta_1 = \theta_2 \Leftrightarrow S(\theta_1, \theta_2) = 1, S(\theta_2, \theta_1) = 1 \Leftrightarrow A = B, and \ \forall t \in A, F(t) = G(t). \tag{8}$$

Thus $\langle\langle LS(X), \approx, \rangle, \preceq\rangle$ is an **L**-order set [2]. When **L**=**2**, the above definitions coincide with [11]; when **L**=[0, 1], the above definition coincides with [1, 8].

Example 4. Follows Example 2, (F_1, A_1) is a **L**-soft set. Let $A_3 = A_1$, $F_3 : A_3 \to L^X$, where $F_3(t_1) = \{0.4/x_1\}$, $F_3(t_2) = \{0.9/x_1, 0.5/x_2\}$, $F_3(t_3) = \{0.4/x_1, 0.2/x_2, 0.5/x_3\}$, thus (F_3, A_3) is also an **L**-soft set, and $(F_3, A_3)\tilde{\subseteq}(F_1, A_1)$.

Third, we introduce the union and the extended (restricted) intersection of two **L**-soft sets. In [8], P.K.Maji et al. defined the union of two fuzzy soft sets as follows.

Definition 5. *Suppose* $(F, A), (G, B) \in LS(X)$ *are two* **L**-soft sets, the union of (F, A) and (G, B) is an **L**-soft set (H, C), where $C = A \cup B$, and for $t \in C$,

$$H(t) = \begin{cases} F(t) & if \ t \in A - B \\ G(t) & if \ t \in B - A \\ F(t) \vee G(t) & if \ t \in A \cap B \end{cases} \tag{9}$$

and written as $(F, A)\tilde{\cup}(G, B) = (H, C)$.

About some properties of the union, we combine [1] Proposition 1, [8] Proposition 3.2 and [13] Proposition 2 as follows.

Proposition 1. *(1)* $(F, A)\tilde{\cup}(F, A) = (F, A)$,

(2) $(F, A)\tilde{\cup}(G, B) = (G, B)\tilde{\cup}(F, A)$,

(3) $((F, A)\tilde{\cup}(G, B))\tilde{\cup}(H, C) = (F, A)\tilde{\cup}((G, B)\tilde{\cup}(H, C))$

(4) $(F, A)\tilde{\subseteq}(F, A)\tilde{\cup}(G, B)$, and $(G, B)\tilde{\subseteq}(F, A)\tilde{\cup}(G, B)$

(5) $(F, A)\tilde{\subseteq}(G, B) \Rightarrow (F, A)\tilde{\cup}(G, B) = (G, B)$,

(6) $(F, A)\tilde{\cup}\Phi_A = (F, A)$, $(F, A)\tilde{\cup}\Gamma_A = \Gamma_A$.

In [8], P.K.Maji et al. also defined the intersection of two fuzzy soft sets, i.e., suppose $(F, A), (G, B) \in LS(X)$ are two **L**-soft sets, the intersection of (F, A) and (G, B) is also an **L**-soft set (K, D), where $D = A \cap B$, and for $t \in D$, $K(t) = F(t)$ or $G(t)$ (as both are the same **L**-set). But generally $F(t) = G(t)$ does not hold, and $A \cap B$ may be a empty set. So B. Ahmad, Athar Kharal, M.I.Ali and M.Shabir introduced a new definition, see [1] Definition 7, [12] Definition 3.3 and [13] Definition 10. In [12, 13], it is called the restricted intersection($\widetilde{\cap}$).

Definition 6. *Suppose* $(F, A), (G, B) \in LS(X)$ *are two* **L***-soft sets, such that* $A \cap B \neq \emptyset$, *the (restricted) intersection of* (F, A) *and* (G, B) *is also an* **L***-soft set* (K, D), *where* $D = A \cap B$, *and for* $t \in D$, $K(t) = F(t) \wedge G(t)$. *It denoted by* $(F, A) \widetilde{\cap} (G, B) = (K, D)$.

Example 5. Follows Example 3, we obtain $(H, C) = (F_1, A_1) \widetilde{\cup} (F_2, A_2)$, where $C = A_1 \cup A_2 = \{t_1, t_2, t_3, t_4\}$, and $H(t_1) = F_1(t_1) \vee F_2(t_1) = \{0.7/x_1\}$, $H(t_2) = F_1(t_2) \vee F_2(t_2) = \{1/x_1, 0.5/x_2, 0.3/x_3\}$, $H(t_3) = F_1(t_3) \vee F_2(t_3) = \{0.6/x_1, 0.2/x_2, 0.7/x_3\}$, $H(t_4) = F_2(t_4) = \{1/x_1, 0.7/x_2, 0.6/x_3\}$. Similarly, $(F_4, A_4) = (F_1, A_1) \widetilde{\cap} (F_2, A_2)$, where $A_4 = A_1 \cap A_2 = \{t_1, t_2, t_3\}$, and $F_4(t_1) = F_1(t_1) \wedge F_2(t_1) = \{0.4/x_1\}$, $F_4(t_2) = F_1(t_2) \wedge F_2(t_2) = \{0.9/x_1, 0.5/x_2\}$, $F_4(t_3) = F_1(t_3) \wedge F_2(t_3) = \{0.4/x_1, 0.2/x_2, 0.5/x_3\}$.

In [12, 13], M.I.Ali et al. defined a new intersection, which is called the extended intersection.

Definition 7. *Suppose* $(F, A), (G, B) \in LS(X)$ *are two* **L***-soft sets, the extended intersection of* (F, A) *and* (G, B) *is also an* **L***-soft set* (J, C), *where* $C = A \cup B$, *and for* $t \in C$,

$$
J(t) = \begin{cases} F(t) & \text{if } t \in A - B \\ G(t) & \text{if } t \in B - A \\ F(t) \wedge G(t) & \text{if } t \in A \cap B \end{cases} \tag{10}
$$

and written as $(F, A) \sqcap (G, B) = (J, C)$.

Example 6. Follows Example 3, we obtain $(J, C) = (F_1, A_1) \sqcap (F_2, A_2)$, where $C = A_1 \cup A_2 = \{t_1, t_2, t_3, t_4\}$, and $J(t_1) = F_1(t_1) \wedge F_2(t_1) = \{0.4/x_1\}$, $J(t_2) = F_1(t_2) \wedge F_2(t_2) = \{0.9/x_1, 0.5/x_2\}$, $J(t_3) = F_1(t_3) \wedge F_2(t_3) = \{0.4/x_1, 0.2/x_2, 0.5/x_3\}$, $J(t_4) = F_2(t_4) = \{1/x_1, 0.7/x_2, 0.6/x_3\}$.

Now, we introduce a new operator $\widetilde{\otimes}$ on LS(X).

Definition 8. *Suppose* $(F, A), (G, B) \in LS(X)$, *with* $A \cap B \neq \emptyset$, *the (restricted)*$\otimes$ *of* (F, A) *and* (G, B) *is an* **L***-soft set* (Y, D), *where* $D = A \cap B$, *and* $\forall t \in D$, $Y(t) = F(t) \otimes G(t)$, *denoted by* $(F, A) \widetilde{\otimes} (G, B) = (Y, D)$.

Example 7. Suppose $X = \{x_1, x_2, x_3\}$, and $L = [0, 1]$ with $a \otimes b = max(a + b - 1, 0)$, $a \to b = min(1 - a + b, 1)$ (Lukasiewicz Structure). (F_1, A_1) and (F_2, A_2) are two **L**-soft sets, see Example 3. $(F_1, A_1) \widetilde{\otimes} (F_2, A_2) = (F_5, A_5)$, where $A_5 = A_1 \cap A_2 = \{t_1, t_2, t_3\}$; and $F_5(t_1) = \{0.1/x_1\}$, $F_5(t_2) = \{0.9/x_1, 0/x_2\}$, $F_5(t_3) = \{0/x_1, 0/x_2, 0.2/x_3\}$.

Similarly, we also define the extended $\otimes$ of $(F, A), (G, B)$, it is an **L**-soft set (Z, C), where $C = A \cup B$, and $\forall t \in C$,

$$Z(t) = \begin{cases} F(t) & \text{if } t \in A - B \\ G(t) & \text{if } t \in B - A \\ F(t) \otimes G(t) & \text{if } t \in A \cap B \end{cases} \tag{11}$$

written as $(F, A) \underline{\otimes} (G, B) = (Z, C)$.

In Example 7, $(F_1, A_1) \underline{\otimes} (F_2, A_2) = (Z, C)$, where $C = A_1 \cup A_2 = \{t_1, t_2, t_3, t_4\}$, and $Z(t_1) = F_4(t_1)$, $Z(t_2) = F_4(t_2)$, $Z(t_3) = F_4(t_3)$, $Z(t_4) = F_2(t_4)$.

When $\otimes = \wedge$, the extended (restricted) $\otimes$ coincides with the extended (restricted) intersection, respectively. Clearly, we have,

Proposition 2. *(1)* $(F, A) \tilde{\cap} (F, A) = (F, A) \sqcap (F, A) = (F, A)$,

(2) $(F, A) \tilde{\cap} (G, B) = (G, B) \tilde{\cap} (F, A)$, $(F, A) \sqcap (G, B) = (G, B) \sqcap (F, A)$,

$\quad (F, A) \tilde{\otimes} (G, B) = (G, B) \tilde{\otimes} (F, A)$, $(F, A) \underline{\otimes} (G, B) = (G, B) \underline{\otimes} (F, A)$,

(3) $(F, A) \tilde{\otimes} (G, B) \tilde{\subseteq} (F, A) \tilde{\cap} (G, B) \tilde{\subseteq} (F, A) \sqcap (G, B) \tilde{\subseteq} (F, A) \tilde{\cup} (G, B)$,

$\quad (F, A) \tilde{\otimes} (G, B) \tilde{\subseteq} (F, A) \underline{\otimes} (G, B) \tilde{\subseteq} (F, A) \sqcap (G, B) \tilde{\subseteq} (F, A) \tilde{\cup} (G, B)$,

(4) $(F, A) \tilde{\cap} \Gamma_A = (F, A)$, $(F, A) \sqcap \Gamma_A = (F, A)$,

$\quad (F, A) \tilde{\otimes} \Gamma_A = (F, A)$, $(F, A) \underline{\otimes} \Gamma_A = (F, A)$,

(5) $(F, A) \tilde{\cap} \Phi_A = \Phi_A$, $(F, A) \sqcap \Phi_A = \Phi_A$,

$\quad (F, A) \tilde{\otimes} \Phi_A = \Phi_A$, $(F, A) \underline{\otimes} \Phi_A = \Phi_A$,

(6) $(F, A) \tilde{\otimes} (G, B) \tilde{\subseteq} (F, A) \tilde{\cap} (G, B) \tilde{\subseteq} (F, A)$,

$\quad (F, A) \tilde{\otimes} (G, B) \tilde{\subseteq} (F, A) \tilde{\cap} (G, B) \tilde{\subseteq} (G, B)$.

In [1], B. Ahmad and Athar Kharal defined arbitrary union and intersection of a family of fuzzy soft sets. As generalizations of the above operators, we define infinitely union, infinitely extended intersection operators on LS(X).

Definition 9. *On LS(X), the infinitely union of a system of* **L**-*soft sets* $\theta_i = (F_i, A_i)$ *is an* **L**-*soft set* (H, C), *where* $C = \bigcup_i A_i$, $H : C \to L^X$, *for every* $t \in C$,

$$H(t) = \bigvee_{i \in I_t} F_i(t), \text{ where } I_t = \{i \mid t \in A_i\}, \text{ denoted by } \tilde{\bigcup} \theta_i.$$

The infinitely extended intersection of a system of **L**-*soft sets* $\theta_i = (F_i, A_i)$ *is an* **L**-*soft set* (K, C), *where* $C = \bigcup_i A_i$, $K : C \to L^X$, *for every* $t \in C$,

$$K(t) = \bigwedge_{i \in I_t} F_i(t), \text{ where } I_t = \{i \mid t \in A_i\}, \text{ denoted by } \sqcap \theta_i.$$

Example 8. Follows Example 3. Suppose (F_i, A_i) is a system of **L**-soft sets, $i = 2k$, $A_i = \{t_1, t_2\}$, and $F_i : A_i \to L^X$, $F_i(t_1) = \{\underbrace{0.99 \cdots 9}_{i}/x_1, 0.7/x_2\}$,

$F_i(t_2) = \{0.5/x_1, 1/i x_2\}$; $i = 2k + 1$, $A_i = \{t_1, t_3\}$, and $F_i : A_i \to L^X$, $F_i(t_1) = \{1/i/x_1, 0.4/x_2, \underbrace{0.99 \cdots 9}_{i}/x_3\}$, $F_i(t_3) = \{0.5/x_1, \underbrace{0.99 \cdots 9}_{i}/x_2, 0.6/x_3\}$, where $k = 1, 2, 3, \cdots$. Then we have $\tilde{\bigcup}_i(F_i, A_i) = (F, A)$, where $A = \{t_1, t_2, t_3\}$, and $F : A \to L^X$, $F(t_1) = \{1/x_1, 0.7/x_2, 1/x_3\}$, $F(t_2) = \{0.5/x_1, 1/x_2\}$, $F(t_3) = \{0.5/x_1, 1/x_2, 0.6/x_3\}$. $\sqcap(F_i, A_i) = (G, B)$, where $B = \{t_1, t_2, t_3\}$, and $G : B \to L^X$, $G(t_1) = \{0.4/x_2\}$, $G(t_2) = \{0.5/x_1\}$ $G(t_3) = \{0.5/x_1, 0.9/x_2, 0.6/x_3\}$.

We obtain the following proposition about the infinity union, and the infinity extended intersection.

Proposition 3. *(De Morgan Law) Suppose* **L** *satisfies the law of double negation,* (F_i, A_i) *is a system of* **L**-*soft sets over* X, *then*

(1) $\sqcap_i(F_i, A_i)^c = (\tilde{\bigcup}_i(F_i, A_i))^c$, *(2)* $(\sqcap_i(F_i, A_i))^c = \tilde{\bigcup}_i(F_i, A_i)^c$.

We close the section by considering the influence of the distributivity law of **L** on the above two operators. If **L** satisfies the distributivity law, i.e., $a \wedge (b \vee c) = (a \wedge b) \vee (a \wedge c)$, $a \vee (b \wedge c) = (a \vee b) \wedge (a \vee c)$ hold for $a, b, c \in L$, we have

$$(F, A) \sqcap ((G, B) \tilde{\bigcup} (H, C)) = ((F, A) \sqcap (G, B)) \tilde{\bigcup} ((F, A) \sqcap (H, C)),$$

$$(F, A) \tilde{\bigcup} ((G, B) \sqcap (H, C)) = ((F, A) \tilde{\bigcup} (G, B)) \sqcap ((F, A) \tilde{\bigcup} (H, C)). \quad (12)$$

If **L** satisfies the join infinity distributivity law, i.e., $a \wedge \bigvee_{i \in I} b_i = \bigvee_{i \in I} a \wedge b_i$, where $a \in L$, $b_i \in L$ for $i \in I$, we have

$$(F, A) \sqcap \tilde{\bigcup}_{i \in I}(G_i, B_i) = \tilde{\bigcup}_{i \in I}(F, A) \sqcap (G_i, B_i). \quad (13)$$

Furthermore, If **L** satisfies the completely distributivity law, i.e., $\bigvee_{i \in I} \bigwedge_{j \in J} a_{ij} = \bigwedge_{\alpha:I \to J} \bigvee_{i \in I} a_{i\alpha(i)}$, where $a_{ij} \in L$ for $i \in I$, $j \in J$, we may obtain

$$\tilde{\bigcup}_{i \in I} \sqcap_{j \in J} (F_{ij}, A_{ij}) = \sqcap_{\alpha:I \to J} \tilde{\bigcup}_{i \in I}(F_{i\alpha(i)}, A_{i\alpha(i)}). \quad (14)$$

Note that I, J are two index sets.

In conclusion, suppose **L** satisfies the distributivity law (the join infinity distributivity law, the completely distributivity law), then LS(X) also satisfies the corresponding one with respect to the operators $\tilde{\bigcup}$ and $\sqcap$.

4   Rough L-Soft Sets

In the section, we define the two rough operators on LS(X) by means of N_φ, H_φ on L^X, and investigate some of their properties.

First, we define two rough operators N, H on LS(X) in following manner.

Definition 10. *For every $\theta = (F, A)$, for every $t \in A$, let*

$$F_*(t) : A \to L^X, \; F_*(t) = N_\varphi(F(t)),$$

$$F^*(t) : A \to L^X, F^*(t) = H_\varphi(F(t)). \tag{15}$$

*Then we obtain two **L**-soft sets $N(\theta) = (F_*, A)$, $H(\theta) = (F^*, A)$. The operators N, H are called the lower and upper rough approximations of **L**-soft sets. If $N(\theta) = H(\theta)$, the **L**-soft set θ is said to be definable; otherwise $(N(\theta), H(\theta))$ is called a pair of rough **L**-soft set.*

We present the following example.

Example 9. Let $(F, A) = (F_1, A_1)$ defined in Example 2, we may obtain,
$F_*(t_1) = N_\varphi(F(t_1)) = \{1/x_1\}, \; F_*(t_2) = N_\varphi(F(t_2)) = \{1/x_1, 1/x_2\},$
$F_*(t_3) = N_\varphi(F_1(t_3)) = \{1/x_1, 0.2/x_2, 1/x_2\};$ and
$F^*(t_1) = H_\varphi(F_1(t_1)) = \{0.5/x_1\}, F^*(t_2) = H_\varphi(F_1(t_2)) = \{0.5/x_1, 0.5/x_2\},$
$F^*(t_3) = H_\varphi(F_1(t_3)) = \{0.5/x_1, 0.2/x_2, 0.5/x_2\}.$ (See Example 1).

Then $N(\theta) = (F_*, A)$, $H(\theta) = (F^*, A)$ is the lower and upper approximations of $\theta = (F, A)$.

Next, we investigate some properties about N, H on LS(X).

In the classical case, N and H are monotone increasing, i.e., if $A \subseteq B$, $N(A) \subseteq N(B)$ and $H(A) \subseteq H(B)$ hold. For LS(X), from the point of view of graded approach, we will prove the two rough operators N, H are monotone increasing for the subsethood degrees, see Proposition 4(2) and (3).

Proposition 4. *(1) $N(\Gamma_A) = \Gamma_A, H(\Phi_A) = \Phi_A$,*
(2) $S(\theta_1, \theta_2) \leq S(N(\theta_1), N(\theta_2))$, (3) $S(\theta_1, \theta_2) \leq S(H(\theta_1), H(\theta_2))$.

Proof. We prove (2) only.

(2) For $\theta_1 = (F, A), \theta_2 = (G, B) \in$ LS(X), and every $x \in X$,

$$S(\theta_1, \theta_2) \otimes F_*(t)(x) = S(\theta_1, \theta_2) \otimes N_\varphi(F(t))(x)$$

$$= [\bigwedge_{t \in A} S(F(t), G(t))] \otimes [\bigvee_{\{a/x\} \in M} a \otimes S(\varphi(\{a/x\}), F(t))]$$

$$\leq S(F(t), G(t)) \otimes [\bigvee_{\{a/x\} \in M} a \otimes S(\varphi(\{a/x\}), F(t))]$$

$$= \bigvee_{\{a/x\} \in M} S(F(t), G(t)) \otimes a \otimes S(\varphi(\{a/x\}), F(t))$$

$$\leq \bigvee_{\{a/x\} \in M} a \otimes S(\varphi(\{a/x\}), G(t))$$

$$= N_\varphi(G(t))(x)$$

$$= G_*(t)(x). \tag{16}$$

So $S(\theta_1, \theta_2) \leq F_*(t)(x) \to G_*(t)(x)$, thus $S(\theta_1, \theta_2) \leq S(N(\theta_1), N(\theta_2))$ holds.

Clearly, for $\theta_1 = (F, A), \theta_2 = (G, B) \in LS(X)$, if $\theta_1 \tilde{\subseteq} \theta_2$, we also have $N(\theta_1) \tilde{\subseteq} N(\theta_2), H(\theta_1) \tilde{\subseteq} H(\theta_2)$.

In [7], the set of all lower approximations and the set of all upper approximations form complete lattices. For LS(X), we also obtain the following propositions.

Proposition 5. *Suppose $\{\theta_i \mid i \in I\} \subseteq LS(X)$, we have*

(1) $\tilde{\bigcup}_{i \in I} N(\theta_i) \tilde{\subseteq} N(\tilde{\bigcup}_{i \in I} \theta_i)$, *(2)* $N(\sqcap_{i \in I} \theta_i) \tilde{\subseteq} \sqcap_{i \in I} N(\theta_i)$,

(3) $H(\sqcap_{i \in I} \theta_i) \tilde{\subseteq} \sqcap_{i \in I} H(\theta_i)$.

Proposition 6. *Suppose **L** satisfies idempotency [2], then for $\theta_1, \theta_2 \in LS(X)$, we have*

(1) $H(\theta_1) \tilde{\otimes} H(\theta_2) = H(\theta_1 \tilde{\otimes} \theta_2)$, *(2)* $H(\theta_1) \underline{\otimes} H(\theta_2) = H(\theta_1 \underline{\otimes} \theta_2)$.

Proposition 7. $Q = \{H(\theta) \mid \theta \in LS(X)\}$ *is closed for $\tilde{\bigcup}$, that is, suppose* $\{\theta_i \mid i \in I\} \subseteq Q$, *then* $H(\tilde{\bigcup}_{i \in I} \theta_i) \in Q$.

Proof. Suppose $\{\theta_i = (F_i, A_i) \mid i \in I\} \subseteq Q$, let $\theta = (F, A) = \tilde{\bigcup}_{i \in I} \theta_i$, where $A = \bigcup_{i \in I} A_i$, and for every $t \in A$, $F(t) = \bigvee_{i \in I_t} F_i(t)$, where $I_t = \{i \mid t \in A_i\}$.

Then for every $x \in X$, and $t \in A = \bigcup_{i \in I} A_i$, we have

$$
\begin{aligned}
F^*(t)(x) &= H_\varphi(F(t))(x) \\
&= H_\varphi(\bigvee_{i \in I_t} F_i(t))(x) \\
&= [\bigvee_{\{a/x\} \in M} a \otimes \rho(\varphi(\{a/x\}), \bigvee_{i \in I_t} F_i(t))](x) \\
&= [\bigvee_{\{a/x\} \in M} a \otimes \bigvee_{i \in I_t} \rho(\varphi(\{a/x\}), F_i(t))](x) \\
&= [\bigvee_{i \in I_t} \bigvee_{\{a/x\} \in M} a \otimes \rho(\varphi(\{a/x\}), F_i(t))](x) \\
&= \bigvee_{i \in I_t} H_\varphi(F_i(t))(x) \\
&= \bigvee_{i \in I_t} F_i^*(t)(x). \qquad (17)
\end{aligned}
$$

Note 1. By the above propositions, we know that if **L** satisfies idempotency, Q is a semilattice with respect to $\tilde{\bigcup}$, and the minimal element is Φ_E, the greatest element is Γ_E.

In fact, Definition 3 may be seen as a special case of Definition 10. As shown in Section 2.2, M is the set of all singletons. If we choose M as the collection of all possible parameters, i.e., $E = M$; for every **L**-set $\tilde{A}$, $\tilde{A} = \{A(x)/x \mid x \in X\} \subseteq M$. For $\tilde{A}$, we consider $F : \tilde{A} \to L^X$, then $(F, \tilde{A})$ is an **L**-soft set, where for $t \in \tilde{A}$, $F(t) = \tilde{A}$. According to Definition 10, we also obtain Definition 3.

5   Conclusions

In the paper, first, we generalized the notion of a soft set in fuzzy setting, and defined **L**-order, **L**-equivalence relation, the union, the extended (restricted) intersection, the infinite union, the extended (restricted)infinite intersection, $\tilde{\otimes}$, $\otimes$ and complement on the collection of all **L**-soft sets. Second, we defined two rough operators on LS(X) induced by the two rough operators [3] on L^X, and investigated some of their properties.

References

1. Ahmad, B., Kharal, A.: On Fuzzy Soft Sets. Advances in Fuzzy Systems 2009, 1–6
2. Bĕlohlávek, R.: Fuzzy relational systems, Foundations and Principles. Kluwer, New York (2002)
3. Chen, X., Li, Q.: Construction of Rough Approximations in Fuzzy Setting. Fuzzy Sets and Systems 158, 2641–2653 (2007)
4. Dubois, D., Prade, H.: Rough fuzzy sets and fuzzy rough sets. International Journal of General Systems 17, 191–209 (1990)
5. Feng, F., Li, C., Davvaz, B., Ali, M.I.: Soft sets combined with fuzzy sets and rough sets a tentative approach. Soft Comput. 14(6), 899–911 (2010)
6. Feng, F., Liu, X.Y., Leoreanu-Fotea, V., Jun, Y.B.: Soft sets and soft rough sets. Information Sciences 181, 1125–1137 (2011)
7. Järvinen, J.: On the structure of rough approximations. In: Alpigini, J.J., Peters, J.F., Skowron, A., Zhong, N. (eds.) RSCTC 2002. LNCS (LNAI), vol. 2475, pp. 123–130. Springer, Heidelberg (2002)
8. Maji, P.K., Biswas, R., Roy, A.R.: Fuzzy soft sets. The Journal of Fuzzy Mathematics 9(3), 589–602 (2001)
9. Maji, P.K., Biswas, R., Roy, A.R.: Soft Set Theory. Computers and Mathematics with Applications 45, 555–562 (2003)
10. Meng, D., Zhao, X., Qin, K.: Soft rough fuzzy sets and soft fuzzy rough sets. Computers and Mathematics with Applications 62, 4635–4645 (2011)
11. Molodtsov, D.: Soft Set Theory First Results. Computers and Mathematics with Applications 37, 19–31 (1999)
12. Ali, M.I., Feng, F., Liu, X., Min, W.K., Shabir, M.: On some new operations in soft set theory. Computers and Mathematics with Applications 57, 1547–1553 (2009)
13. Ali, M.I., Shabir, M.: Comments on De Morgan's law in fuzzy soft sets. The Journal of Fuzzy Mathematics 18(3), 679–686 (2010)
14. Ali, M.I.: A note on soft sets, rough sets and fuzzy sets. Appl. Soft Comput. 11, 3329–3332 (2011)
15. Pawlak, Z.: Rough sets. International Journal of Computer and Information Science 11, 341–356 (1982)

A Circuit Generating Mechanism with Evolutionary Programming for Improving the Diversity of Circuit Topology in Population-Based Analog Circuit Design

Mei Xue and Jingsong He[*]

Department of Electronic Science and Technology,
University of Science and Technology of China, Hefei, China, 230027
jxue@mail.ustc.edu.cn, hjss@ustc.edu.cn

Abstract. This paper presents an analog circuit generating mechanism based on connecting point guidance existing in circuit netlist. With the proposed mechanism, the initial circuit topology can be a random netlist, and the evolutionary operation can be executed directly on connecting point. Also, the knowledge of graph theory is introduced for evaluating the degree of diversity of circuit structures. Experimental results show that the proposed mechanism is beneficial to improve the diversity of topology in population. In the case of no robustness evolution mechanism, the diversity of topology in population can improve the fault tolerance of population.

Keywords: Analog circuit, circuit topology, evolutionary programming.

1 Introduction

Analog circuit module plays an important role in many electronic systems. Unlike digital circuit design, analog circuit design doesn't have nature CAD tools. The task of analog circuit design is very complicate and the final result largely depends on the designers' knowledge and experience. In the past, researchers pay many attentions to analog circuit design automation with the application of artificial intelligence [1-9]. Koza et al. [1] propose a tree-coded scheme with genetic programming, by which the circuit topological structure is determined by the tree storage structure and connection-modifying functions. Lohn at el. [2] propose a linear-coded mechanism combining with genetic algorithm, which has been successfully used in evolutionary design of analogy filters and amplifiers. Grimbleby [3] propose a kind of netlist-based representation method which can directly generate circuit with a few restriction on the topology structures of circuits. Zebulum at el. [4] apply the netlist-based representation to the synthesis of circuits with three-terminal component. [5-6] limit the number of components, where the number of components is different from the length of chromosomes.

---

[*] Corresponding author.

Y. Tan, Y. Shi, and H. Mo (Eds.): ICSI 2013, Part I, LNCS 7928, pp. 540–547, 2013.

In all the above methods, encoding method presented in [1] is complicated, the connection type of component is set in advance in [2], and the circuit topological structure is determined by connection type. Since a certain amount of connections cannot include all kinds of topologies, the encoding method has limitation on circuit topological structure. The netlist-based encoding mechanism directly generates the topology of circuits by connecting points. Topologies generated by this method are more flexible, thereby the method is helpful to get rich topologies. However, the length of chromosomes in [3-6] is fixed. This situation is not conducive to the diversity of population.

This paper presents a connecting point guidance circuit generating mechanism. It constructs the topology of a circuit by all its connecting points. The number of components in a circuit is determined both by the number of connecting points and the number of components connected to every connecting point. It directly operates the connecting points by mutation operators during evolution. These mutation operators offer a more flexible approach to obtain rich topologies. Experimental results show that the connecting point guidance circuit generating mechanism is beneficial to improve the diversity of topologies in population. In the case of no robustness evolution mechanism, the diversity of circuit topologies in population can improve the fault-tolerance of population.

2   Circuit Generating

This new circuit generating mechanism is based on connecting point of circuits, where the number of components in a circuit is determined both by the number of connecting points and the number of components connected to every connecting point. In this way, the number of components of the candidate circuits will be changeable and flexible, on condition that these two parameters have not been fixed at the stage of initialization. Genetic operators are directly associated with encoding method, we adopt five kinds of mutation operators which specially designed for connecting point guidance circuits generating mechanism. These five mutation operators are complete to the evolutionary process, and can produce any circuit structure.

2.1   Netlist Generating

Netlist-based encoding method of circuit was firstly proposed by Grimbleby in [3]. The netlist-based encoding method we used here is different from previous ones. Our method is to control the number of components in a circuit by the number of connecting points and the number of components connected to every connecting point. It is clear, if these two parameters are fixed, the number of components in circuit are same in initialized population. As a solution, we can get diverse individuals by setting the range of the two parameters. We set a parameter MC to control the number of connecting point, and another parameter N_i to control the number of components that the i-th connecting point connected. When these two parameters are variable, the number of component in every circuit is unfixed. On the basic of embryonic circuit (see left in Fig. 1), 0 represents ground, 1 represents input and 255 represents output, connecting points in a circuit are [0,1,2,..., MC,255]. The length of the circuit is L ,

$$L = N * \left\lceil \frac{MC+2}{2} \right\rceil .$$
(1)

Assume that N of a circuit is from 2 to 5, then

$$2 * \left\lceil \frac{MC+2}{2} \right\rceil \leq L \leq 5 * \left\lceil \frac{MC+2}{2} \right\rceil .$$
(2)

If the range of MC is from 4 to 10, the length of the circuit is from 6 to 30.

The right figure in Figure 1 shows the map of a random netlist. We take a connecting point connect at least two not parallel components in whole circuit as guiding principle to avoid invalid circuits as much as possible. We use real number encoding technique to initialize population, e.g., 1 represents resistance, 2 represents capacitance, and 3 represents inductance. Thus, the value of every component can randomly be selected by its type and the corresponding range of value.

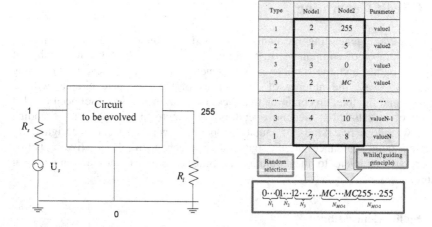

Fig. 1. Left: the embryonic circuit of analog filter ; right: the process of netlist generating

2.2 Mutation Operations

We adopt five mutation operators according to the actual requirements. These five mutation operators are enough to generating any kinds of topologies.

1. parameter change: select a component randomly, replace its value with a new one which randomly select from the range of parameter.

2. type change: select a component randomly, the type is changed to a different one and the parameter is changed simultaneously.
3. point change: randomly select two different connecting point in circuit netlist, then swap them.
4. component adding by point: select two different connecting points randomly in circuit netlist, connect the two connecting points by one component, its type and value are randomly selected.
5. component deletion: delete a component which is randomly selected from circuit netlist by merging the two connecting points into one.

In these five mutation operators, the last three are specially designed for connecting point, they can flexibly operate circuit topology, and can produce circuits with any topology. Mutation operation is likely to produce invalid individuals in evolutionary process , we take a connecting point connect at least two non-parallel components in an whole circuit as guiding principle to avoid invalid circuits as much as possible.

2.3   The Measurement of Circuit Topology

It is said that small world patterns exist in electronic circuits in [10]. Taking circuits as graphs, graphs with a small world structure are highly clustered but path length will be small [10]. Here, we choose clustering coefficient to describe circuit topology. We simplify circuits by series and parallel until their topologies contains only nodes, the node is the connecting point which connect three or more than three components.

Assume that node i has k edges to other nodes, the maximum number of edges between the k_i nodes is $C_{k_i}^2$, and the actual number of edges is E_i. The clustering coefficient of a circuit is the average of all nodes, i.e.,

$$C_3 = \frac{1}{N}\sum_{i=1}^{N}\frac{E_i}{C_{k_i}^2} \ . \tag{3}$$

From the perspective of geometry, the clustering coefficient of node i is the number of triangles connected to node i relative to the number of ternary group with the center of node i.

It can be found that the clustering coefficient above can only reflect the topology of triangle, not include the quadrilateral topology. For this reason, we have

$$C_4 = \frac{1}{N}\sum_{i=1}^{N}\frac{E2_i}{C_{k_i}^3} \ . \tag{4}$$

where, $E2_i$ denotes the number of quadrilaterals connected to node i. The measurement coefficient of topologies can be $C = \alpha C_3 + \beta C_4$, and $\alpha + \beta = 1$. In this paper, we have $\alpha = \beta = 0.5$.

3   Evolutionary Method

The search engine we adopt in this paper is evolutionary programming [11], it is a kind of evolutionary algorithm based on population. The number of initialized individuals is NP, every individual produce a child after mutation operation, every time one mutation operator is randomly selected to operate a individual. Each iteration we select ten percent of optimal individuals from parents and children, others are selected by stochastic tournament mechanism. The process of analog circuit design is described as follow:

1. Randomly generate the initial population of NP individuals with the connecting point guidance circuit generating mechanism, each circuit is stored in a two-dimensional array just like the right figure in Figure 1, X_n denote the n-th individual, $\forall n \in \{1, \cdots, NP\}$.

2. Evaluate the fitness of each individual in the population based on the fitness function in [1].

3. Each individual X_n, n = 1,...,NP, produce a offspring X_n by a kind of mutation operators which are randomly selected from five mutation operators with equal probability.

4. Calculate the fitness of each offspring X_n' as step 2, $\forall n \in \{1, \cdots, NP\}$.

5. Select $0.1 * NP$ optimal individuals out of $X_n \cup X_n'$, $\forall n \in \{1, \cdots, NP\}$, the other $0.9 * NP$ individuals are selected by q stochastic tournament mechanism in remaining population. For each individual a_i, $\forall i \in \{1, \cdots, 1.9 * NP\}$ in remaining population, q opponents are chosen randomly with equal probability, $score_i$ is the number of individuals whose fitness are bigger than a_i. After the process of comparison, sort the individuals by the rising trend of their scores. Then select $0.9 * NP$ individuals with high scores to the parents of the next generation.

6. Stop if the halting criterion is satisfied; otherwise, go to step 3.

4   The Experiments and Results

We take analog filter as experimental observation object. The goal of experiments is to automatically design analog lowpass filters using evolutionary programming. The transition zone is from $1000\ Hz$ to $2000Hz$, the ripple of passband is $30mV$ or less and the ripple of stopband is $1mV$ or less.

There are many kinds of faults, they can be roughly divided into two classes, parameter perturbations and topology failures. We use topology failure model to test every component in circuits with single point of short and disconnection damages[8]. A crucial-component in a circuit is the component whose failure will result in the losing of its original function.

4.1 Filter Design

Population size is set to 200 individuals, and each run proceeds for 400 generations. The results shown in Figure 2 are chosen from ten runs. Top left shows the best fitness through generation, fitness of four runs decrease gradually indicate that populations are convergent. Top right shows a schematic of lowpass filter, it is satisfied the specification and randomly selected from four runs. From the schematic we can see that the circuit evolved by the connecting point guidance circuit generating mechanism contains novel topology, rather than the traditional filters contain only T-shaped and π-shaped topology. Bottom left shows the change of circuits' length in population in four runs, we can see that the number of component in population is changeable and flexible. Bottom right shows the change of circuit structure in population in four runs, we can see that there are a variety of topologies in population. These experiments illustrate that the connecting point guidance circuits generating mechanism proposed by this paper can be a potential assistant method to design analog circuits. It is beneficial to improve the diversity of topology in population and produce novel circuit topology.

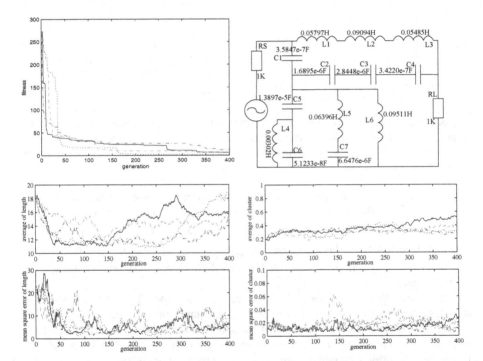

Fig. 2. The results of four runs which chosen from ten runs in lowpass filter design. Top left: Fitness of best circuit through generation in four runs; Top right: a circuit randomly selected which satisfy the specification. Bottom left: the average and the mean square error of circuits' length in four populations; Bottom right: the average and the mean square error of measurement coefficient in four populations; The same color represents the same population

4.2   Fault Tolerance Test

The statistics of fault tolerant performance of individuals in population can better indicate the fault tolerance of population, so we can choose some individuals which meet the design requirements from populations, then test their fault tolerance. Suppose $f_{tolerance}$ is the fault tolerance, it is the sum of single point fault model test in Eq. (5). Assume that the number of component in a circuit is n, and T is a threshold, if the fitness of single point fault model test is bigger than T, the weight of fitness is 0.1, else the weight is 0.01.

$$f_{tolerance} = \frac{w}{n} \sum_{i=1}^{n} fitness\ (i)\ .\tag{5}$$

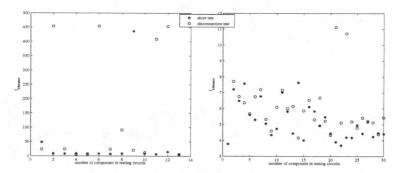

Fig. 3. The results of fault tolerance test. Left: the fault-tolerance performance of top right with single point of shorts and disconnection damages; Right: the fault tolerance of 30 individuals randomly selected in ten runs with single point of shorts and disconnection damages, and they all match the specification.

In Figure 3, left reflects the fault tolerance of top right in Figure 2, we can see that there is one crucial-component in circuit when we have single point short damage, and four crucial-components with disconnection damage. The $f_{tolerance}$ of single point short damage test is 3.7770, and the $f_{tolerance}$ of single point disconnection damage test is 11.6282, the circuit's fault tolerance to disconnection damage is worse than short damage.

We choose 30 different individuals which meet the specification from ten runs, then test them with single point of short and disconnection damages respectively. It can be seen from right in Figure 3 that most individuals have some degree of fault tolerance. The circuits are randomly selected, so we have reasons to believe that rich topologies can bring a certain degree of fault tolerance to population.

5   Conclusion

In this paper, we propose a connecting point guidance circuit generating mechanism, this mechanism coordinates with evolutionary programming can evolve analog circuits. This circuit generating mechanism almost has no limit on circuit topology, circuits evolved by this method have various number of components. The experimental results shows that the connecting point guidance circuit generating mechanism benefits to improve the diversity of topologies in population. In the case of no robustness evolution mechanism, rich topologies can bring a certain degree of fault tolerance to population. In the following work, we would use this circuit generating mechanism in the evolution of circuits which contain three-terminal components, even circuits with multiport modularized circuit, and make connecting point guidance circuit generating mechanism become a universal assist to netlist-based representation in analog design automation.

Acknowledgments. This work is supported by National Nature Science Foundation of China under Grant 60975051 and 61273315.

References

1. Koza, J.R., Benett III, F.H., Andre, D., et al.: Automated synthesis of analog electrical circuits by mean of genetic programming. IEEE Transactions on Evolutionary Computation 1(2), 109–128 (1997)
2. Lohn, J.D., Colombano, S.P.: A Circuit Representation Technique for Automated Circuit Design. IEEE Transactions on Evolutionary Computation 3(3), 205–219 (1999)
3. Grimbleby, J.B.: Automatic Analogue Network Synthesis Using Genetic Algorithms. In: The First Conference on Genetic Algorithms in Engineering Systems: Innovations and Applications, Sheffield, UK, pp. 53–58 (1995)
4. Zebulum, R.S., Pacheco, M.A., Vellasco, M.: Artificial Evolution of Active Filters: A Case Study. In: The First NASAIDOD Workshop on Evolvable Hardwar (1999)
5. Ando, S., Iba, H.: Analog circuit design with a variable length chromosome. In: Proceedings of the 2000 Congress on Evolutionary Computation, vol. 2, pp. 994–1001 (2000)
6. Goh, C., Li, Y.: GA automated design and synthesis of analog circuits with practical constraints. In: Proc. of the 2001 Congress on Evolutionary Computation, vol. 1, pp. 170–177 (2001)
7. Sapargaliyev, Y., Kalganova, T.: Constrained and unconstrained evolution of "LCR" low-pass filters with oscillating length representation. In: Proceedings of IEEE Congress on Evolutionary Computation, pp. 1529–1536 (2006)
8. Kim, K.J., Wong, A., Lipson, H.: Automated synthesis of resilient and tamper-evident analog circuits without a single point of failure. Genet Program Evolvable Mach 11, 34–59 (2010)
9. He, J., Zou, K., Liu, M.: Section-representation scheme for evolutionary analog filter synthesis and fault tolerance design. In: Advanced Computational Intelligence (IWACI), vol. 3, pp. 265–270 (2010)
10. Ferrer, I., Cancho, R., Janssen, C., Sole, R.V.: Topology of technology graphs: Small world patterns in electronic circuits. Physical Review 64, 461 (2001)
11. Yao, X., Liu, Y.: Fast evolutionary programming. In: The Fifth Annual Conference on Evolutionary Programming (1996)

An Evolutionary Game Model
of Organizational Routines on Complex Networks

Dehua Gao, Xiuquan Deng, and Bing Bai

School of Economics and Management, Beihang University, Beijing 100191, China
{smqinghua182,szzxbb}@163.com, dengxiuquan@buaa.edu.cn

Abstract. Organizational routines are collective phenomena with multiple actors involved in. In this paper, we introduce the evolutionary game theory into the study of organizational routines, and build up an evolutionary game model of organizational routines on complex networks. On the bases of this model, we provide a multi-agent based simulation via Swarm package. The results of our research show that: the evolutionary game theory – with the aid of multi-agent simulation as well – can afford us a general framework for formalized quantitative analysis, and provide an absolutely novelty directions for the researches of organizational routines based on mathematical methods.

Keywords: organizational routines, individual habits, small-world network, evolutionary game theory, agent-based simulation.

1 Introduction

Organizational routines are some specific generative systems that may produce "repetitive and recognizable patterns of interdependent actions", carried out by multiple actors within a pre-existing social context [1], [2], [3]. Some recent efforts have been paid to identify the micro-foundations of organizational routines that focused on the role of individual actors [4], and argued that there is a recursive relation between individual habits and routines in the organizational-level. On one hand, organizational routines are virtually some special structures of interlocking individual habits and they operate through the triggering of these individual habits. On the other hand, individual habits are the fundamental building blocks in the individual-level from which organizational routines emerge. They are the basis and individual analogue of organizational routines [5], [6].

While from the perspective of the evolutionary game theory, an organizational routine can be regarded as some evolutionary process for which all the individual actors devote to reaching a specific consensus or equilibrium among their interaction activities [7]. However, what do the internal dynamics work during the evolution of organizational routines? In this paper, we try to introduce the Evolutionary Game Theory (EGT) into our study, and discuss the evolutionary trajectories of organizational routines on one of a specific type of complex networks – i.e., the small-world networks. The paper is organized as follows: In the second section, an

Y. Tan, Y. Shi, and H. Mo (Eds.): ICSI 2013, Part I, LNCS 7928, pp. 548–555, 2013.
© Springer-Verlag Berlin Heidelberg 2013

evolutionary game model of organizational routines is given as the theoretical foundation of our study. In the third section, an agent-based simulation model is build up, and the simulation results are discussed on the small-world networks. And the forth section is the conclusion.

2   Design of the Model

2.1   Complex Networks for Individual Actors Involved in the Routine

Organizational routines are collective phenomena carried out by distributed multiple actors that may be located in different places and belong to different organizational units. These actors act depending on their own knowledge contexts, capabilities, and the roles they play during the implementation of the routines, and linked with each other by interactions – i.e., all the individual actors involved in the routine are interacting with each other on some type of complex networks. And network topology plays a crucial role in determining the system's dynamic properties [8], [9], [10]. In general, there are mainly three kinds of complex network to be considered – namely, local, small-world, and random networks. These different types of networks may share the same number of actors (i.e., nodes), but differ in terms of the pattern of connections, sees as in fig. 1[8]. The first is the regular network, which is indeed a lattice that consists of the undetermined amount of actors who are placed on the ring-shaped networks. For example, the most frequently used structure is that within which every actor located has two neighbors in both directions that they are certain to play with. That is, there are four connections that are certain for each of the actors playing on the regular lattice.

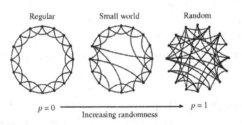

Fig. 1. Regular, small-world, and random networks

The second is the random network which is placed at the opposite end of the spectrum from the regular network. On a random network, individual actors are randomly connected with each other, and the other actors they interact are not limited to the immediate neighborhood but are randomly drawn from the entire population. That is, Actors play this type of network structure are not clustered but may have short average path lengths.

The third is the small-world network that can be created from a regular lattice by randomly rewiring each of the links with some probability p. The effect rewiring is to substitute some short-range connections with long-range ones, which may reduce drastically the average distance between randomly chosen pair of sits and produce a small world phenomenon characteristic to many social networks. Thus, small-world networks are highly clustered and the minimum path length is relatively short. This means that most of the actors are still connected to their nearest neighbors, but now a very small number of new links can directly connect two actors that are far apart from each other. However, the probability p plays an important role in dominating the structure of the networks. That is, if $p = 1$ (which means that all connections are replaced), the rewiring process may yield the well-investigated random graphs. While in the limit $p \rightarrow 0$, the depicted structure is equivalent to the regular lattice.

2.2   Theoretical Foundation of the Evolutionary Game

We suppose that there are only two groups of actors involved in a routine – namely, group A and group B – the number of actors each of the group consists is *agent1Num* and *agent2Num*, respectively. Actors from each of the groups have only two behavioral strategies – i.e., the one is of maintaining the current habitual behavior, and the other is of searching for a new way of behaviors to follow. We may denote the former behavioral strategy as $x1$, and the latter as $x2$. Further, we have that:

If actors of both the two groups play behavioral strategy $x1$, there is no additional efforts needed, and they get the payoff as $V1$ and $V2$, respectively. Similarly, if they both play strategy $x2$, then an excess return ΔV would be attained (that is, $\alpha \Delta V$ for actors from group A and $(1-\alpha)\Delta V$ for actors from group B, respectively, where α is a coefficient, $0 \le \alpha \le 1$), and a cost C should be paid for their searching activities (that is, βC for actors from group A and $(1-\beta)C$ for actors from group B, respectively, where β is a coefficient, $0 \le \beta \le 1$).

However, If actors of both the two groups play different behavioral strategies, then the ones searching for new behaviors (playing strategy $x2$) may gain an additional return (denoting as ΔV_1 for actors from group A and ΔV_2 for actors from group B, respectively), and pay for the cost (βC for actors from group A and $(1-\beta)C$ for actors from group B, respectively). While the other ones adopting strategy $x1$ would gain an excess return (denoting as $EV1$ for actors from group A and $EV2$ for actors from group B, respectively), but no costs should be paid – i.e., the hitchhike phenomenon occurring. The payoff matrix of this game is given as in table 1.

Table 1. The payoff matrix of the evolutionary game

		Actors of group B	
		$x1$	$x2$
Actors of group A	$x1$	$V_1,\ V_2$	$V_1 + EV_1,\ V_2 + \Delta V_2 - (1-\beta)C$
	$x2$	$V_1 + \Delta V_1 - \beta C,\ V_2 + EV_2$	$V_1 + \alpha\Delta V - \beta C,\ V_2 + (1-\alpha)\Delta V - (1-\beta)C$

Note: all the parameters in table 1 are bigger than 0, and there are relations as that: $\Delta V > C$, $\Delta V_1 > \beta C$, and $\Delta V_2 > (1-\beta)C$.

Let θ_A $(0 \le \theta_A \le 1)$ and θ_B $(0 \le \theta_B \le 1)$ be the proportion of actors of both the group A and B that play behavioral strategy $x2$, respectively. Then, the proportion of actors of these two groups that play strategy $x1$ is $(1 - \theta_A)$ and $(1 - \theta_B)$, respectively. Then, we can get the expect payoffs for actors of group A that playing strategy $x1$ and $x2$ as that:

$$\pi_A^{(x_1)} = (1 - \theta_B)V_1 + \theta_B(V_1 + EV_1) \tag{1}$$

$$\pi_A^{(x_2)} = (1 - \theta_B)(V_1 + \Delta V_1 - \beta C) + \theta_B(V_1 + \alpha \Delta V - \beta C) \tag{2}$$

And the average payoff for actors from group A is as that:

$$\begin{aligned}
\bar{\pi}_A &= (1 - \theta_A)\pi_A^{(x_1)} + \theta_A \pi_A^{(x_2)} \\
&= V_1 + \theta_B EV_1 + \theta_A[\Delta V_1 - \beta C + \theta_B(\alpha \Delta V - \Delta V_1 - EV_1)]
\end{aligned} \tag{3}$$

Accordingly, the expect payoffs for actors of group B that playing strategy $x1$ and $x2$ are as that:

$$\pi_B^{(x_1)} = (1 - \theta_A)V_2 + \theta_A(V_2 + EV_2) \tag{4}$$

$$\pi_B^{(x_2)} = (1 - \theta_A)(V_2 + \Delta V_2 - (1 - \beta)C) + \theta_A(V_2 + (1 - \alpha)\Delta V - (1 - \beta)C) \tag{5}$$

And the average payoff for actors of group B is as that:

$$\begin{aligned}
\bar{\pi}_B &= (1 - \theta_B)\pi_B^{(x_1)} + \theta_B \pi_B^{(x_2)} \\
&= V_2 + \theta_A EV_1 + \theta_B[\Delta V_2 - (1 - \beta)C + \theta_A((1 - \alpha)\Delta V - \Delta V_2 - EV_2)]
\end{aligned} \tag{6}$$

2.3   Evolutionary Mechanisms of the Routine

There are typically three mechanisms for actors to evolve their behavioral strategies. The first is by incentives. All the actors involved in the routine are bounded rational – i.e., they only have limited information and constrained computational capacities [11], and they consider the pursuit of maximum returns as their basic motives for interacting with each other. This means that an actor may change his behavioral strategy if and only if the payoff he gains is less than both of the expected and the average one of the whole group he belongs to. That is, for any actor (i, j) from group i ($i = A, B$; $1 \le j \le agent1Num$ while $i = A$, or $1 \le j \le agent2Num$ while $i = B$), at the time t, he would choose a partner from the other group randomly, and earn his payoff as $U_{i,j}^{(t)} = \pi_i^{(x_k)}$ (where, $k = 1, 2$). And we have that:

(1) If the current behavioral strategy is $x1$, then the actor would change to $x2$ at the time $t + 1$ if and only $U_{i,j}^{(t)} < \min\{\pi_i^{(x_1)}, \bar{\pi}_i\}$

(2) If the current behavioral strategy is $x2$, then he would change to $x1$ at the time $t + 1$ if and only if $U_{i,j} < \min\{\pi_i^{(x_2)}, \bar{\pi}_i\}$

The second is by imitation [6]. According to [12], we adopt the following formula to describe this mechanism as that:

$$p_{i,jk}^{(t)} = \frac{1}{1 + \exp\left[\left(U_{i,j}^{(t)} - U_{i,k}^{(t)}\right)/\tau\right]} \tag{7}$$

Where, the coefficient $p_{i,jk}^{(t)}$ represents the probability of actor j from group i to imitate the behavioral strategy from its nearest neighbor k at the time $t + 1$; Ui,j and Ui,k are the payoffs of the actor j and k of the group i, respectively; and the parameter τ is a constant given arbitrarily.

The third is by improvisation. Organizational routines are not just simply followed or reproduced – rather, individual actors have choices between whether to do so, or whether to amend the routine [1], [13], [14], [15]. That is, actors involved in organizational routines are allowed to take up one of their behavioral strategies randomly with a very small given probability *improve_P*.

3   A Multi-Agent Based Simulation Study

3.1   Why the Multi-Agent Based Simulation

As the complex and dynamic characteristics inherited in social and economic phenomena, traditional tools – such as mathematics and experimental techniques *etc.* – are often becoming powerless in their applications. While the simulation method – the multi-agent based simulation especially– with its arising and development in the past few years, is being more and more popular in studies of social complex and dynamic problems, and earns vast applications [16], [17].

The multi-agent based simulation method takes the systems' macro-phenomena as the results of interactions between and within the micro-individuals [17], [18]. By defining the active agents and their environments, as well as describing rules of the interactions between and within these agents, we can execute simulation experiments via computer and reveal some emergences in the macro-level. Based on the overview about lots of works in which multi-agent based models have been used, [17] showed that there were associated relationships between multi-agent based models and the simulation of social phenomena with complex and dynamic characteristics inherited. Therefore, multi-agent based simulation can be regarded as an effective method for studying the emergence process from individuals to social organizations.

3.2   Simulation Results and Discussion

In this paper, we assume that some fixed relationships have long been formed among all the individual actors involved in the routine – that is, actors from each of the groups have connections only with a certain number of others (which we call as their nearest neighbors), and are organized with a given network typology – especially, the small-world network is considered in our model (where, we let the coefficient p that dominating network typologies in fig. 1 as that $p = 0.005$). While every actor of group A would be randomly linked with 1 to *agent1CopNum* numbers of actors from group B, (which we call as pairs of copartners), and every actor in group B would be randomly linked with 1 to *agent2CopNum* number of actors from group A, likewise.

Then, by considering individual actors involved in the routine as "active agents", we realize the multi-agent based simulation of this evolutionary game model via the Swarm Package [19]. Let the coefficients of the payoff matrix in table 1 as follows: $agent1Num = agent2Num = 120$, $agent1CopNum = agent2CopNum = 4$, $V1 = V2 = 10$, $\Delta V = 3.4$, $\Delta V_1 = \Delta V_2 = 2$, $EV1 = EV2 = 1.2$, $C = 1$, $\alpha = \beta = 0.5$, $improve\_P = 0.001$, $\tau = 0.01$, we can obtain the simulation results as shown in fig. 2 and fig. 3.

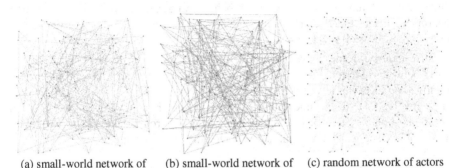

(a) small-world network of   (b) small-world network of   (c) random network of actors
individual actors of group A   individual actors of group B   between group A and B

Fig. 2. Network typology of the individual actors involved in the routine

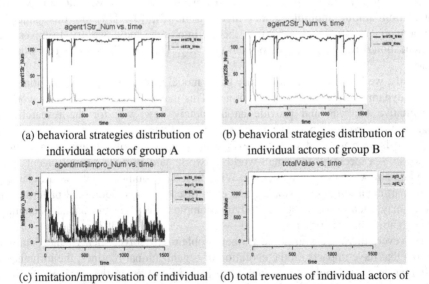

(a) behavioral strategies distribution of   (b) behavioral strategies distribution of
individual actors of group A   individual actors of group B

(c) imitation/improvisation of individual   (d) total revenues of individual actors of
actors of both group A and B   both group A and B

Fig. 3. The simulation results with the given scenario

From the simulation results, we have the conclusion that: at the beginning of the simulation, as all the individual actors of both the group A and B are taking the behavioral strategy $x1$, the game is at a relatively stable state. However, some actors may search for and adopt the behavioral strategy $x2$ occasionally due to the improvisation activities, and obtain much more revenues (when $\Delta V_1 > \beta C$ and $\Delta V_2 > (1 - \beta)C$). Further, the nearest neighbors associated with these actors would imitate and learn to adopt the new strategy with the motivation of pursuing much more payoffs, which may lead to the spread of behavioral strategy $x2$ within the whole group of both A and B. Simultaneously, if there are some actors of one group that take the behavioral strategy $x1$, then, their copartners from the other group are also tending to adopt the strategy $x2$ (when $\Delta V_1 > \beta C$ and $\Delta V_2 > (1 - \beta)C$). While actors of this group change their behavioral strategies from $x1$ to $x2$, their copartners from the other group would also tend to choose $x2$ (when $\alpha\Delta V - \beta C > EV_1$ and $(1 - \alpha)\Delta V - (1 - \beta)C > EV_2$).

Thus, underlying these two dynamical mechanisms – i.e., the one is of playing games between individual actors of both the two groups A and B, and the other is of imitating and learning between these actors within the same group – all the individual actors involved in the routine interact with each other, and mutually result in an evolutionary stable state that actors of both the two groups take $x2$ as a common strategy. This implies that the routine is evolved in the organizational-level.

4   Conclusions

Organizational routines are collective phenomena that come out through repeated interaction among multiple actors involved in [2], [3], [14]. In this paper, by introducing the evolutionary game theory into our study, we try to investigate organizational routines from a "bottom-up" way. The contributions of our work are that:

Firstly, we build up an evolutionary game model as a formalized description of the micro-dynamics of organizational routines. This may afford a general framework for quantitative analysis and provide an absolutely new way for the researches of organizational routines based on mathematical methods.

Secondly, we try to discuss the evolution of organizational routines on complex networks – especially, the small-world network, and provide a multi-agent based simulation model via Swarm package. With the aid of computer tools, we can develop some particular simulation studies, and even discover some additional patterns. And these, may shed some light on studying complex and dynamic social phenomena through artificial experiments.

However, there are still some further problems to solve, such as the details of factors that have impacts on the evolution of organizational routines, the influence of different network typologies of individual actors involved in the routines, the inter-relationships between different kinds of individual behaviors and their complicated interactions, and so forth. All of these may hint some directions of our future work.

Acknowledgements. The research work in this paper is sponsored by National Natural Science Foundation of China (No. 70872008).

References

1. Feldman, M.S., Pentland, B.T.: Reconceptualizing Organizational Routines as a Source of Flexibility and Change. Admin. Sci. Quart. 48, 94–118 (2003)
2. Pentland, B.T., Feldman, M.S.: Organizational Routines as a Unit of Analysis. Ind. Corp. Change 14, 793–815 (2005)
3. Pentland, B.T., Feldman, M.S.: Designing Routines: on the Folly of Designing Artifacts, While Hoping for Patterns of Action. Inf. Organ 18, 235–250 (2008)
4. Abell, P., Felin, T., Foss, N.J.: Building Micro-Foundations for the Routines, Capabilities, and Performance Links. Manag. Decis. Econ. 29, 489–502 (2008)
5. Knudsen, T.: Organizational Routines in Evolutionary Theory. In: Becker, M.C. (ed.) Handbook of Organizational Routines, pp. 125–151. Edward Elgar, Cheltenham (2008)
6. Hodgson, G.M.: The Concept of a Routine. In: Becker, M.C. (ed.) Handbook of Organizational Routines, pp. 15–30. Edward Elgar, Cheltenham (2008)
7. Gao, C., Chen, Y.: Research on the Evolution of Enterprise Routines: Motives, Paths and Model Construction. J. Bus. Econ. 3, 19–26 (2012) (in Chinese)
8. Watts, D.J.: Small Worlds: The Dynamics of Networks between Order and Randomness. Princeton University Press, Princeton (1999)
9. Barabâsi, A.L., Albert, R., Jeong, H.: Mean-Field Theory for Scale Free Random Networks. Phys. A. 272, 173–187 (1999)
10. Dorogovtsev, S.N., Mendes, J.F.F.: Evolution of Networks: From Biological Nets to the Internet and WWW. Oxford University Press, Oxford (2003)
11. Cybert, R.M., March, J.G.: A Behavioral Theory of the Firm. Prentice-Hall, New York (1963)
12. Hauert, C., Szabo, G.: Game Theory and Physics[J]. Amer. J. Phys. 73, 405–414 (2005)
13. Feldman, M.S.: Organizational Routines as a Source of Continuous Change. Organ. Sci. 11, 611–629 (2000)
14. Becker, M.C.: Organizational Routines: A Review of the Literature. Ind. Corp. Change 13, 643–677 (2004)
15. Becker, M.C.: The Concept of Routines: Some Clarifications. Camb. J. Econ. 29, 249–262 (2005)
16. Axelrod, R.: The Complexity of Cooperation: Agent-based Models of Competition and Collaboration. Princeton University Press, New Jersey (1997)
17. Gilbert, N.: Agent-based Social Simulation: Dealing with Complexity (2004), http://cress.soc.surrey.ac.uk/.../ABSS%20%20dealing%20with%20complexity-1-1.pdf
18. Gilbert, N., Terna, P.: How to Build and Use Agent-Based Models in Social Science. Mind Soc.: Cognitive Stud. Econ. Soc. Sci. 1, 57–72 (2000)
19. Swarm Development Group (SDG), http://www.swarm.org

A Novel Negative-Correlation Redundancy Evolutionary Framework Based on Stochastic Ranking for Fault-Tolerant Design of Analog Circuit

Chao Lin and Jingsong He[*]

Department of Electronic Science and Technology, University of Science
and Technology of China, Hefei, China
chaolin@mail.ustc.edu.cn, hjss@ustc.edu.cn

Abstract. The fault-tolerant evolutionary design based on negative-correlation redundancy technique is an effective way to improve the fault-tolerance of analog circuits with uncertain faults. In the existing negative-correlation redundancy evolutionary framework (ENCF), the negative-correlation penalty coefficient plays an important role, and it affects the performance of ENCF greatly. However, the value of the negative-correlation penalty coefficient is heavily dependent on the experience of designers. In this paper, we propose a new negative-correlation redundancy evolutionary framework based on stochastic ranking strategy. In order to make comparisons with the existing researches, we employ analog filter as a design example. Experimental results show that the framework proposed in this paper can generate negatively correlated redundancies without specifying the penalty coefficient, and it shows a relatively high ability to convergence compared to ENCF.

Keywords: Negative-correlation, analog circuit, fault-tolerant, stochastic ranking, genetic algorithm.

1 Introduction

Analog circuit is a necessary part of modern electronic systems. The fault-tolerant design of analog circuit is especially important when it is working in harsh environment, such as the battle field and outer space. Usually, designers need to prepare multiple redundant circuits instead of single circuit to an electronic system to increase its fault-tolerant performance. One of the key factors of this heterogeneous redundancy approach is how to generate diverse analog redundant modules with different structures. Therefore, the research of designing diverse analog redundant modules effectively has the potential value.

There are only a few works on the fault-tolerant circuits evolutionary design based on redundancy technique. T. Schnier and X. Yao [1] used negative correlation approach to make individuals of population as diverse as possible and use majority mechanism to combine the evolved multiple digital circuits to generate fault-tolerant

---

[*] Corresponding author.

Y. Tan, Y. Shi, and H. Mo (Eds.): ICSI 2013, Part I, LNCS 7928, pp. 556–563, 2013.

circuit. Liu and He [2], [3], [4] proposed ENCF to design negatively correlated analog circuits by the negative correlation information communication among three different populations, and chosen the best circuit as the output of the ensemble, experimental results show that this method can improve the robustness-generation ability of analog circuits with uncertain faults. Kim et al. [5], [6] employed evolutionary strategy (single population and multi-population, respectively) to design multiple redundant circuits and used the weight summing circuit to build fault-tolerant ensemble circuits. Chang and He [7] added structure space crowding factor and genotype similarity penalty into fitness function to evolve multiple analog redundant modules diversely. The negatively correlated redundancy evolutionary technique proposed in [3] is a promising way for fault-tolerant evolutionary design of analog circuits with uncertain faults.

The task of negative-correlation redundancy evolutionary framework is generating analog redundant modules, and they are negatively correlated each other in the frequency region. The existing ENCF framework proposed by Liu and He [3] is able to evolve the negatively correlated analog redundancies automatically. In this framework, an interactive-evaluate strategy is used, and the interactive-evaluation fitness function is defined by the sum of the circuit function and the negative-correlation penalty function. A negative-correlation penalty coefficient is used to adjust the weight of negative-correlation degree to the circuit function, and it influences the performance of ENCF greatly [2], [4]. But its value is hard to determine, in order to find a proper negative-correlation penalty coefficient value, a predefined monotonically non-decreasing sequence [0.25 0.5 0.75 1] has to be used in ENCF, and a trial-and-error process must be used in this situation. Although this penalty function method used in ENCF is a suitable approach, but the value of negative-correlation penalty coefficient is heavily dependent on the experience of the designers.

In this paper, we try to introduce the idea of stochastic ranking [8] into the problem of designing the negatively correlated analog redundancies, and propose a new negative-correlation redundancy evolutionary framework based on stochastic ranking strategy. Our experimental results on the low-pass filter show that the framework can generate negative-correlation analog redundancies without specifying the negative-correlation penalty coefficient. Compare to ENCF, this framework can avoid the problem of determine the hard-to-set value of penalty coefficient, and it shows a relatively high ability to convergence to the negatively-correlated analog redundancies.

2   Negatively Correlated Analog Redundancies Design

There are two parts in this section. Firstly, we briefly analyze the existing ENCF framework; Then we will explain why we introduce the stochastic ranking into the problem of designing negatively correlated analog redundancies and show the detail description of the new negative-correlation evolutionary framework proposed in this paper.

2.1   Analysis of the Existing ENCF Framework

Inspired by the advantages of negative correlation strategy in the neural network ensemble systems [9], [10], Liu and He [3] firstly introduced negative correlation into the field of fault-tolerant of analog circuit. The negative-correlation define in analog circuit is based on the outputs of analog circuits, that is using the output error between the under-design circuit and the design goal to define the negative-correlation among three different circuits. In ENCF, the fitness function given by Eq. (1) guides the evolution of the negative-correlation analog modules. A negative-correlation penalty coefficient λ is used to adjust the weight of negative-correlation degree to the circuit function. The value of the negative-correlation is hard to determine. If the negative-correlation penalty coefficient is not chosen well, analog redundancies may meet their functional requirements but they are not negatively correlated.

$$Fitness = Fit + \lambda * P .\tag{1}$$

Where *Fit* denotes the circuit function as the objective value given by Eq. (2).

$$Fit = \sum_{n=1}^{N}(f(x_n) - F(x_n)) .\tag{2}$$

P denotes the negative-correlation degree as the penalty term given by Eq. (3).

$$P = \sum_{i=1}^{N}((f_i(x_n) - F_i(x_n)) * \sum_{j=1,i\neq j}^{M}(f_j(x_n) - F_j(x_n))) .\tag{3}$$

Where N is the number of the whole sampling points, M is the number of the analog redundancy, $f_i(x_n)$ denotes the actual circuit output of the *ith* sampling, $F_i(x_n)$ denotes the design goal of the *ith* sampling.

According to Eq. (1), different negative-correlation penalty coefficients define different fitness functions. A fit individual under one fitness function may not be fit under another different fitness function. Different values of negative-correlation penalty coefficients represent different weights in objective value and negative-correlated degree. When the objective value of the best individual in the population falls down to a very small value, this weighted summing method could bring bad influence. The individuals with better performance cannot be separated from those individuals with worse performance but higher negative-correlation value. In the researches on numerical constrained optimization, researchers use stochastic ranking [8] method to solve this problem. Finding a near-optimal negative-correlation penalty coefficient is equivalent to ranking individuals in a population adaptively.

Thus, we transform the problem of designing negatively correlated redundant modules for robust analog circuit into how to rank individuals according to the function requirements of the analog redundancies and their negative-correlation degree in the frequency region. Thus, the fitness value representing the circuit function can be isolated with the negative-correlation degree, the mutual interference between the two variables can be prevented, and the algorithm performance can become better. We propose a new evolutionary framework based on the stochastic ranking strategy, and the framework is described in the next section.

2.2   Detail Description of the New Framework

Designing negatively-correlated analog redundancies is based on analog circuit evolution design. Before evolving analog redundancies, we should encode analog circuits. We use the linear representation method proposed in [11], which is an effective circuit representation for analog circuit automatic design. We choose genetic algorithm as the evolution engine, which has been demonstrated to be an efficient evolutionary algorithm [3], [11]. In the evolutionary framework proposed in this paper, the key strategies are double population strategy, candidate strategy and the stochastic ranking strategy.

The process of the new negative-correlation evolutionary algorithm is described as follows:

Step1: Initial two populations randomly. Then evaluate all the individuals according to Eq. (2), and choose the individual with best fit as the candidate individual for each population.

Step2: The two populations evolve simultaneously by the genetic operations (selection, crossover and mutate), evaluate the new populations and compute the correlation degree between all the individuals and the candidate in another population according to Eq. (3), respectively .

Step3: Use the stochastic ranking to sort all the individuals in the two population simultaneously.

Step4: Choose some individuals into the next generation, renew the two populations' candidates, respectively. Go back to Step2.

In Step 3: Stochastic ranking is the most important factor in the new evolutionary framework proposed in this paper. The detail usage of stochastic ranking is that given any pair of two adjacent individuals, the ratio of comparing them (in order to determine which is fitter) according to the function of the circuit is 1, if they are negatively-correlated; Otherwise, it is P_f. P_f is introduced here to use only the circuit function for comparisons in ranking in the feasible regions (that is the negative correlation region) of the search space. That is when the analog redundancies are negatively-correlated, we just compare their circuit function, the smaller the circuit function fitness value the better; Otherwise, if rand() < P_f, we still compare the function of the circuits; If rand() > P_f, we compare the correlation degree of the analog redundancies, the more negative the better.

Compared to ENCF, the most outstanding feature of the new framework proposed in this paper is that it employs stochastic ranking strategy to balance the circuit function and the negative correlation degree stochastically, and it don't need to determine the value of negative-correlation penalty coefficient.

3   Experiments and Result Discussions

Experiments are employed to value whether the framework in this paper can generate negative-correlation analog redundant modules without a negative-correlation penalty coefficient. In order to make comparisons with ENCF, we use the typical low-pass filter design problem shown in [3] as a case. Its design specification is shown in Table 1. A platform with Matlab and Winspice is employed. In the evolutionary process, the population number is 2, population size is 200, crossover rate is 0.8, mutation rate is 0.1, the maximum number of components in a circuit is 20, the best value of P_f is 0.45 and the number of generation is 200.

We make experiments for 40 independent runs. Fig. 1-Fig. 3 show the corresponding results. From the amplitude-frequency curves of the best pair of circuits shown in Fig.1, we can see that the two low-pass filters both meet the functional requirements and they are negatively correlated visually. Fig.2 (a) shows the fitness track of the two modules. The fit value goes down to zero quickly, it means the both of the analog modules meet the functional requirements. Fig.2 (b) shows the negative-correlation evolutionary track of the two negatively correlated low-pass filters in 40 runs. From it, we can observe that after the 150th generation, they appear to keep the negative-correlation relationship all the time. Fig.3 gives the circuit blocks of the optimal negatively correlated low-pass filters.

From the results, we can see that the stochastic ranking strategy works well, and the evolutionary framework proposed in this paper is capable of generating analog redundancies and the most important is that it needs not specifying the negative-correlation penalty coefficient.

We also make comparisons between the framework proposed in this paper and ENCF [3] in the same conditions to verify the performance of the framework proposed in the paper. Independent experiments on both evolutionary framework runs 40 times, respectively. If circuit function meets and they are negatively correlated at the same time, That means the framework successfully converge. Two key comparison indexes are the number of successful convergence and ratio of convergence. The ratio of convergence is defined by the quotient between numbers of successful convergence and the whole 40 runs. Population size is 200, and the number of generation is 200. Table 2 shows the comparison results in 40 independent statistical runs. The index ENCF_λ_x denotes the ENCF framework when negative-correlation penalty coefficient is x. In ENCF, x is marked by [0.25 0.5 0.75 1].

Table 1. The specifications of the under-design low-pass filters in the experiments

F p (Hz)	Fs (Hz)	K p (dB)	Ks (dB)
1000	2000	-3	-60

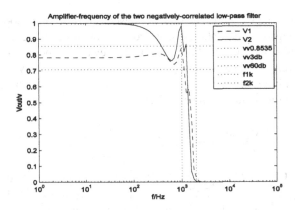

Fig. 1. The amplitude-frequency curves of the best pair of circuits

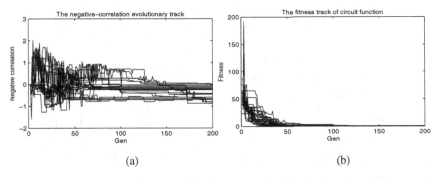

Fig. 2. Statistical results of 40 runs. **a** The fitness track of circuit function; **b** The negative-correlation evolutionary track of the two analog redundancies.

Table 2. Experimental comparison results between our framework and ENCF

Evolutionary Framework	Numbers of convergence	Ratio of convergence
Framework in this paper	35	87.5%
ENCF_λ_0.5	24	60%
ENCF_λ_0.25	16	40%
ENCF_λ_0.75	12	30%
ENCF_λ_1.00	12	30%

From the Table 2, we can see that a trial-and-error process has to be used in ENCF in order to find a proper negative-correlation penalty coefficient. When the negative-correlation penalty coefficient is 0.5 (this value is not known in advance), the number of successful convergence is the most, and the ratio of convergence is 60%. However, compared to the ratio of convergence of the framework proposed in this paper, whose ratio is 87.5%, it is not so good. The new framework don't need to choose the hard-to-set value of the negative-correlation penalty coefficient, thus, saving much resources.

From the results, we can conclude that the framework proposed in this paper can generate negatively correlated analog modules with a relatively high ability to convergence, and it doesn't have to choose the penalty coefficient, thus, the trial-and-error process is not necessary compared to ENCF.

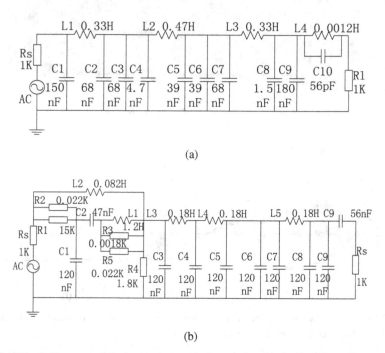

(a)

(b)

Fig. 3. The circuit diagrams of the best pair negatively-correlated redundancies. **a** One evolved filter and **b** another filter negatively correlated with the first circuit in **a**.

4   Conclusion

The performance of the existing evolutionary negative-correlation redundancy framework proposed in [3] is affected greatly by the negative-correlation penalty coefficient, but the value of the negative-correlation penalty coefficient is dependent on the experience of the designers. In this paper, we propose a new negative-correlation redundancy evolutionary framework based on stochastic ranking strategy. Our experiment results suggest that the new framework is capable of evolving the negatively correlated analog redundant modules efficiently. The most important is that it solves the problem of the hard-to-set value of the negative-correlation penalty coefficient used in ENCF. Compared to ENCF, this framework can avoid the use of the negative-correlation penalty coefficient, thus, the trial-and-error process is not needed and it shows a relatively high ability to convergence to the negatively correlated analog redundancies.

Acknowledgments. This work is supported by National Nature Science Foundation of China under Grant 60975051 and 61273315.

References

1. Schnier, T., Yao, X.: Using Negative Correlation to Evolve Fault-Tolerant Circuits. In: Proceedings of the 5th International Conference on Evolvable Systems, pp. 35–46 (2003)
2. Liu, M., He, J.: Negatively-Correlated Redundancy Circuits Evolution: A New Way of Robust Analog Circuit Synthesizing. In: Third International Workshop on Advanced Computational Intelligence, Suzhou, Jiangsu, China, August 25-27 (2010)
3. Liu, M., He, J.: An Evolutionary Negative-Correlation Framework for Robust Analog-Circuit Design under Uncertain Faults. IEEE Transactions on Evolutionary Computation (2012)
4. Liu, M.: Analog circuits evolution design and the study of negative-correlation evolutionary framework for fault tolerant of analog circuits. Science and Technology University of Science and Technology of China, Anhui (2012)
5. Kim, K.-J., Wong, A., Lipson, H.: Automated synthesis of resilient and tamper-evident analog circuits without a single point of failure. Genet Program Evolvable Mac 11(2010), 34–59 (2010)
6. Kim, K.-J., Chob, S.-J.: Automated synthesis of multiple analog circuits using evolutionary computation for redundancy-based fault-tolerance. Applied Soft Computing 12(4), 1309–1321 (2012)
7. Chang, H., He, J.: Structure diversity design of analog circuits by evolution computation for fault-tolerance. In: 2012 International Conference on Systems and informatics (2012)
8. Runarsson, T.P., Yao, X.: Stochastic Ranking for Constrained Evolutionary Optimization. IEEE Transactions on evolutionary computation 4(3) (2012)
9. Liu, Y., Yao, X.: Ensemble Learning Via Negative Correlation. Neural Networks 12, 1399–1404 (1999)
10. Liu, Y., Yao, X., Higuchi, T.: Evolutionary Ensembles with Negative Correlation Learning. IEEE Transactions on Evolutionary Computation 4(4), 380–387 (1999)
11. Lohn, J.D., Colonbano, S.P.: A Circuit Representation Techniques for Automated Circuit Design. IEEE Transactions on Automatic Control 3(3), 205–219 (1999)

Intelligent Modeling and Prediction of Elastic Modulus of Concrete Strength via Gene Expression Programming

Amir Hossein Gandomi[1], Amir Hossein Alavi[2], T.O. Ting[3,*], and Xin-She Yang[4]

[1] Department of Civil Engineering, The University of Akron, Akron, OH 44325-3905, USA
[2] Department of Civil and Environmental Engineering, Engineering Building,
Michigan State University, East Lansing, MI, 48824, USA
[3] Department of Electrical and Electronic Engineering, Xi'an Jiaotong-Liverpool University,
Suzhou, Jiangsu Province, P.R. China
[4] School of Science and Technology, Middlesex University Hendon Campus, London, UK
ag72@uakron.edu, alavi@msu.edu, toting@xjtlu.edu.cn,
x.yang@mdx.ac.uk

Abstract. The accurate prediction of the elastic modulus of concrete can be very important in civil engineering applications. We use gene expression programming (GEP) to model and predict the elastic modulus of normal-strength concrete (NSC) and high-strength concrete (HSC). The proposed models can relate the modulus of elasticity of NSC and HSC to their compressive strength, based on reliable experimental databases obtained from the published literature. Our results show that GEP can be an effective method for deriving simplified and precise formulations for the elastic modulus of NSC and HSC. Furthermore, the comparison study in the present work indicates that the GEP predictions are more accurate than other methods.

Keywords: Tangent elastic modulus, Normal and High strength concrete, Gene expression programming, Compressive strength, Formulation.

1 Introduction

In many civil engineering applications, to estimate the material properties such as elastic modulus is very important to meet design requirements. For example, the elastic modulus of normal and high strength concrete is a key parameter in structural engineering, and this parameter helps to determine the static and time-dependent deformation and system behaviour. It is also related to the assessment of other key processes such as creep, shrinkage, crack propagation and control in both reinforced concrete and prestressed concrete [1,2]. From the slope of a stress-strain curve of a given concrete material, we can estimate the elastic modulus of the sample.

Despite its importance, the elastic modulus is not usually measured in situ as it is time-consuming and expensive. The common practice is to estimate it using empirical relationships, based on various codes of practice. Such models often link the elastic

---

* Corresponding author.

Y. Tan, Y. Shi, and H. Mo (Eds.): ICSI 2013, Part I, LNCS 7928, pp. 564–571, 2013.

modulus with compressive strength, which essentially eliminate the need for going through laborious and time-consuming direct measurements from load-deformation curve [2, 3].

In recent years, techniques such as pattern recognition systems have received much attention in civil engineering applications. These systems are trained based on empirical data and thus can extract various discriminators. Loosely speaking, in the context of engineering applications, Artificial Neural Networks (ANNs), Fuzzy Logic (FL), Adaptive Neuro Fuzzy Inference System (ANFIS), and Support Vector Machine (SVM) can all be referred to as pattern recognition methods. Not surprisingly, these techniques have been used in predicting the elastic modulus of normal and high strength concrete (NSC and HSC) [4-6]. Although ANNs, FL, ANFIS, and SVM are successful in prediction, they cannot produce explicit equations for predictions, and thus limiting their usage.

In this paper, we present an alternative approach to produce explicit equations for elastic modulus of concrete materials by using genetic programming (GP), and this partly overcomes the limitations of ANNs, FL, ANFIS, and SVM for this type of applications. To achieve this goal, we investigate a relatively new variant of GP, namely gene expression programming (GEP) [7] that have been used to solve civil engineering applications such as concrete modeling [2,8,9]. In our predictions and model formulation, we have used reliable databases of previously published test results. A comparative study is carried out between the results obtained by GEP and those obtained from the buildings codes [10-13], compatibility aided [14, 15], FL [4], and ANN [5] models. The rest of the paper is organized as follows: Section 2 provides a brief description of the gene expression programming. In Section 3, a detailed study of model prediction of concrete strength and parameters using GEP is presented. Further, Section 4 provides the performance comparison and analysis and finally we draw brief conclusions in Section 5.

2   Gene Expression Programming

Genetic programming is a branch of artificial intelligence techniques that creates computer programs to solve a problem by mimicking the evolution of living or biological organisms [16]. In essence, the main aim of this method is to use inputs and their corresponding output data samples so as to create a computer program that connects them with the minimum fitting or prediction errors. The major difference between GP and genetic algorithms (GA) is the way of representing the solutions. In GA, a solution is represented by a string of numbers, either binary or real, while in the classical GP, solutions are represented as computer programs in terms of tree structures and are the expressed in a functional programming language (such as LISP) [2, 8]. In GP, a random set or population of individuals (computer programs) are created and evolved in an iterative manner to achieve sufficient diversity. A comprehensive description of GP can be found in Koza (1992) [16]. GEP is a new variant of GP first proposed by Ferreira [17]. GEP has five main components: function set, terminal set, fitness function, control parameters, and termination condition. GEP uses a fixed length of character strings to represent solutions in a domain of interest, which are then expressed as parse trees with different sizes and shapes. These trees are called GEP expression trees (ET). A main advantage of the GEP technique is that its creation of genetic diversity in solution is simplified and carried out using genetic operators that work at the chromosome level. In GEP, individuals are selected and copied into the next generation according to their fitness by the so-called roulette wheel sampling technique, together with elitism. This essentially guarantees the survival and cloning of the best individual to the next generation, which may speed up the overall convergence rate. Variations in the population are introduced by applying single or several genetic operators on selected chromosomes, and these genetic operators include crossover, mutation and rotation [7, 18]. The GEP algorithm has four main steps until it reaches one of the stop criteria [9, 17]:

I.   Randomly generating the fixed-length chromosomes as initial population.
II.   Expressing chromosomes as expression trees and evaluating fitnesses.
III. Selecting the best individuals according to their fitnesss to reproduce with modification.
IV.   Repeating the steps II an III until a termination condition is reached.

3   GEP-Based Modelling of Elastic Modulus of NSC and HSC

The main goal of this study is to obtain the prediction equations for elastic modulus (E_c) of NSC and HSC in terms of compressive strength (f_c) in the following generic form:

$$E_c = f(f_c)$$ (1)

Hence, there is only one parameter that has been used for the GEP models as the input variable. Using reliable databases for the NSC and HSC, two different GEP-based formulas for the elastic modulus of NSC and HSC can be obtained. In this study, basic arithmetic operators and mathematical functions are utilized to obtain the optimum GEP models. The actual number of generation depends on the number of possible solutions and complexity of the problem. However, it must be set properly before the runs. A large number of generations has to be tested so as to find the models with minimum errors. The program is run iteratively until there is no longer significant improvement in the performance of the models, or a specified number of iterations is reached. The values of the other parameters are selected, based on some previously suggested values [7, 18] or determined by a trial and error approach. For the GEP-based analysis, we adopted the computer software known as GeneXproTools [19]. The best GEP model is chosen on the basis of a multi-objective strategy as below:

i.   The simplicity of the model, although this is not a predominant factor.
i.   The goodness of the best fit on the training set of data.
iii. The best fitness value on a test set of unseen data.

The first objective can be controlled by the user through the parameter settings (e.g., head size or number of genes), while for the other two objectives, the following objective function (Obj) is constructed as a measure of how well the model fits the experimental data [2]. The selection criterion of the best GEP model is based on the minimization of the following function:

$$Obj = \left(\frac{N_{Train} - N_{Test}}{N_{All}} \right) \frac{MAE_{Train}}{R_{Train}^2} + \frac{2N_{Test}}{N_{All}} \frac{MAE_{Test}}{R_{Test}^2} \qquad (2)$$

where N_{Train}, N_{Test} and N_{All} are the numbers of training, testing and whole, respectively, of data. R and MAE are the correlation coefficient and mean absolute error, respectively. The above objective function has taken into account the changes of R and MAE together. Higher R values and lower MAE values result in lower Obj and, consequently, corresponds to a more precise model. In addition, the above function has also taken into account the effects of different data divisions between the training and testing data.

3.1   Experimental Database

The experimental database of previously published test results consist of 89 and 70 test results for the elastic modulus of HSC and NSC, respectively [7]. Descriptive statistics of the variables used in the model development are given in Fig. 1.

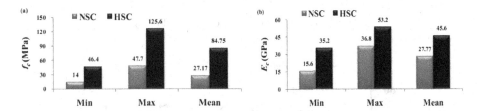

Fig. 1. Descriptive statistics of the variables

For the analysis, the data sets have been divided into training and testing subsets. The training data are applied in the learning process via genetic evolution whereas the validation data were used to measure the prediction capability of the obtained models on data that played no role in building the models. Out of 89 data sets for HSC, 69 values were taken for training of the GEP algorithm and the remaining 20 values are used for the testing and prediction. For NSC, 57 values are taken for the training process and the remaining 13 values are used for testing of the models. Out of a the total 159 data sets for HSC and NSC, 126 values were used for the training, 33 values were used for the testing of the generic model for both HSC and NSC. From these simulation, training and multiple runs, the main results can be summarized in the following sections.

3.2   Explicit Formula for Elastic Modulus of HSC and NSC

The GEP-based formulation of the E_c of HSC in terms of f_c is as given below:

$$E_{c,GEP} = 7\left(\sqrt[3]{7 + f_c} + 2\right)$$ (3)

This proposed model for the E_c of HSC gives a value of 5.462 (Obj=5.462). The expression tree of the above formulation is given in Fig. 2. The comparisons of the GEP predicted values against experimental elastic modulus of HSC are shown in Fig. 3.

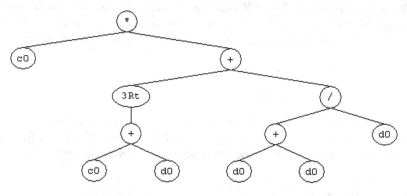

Fig. 2. Expression tree for Ec of HSC (d0 = f_c)

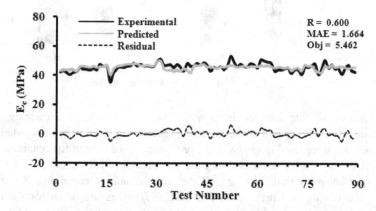

Fig. 3. Predicted versus experimental E_c of HSC using the GEP model

The GEP-based formulation of the E_c of NSC in terms of f_c can be written as

$$E_{c,GEP} = \sqrt[3]{875 f_c - 2100},$$ (4)

which yields an Obj value of 7.841. The expression tree of the above formulation is given in Fig. 4. Comparisons of the GEP predicted values against experimental elastic modulus of NSC are shown in Fig. 5.

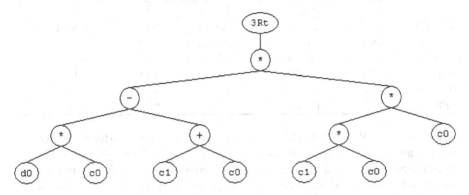

Fig. 4. Expression tree for E_c of HSC and NSC (d0 = f_c)

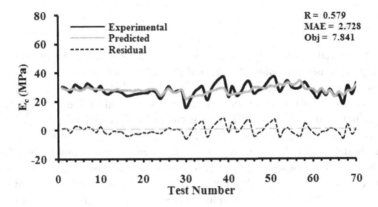

Fig. 5. Predicted versus experimental E_c of NSC using the GEP model

4   Performance Analysis

Table 1 shows the prediction performance of the GEP models, Iranian (NBS) [10], American (ACI 318-95) [11], Norwegian (NS 3473) [12], and Turkish (TS 500) [13] codes, two compatibility aided model [14, 15], FL [6], and ANN [7] models for the E_c of NSC and HSC, respectively. It can be clearly seen from this table that the proposed GEP models provide more accurate predictions than the available codes and models for the elastic modulus of HSC and NSC. However, the exception is the FL and ANN models for HSC provide better results than the GEP models.

Table 1. Comparisons between the GEP models and other models in the literature

HSC			NSC		
Model	MAE (%)	R	Model	MAE (%)	R
FL [6]	0.0368	0.6130	FL [6]	0.1031	0.5536
ANN [7]	0.0365	0.6354	ANN [7]	0.1032	0.5151
ACI [11]	0.1808	0.6024	ACI [11]	0.1327	0.5784
NS [12]	0.2124	0.5916	NBS [10]	0.1057	0.5719
[14]	0.0412	0.5577	TS [13]	0.1411	0.5693
[15]	0.1354	0.6002	[15]	0.1028	0.5839
GEP	0.0374	0.6005	GEP	0.0982	0.5795

Numerically, although the ANN and FL models have a good performance, they do not give any explicit function or formula. ANN has only final synaptic weights to obtain the outcome in a parallel manner. The determination of the fuzzy rules in FL is also a non-trivial task [8]. In addition, the ANN and FL approaches are appropriate to be used as a part of a computer program and may not be suitable for practical calculations such as *in situ* applications.

5   Conclusion

We have adopted a relatively new technique, GEP, to obtain best-fit equations for predicting the elastic modulus of HSC and NSC. Two design formulas for the elastic modulus have been obtained via GEP using a reliable database of previously published elastic modulus test results. The database is used for the training and testing of the prediction models. The GEP models can indeed give reliable estimations of the elastic modulus of HSC and NSC. The obtained formulas and proposed approach can outperform the other existing models in nearly all cases. In addition to the advantages of the acceptable accuracy, the GEP-based prediction equations are really simple to use, and can thus be used reliably for practical pre-planning and pre-design purposes by simple calculations. Such simple models for estimating elastic moduli are advantageous due to the demand in carrying out destructive, sophisticated and time-consuming laboratory tests. Further studies can focus on the extension of the proposed approach to model prediction equations for other time-consuming tasks and key parameters in engineering applications with reliable databases.

References

1. Mesbah, H.A., Lacherni, M., Aitcin, P.C.: Determination of elastic properties of high performance concrete at early age. ACI Material Journal 99(1), 37–41 (2002)
2. Gandomi, A.H., Alavi, A.H., Sahab, M.G., Arjmandi, P.: Formulation of Elastic Modulus of Concrete Using Linear Genetic Programming. Journal of Mechanical Science and Technology 24(6), 1011–1017 (2010)
3. ASTM C 469. Standard test method for static modulus of elasticity and poisson's ratio of concrete in compression. Annual Book of ASTM standards (1994)

4. Demir, F.: A new way of prediction elastic modulus of normal and high strength concrete–fuzzy logic. Cement and Concrete Research 35, 1531–1538 (2005)
5. Demir, F.: Prediction of elastic modulus of normal and high strength concrete by artificial neural networks. Construction and Building Materials 22, 1428–1435 (2008)
6. Yan, K., Shi, C.: Prediction of elastic modulus of normal and high strength concrete by support vector machine. Construction and Building Materials 24(8), 1479–1485 (2010)
7. Alavi, A.H., Gandomi, A.H.: A robust data mining approach for formulation of geotechnical engineering systems. Engineering Computations 28(3), 242–274 (2011)
8. Gandomi, A.H., Babanajad, S.K., Alavi, A.H., Farnam, Y.: A Novel Approach to Strength Modeling of Concrete under Triaxial Compression. Journal of Materials in Civil Engineering-ASCE 24(9), 1132–1143 (2012)
9. Gandomi, A.H., Alavi, A.H.: Expression Programming Techniques for Formulation of Structural Engineering Systems. In: Gandomi, A.H., et al. (eds.) Metaheuristic Applications in Structures and Infrastructures, ch. 18. Elsevier, Waltham (2013)
10. NBS. Analysis and Design of Reinforced Concrete Buildings, National Building Standard, Part 9, Iran (2006)
11. ACI 318-95, Building code requirements for structural concrete. ACI Manual of Concrete Practice Part 3: Use of concrete in Buildings –Design, Specifications, and Related Topics. Detroit, Michigan (1996)
12. NS 3473. Norwegian Council for Building Standardization. Concrete Structures Design Rules. Stockholm (1992)
13. TS 500. Requirements for design and construction of reinforced concrete structures. Ankara: Turkish Standardization Institute (2000)
14. Wee, T.H., Chin, M.S., Mansur, M.A.: Stress–strain relationship of high-strength concrete in compression. Journal of Materials in Civil Engineering-ASCE 8(2), 70–76 (1994)
15. Mostofinejad, D., Nozhati, M.: Prediction of the modulus of elasticity of high strength concrete. Iranian Journal of Science & Technology, Transaction B: Engineering 29(B3), 85–99 (2005)
16. Koza, J.: Genetic programming, on the programming of computers by means of natural selection. MIT Press, Cambridge (1992)
17. Ferreira, C.: Gene expression programming: a new adaptive algorithm for solving problems. Complex Syst. 2001 13(2), 87–129 (1988)
18. Gandomi, A.H., Alavi, A.H., Mirzahosseini, M.R., Moqhadas Nejad, F.: Nonlinear Genetic-Based Models for Prediction of Flow Number of Asphalt Mixtures. Journal of Materials in Civil Engineering-ASCE 23(3), 248–263 (2011)
19. GEPSOFT. GeneXproTools. Version 4.0 (2006), http://www.gepsoft.com

Author Index